Kosuke Izutsu

Electrochemistry in Nonaqueous Solutions

Related Titles

Endres, F., MacFarlane, D., Abbott, A. (eds.)

Electrodeposition from Ionic Liquids

2008
ISBN: 978-3-527-31565-9

Eftekhari, A. (ed.)

Nanostructured Materials in Electrochemistry

2008
ISBN: 978-3-527-31876-6

Hamann, C. H., Hamnett, A., Vielstich, W.

Electrochemistry

Second, Completely Revised and Updated Edition
2007
ISBN: 978-3-527-31069-2

Kosuke Izutsu

Electrochemistry in Nonaqueous Solutions

Second, Revised and Enlarged Edition

WILEY-VCH Verlag GmbH & Co. KGaA

The Author

Prof. Dr. Kosuke Izutsu
4-31-6-208 Kichijoji-honcho
Musashino 180-0004
Japan

All books published by Wiley-VCH are carefully
produced. Nevertheless, authors, editors, and
publisher do not warrant the information contained
in these books, including this book, to be free of
errors. Readers are advised to keep in mind that
statements, data, illustrations, procedural details or
other items may inadvertently be inaccurate.

Library of Congress Card No.: applied for

British Library Cataloguing-in-Publication Data
A catalogue record for this book is available from the
British Library.

**Bibliographic information published by
the Deutsche Nationalbibliothek**
The Deutsche Nationalbibliothek lists this
publication in the Deutsche Nationalbibliografie;
detailed bibliographic data are available on the
Internet at http://dnb.d-nb.de.

© 2009 WILEY-VCH Verlag GmbH & Co. KGaA,
Weinheim

Cover Design   Formgeber, Eppelheim
Typesetting   Thomson Digital, Noida, India
Printing   betz-druck GmbH, Darmstadt
Binding   Litges & Dopf GmbH, Heppenheim

Printed in the Federal Republic of Germany
Printed on acid-free paper

ISBN: 978-3-527-32390-6

Contents

Electrochemistry in Nonaqueous Solutions, Second, Revised and Enlarged Edition. Kosuke Izutsu
Copyright © 2009 WILEY-VCH Verlag GmbH & Co. KGaA, Weinheim
ISBN: 978-3-527-32390-6

Preface to the First Edition

A majority of chemical reactions are carried out in solution. The use of a solvent as reaction medium makes it easy to control reaction conditions such as temperature, pressure, pH, rate of mass transfer and concentration of reactant. Water is the most popular solvent. However, by using appropriate nonaqueous solvents, substances that are insoluble in water can be dissolved, substances that are unstable in water remain stable and chemical reactions that are impossible in water become possible. The reaction environments are markedly wider in nonaqueous solvents than in water.

The widespread use of nonaqueous solvents, especially dipolar aprotic solvents, began in the 1950s in various fields of pure and applied chemistry and has contributed greatly to advances in chemical sciences and technologies. From the very beginning, electrochemistry in nonaqueous solutions has played an important role in exploring new chemical possibilities as well as in providing the methods to evaluate static solvent effects on various chemical processes. Moreover, many new electrochemical technologies have been developed using nonaqueous solvents. Recently, electrochemistry in nonaqueous solutions has made enormous progress: the dynamic solvent effects on electrochemical processes have been greatly elucidated and solvent effects are now understood much better than before. On the other hand, however, it is also true that some useful solvents have properties that are problematic to human health and the environment. Today, efforts are being made, under the framework of 'green chemistry', to find environmentally benign media for chemical processes, including harmless nonaqueous solvents, immobilized solvents, ionic liquids, supercritical fluids, aqueous systems and even solventless reaction systems. For electrochemical purposes, replacing hazardous solvents with harmless solvents, ionic liquids and supercritical fluids appears to be promising.

This book was written to provide readers with some knowledge of electrochemistry in nonaqueous solutions, from its fundamentals to the latest developments, including the current situation concerning hazardous solvents. The book is divided into two parts. Part I (Chapters 1–4) contains a discussion of solvent properties and then deals with solvent effects on chemical processes such as ion solvation, ion complexation, electrolyte dissociation, acid–base reactions and redox reactions. Such

Electrochemistry in Nonaqueous Solutions, Second, Revised and Enlarged Edition. Kosuke Izutsu
Copyright © 2009 WILEY-VCH Verlag GmbH & Co. KGaA, Weinheim
ISBN: 978-3-527-32390-6

solvent effects are of fundamental importance in understanding chemistry in nonaqueous solutions; furthermore, their quantitative evaluations are often carried out by means of electrochemical techniques. Part II (Chapters 5–12) mainly deals with the use of electrochemical techniques in nonaqueous solutions. In Chapter 5, the fundamentals of various electrochemical techniques are outlined in preparation for the following chapters. In Chapters 6–9, the applications of potentiometry, conductimetry, polarography, voltammetry and other new electrochemical techniques in nonaqueous solutions are discussed by focusing on the chemical information they provide. Chapters 10 and 11 examine methods of selecting and purifying the solvents and electrolytes of electrochemical importance. Finally, in Chapter 12, some practical applications of nonaqueous solvents in modern electrochemical technologies are discussed. These include their use in batteries, capacitors and display devices, and such processes as electrolytic refining, plating, synthesis and polymerization. The applicability of ionic liquids and supercritical fluids as environmentally benign media for electrochemical technology is also dealt with.

Most chemists are familiar with chemistry in aqueous solutions. However, the common sense in aqueous solutions is not always valid in nonaqueous solutions. This is also true for electrochemical measurements. Thus, in this book, special emphasis is placed on showing which aspects of chemistry in nonaqueous solutions are different from chemistry in aqueous solutions. Emphasis is also placed on showing the differences between electrochemical measurements in nonaqueous systems and those in aqueous systems. The importance of electrochemistry in nonaqueous solutions is now widely recognized by nonelectrochemical scientists – for example, organic and inorganic chemists often use cyclic voltammetry in aprotic solvents to determine redox properties, electronic states and reactivities of electroactive species, including unstable intermediates. This book will therefore also be of use to such nonelectrochemical scientists.

I obtained most of the information included in this book from the publications of many scientists in this field. I would like to express my sincere thanks to all of them. I also would like to thank my coworkers for their cooperation, the editorial and production staff of Wiley-VCH for their help and support and my wife for her assistance and patience.

Matsumoto   *Kosuke Izutsu*
December 2001

Preface to the Second Edition

The second edition consists of three parts: Part I (Chapters 1–4) is for electrochemical aspects of the fundamentals of chemistry in nonaqueous solutions, Part II (Chapters 5–12) deals with the electrochemical techniques and their applications in nonaqueous solutions and Part III (Chapters 13 and 14) is concerned with the electrochemistry in new solvent systems. Chapter 13 is on the electrochemistry in clean solvents and special emphasis is placed on supercritical fluids and (room-temperature) ionic liquids. Chapter 14 is on the electrochemistry at the liquid–liquid interfaces and, in addition to charge transfers at the interface between water and organic solvents, those at the interface between water and ionic liquid are also discussed. Since the publication of the first edition, considerable efforts have been made in pursuing green solvents that are benign to human health and the environment. Especially, the movement in the field of ionic liquids has been remarkable and the rapid progress is still continuing. Most of the ionic liquids are nonvolatile, nonflammable, less toxic, chemically and thermally stable and good solvents for both organic and inorganic materials. Due to their aprotic properties, many ionic liquids can replace aprotic organic solvents, particularly for use in electrochemistry. Moreover, the immiscibility of ionic liquids with water makes possible their applications to the water/ionic liquid interfaces in various ways, including electrochemical uses.

Besides the addition of the two new chapters, each of the Chapters 1–12 has been updated and revised. Especially, the revision in Chapter 12 is considerable because the use of nonaqueous solutions in modern electrochemical technologies is steadily increasing.

Although some nonaqueous solvents that are hazardous to human health and the environment cannot be used or should be used with great care, the needs for the electrochemical use of less hazardous solvents, including supercritical fluids and ionic liquids, are increasing. Thus, the knowledge of electrochemistry in nonaqueous solutions is also increasing its importance.

Finally, I wish to thank the editorial and production staff at Wiley-VCH Verlag GmbH, for their help and support in making this edition.

Musashino, Tokyo   *Kosuke Izutsu*
August 2009

Electrochemistry in Nonaqueous Solutions, Second, Revised and Enlarged Edition. Kosuke Izutsu
Copyright © 2009 WILEY-VCH Verlag GmbH & Co. KGaA, Weinheim
ISBN: 978-3-527-32390-6

Books, reviews and data compilations on nonaqueous solvents and the chemistry, especially electrochemistry, in nonaqueous solutions are shown below:

1 Lagowski, J.J. (ed.) (1966) *The Chemistry of Non-Aqueous Solvents*, Academic Press, New York, vol. 1; 1967, vol. 2; 1970, vol. 3; 1976, vol. 4; 1978, vol. 5A; 1978, vol. 5B. Includes many reviews.

2 Charlot, G. and Tremillon, B. (1969) *Chemical Reactions in Solvents and Melts*, Pergamon Press, Oxford.

3 Bard, A.J. (ed.) (1969) *Electroanalytical Chemistry*, vol. 3, Marcel Dekker, New York, p. 57; 1975, vol. 8, p. 281, etc.

4 Coetzee, J.F. and Ritchie, C.D. (eds) (1969) *Solute–Solvent Interactions*, Marcel Dekker, New York, vol. I; 1976, vol. II, includes reviews.

5 Mann, C.K. and Barnes, K.K. (1970) *Electrochemical Reactions in Nonaqueous Solvents*, Marcel Dekker, New York.

6 Janz, G.J. and Tomkins, R.P.T. (eds) (1972) *Nonaqueous Electrolytes Handbook*, vol. 1, Academic Press, New York; 1973, vol. 2.

7 Covington, A.K. and Dickinson, T. (eds) (1973) *Physical Chemistry in Organic Solvent Systems*, Plenum Press, New York, includes reviews and data compilations.

8 Fritz, J.S. (1973) *Acid–Base Titrations in Nonaqueous Media*, Allyn & Bacon, Needham Heights, MA.

9 Tremillon, B. (1974) *Chemistry in Nonaqueous Solvents*, D. Reidel, Dordrecht, the Netherlands.

10 (a) Meites, L. and Zuman, P. (eds) (1977–1983) *CRC Handbook Series in Organic Electrochemistry*, vols. I–VI, CRC Press, Boca Raton, FL; (b) (1981–1988) *CRC Handbook Series in Inorganic Electrochemistry*, vols. I–VIII, CRC Press, Boca Raton, FL. Compilations of potential data.

11 Marcus, Y. (1977) *Introduction to Liquid State Chemistry*, John Wiley & Sons, Inc., New York.

12 Burgess, J. (1978) *Metal Ions in Solutions*, Ellis Horwood, Chichester.

13 Gutmann, V. (1978) *Donor–Acceptor Approach to Molecular Interactions*, Plenum Press, New York.

14 Kolthoff, I.M. and Elving, P.J. (eds) (1979) *Treatise on Analytical Chemistry*, 2nd edn, Part I, vol. 2, John Wiley & Sons, Inc., New York, Chapter 19. Excellent reviews on acid–base reactions in non-aqueous systems.

15 Popovych, O. and Tomkins, R.P.T. (1981) *Nonaqueous Solution Chemistry*, John Wiley & Sons, Inc., New York. Treats electrochemical aspects in detail.

16 Coetzee, J.F. (ed.) (1982) *Recommended Methods for Purification of Solvents and Tests for Impurities*, Pergamon Press, Oxford. Reports from IUPAC.

17 Burger, K. (1983) *Solvation, Ionic and Complex Formation Reactions in Nonaqueous Solvents*, Elsevier, Amsterdam.

18 Marcus, Y. (1985) *Ion Solvation*, John Wiley & Sons, Inc., New York. Includes large amounts of data.

19 Riddick, A., Bunger, W.R. and Sakano, T.K. (1986) *Organic Solvents, Physical Properties and Methods of Purification*, 4th edn, John Wiley & Sons, Inc., New York. Includes detailed data on solvent properties and methods of solvent purification.

20 Safarik, L. and Stransky, Z. (1986) *Titrimetric analysis in organic solvents*, Comprehensive Analytical Chemistry, vol. 22, Elsevier, Amsterdam.

21 Izutsu, K. (1990) *Acid–Base Dissociation Constants in Dipolar Aprotic Solvents*, IUPAC Chemical Data Series No. 35, Blackwell Science, Oxford. Data compilation.

22 Krestov, G.A. (1991) *Thermodynamics of Solvation, Solution and Dissolution, Ions and Solvents, Structure and Energetics*, Ellis Horwood, New York.

23 Mamantov, G. and Popov, A.I. (eds) (1994) *Chemistry of Nonaqueous Solutions, Current Progress*, Wiley-VCH Verlag GmbH, Weinheim.

24 Galus, Z. (1994) in *Advances in Electrochemical Science and Engineering*, vol. 2 (eds H. Gerischer and C.W. Tobias), Wiley-VCH Verlag GmbH,

Weinheim, pp. 217–295. Thermodynamics and kinetics of electrode reactions in nonaqueous and mixed solvents.

25 Gutmann, V. and Resch, G. (1995) *Lecture Notes on Solution Chemistry*, World Science, Singapore.

26 Sawyer, D.T., Sobkowiak, A. and Roberts, J.L., Jr (1995) *Electrochemistry for Chemists*, 2nd edn, John Wiley & Sons, Inc., New York. Useful references on electrochemical techniques in nonaqueous solutions.

27 Kissinger, P.T. and Heineman, W.R. (eds) (1996), *Laboratory Techniques in Electroanalytical Chemistry*, Marcel Dekker, New York. Includes many chapters on electrochemical techniques in nonaqueous solutions.

28 Marcus, Y. (1997) *Ion Properties*, Marcel Dekker, New York.

29 Tremillon, B. (1997) *Reactions in Solution: An Applied Analytical Approach*, John Wiley & Sons, Inc., New York.

30 Barthel, J.M.G., Krienke, H. and Kunz, W. (1998) *Physical Chemistry of Electrolyte Solutions: Modern Aspects, Topics in Physical Chemistry*, vol. 5, Springer, Berlin.

31 Marcus, Y. (1998) *The Properties of Solvents*, John Wiley & Sons, Inc., New York.

32 Chipperfield, J.R. (1999) *Non-Aqueous Solvents*, Oxford University Press, Oxford.

33 Aurbach, D. (ed.) (1999) *Nonaqueous Electrochemistry*, Marcel Dekker, New York. Mainly concerned with lithium batteries.

34 Burgess, J. (1999) *Ions in Solution, Basic Principles of Chemical Interactions*, Horwood Publishing, Chichester.

35 Wypych, G. (ed.) (2001) *Handbook of Solvents*, ChemTec Publishing, Toronto.

36 (a) Lund, H. and Hammerich, O. (eds) (2001) *Organic Electrochemistry*, 4th edn, Marcel Dekker, New York; (b)Lund, H. and Baizer, M.M. (eds) (1991) *Organic Electrochemistry*, 3rd edn, Marcel Dekker, New York. Detailed treatment of electrochemical techniques and electrode processes of organic substances.

37 Marcus, Y. (2002) *Solvent Mixtures: Properties and Selective Solvation*, Marcel Dekker, New York.

38 Buncel, E. (2003) *The Role of the Solvent in Chemical Reactions (Oxford Chemistry Master 6)*, Oxford University Press, Oxford.

39 Fawcett, W.R. (2004) *Liquids, Solutions, and Interfaces: From Classical Macroscopic Descriptions to Modern Microscopic Details*, Oxford University Press, Oxford.

40 Reichardt, C. (2005) *Solvents and Solvent Effects in Organic Chemistry*, 3rd edn, Wiley-VCH Verlag GmbH, Weinheim.

41 Wasserscheid, P. and Welton, T. (eds) (2008) *Ionic Liquids in Synthesis*, 2nd edn, vols. 1–2, Wiley-VCH Verlag GmbH, Weinheim.

42 Endres, F., Abbott, A.P. and MacFarlane, D.R. (eds) (2008) *Electrodeposition from Ionic Liquids*, Wiley-VCH Verlag GmbH, Weinheim.

Examples of books dealing with the fundamentals of electrochemistry:

1 Rossiter, B.W. and Hamilton, J.F. (eds) (1986) *Electrochemical Methods, Physical Methods of Chemistry*, vol. II, 2nd edn, John Wiley & Sons, Inc., New York.

2 (a) Brett, C.M.A. and Brett, A.M.O. (1993) *Electrochemistry: Principles, Methods and Applications*, Oxford University Press, Oxford; (b) (1998) *Electroanalysis*, Oxford University Press, Oxford.

3 Koryta, J., Dvorak, J. and Kavan, L. (1993) *Principles of Electrochemistry*, 2nd edn, John Wiley & Sons, Inc., New York.

4 Oldham, H.B. and Myland, J.C. (1994) *Fundamentals of Electrochemical Science*, Academic Press, New York.

5 Galus, Z. (1994) *Fundamentals of Electrochemical Analysis*, 2nd edn, John Wiley & Sons, Inc., New York.

6 Rubinstein, I. (ed.) (1995) *Physical Electrochemistry: Principles, Methods, and Applications*, Marcel Dekker, New York.

7 Fisher, A.C. (1996) *Electrode Dynamics*, Oxford University Press, Oxford.

8 Bockris, J.O'M. and Reddy, A.N. (1998) *Modern Electrochemistry*, 2nd edn, vol. 1, *Ionics*; Plenum Press, New York; (2000) Vol. 2A, *Fundamentals of Electrodics*; (2000) vol. 2B, *Electrodics in Chemistry*,

Engineering, Biology and Environmental Science.

9 Bard, A.J. and Faulkner, L.R. (2001) *Electrochemical Methods: Fundamentals and Applications,* 2nd edn, John Wiley & Sons, Inc., New York.

10 Bond, A.M. (2002) *Broadening Electrochemical Horizons: Principles and Illustration of Voltammetric and Related Techniques,* Oxford University Press, Oxford.

11 Bard, A.J. and Stratmann, M., *et al.* (2002–2007) *Encyclopedia of Electrochemistry,* vols. 1–11, Wiley-VCH Verlag GmbH, Weinheim.

12 Bagotsky, V.S. (2005) *Fundamentals of Electrochemistry (The ECS Series of Text and Monographs),* 2nd edn, Wiley-Interscience, New York.

13 Savéant, J.-M. (2006) *Elements of Molecular and Biomolecular Electrochemistry: An Electrochemical Approach to Electron Transfer Chemistry,* John Wiley & Sons, Inc., NJ.

14 Wang, J. (2006) *Analytical Electrochemistry,* 3rd edn, Wiley-VCH Verlag GmbH, New York.

15 Compton, R.G. and Banks, C.E. (2007) *Understanding Voltammetry,* World Scientific, London.

16 Hamann, C.H., Hamnett, A. and Vielstich, W. (2007) *Electrochemistry,* 2nd edn, Wiley-VCH Verlag GmbH, Weinheim.

Part One
Fundamentals of Chemistry in Nonaqueous Solutions: Electrochemical Aspects

Electrochemistry in Nonaqueous Solutions, Second, Revised and Enlarged Edition. Kosuke Izutsu
Copyright © 2009 WILEY-VCH Verlag GmbH & Co. KGaA, Weinheim
ISBN: 978-3-527-32390-6

1
Properties of Solvents and Solvent Classification

Three types of liquid substances, i.e. molecular liquids, ionic liquids and atomic liquids, can serve as *solvents*. They dissolve *solutes* that are solid, liquid or gaseous and form *solutions*. Molecular liquid solvents are the most common and include, besides water, many organic solvents and some inorganic solvents, such as hydrogen fluoride, liquid ammonia and sulfur dioxide. Ionic liquid solvents are mostly molten salts and usually used at relatively high temperatures. Nowadays, however, various room-temperature ionic liquids (or simply called 'ionic liquids') are being designed and used as a kind of 'green solvents'.[1] There are only a few atomic liquid solvents at room temperatures, metallic mercury being a typical example. Besides these liquid solvents, supercritical fluids are sometimes used as media for chemical reactions and separations (see footnote 1).

Apart from Chapter 13, which deals with electrochemistry in new solvents (supercritical fluids, ionic liquids, etc.), this book mainly considers molecular liquid solvents. Thus, the term 'solvents' means molecular liquid solvents. Water exists abundantly in nature and has many excellent solvent properties. If water is appropriate for a given purpose, it should be used without hesitation. If water is not appropriate, however, some other solvent (or ionic liquids) must be employed. Molecular solvents other than water are generally called *nonaqueous solvents*. Nonaqueous solvents are often mixed with water or some other nonaqueous solvents, to obtain desirable solvent properties. These mixtures of solvents are called *mixed solvents*.

---

1) 'Green' chemistry is the utilization of a set of principles that reduces or eliminates the use or generation of hazardous substances in the design, manufacture and application of chemical products (Anastas, P.T., Warner, J.C. (1998) *Green Chemistry, Theory and Practice*, Oxford University Press, New York, p. 11). Under the framework of green chemistry, efforts are being made to find environmentally benign media (green solvents) for chemical processes; among such media are harmless nonaqueous solvents, immobilized solvents, ionic liquids, supercritical fluids, aqueous reaction systems and solvent-free reaction systems. For the recent situation, see, for example, Abraham, M.A., Moens, L. (eds) (2002) *Clean Solvents: Alternative Media for Chemical Reactions and Processing*, Oxford University Press, New York; Nelson, W.M. (2003) *Green Solvents for Chemistry: Perspectives and Practice* (Green Chemistry Series), Oxford University Press, New York.

Electrochemistry in Nonaqueous Solutions, Second, Revised and Enlarged Edition. Kosuke Izutsu
Copyright © 2009 WILEY-VCH Verlag GmbH & Co. KGaA, Weinheim
ISBN: 978-3-527-32390-6

Table 1.1 Physical properties of organic solvents and some inorganic solvents of electrochemical importance.

Solvent	Abbreviated symbol	Bp (°C)	Fp (°C)	Vapor pressure[a] (mmHg)	Density[a] (g cm^{-3})	Viscosity[a] (cP)	Conductivity[a] (S cm^{-1})	Relative permittivity[a]	Dipole moment[a] (D)	Toxicity[b]
(1) Water		100	0	23.8	0.9970	0.890	6×10^{-8}	78.39	1.77	
Acids										
(2) Hydrogen fluoride		19.6	−83.3	—	0.9529	0.256	1×10^{-4}	84.0	1.82	0.5
(3) Formic acid		100.6	8.27	43.1	1.2141	1.966	6×10^{-5}	58.5_{16}	1.82_{30}	5
(4) Acetic acid	HOAc	117.9	16.7	15.6	1.0439	1.130	6×10^{-9}	6.19	1.68_{30}	10
(5) Acetic anhydride		140.0	−73.1	5.1	1.0749	0.783_{30}	5×10^{-9}	20.7_{19}	2.82	5
Alcohols										
(6) Methanol	MeOH	64.5	−97.7	127.0	0.7864	0.551	1.5×10^{-9}	32.7	1.71	200,T
(7) Ethanol	EtOH	78.3	−114.5	59.0	0.7849	1.083	1.4×10^{-9}	24.6	1.74	1000,C(A4)
(8) 1-Propanol	1-PrOH	97.2	−126.2	21.0	0.7996	1.943	$9 \times 10^{-9}_{18}$	20.5	1.71	200,C(A3)
(9) 2-Propanol	2-PrOH	82.2	−88.0	43.3	0.7813	2.044	6×10^{-8}	19.9	1.66_{30}	200,C(A4)
(10) Methyl cellosolve (i)		124.6	−85.1	9.7	0.9602	1.60	1.1×10^{-6}	16.9	2.04	5
(11) Cellosolve (ii)		135.6	<−90	5.3	0.9252	1.85	9×10^{-8}	29.6_{24}	2.08	5
Ethers										
(12) Anisole (methoxybenzene)	PhOMe	153.8	−37.5	3.54	0.9893	0.895	1×10^{-13}	4.33	1.245	NE
(13) Tetrahydrofuran (iii)	THF	66.0	−108.4	162	0.8892_{20}	0.460	—	7.58	1.75	50,C(A3)
(14) 1,4-Dioxane (iv)		101.3	11.8	37.1	1.028	1.087_{30}	5×10^{-15}	2.21	0.45	20,C(A3),T
(15) Monoglyme (1,2-dimethoxyethane) (v)	DME	84.5	−69	48_{20}	0.8637	0.455	—	7.20	1.71	NE
(16) Diglyme (vi)		159.8	−64	3.4	0.9384	0.989	—	—	1.97	NE

Ketones

(17) Acetone	Ac	56.1	−94.7	231	0.7844	0.303	5×10^{-9}	20.6	2.7_{20}	500,C(A4)
(18) 4-Methyl-2-pentanone	MIBK	117.4	−84	18.8	0.7963	0.546	$<5\times10^{-8}$	13.1_{20}	—	50,T
(19) Acetylacetone	Acac	138.3	−23.2	8.6_{23}	0.9721	0.694	1×10^{-8}	25.7_{20}	2.78_{22}	NE

Nitriles

(20) Acetonitrile	AN	81.6	−43.8	88.8	0.7765	0.341_{30}	6×10^{-10}	35.9	3.53	20,C(A4),T
(21) Propionitrile	PrN	97.4	−92.8	44.6	0.7768	0.389_{30}	8×10^{-8}	28.9_{20}	3.50	Very toxic
(22) Butyronitrile	BuN	117.6	−111.9	19.1	0.7865	0.515_{30}	—	24.8_{20}	3.50	Very toxic
(23) Isobutyronitrile		103.8	−71.5	—	0.7656	0.456_{30}	—	20.4_{24}	3.61	Very toxic
(24) Benzonitrile	BN	191.1	−12.7	$1_{28.2}$	1.0006	1.237	5×10^{-8}	25.2	4.01	Very toxic

Amines

(25) Ammonia		−33.4	−77.7	—	0.681_{-34}	0.25_{-34}	$5\times10^{-11}{}_{-34}$	23.0_{-34}	0.93	25
(26) Ethylenediamine	en	116.9	11.3	$13.1_{26.5}$	0.8931	1.54	9×10^{-8}	12.9	1.90	10,C(A4)
(27) Pyridine	Py	115.3	−41.6	20	0.9782	0.884	4×10^{-8}	12.9	2.37	1,C(A3)

Amides

(28) Formamide	FA	210.5	2.5	1_{70}	1.1292	3.30	$<2\times10^{-7}$	111_{20}	3.37_{30}	10
(29) N-Methylformamide (vii)	NMF	180–185	−3.8	0.4_{44}	0.9988	1.65	8×10^{-7}	182.4	3.86	NE
(30) N,N-Dimethylformamide (viii)	DMF	153.0	−60.4	3.7	0.9439	0.802	6×10^{-8}	36.7	3.24	10,C(A4),T
(31) N-Methylacetamide (ix)	NMA	206	30.5	1.5_{56}	0.9500_{30}	3.65_{30}	$2\times10^{-7}{}_{40}$	191.3_{32}	4.27_{30}	10,C(A4),T
(32) N,N-Dimethylacetamide (x)	DMA	166.1	−20	1.3	0.9363	0.927	10^{-7}	37.8	3.79	10,C(A4)
(33) N-Methylpropionamide		104_{16mm}	−30.9	94_{10}	0.9305	5.22	8×10^{-8}	176	—	
(34) Hexamethylphosphoric triamide (xi)	HMPA	233	7.2	0.07_{30}	1.020	3.10	2×10^{-7}	29.6	5.37	Toxic, C(A3),T
(35) N-Methyl-2-pyrrolidinone (xii)	NMP	202	−24.4	0.3	1.026	1.67	1×10^{-8}	32.2	4.09_{30}	NE
(36) 1,1,3,3-Tetramethylurea	TMU	175.2	−1.2		0.9619	1.395	$<6\times10^{-8}$	23.60	3.50	

(Continued)

Table 1.1 (*Continued*)

Solvent	Abbreviated symbol	Bp (°C)	Fp (°C)	Vapor pressure[a] (mmHg)	Density[a] (g cm⁻³)	Viscosity[a] (cP)	Conductivity[a] (S cm⁻¹)	Relative permittivity[a]	Dipole moment[a] (D)	Toxicity[b]
Sulfur compounds										
(37) Sulfur dioxide		−10.01	−75.46		1.46_{-10}	0.429_0		15.6_0	1.62	2, C(A4)
(38) Dimethyl sulfoxide (xiii)	DMSO	189.0	18.5	0.60	1.095	1.99	2×10^{-9}	46.5	4.06	NE
(39) Sulfolane (xiv)	TMS	287.3	28.5	5.0_{118}	1.260_{30}	10.3_{30}	$<2 \times 10^{-8}{}_{30}$	43.3_{30}	4.7_{30}	
(40) Dimethylthioformamide	DMTF	70_{1mm}	−8.5		1.024_{27}	1.98	—	47.5	4.4	
(41) N-Methyl-2-thiopyrrolidinone	NMTP	145_{15mm}	19.3		1.084	4.25	—	47.5	4.86	
Others										
(42) Hexane		68.7	−95.3	151.3	0.6548	0.294	$<10^{-16}$	1.88	0.085	50,T
(43) Benzene		80.1	5.5	95.2	0.8736	0.603	4×10^{-17}	2.27	0	0.5,C(A1),T
(44) Toluene		110.6	−95.0	28.5	0.8622	0.553	8×10^{-16}	2.38	0.31	50,C(A4),T
(45) Nitromethane	NM	101.2	−28.6	36.7	1.1313	0.614	5×10^{-9}	36.7	3.17	20,C(A3)
(46) Nitrobenzene	NB	210.8	5.76	0.28	1.1983	1.62_{30}	2×10^{-10}	34.8	4.00	1,C(A3),T
(47) Dichloromethane		39.6	−94.9	436	1.3168	0.393_{30}	4×10^{-11}	8.93	1.55	50,C(A3)
(48) 1,2-Dichloroethane	DCE	83.5	−35.7	83.4_{20}	1.2464	0.73_{30}	4×10^{-11}	10.37	1.86	10,C(A4)
(49) γ-Butyrolactone (xv)	γ-BL	204	−43.4	3.2	1.1254	1.73		39.1	4.12	
(50) Propylene carbonate (xvi)	PC	241.7	−54.5	1.2_{55}	1.195	2.53	1×10^{-8}	64.92	4.94	
(51) Ethylene carbonate	EC	248.2	36.4	3.4_{95}	1.3383	1.9_{40}	$5 \times 10^{-8}{}_{40}$	89.8_{40}	4.9	
(52) Methyl acetate	MA	56.9	−98.0	216.2	0.9279	0.364	$3 \times 10^{-6}{}_{20}$	6.68	1.72	200
(53) Ethyl acetate		77.1	−83.6	94.5	0.8946	0.426	$<1 \times 10^{-9}$	6.02	1.82	400

Except for the column of 'Toxicity', the data in this table are mainly from Ref. [3], though some are from Ref. [1].

[a]Unless otherwise stated, the data are at 25°C. The temperatures other than 25°C are shown as subscript.

[b]The numerical value shows the TLV as time-weighted average (TWA), i.e. the maximum permissible vapor concentration that the average person can be exposed for 8 h per day, 5 days per week without harm, in ppm [cm³ of solvent vapor per 1 m³ of air] [29, 30a, 31]. The mark 'C' shows that the solvent is or is suspected to be carcinogenic and A1,

A3 and A4 show 'confirmed human carcinogen', 'confirmed animal carcinogen with unknown relevance to humans' and 'not classifiable as a human carcinogen', respectively. The mark 'T' shows the solvent has been listed in Title III of the Clean Air Act Amendments of 1990 as a hazardous air pollutant. NE stands for 'not established'.

(i) $CH_3OCH_2CH_2OH$

(ii) $C_2H_5OCH_2CH_2OH$

(iii) 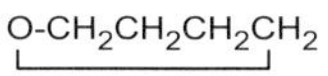 $O\text{-}CH_2CH_2CH_2CH_2$

(iv) 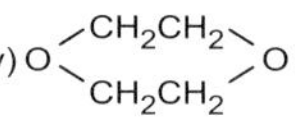 structure: O with CH_2CH_2 / CH_2CH_2 / O

(v) $CH_3OCH_2CH_2OCH_3$

(vi) $(CH_3OCH_2CH_2)_2O$

(vii) $HCONH(CH_3)$

(viii) $HCON(CH_3)_2$

(ix) $CH_3CONH(CH_3)$

(x) $CH_3CON(CH_3)_2$

(xi) 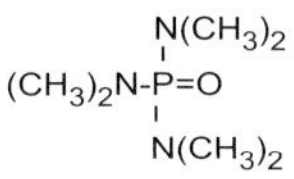 $(CH_3)_2N\text{-}P{=}O$ with $N(CH_3)_2$ and $N(CH_3)_2$

(xii) 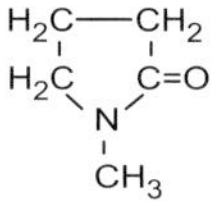 structure: $H_2C{-}CH_2$ / H_2C $C{=}O$ / N / CH_3

(xiii) $(CH_3)_2S{=}O$

(xiv) 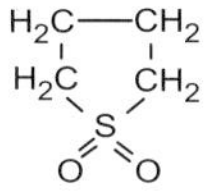 structure: $H_2C{-}CH_2$ / H_2C CH_2 / S / $O{=}$ O

(xv) 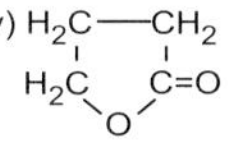 structure: $H_2C{-}CH_2$ / H_2C $C{=}O$ / O

(xvi) 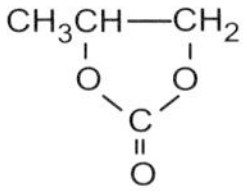 $CH_3CH{-}CH_2$ / O O / $C{=}O$

There are a great many kinds of neat nonaqueous solvents. Substances that are solid or gaseous at ambient temperatures also work as solvents, if they are liquefied at higher or lower temperatures. For mixed solvents, it is possible to vary the mixing ratio and thus the solvent properties continuously. Therefore, if both nonaqueous and mixed solvents are included, the number of solvents really is infinite.

When a nonaqueous solvent is to be used for a given purpose, a suitable one must be selected from the infinite number available. This is not easy, however, unless there are suitable guidelines available on how to select solvents. In order to make solvent selection easier, it is useful to classify solvents according to their properties. The properties of solvents and solvent classification have been dealt with in detail in the literature [1, 2]. In this chapter, these problems are briefly discussed in Sections 1.1 and 1.2, and then the influences of solvent properties on reactions of electrochemical importance are outlined in Section 1.3.

Organic solvents and some inorganic solvents for use in electrochemical measurements are listed in Table 1.1, with their physical properties.

1.1
Properties of Solvents

Physical and chemical properties that are important in characterizing solvents as reaction media are listed in Table 1.2 and are briefly discussed in Sections 1.1.1 and 1.1.2. The data of solvent properties have been compiled in Refs [2–4] for a number of solvents. In addition to these properties, structural aspects of solvents are outlined in Section 1.1.3 and the effects of toxicity and the hazardous properties of solvents are considered in Section 1.1.4.

1.1.1
Physical Properties of Solvents

Each of the physical properties in Table 1.2 has its own significance.[2] The boiling point, T_b, and the melting (or freezing) point determine the liquid range of solvents. The vapor pressure is a fundamental vaporization property, and it is also important when considering the problem of toxicity and other hazards of solvents (Section 1.1.4). The heat of vaporization, $\Delta_v H$, determines the cohesive energy density, c, defined by $c = (\Delta_v H - RT)/V_m$, and the solubility parameter, δ, is defined by $\delta = c^{1/2} = [(\Delta_v H - RT)/V_m]^{1/2}$, where V_m is the molar volume. The cohesive energy density is a measure of the 'stickiness' of a solvent and is related to the work necessary to create 'cavities' to accommodate solute particles in the solvent. Conversely, the solubility parameter proposed by Hildebrand is useful in predicting the solubilities of nonelectrolyte solutes in low-polarity solvents. In many cases, two liquid substances with similar δ values are miscible, while those with dissimilar

---

2) See Refs [1–3] or advanced textbooks of physical chemistry.

Table 1.2 Physical and chemical properties of solvents.

Physical properties	*Bulk properties*: boiling point, melting (or freezing) point, molar mass, density, viscosity, vapor pressure, heat capacity, heat of vaporization, refractive index, relative permittivity, electric conductivity; *molecular properties*: dipole moment, polarizability
Chemical properties	Acidity (including the abilities as proton donor, hydrogen bond donor, electron pair acceptor and electron acceptor)[a]; basicity (including the abilities as proton acceptor, hydrogen bond acceptor, electron pair donor and electron donor)[a]

[a]The terms 'acidity' and 'basicity' are used in somewhat wider ways than usual (see text).

δ values are immiscible.[3] The heat of vaporization at the boiling point, $\Delta_v H(T_b)$ in kJ mol^{-1}, determines Trouton's constant, $(\Delta_v S(T_b)/R)$, which is equal to $\Delta_v H(T_b)/T_b$. Solvents with $\Delta_v S(T_b)/R \leq 11.6$ are usually nonstructured (e.g. $\Delta_v S(T_b)/R = 7.2$ for acetic acid, 10.2 for hexane, 10.5 for benzene and 10.9 for acetone), while those with $\Delta_v S(T_b)/R \geq 12$ are structured (e.g. $\Delta_v S(T_b)/R = 12.5$ for methanol and 13.1 for water). The viscosity (η) influences the rate of mass transfer in the solvent and therefore the conductivity of electrolyte solutions.

The relative permittivity, ε_r, influences the electrostatic interactions between electric charges. If two charges, q_1 and q_2, are placed in a vacuum at a distance r from each other, the electrostatic force F_{vac} between them is expressed by Eq. (1.1):

$$F_{vac} = \frac{q_1 q_2}{4\pi\varepsilon_0 r^2} \tag{1.1}$$

where ε_0 is the permittivity of a vacuum and $\varepsilon_0 = 8.854 \times 10^{-12}\,\mathrm{F\,m^{-1}}$. F_{vac} is a repulsive force if q_1 and q_2 are of the same sign, while F_{vac} is an attractive force if they

3) The primary role of solvents is to 'dissolve' substances. There is an old principle – *'like dissolves like'*. In general, polar solvents can dissolve polar substances, while nonpolar solvents can dissolve nonpolar substances. The following table shows the relationship between the polarities of solvents and solutes and their mutual solubilities:

obtained by dissolution. The energetic stabilization depends on the energies of three interactions, i.e. solute–solvent, solute–solute and solvent–solvent interactions. When both the solvent and the solute are nonpolar, all of the three interactions are weak. In that case, the energy gained by the entropy of mixing of the solvent and the solute plays an important role

Solvent A	Solute B	Interaction			Mutual solubility
		A $\cdots$ A	B $\cdots$ B	A $\cdots$ B	
Nonpolar	Nonpolar	Weak	Weak	Weak	High
Nonpolar	Polar	Weak	Strong	Weak	Low
Polar	Nonpolar	Strong	Weak	Weak	Low
Polar	Polar	Strong	Strong	Strong	High

The necessary condition for dissolution of a substance is that energetic stabilization is

in the high mutual solubility. For the dissolution of electrolytes, see Section 2.1.

are of opposite sign. If the two charges are placed in a solvent of relative permittivity ε_r and at a distance r, the electrostatic force F_{solv} between them is expressed by Eq. (1.2):

$$F_{solv} = \frac{q_1 q_2}{4\pi\varepsilon_0\varepsilon_r r^2} = \frac{F_{vac}}{\varepsilon_r} \tag{1.2}$$

Because ε_r is larger than $\sim$1.8 for most solvents (1.84 for *n*-pentane and 1.88 for *n*-hexane are examples of lowest ε_r values), the electrostatic interaction between charges is always weakened by solvents. As discussed in Chapter 2, the relative permittivity of a solvent has a decisive influence on the electrostatic solute–solute and solute–solvent interactions as well as on the dissolution and dissociation of electrolytes. Thus, it is used in classifying solvent polarity or solvating capability. Solvents of high permittivities ($\varepsilon_r \geq 15$ or 20) are called *polar solvents*, while those of low permittivities are called *apolar* or *nonpolar solvents* (Section 1.2). Many of the solvents listed in Table 1.1 are polar solvents, because the solvents for electrochemical use must dissolve and dissociate electrolytes. The relative permittivities of *N*-methylformamide (NMF) and *N*-methylacetamide (NMA) are exceptionally high, at 182 and 191, respectively. This is because the molecules of these solvents mutually interact by hydrogen bonding and are linearly arranged, causing high permittivities (Section 1.1.3). However, some nonpolar solvents, e.g. hexane and benzene ($\varepsilon_r \sim 2$), are now also used in electrochemical measurements, as will be discussed in Section 8.4.

If a solvent is placed in a low-frequency electric field ($<10^7$ Hz), its molecules are polarized in two ways: one is the *induced polarization*, which is due to the atomic and electronic displacements, and the other is the *orientational polarization*, which is due to the alignment of the permanent dipoles. They both contribute to the *static permittivity*, ε_s, which is equal to ε_r in Table 1.1. However, if the frequency of the electric field is increased, the orientational polarization is lost in the microwave region (10^9–10^{11} Hz) because the permanent dipoles need some time to rotate or reorient. The permittivity after this Debye (rotational) relaxation is the *infinite frequency permittivity* and is denoted by ε_∞ (Figure 1.1). Then, after the resonant transition in the IR region, the polarization occurs only due to electronic displacement. The permittivity then obtained is the *optical permittivity* and is denoted by ε_{op}. After the transition in the UV region, no polarization occurs and the permittivity becomes equal to unity. Table 1.3 shows the values of ε_s, ε_∞ and ε_{op} for some solvents. It also shows the values of the Debye relaxation time, τ_D, and the longitudinal relaxation time, τ_L; τ_D is obtained experimentally by such methods as dielectric relaxation spectroscopy [5] and τ_L is obtained by the relation $\tau_L = (\varepsilon_\infty/\varepsilon_s)\tau_D$ [6]. For H-bonding solvents such as alcohols and water, the Debye relaxation process is more complicated. Table 1.4 shows the data for the sequential relaxation of such solvents. For example, monoalcohols give three relaxation processes: the first (slowest) one (τ_1) is attributed to the winding chain formed by association, the second one (τ_2) is attributed to the rotation of monomers and molecules situated at the chain end and the third one (τ_3) is attributed to the hindered rotation of molecules within the H-bonded system. Solvents that undergo one Debye relaxation are called 'Debye' solvents, while those that undergo sequential relaxations are called 'non-Debye' solvents. For PC and DMF, some confusion is observed whether their behavior is Debye or non-Debye. According to the recent studies, these dynamic properties of solvents have remarkable influences on various electrochemical processes such as

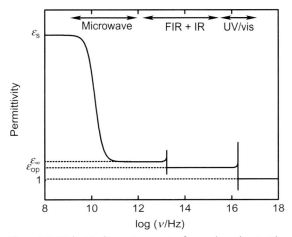

Figure 1.1 Dielectric dispersion spectra for a polar solvent with a single Debye relaxation process in the microwave region and two resonant transmissions in the IR and UV ranges [5b].

ion solvation, homogeneous and heterogeneous electron-transfer reactions and ionic migrations, as discussed in Sections 2.2.2, 4.13, 7.2.1, 8.2.2 and 8.3.1.

The refractive index, n_D, defined as the ratio of light speed at the sodium D-line in a vacuum to that in the medium, is used in obtaining the polarizability, α, of

Table 1.3 Solvent dielectric and related properties at $25\,^{\circ}C$.[a]

Solvent	$\varepsilon_s\ (= \varepsilon_r)$	ε_{op}	ε_∞	$\varepsilon_{op}^{-1} - \varepsilon_s^{-1}$	τ_D (ps)	τ_L (ps)
Debye solvents						
AN	37.5	1.80	2	0.528	3.3	~0.2
Ac	21	1.84	2	0.495	3.3	0.3
DMSO	46.7	2.18	5.7	0.438	19.5	2.4
HMPA	29.6	2.12	3.3	0.438	80	8.9
NB	35.7	2.40	4.1	0.389	45.6	5.2
Py	13.3	2.27	2.3	0.365	6.9	1.2
THF	7.58	1.97	2.3	0.376	3.3	1.0
Non-Debye solvents						
EtOH	24.5	1.85	4.2	0.499	130	22
FA	110	2.09	7.0	0.469	37	2.35
MeOH	32.7	1.76	5.6	0.628	48	8.2
NMF	182	2.04	5.4	0.485	123	3.7
1-PrOH	20.4	1.92	2.2	0.472	390	42
Debye or non-Debye (?) solvents (different viewpoints exist)						
DMF	36.7	2.04	4.5	0.472	11.0	1.3
PC	65	2.02	4.1	0.480	43	2.7

[a]From McManis, G.E., Golovin, M.N. and Weaver, M.J. (1986) *J. Phys. Chem.*, **90**, 6563; Barthel, J., Bachhuber, K., Buchner, R., Gill, J.B. and Kleebauer, M. (1990) *Chem. Phys. Lett.*, **167**, 62; Galus, Z. (1995) *Advances in Electrochemical Science and Engineering*, vol. 4 (eds H. Gerischer and C.W. Tobias), Wiley-VCH Verlag GmbH, Weinheim, p. 222.
ε_s: static permittivity; ε_{op}: optical permittivity; ε_∞: infinite frequency permittivity; $(\varepsilon_{op}^{-1} - \varepsilon_s^{-1})$: solvent Pekar factor; τ_D: Debye relaxation time and τ_L: longitudinal relaxation time.

Table 1.4 Dielectric relaxation parameters of water and lower alcohols determined by femtosecond terahertz pulse spectroscopy at 25 °C.[a]

Solvent	ε_s	τ_1 (ps)	ε_2	τ_2 (ps)	ε_3	τ_3 (ps)	ε_∞
Water	78.36	8.24	4.93	0.18			3.48
MeOH	32.63	48	5.35	1.25	3.37	0.16	2.10
EtOH	24.35	161	4.15	3.3	2.72	0.22	1.93
1-PrOH	20.44	316	3.43	2.9	2.37	0.20	1.97

[a]From Kindt, J.T. and Schmuttenmaer, C.A. (1996) *J. Phys. Chem.*, **100**, 10373.

solvent molecules. The relationship between α and n_D is given by $\alpha = (3V_m/4\pi N_A)$ $(n_D^2 - 1)/(n_D^2 + 2)$, where N_A is the Avogadro constant and V_m is the molar volume.[4] Solvent molecules with high α values tend to interact easily with one another or with other poralizable solute particles by dispersion forces.[5]

Most solvents consist of molecules that are intrinsic dipoles and have permanent dipole moments (μ). If such molecules are placed between the two plates of a capacitor as a vapor (or as a dilute solution in a nonpolar liquid), they are oriented by the electric field. Then, the orientational polarization and the induced polarization occur simultaneously, as described above. If ε_r is the relative permittivity of the vapor, there is a relationship

$$\frac{\varepsilon_r - 1}{\varepsilon_r + 2} = \frac{4\pi N_A}{3V_m}\left(\alpha + \frac{\mu^2}{3k_BT}\right) \tag{1.3}$$

where k_B is the Boltzmann constant. By plotting the relation between $V_m(\varepsilon_r - 1)/(\varepsilon_r + 2)$ and $1/T$, the value of μ is obtained simultaneously with the value of α, although a more accurate value of μ is obtainable from the Stark splitting of microwave lines.

The dipole moment is also used to assess the solvent polarity: solvents with high dipole moments (e.g. $\mu \geq 2.5$ D, 1 D = 3.33564×10^{-30} C m) are called *dipolar solvents*, while those with low dipole moments are called *apolar* or *nonpolar solvents*. Many solvents with high ε_r values also have high μ values (see Table 1.1). However, the μ value of water (1.77 D) is lower than expected from its high solvating abilities. The dipole moment

4) Examples of n_D values: methanol 1.326, water 1.332, AN 1.341, hexane 1.372, PC 1.419, DMF 1.428, DMSO 1.477, benzene 1.498, Py 1.507, NB 1.550 and DMTF 1.576 (Table 3.5 in Ref. [2a]). For all solvents, the value of n_D is between 1.2 and 1.8. There is a relationship $\varepsilon_{op} \sim n_D^2$.

5) Dispersion forces (instantaneous dipole–induced dipole interactions): Even in atoms and molecules with no permanent dipole moment, the continuous movement of electrons results, at any instant, in small dipole moments, which fluctuatingly polarize the electronic system of the neighboring atoms or molecules. This coupling causes the electronic movements to be synchronized in such a way that a mutual

attraction results (Ref. [1a], p. 13). The dispersion forces, which are universal for all atoms and molecules, are proportional to the products of the polarizabilities (α) of the two interacting species but are short range in action. Among the intermolecular forces, the dispersion forces are often stronger than the dipole–dipole and dipole–induced dipole forces, though they are weaker than the hydrogen bonding. Due to the dispersion forces, benzene exists as a liquid at normal temperatures, and hydrogen and argon are condensed to liquids at low temperatures. See, for example, Israelachvili, J.N. (1992) *Intermolecular and Surface Forces*, 2nd edn, Academic Press, London, Chapter 6.

tends to underestimate the polarity of small solvent molecules because it depends on the distance between the positive and negative charge centers in the molecule.

Many efforts have been made to correlate solute–solvent and solute–solute interactions in solutions with such polarity scales as relative permittivity and dipole moment, but these have often been unsuccessful. The chemical properties of solvents, as described below, often play more important roles in such interactions.

1.1.2
Chemical Properties of Solvents

Here, we mean by 'chemical properties' the acidity and basicity of solvents. Furthermore, we use the terms 'acidity' and 'basicity' in a somewhat broader sense than usual. The ability to accept an electron is included in the acidity of solvents, in addition to the abilities to donate a proton and a hydrogen bond and to accept an electron pair, while the ability to donate an electron is included in the basicity of solvents, along with the abilities to accept a proton and a hydrogen bond and to donate an electron pair. Conventionally, acidity and basicity are defined by the proton-donating and proton-accepting capabilities by the Brønsted acid–base concept and the electron pair-accepting and electron pair-donating capabilities by the Lewis acid–base concept. However, a solvent with a strong proton-donating ability usually has strong hydrogen bond-donating, electron pair-accepting and electron-accepting abilities. Moreover, a solvent with a strong proton-accepting ability usually has strong hydrogen bond-accepting, electron pair-donating and electron-donating abilities. Inclusion of electron-accepting and electron-donating abilities in acidity and basicity, respectively, is also justified by the fact that the energies of the highest occupied molecular orbital (HOMO) and the lowest unoccupied molecular orbital (LUMO) for molecules of various solvents are linearly correlated with the donor and acceptor numbers (see below), respectively [7].

As outlined in Section 1.3, the solvent acidity and basicity have a significant influence on the reactions and equilibria in solutions. In particular, differences in reactions or equilibria among the solvents of higher permittivities are often caused by the differences in solvent acidity and/or basicity. Because of the importance of solvent acidity and basicity, various empirical parameters have been proposed to express them quantitatively [1a, 2].[6] Examples of the solvent acidity scales are Kosower's Z values [8], Dimroth and Reichardt's E_T scale [1a, 9], Mayer, Gutmann and Gerger's acceptor number (AN) [10, 11], Kamlet and Taft's α parameter [12] and Catalán and Díaz's SA parameter [13]. On the other hand, examples of the solvent basicity scales are Gutmann's donor number (DN) [11, 14], Kamlet and Taft's β parameter [12] and Catalán et al.'s SB parameter [15]. Besides the acidity and basicity parameters, empirical solvent polarity/polarizability parameters such as the π^* scale [16] have also been proposed. The correlations between these empirical parameters have been

---

6) The parameters for solvent acidity and basicity are included in the parameters of solvent 'polarity'. Here, the solvent polarity is defined as solvent's overall solvation capability (solvation power) for solutes, which in turn depends on the action of all possible, specific and nonspecific, intermolecular interactions between solute ions or molecules and solvent molecules [17]. As reviewed in Ref. [18], the quantitative measures of solvent polarity are numerous.

studied in detail [2, 19]. Moreover, to relate these parameters to solvent effects on various physicochemical quantities in solutions, linear solvation energy relationships (LSER) [20], as expressed by Eq. (1.4), are often employed:

$$XYZ = XYZ_0 + a \cdot \alpha + b \cdot \beta + s \cdot \pi^* + \cdots \tag{1.4}$$

where XYZ is the given quantity, XYZ_0 is the quantity at $\alpha = \beta = \pi^* = 0$ and a, b and s are the coefficients for α, β and π^*, respectively. In the same way, for AN and DN, a semiempirical multiparameter relationship has been given for the solvent effects on physicochemical quantities:

$$\Delta G_S - \Delta G_R = a(DN_S - DN_R) + b(AN_S - AN_R) + c(\Delta G_{vp}^{\circ S} - \Delta G_{vp}^{\circ R})$$

Here, ΔG is the Gibbs energy for the quantity under consideration and ΔG_{vp}° is the standard molar Gibbs energy of vaporization; the sub- and superscripts S and R denote solvents S and R, respectively, and a, b and c are coefficients.

In this book, we only use the acceptor number, AN, and the donor number, DN, because they are the most popular in the field of electrochemistry and convenient to use.[7]

---

7) Kamlet and Taft's parameters (α for solvent hydrogen bond donor (HBD) acidities, β for solvent hydrogen bond acceptor (HBA) basicities and π^* for solvent dipolarity/polarizability) are dealt with in Chapter 13, in sections on supercritical fluids and ionic liquids. These parameters are evaluated by using UV/vis spectral data of solvatochromic compounds as follows: α, β and π^* values for cyclohexane (c-C_6H_{12}) being defined as equal to zero [1a]. (1) β scale: The enhanced solvatochromic shifts, $\Delta\Delta\tilde{\nu}$, are determined in HBA solvents for 4-nitroaniline relative to homomorphic N,N-diethyl-4-nitroaniline. Both standard compounds can act as HBA substrates (at the nitro-oxygen) in HBD solvents, but only 4-nitroaniline can act as an HBD substrate (at NH_2 group) in HBA solvents. Taking the $\Delta\Delta\tilde{\nu}$ value at 2800 cm^{-1} for HMPA (a strong HBA solvent) as $\beta = 1.00$, β values for HBA solvents are

obtained. (2) α scale: The enhanced solvatochromic shifts, $\Delta\Delta\tilde{\nu}$, are determined in HBD solvents for 4-nitroanisole and the pyridinium-N-phenolate betaine dye, and taking the $\Delta\Delta\tilde{\nu}$ value of 6240 cm^{-1} for methanol (a strong HBD solvent) as $\alpha = 1.00$, α values for HBD solvents are obtained. (3) π^* scale: Solvent effects on the $\pi \rightarrow \pi^*$ electronic transitions of positively solvatochromic nitroaromatics of the type D–C_6H_4–A, where D and A stand for electron donor (e.g. NMe_2) and electron acceptor (e.g. NO_2) groups, respectively (e.g. 4-nitroanisole and N,N-dimethylamino-4-nitroaniline). The relation $\pi^*(S) = [\tilde{\nu}(S) - \tilde{\nu}(c\text{-}C_6H_{12})]/[\tilde{\nu}(DMSO) - \tilde{\nu}(c\text{-}C_6H_{12})]$, where $\tilde{\nu}(S)$ is the wavenumber of the maximum of the long-wavelength solvatochromic absorption band of the indicator in solvent S ($\pi^*(DMSO) = 1.00$), is used. The values obtained by somewhat modified methods are shown below.

Solvents	α	β	π^*	Solvents	α	β	π^*
Gas phase	0.00	0.00	−1.23	NM	0.22	0.06	0.75
Cyclohexane (c-C_6H_{12})	*0.00*	*0.00*	*0.00*	HMPA	0.00	*1.00*	0.87
Dichloromethane	0.13	0.10	0.73	DMF	0.00	0.69	0.88
NB	0.00	0.30	0.86	NMF	0.62	0.80	0.90
Py	0.00	0.64	0.87	FA	0.71	0.48	0.97
THF	0.00	0.55	0.55	EtOH	0.86	0.75	0.54
Ac	0.08	0.48	0.62	MeOH	0.98	0.66	0.60
PC	0.00	0.40	0.83	Acetic acid	1.12	0.45	0.64
DMSO	0.00	0.76	*1.00*	2,2,2-Trifluoroethanol	1.51	0.00	0.73
AN	0.19	0.40	0.66	Water	1.17	0.47	1.09

The donor number, DN [11, 14], of solvent D (Lewis base) is determined calorimetrically as the negative value of the standard enthalpy change, $-\Delta H°$ (in kcal mol^{-1}), for the $1:1$ adduct formation between solvent D and antimony pentachloride (SbCl$_5$), both being dilute, in 1,2-dichloroethane (DCE) at 25°C (Eq. (1.5)).

$$\text{D}: +\text{SbCl}_5 \rightleftarrows \text{D} - \text{SbCl}_5 \qquad DN = -\Delta H° \,(\text{kcal mol}^{-1}) \qquad (1.5)$$

The values of DN are listed in Table 1.5 in increasing order. The solvent basicity increases with the increase in the DN value. The DN value for DCE (reference solvent) is zero.

The acceptor number, AN [10, 11], of solvent A (Lewis acid) is obtained by measuring the ^{31}P-NMR chemical shift ($\Delta\delta$, ppm) of triethylphosphine oxide (Et$_3$P=O, a strong Lewis base) in solvent A:

$$(\text{Et}_3\text{P} = \text{O} \leftrightarrow \text{Et}_3 \overset{\oplus}{\text{P}} - \overset{\ominus}{\text{O}}) + \text{A} \rightleftarrows \text{Et}_3 \overset{\delta+}{\text{P}} \cdots \overset{\delta-}{\text{O}} - \text{A}$$

The ^{31}P-NMR chemical shift of Et$_3$P=O is also measured in hexane ($\Delta\delta$(hexane)) and in DCE containing SbCl$_5$ ($\Delta\delta$(SbCl$_5$ in DCE)). Here, by definition, $AN = 0$ for hexane and $AN = 100$ for SbCl$_5$ in DCE. Then, the AN of solvent A is obtained by Eq. (1.6).

$$AN = 100 \times \frac{\Delta\delta(A) - \Delta\delta(\text{hexane})}{\Delta\delta(\text{SbCl}_5 \text{ in DCE}) - \Delta\delta(\text{hexane})} = 2.348[\Delta\delta(A) - \Delta\delta(\text{hexane})]$$

$$(1.6)$$

The values of AN are also included in Table 1.5.[8] The solvent acidity increases with the increase in the AN value. Here, it should be noted that neither DN nor AN can be correlated with the relative permittivity of the corresponding solvents.

Lewis acids are electron pair acceptors and Lewis bases are electron pair donors. However, according to the hard and soft acids and bases (HSAB) concept [21], Lewis acids are classified into hard and soft acids, while Lewis bases are classified into hard and soft bases. Hard acids interact strongly with hard bases, soft acids with soft bases.

The HSAB concept also applies to solvent–solute interactions. Therefore, we have to know whether the solvent is hard or soft as a Lewis acid and a Lewis base. Water is a hard acid and a hard base. In general, hydrogen bond donor solvents are hard acids and solvate strongly to hard-base anions (i.e. small anions such as OH$^-$, F$^-$, Cl$^-$ and anions with a negative charge localized on a small oxygen atom (CH$_3$O$^-$, CH$_3$COO$^-$, etc.)). On the other hand, for solvents having electron pair donor atoms such as O, N and S, the softness increases in the order O < N < S. Here, examples of solvents with an O atom are water, alcohols, ketones and amides, those with an N atom are nitriles, amines and pyridine and those with an S atom are thioethers and

8) Riddle and Fowkes [22] considered that the acceptor numbers determined by the NMR method are partly due to the van der Waals forces between Et$_3$P=O and solvent molecules and attributed somewhat large AN values of strongly basic solvents such as pyridine to it. They proposed new acceptor numbers, cor-rected for the influence of the van der Waals forces, as follows: hexane 0.0, Py 0.5, THF 1.9, DMA 5.7, DMF 6.6, Ac 8.7, DMSO 10.8, NM 14.8, AN 16.3, FA 32.2, MeOH 41.7, CH$_3$COOH 49.3, W 52.4, CF$_3$COOH 111.0 (compare with the AN values in Table 1.5).

Table 1.5 Chemical properties of organic solvents of electrochemical interest.

Solvent[a]	DN	AN	pK_{SH}	ε_r	Solvent	DN	AN	pK_{SH}	ε_r
(48) 1,2-Dichloroethane (DCE)	0	16.7		10.4	(6) Methanol (MeOH)	(19)	41.3	17.2	32.7
(42) Hexane	(0)	0.0		1.88	(3) Formic acid	(19)	83.6	6.2	58.5_{16}
(43) Benzene	0.1	8.2		2.27	(13) Tetrahydrofuran (THF)	20.0	8.0		7.6
(45) Nitromethane (NM)	2.7	20.5		36.7	(4) Acetic acid (HOAc)	(20)	52.9	14.45	6.2
(46) Nitrobenzene (NB)	4.4	14.8		34.8	(15) 1,2-Dimethoxyethane (DME)	23.9	10.2		7.2
(5) Acetic anhydride	10.5	—	14.5	20.7_{19}	(28) Formamide (FA)	(24)	39.8	16.8_{20}	111.0
(24) Benzonitrile	12.0	—		25.2	(30) N,N-Dimethylformamide (DMF)	26.6	16.0	29.4	36.7
(20) Acetonitrile (AN)	14.1	18.9	33.3	35.9	(35) N-Methyl-2-pyrrolidinone (NMP)	27.3	13.3	25.6	32.2
(39) Sulfolane (TMS)	14.8	—	25.5	43.3	(32) N,N-Dimethylacetamide (DMA)	27.8	13.6	23.9	37.8
(14) 1,4-Dioxane	14.8	10.8		2.21	(36) Tetramethylurea (TMU)	29.6			23.6
(50) Propylene carbonate (PC)	15.1	18.3		66.1	(38) Dimethyl sulfoxide (DMSO)	29.8	19.3	33.3	46.5
Diethyl carbonate (DEC)	16.0	—		2.8	(27) Pyridine (Py)	33.1	14.2		12.9
(51) Ethylene carbonate (EC)	16.4	—		89.6	(34) Hexamethylphosphoric triamide (HMPA)	38.8	10.6	20.6	29.6
(52) Methyl acetate (MA)	16.5	10.7		6.7	(7) Ethanol (EtOH)	(32?)	37.9	19.1	24.6
(23) Butyronitrile (BuN)	16.6	—		20.3	(8) 1-Propanol (1-PrOH)		37.3	19.4	20.5
(17) Acetone (Ac)	17.0	12.5	32.5	20.7	(9) 2-Propanol (2-PrOH)	(36?)	33.6	21.1	19.9
(53) Ethyl acetate	17.1	9.3	22.8	6.0	(29) N-Methylformamide (NMF)	(49?)	32.1	10.74	182.4
(49) γ-Butyrolactone (γ-BL)	(18)	17.3		39	Trifluoroacetic acid		105.3		8.55
(1) Water	18(G)–33(L)[b]	54.8	14.0	78.4					

Donor number (DN), acceptor number (AN) and autoprotolysis constants (pK_{SH}) (with relative permittivities (ε_r)).
[a] For the numbers before the solvent names, see Table 1.1.
[b] G means gas and L means liquid.

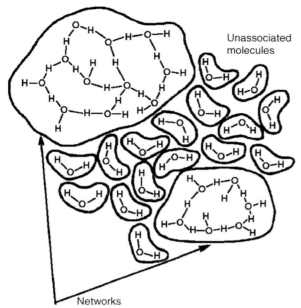

Figure 1.2 The three-dimensional structure of water. (Nemethy, G. and Scheraga, H.A. (1962) *J. Chem. Phys.*, **36**, 3382.)

thioamides. Hard-base solvents solvate strongly to hard-acid cations (Na$^+$, K$^+$, etc.), while soft-base solvents easily solvate to soft-base cations (Ag$^+$, Cu$^+$, etc.). Antimony pentachloride (SbCl$_5$), used in determining the donor number of solvents, is in between a hard acid and a soft acid. However, the donor number is considered to be the scale of solvents as hard bases. Recently, some scales have been proposed for the softness of solvents [23].[9]

1.1.3
Structural Aspects of Solvents

The physical and chemical properties of solvents are closely related to their structures. Water molecules have strong hydrogen-bonding ability and considerable parts of them are combined with one another to form three-dimensional networks (Figure 1.2) [24]. A water molecule held in a network does not stay there long and is liberated as a free molecule; the lifetime of an individual hydrogen bond is of

9) For example, Marcus [23] proposed the μ scale for the softness of solvents. If the Gibbs energy of transfer of species *i* from water to solvent S is expressed by $\Delta G_t^\circ(i, W \rightarrow S)$ (in kJ mol^{-1}), μ is defined by $\mu = \{(1/2)[\Delta G_t^\circ(Na^+, W \rightarrow S) + \Delta G_t^\circ(K^+, W \rightarrow S)] - \Delta G_t^\circ(Ag^+, W \rightarrow S)\}/ 100$. This scale is based on the fact that the size of the soft acid, Ag$^+$ (0.115 nm in radius), is

between the sizes of the hard acids, Na$^+$ (0.102 nm) and K$^+$ (0.138 nm). Examples of the μ values are as follows [2]: AN 0.34; DMF 0.11, DMTF 1.35; DMSO 0.22; FA 0.09; NB 0.23; NM 0.03; NMP 0.13, NMTP 1.35; PC −0.09; Py 0.64. Here, DMTF and NMTP are soft bases, and Py and AN are between soft and hard.

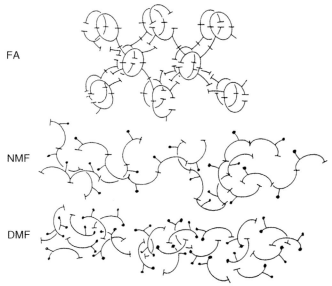

FA

NMF

DMF

Figure 1.3 The liquid structures of FA, NMF and DMF [27a].
•– Methyl group; ✗ H-bond (−NH ··· O=CH−).

the order of 0.6 ps [25]. However, the network formation by hydrogen bonding is responsible for various anomalous physical properties of liquid water, including high boiling and melting points, high values of heat of vaporization, surface tension, compressibility and viscosity and peculiar density change with temperature [26]. Due to the network formation, molecules and ions, which are large in size, are often difficult to dissolve in water unless they have hydrophilic site(s).

Studies of solvent structure are usually carried out by analyzing radial distribution functions that are obtained by X-ray or neutron diffraction methods. Monte Carlo (MC) or molecular dynamics (MD) calculations are also used. Studies of the structure of nonaqueous and mixed solvents are not extensive yet, but some of the results have been reviewed. Pure and mixed solvents included in the reviews [27] are FA, NMF, DMF, DMSO, AN, 2,2,2-trifluoroethanol, EtOH, DMF/AN and 2,2,2-trifluoroethanol/DMSO. For example, Figure 1.3 schematically shows the liquid structures of FA, NMF and DMF. In FA, chain structure and ring-dimer structure are combined by hydrogen bonding to form three-dimensional networks, causing high melting and boiling points and high viscosity of FA.[10] In NMF, linear but short chain structures predominate, giving it a high permittivity. DMF is not hydrogen bonding and most DMF molecules exist as monomers. Thus, the melting and boiling points and the viscosity of DMF are lower than those of FA and NMF. For other solvents, see Ref. [27].

10) The results of X-ray diffraction and molecular dynamics at high pressure and high temperature show that the pressure increase enhances ring dimmer formation at the expense of linear chain structure, but the temperature increase gives an opposite effect [28].

1.1.4
Toxicity and Hazardous Properties of Solvents

There have been recent concerns that many solvents are toxic or hazardous to human health and/or the environment. The latest situation has been discussed in detail in Refs [29–31]. Especially, the *International Chemical Safety Cards (ICSCs)* [31], which is freely available on Web, provides very useful information on the safety and toxicity of many individual solvents.

Effects on human health usually occur by exposure to solvents or by the uptake of solvents through the lungs or skin. General effects that are caused by acute exposure to high solvent concentrations are dysfunctions of the central nervous system (CNS); symptoms such as dizziness, euphoria, confusion, nausea, headache, vomiting, paresthesia, increased salivation, tachycardia, convulsions, and coma can occur, depending on the situation. Besides these general effects, specific effects by particular solvents are also observed. Among these are nonimmunological hepato-toxicity (halogenated hydrocarbons, EtOH, DMF), nephrotoxicity (halogenated hy-drocarbons, toluene, dioxane, ethylene glycol), reproductive toxicity (CS_2, benzene, nitrobenzene), hemopoietic toxicity (benzene metabolites), neurotoxicity (hexane, EtOH, styrene), ocular toxicity (MeOH) and immunological allergies to various solvents. More seriously, carcinogenic solvents are considered to induce malignant tumor; even among the solvents listed in Table 1.1; benzene (A1: confirmed human carcinogen) and dichloromethane, HMPA, NB, NM, 1-PrOH, Py and THF (A3: confirmed animal carcinogen with unknown relevance to humans) are suspected to have carcinogenic effect on humans. Threshold limit values (TLVs), listed in the last column of Table 1.1, are the values of time-weighted average (TWA, the definition of which is in the footnote of the table) taken from Ref. [31], in which other information on the safety and toxicity of solvents is also available. Because of the complicated nature of carcinogenesis, it is often difficult to define TLVs for carcinogens.

Many solvents in common use are volatile organic compounds (VOCs),[11] and various environmental problems are caused by their evaporation. In the lower atmosphere, VOCs participate in photochemical reactions to form, to varying degrees, ground-level ozone and other oxidants that affect health, as well as cause damage to materials, crops and forests. Ozone impairs normal functioning of the lungs and reduces the ability to perform physical activity. Some solvents are listed as hazardous air pollutants (HAPs): they are toxic and/or carcinogenic and are associated with serious health effects such as cancer, liver or kidney damage, reproductive disorders and developmental or neurological problems. They also have detrimental environmental effects on wildlife and degrade water or habitat quality. The 'T' symbol in the last column of Table 1.1 shows that the solvent has been listed as an HAP. The solvents known as chlorofluorocarbons (CFCs) generally do not contribute to ground-level ozone formation but cause stratospheric ozone depletion.

---

11) A volatile organic compound is defined by the Environmental Protection Agency (EPA) as any compound of carbon, excluding carbon monoxide, carbon dioxide, carbonic acid, metallic carbides or carbonates and ammonium carbonate, which is emitted or evaporated into the atmosphere.

In the stratosphere, these gradually release chlorine and other halogens into the atmosphere; these are effective in destroying the ozone layer that protects us from damage by ultraviolet light. The production and use of many CFCs have been banned and new chemicals are used instead. Recently, there has been serious contamination of water and soil with hazardous solvents, but this is not discussed here. In the laboratory, there are potential hazards of accidental spillage of organic solvent of low boiling point, which may be highly inflammable.

In recent years, many efforts have been made to avoid problematic effects of solvents, and many international and national regulations have been established (see Chapter 19 of Ref. [29]); toxic solvents are being replaced with nontoxic or less toxic ones and environmentally hazardous solvents with harmless ones. The use of 1,3-dimethyl-2-imidazolidinone, 1,3-dimethylpropyleneurea or 1,1,3,3-tetramethylurea instead of HMPA is one such example. The search for new, environmentally benign reaction media is the subject of current research. As such media, electrochemical and nonelectrochemical uses of supercritical fluids, room-temperature ionic liquids and fluorous solvents are becoming frequent (see Chapter 13). In nonelectrochemical fields, many reactions that have been carried out in nonaqueous solvents are now performed in aqueous systems or nonsolvent systems.

1.2
Classification of Solvents

The classification of solvents has been dealt with in various books on nonaqueous solvents [32, 33]. In the classification of solvents, it is usual to use some solvent properties as criteria. In order to discuss solvent effects on chemical reactions, it is convenient to use relative permittivities and acid–base properties as criteria.

Type	1	2	3	4	5	6	7	8
Relative permittivity	+	+	+	+	−	−	−	−
Acidity	+	+	−	−	+	+	−	−
Basicity	+	−	+	−	+	−	+	−

In 1928, Brønsted [34] used these criteria and classified solvents into the above eight types. In the table, plus (+) means high or strong and minus (−) means low or weak. Various improved methods of classification have been proposed since; in this book, we follow the classification by Kolthoff [32] (Table 1.6).[12] According to his

12) In Ref. [33c], solvents are classified as follows: *protic solvents* (amphiprotic hydroxylic solvents such as water, methanol and glycols; amphiprotic protogenic solvents such as CH_3COOH and HF; protophilic H-bond donor solvents such as FA, NMF and NH_3); *dipolar aprotic solvents* (aprotic protophilic solvents such as DMF, DMSO and Py; aprotic protophobic solvents such as AN, Ac, NM and PC; low-permittivity electron donor solvents such as diethyl ether, dioxane and THF); *low-polarity and inert solvents* (low-polarity solvents of high polarizability such as CH_2Cl_2, $CHCl_3$ and benzene; inert solvents such as *n*-hexane and cyclohexane).

Table 1.6 Classification of solvents (Kolthoff) [32].

	No.	ε_r, μ	Acidity[b]	Basicity[b]	Examples[a]
Amphiprotic solvents					
Neutral	1a	+	+	+	Water (78); MeOH (33); ethylene glycol (38)
	1b	−	+	+	t-BuOH (11)
Protogenic	2a	+	++	±	H_2SO_4; HF; HCOOH (58)
	2b	−	++	±	CH_3COOH (6)
Protophilic	3a	+	±	++	NMF (182); DMSO (46)[c]; tetra-methylurea (24); FA (111); NH_3 (23)
	3b	−	±	++	En (13); tetramethylguanidine (12)
Aprotic solvents					
Dipolar protophilic[d]	4a	+	− (±)	++ (+)	DMF (37); DMSO (46)[c]; NMP (32); HMPA (30)
	4b	−	−	++ (+)	Py (13); THF (8); ether (4)
Dipolar protophobic[d]	5a	+	− (±)	−	AN (36); PC (65); NM (37); TMS (43); Ac (21)
	5b	−	−	−	MIBK (13); methylethylketone (17)
Inert	5c	−	−	−	Aliphatic hydrocarbons (∼2); benzene (2); CCl_4 (3); DCE (10)

[a] The symbol + is for $\varepsilon_r \geq 15$ or 20, $\mu \geq 2.5$ D and − is for $\varepsilon_r < 15$ or 20, $\mu < 2.5$ D. In parentheses in the column 'Examples' are shown approximate values of ε_r.
[b] The symbol + is for the case comparable to water, ++ for the case much stronger than water, ± for the case somewhat weaker than water and − for the case much weaker than water.
[c] DMSO is an amphiprotic solvent because its autoprotolysis occurs slightly ($pK_{SH} \sim 33$) and the lyate ion ($CH_3SOCH_2^-$) is somewhat stable. However, DMSO is classified as an aprotic solvent. The rough criteria for aprotic solvents are $pK_{SH} > 22$ and $AN < 20$.
[d] Some solvents with $\varepsilon_r < 15$ (or $\mu < 2.5$ D) are also classified as 'dipolar'. For the reason, see text.

classification, solvents are roughly divided into two groups, *amphiprotic solvents* and *aprotic solvents*.[13]

Amphiprotic solvents have both acidic and basic properties in terms of the Brønsted acid–base concept. If we denote an amphiprotic solvent by SH, it donates a proton by $SH \rightleftarrows S^- + H^+$ and accepts a proton by $SH + H^+ \rightleftarrows SH_2^+$. Overall, the autoprotolysis (autoionization) occurs by $2SH \rightleftarrows SH_2^+ + S^-$. The extent of autoprotolysis is expressed by the autoprotolysis constant, $K_{SH} = a_{SH_2^+} \times a_{S^-}$, the values of which are also included in Table 1.5 as pK_{SH} values (for more details, see Table 6.6).

Using water as reference, an amphiprotic solvent with an acidity and a basicity comparable to those of water is called a *neutral solvent*, the one with a stronger acidity and a weaker basicity than water is called a *protogenic solvent* and the one with a weaker acidity and a stronger basicity than water is called a *protophilic solvent*. The solvent with relatively strong acidity usually has in its molecule a hydrogen atom that is joined to

13) There is an opinion that the term 'aprotic' should be reserved for solvents with no hydrogen atom (e.g. SO_2 and BF_3). However, it is more popular to use 'aprotic' for solvents that are very weak in proton-donating and hydrogen bond-donating abilities.

an electronegative atom such as oxygen (O), nitrogen (N) or halogen (X). Because of the electron pair donor capacity of the electronegative atom, a solvent with relatively strong acidity also has some basicity. Actually, there are no acidic solvents without some basicity.

Aprotic solvents, on the other hand, do not have a hydrogen atom joined to electronegative atom. Generally, the hydrogen atom(s) of an aprotic solvent is joined only to a carbon atom. Therefore, aprotic solvents have very weak proton-donating and hydrogen bond-donating abilities. Concerning the basicity, however, some aprotic solvents are stronger, although some are much weaker, than water. Aprotic solvents with strong basicity are said to be *protophilic*, while those with very weak basicity are said to be *protophobic*. The molecules of protophilic aprotic solvents have an oxygen atom or a nitrogen atom, on which negative charge is located. Among the aprotic solvents, those with relatively high permittivities ($\varepsilon_r \geq 15$ or 20) or large dipole moments ($\mu \geq 2.5$ D) are often called *dipolar aprotic solvents*. As shown in Table 1.6, some aprotic solvents with $\varepsilon_r < 15$ or $\mu < 2.5$ D (e.g. Py, THF, diethyl ether, MIBK) are classified as dipolar solvents. This is because, due to their acidic or basic properties, they behave like dipolar solvents. Solvents with low relative permittivities (or dipole moments) and very weak acidic and basic properties are called *inert solvents*.

The distinction between amphiprotic and aprotic solvents is not always clear. For instance, dimethyl sulfoxide (DMSO) is usually considered aprotic, but it undergoes an autoprotolysis as follows:

$$2CH_3SOCH_3 \rightleftharpoons (CH_3SOCH_3)H^+ + CH_3SOCH_2^- \quad (pK_{SH} \approx 33)$$

where $(CH_3SOCH_3)H^+$ is a lyonium ion and $CH_3SOCH_2^-$ is a lyate ion. Thus, DMSO may be considered to be an amphiprotic solvent.[14] It is usual, however, to include solvents with $pK_{SH} > 22$ as aprotic solvents. On the other hand, the values of acceptor number are often less than 10 for inert solvents, between 10 and 20 for dipolar aprotic solvents and 25 or more for neutral or protogenic amphiprotic solvents.

1.3
Effects of Solvent Properties on Chemical Reactions: An Outline

Chemical reactions in solutions are often affected drastically by the solvents used. The main objective of this book is to correlate the properties of solvents and the solvent effects on various chemical processes relevant to electrochemistry. The most important solvent properties in considering solvent effects are the solvent permittivity and the solvent acidity and basicity. If the permittivity of one solvent is high ($\varepsilon_r > 30$) and that of the other is low ($\varepsilon_r < 10$), the difference in a chemical process in the two solvents is usually attributable to the influence of permittivity. However, the

14) The lyate ion of DMSO ($CH_3SOCH_2^-$) is called *dimsyl ion*. Its alkali metal salts are fairly stable and have been used as titrant in DMSO [32a].

Table 1.7 Acid–base properties of solvents and the characteristics of reactions

Solvents with weak (strong) acidity	Solvents with weak (strong) basicity
(1) Solvation to small anions is difficult (easy) • Small anions are reactive (nonreactive)	(1) Solvation to small cations is difficult (easy) • Small cations are reactive (nonreactive)
(2) Proton donation from solvent is difficult (easy) • pH region is wide (narrow) on the basic side • Strong bases are differentiated (leveled) • Very weak acids can (cannot) be titrated	(2) Proton acceptance by solvent is difficult (easy) • pH region is wide (narrow) on the acidic side • Strong acids are differentiated (leveled) • Very weak bases can (cannot) be titrated
(3) Reduction of solvent is difficult (easy) • Potential region is wide (narrow) on negative side • Strong reducing agent is stable (unstable) in the solvent • Substances difficult to reduce can (cannot) be reduced	(3) Oxidation of solvent is difficult (easy) • Potential region is wide (narrow) on positive side • Strong oxidizing agent is stable (unstable) in the solvent • Substances difficult to oxidize can (cannot) be oxidized

difference in a chemical process in two high-permittivity solvents (e.g. $\varepsilon_r > 30$) is often attributable to the influence of the acidity or basicity of the two solvents rather than the influence of permittivity. General tendencies of the effects of solvent acid–base properties on chemical processes are summarized in Table 1.7. For example, the items on the left-hand column of the table should be read as follows:

1. A solvent with weak acidity is a weak hydrogen bond donor and solvates only very weakly to small anions (F^-, Cl^-, OH^-, CH_3COO^-, etc.). Thus, small anions are very reactive in it. In contrast, a solvent with strong acidity easily solvates to small anions by hydrogen bonding and weakens their reactivity.

2. In a solvent with weak acidity, the solvent molecule cannot easily release a proton. Thus, the pH region is wider on the basic side than in water; some strong bases, whose strengths are leveled in water, are differentiated; some very weak acids, which cannot be determined by neutralization titration in water, can be determined. In contrast, in a solvent with strong acidity, a proton is easily released from the solvent molecule. Thus, the pH region is narrow on the basic side; strong bases are easily leveled; neutralization titrations of very weak acids are impossible.

3. A solvent with weak acidity is a weak electron acceptor and is more difficult to reduce than water. Thus, in it, the potential window is wider on the negative side than in water; some strong reducing agents that are not stable in water can survive; some substances that are difficult to reduce in water can be reduced. In contrast, a solvent with strong acidity easily accepts electrons and is reduced. Thus, in it, the potential window is narrow on the negative side; strong reducing agents easily reduce the solvent; some substances, which can be reduced in water, cannot be reduced until the reduction of the solvent.

Water has high permittivity and moderate acidity and basicity. Thus, in water, many cations and anions are easily solvated (hydrated) and many electrolytes are highly soluble and dissociate into ions. Water has fairly wide pH and potential ranges and a convenient liquid temperature range. Of course, water is an excellent solvent. However, as in Table 1.7, the reaction environment can be expanded much wider than in water by using a solvent with weak acidity and/or basicity. This is the reason why dipolar aprotic solvents, which are either protophilic or protophobic, are used in a variety of ways in modern chemistry.

Although water is an excellent solvent and the most popular, it has somewhat anomalous properties that come from the hydrogen-bonding ability of water to form three-dimensional networks (Figure 1.2, Section 1.1.3). Large molecules and ions are often difficult to dissolve in water, unless they have hydrophilic site(s). Therefore, water is not suitable as a medium for reactions involving large hydrophobic molecules or ions. In contrast, most dipolar aprotic solvents are nonstructured or only weakly structured and can dissolve many large hydrophobic molecules and ions. This is another major reason why dipolar aprotic solvents are often used instead of water.

References

1 (a) Reichardt, C. (2003) *Solvents and Solvent Effects in Organic Chemistry*, 3rd edn, Wiley-VCH Verlag GmbH, Weinheim; (b) Chipperfield, J.R. (1999) *Non-Aqueous Solvents*, Oxford University Press, Oxford.

2 (a) Marcus, Y. (1998) *The Properties of Solvents*, John Wiley & Sons, Inc., New York; (b) Marcus, Y. (1985) *Ion Solvation*, John Wiley & Sons, Inc., New York.

3 Riddick, J.A., Bunger, W.B. and Sakano, T.K. (eds) (1986) *Organic Solvents: Physical Properties and Methods of Purification*, 4th edn, John Wiley & Sons, Inc., New York.

4 Abboud, J.-L.M. and Notario, R. (1999) *Pure Appl. Chem.*, **71**, 645.

5 (a) Buchner, R. and Barthel, J. (1994) *Annu. Rep. Prog. Chem., Sect. C, Phys. Chem.*, **91**, 71; (b) Barthel, J.M.G., Krienke, H. and Kunz, W. (1998) *Physical Chemistry of Electrolyte Solutions, Modern Aspects*, Springer, Darmstadt, p. 88.

6 (a) Fröhlich, H. (1958) *Theory of Dielectrics*, 2nd edn, Clarendon Press, Oxford, p. 72;

(b) Kivelson, D. and Friedman, H. (1989) *J. Phys. Chem.*, **93**, 7026.

7 Sabatino, A., LaManna, G. and Paolini, I. (1980) *J. Phys. Chem.*, **84**, 2641.

8 Kosower, E.M. (1956) *J. Am. Chem. Soc.*, **78**, 5700; 1958, **80**, 3253, 3261, 3267.

9 (a) Dimroth, K., Reichardt, C., Siepmann, T. and Bohlmann, F. (1963) *Ann. Chem.*, **661**, 1; (b) Reichardt, C. (1994) *Chem. Rev.*, **94**, 2319 (A review on solvochromic solvent polarity indicators).

10 Mayer, U., Gutmann, V. and Gerger, W. (1975) *Monatsh. Chem.*, **106**, 1235.

11 (a) Gutmann, V. (1978) *The Donor–Acceptor Approach to Molecular Interactions*, Plenum Press, New York; (b) Gutmann, V. and Resch, G. (1995) *Lecture Notes on Solution Chemistry*, World Scientific, Singapore.

12 (a) Kamlet, M.J. and Taft, R.W. (1976) *J. Am. Chem. Soc.*, **98**, 377; (b) Taft, R.W. and Kamlet, M.J. (1976) *J. Am. Chem. Soc.*, **98**, 2886; (c) Kamlet, M.J., Abboud, J.-L.M., Abraham, M.H. and Taft, R.W. (1983) *J. Org. Chem.*, **48**, 2877; (d) Taft, R.W., Abboud, J.-L.M., Kamlet, M.J. and

Abraham, M.H. (1985) *J. Solution Chem.*, **14**, 153.

13 Catalán, J. and Díaz, C. (1997) *Liebigs Ann. Recl.* (9), 1941.

14 Gutmann, V. and Vychere, E. (1966) *Inorg. Nucl. Chem. Lett.*, **2**, 257.

15 Catalán, J., Díaz, C., López, V., Pérez, P., De Paz, J.L.G. and Rodríguez, J.G. (1996) *Liebigs Ann.* (11), 1785.

16 (a) Kamlet, M.J., Abboud, J.-L.M. and Taft, R.W. (1977) *J. Am. Chem. Soc.*, **99**, 6027; (b) Abboud, J.-L.M., Kamlet, M.J. and Taft, R.W. (1977) *J. Am. Chem. Soc.*, **99**, 8325.

17 (a) Ref. [1a], p. 389; (b) Müller, P. (1994) *Pure Appl. Chem.*, **66**, 1077.

18 Katritzky, A.R., Fara, D.C., Yang, H. and Tämm, K. (2004) *Chem. Rev.*, **104**, 175.

19 Marcus, Y. (1993) *Chem. Soc. Rev.*, **22**, 409.

20 Kamlet, M.J. and Taft, R.W. (1985) *Acta Chem. Scand.*, **B39**, 611.

21 Pearsons, R.G. (1963) *J. Am. Chem. Soc.*, **85**, 3533.

22 Riddle, F.L., Jr and Fowkes, F.M. (1990) *J. Am. Chem. Soc.*, **112**, 3259.

23 Marcus, Y. (1987) *J. Phys. Chem.*, **91**, 4422.

24 For example, Bockris, J.O'M. and Reddy, A.K.N. (1998) *Modern Electrochemistry, vol. 1, Ionics*, 2nd edn, Plenum Press, New York, p. 41.

25 Bertolini, D., Cassettari, M., Ferrario, M., Grigolini, P. and Salvetti, G. (1985) *Adv. Chem. Phys.*, **62**, 277.

26 For example, Ref. [11b], Chapter 6.

27 (a) Ohtaki, H. and Ishiguro, S. (1994) *Chemistry of Nonaqueous Solutions: Current Progress* (eds G. Mamantov and A.I. Popov), Wiley-VCH Verlag GmbH, New York, Chapter 3; (b) Yamaguchi, T. (1995) *Molecular Pictures of Solutions*, Gakkai Shuppann Center, Tokyo, Chapter 2 (in Japanese).

28 Radnai, T., Megyes, T., Bakó, I., Kosztolányi, T., Pálinkás, G. and Ohtaki, H. (2004) *J. Mol. Liq.*, **110**, 123.

29 Wypych, G. (ed.) (2001) *Handbook of Solvents*, ChemTec Publishing, Toronto.

30 (a) TLVs and BEIs: Based on the Documentation of the Threshold Limit Values for Chemical Substances and Physical Agents & Biological Exposure Indices, 2008; (b) Greim, H. (ed.) (2006) *MAK Value Documentations*, Wiley–VCH Verlag GmbH, Weinheim.

31 International Chemical Safety Cards (ICSCs), International Occupational Safety and Health Information Centre, International Labor Organization (ILO).

32 (a) Kolthoff, I.M. and Chantooni, M.K., Jr (1979) *Treatise on Analytical Chemistry, Part I*, 2nd edn, vol. 2 (eds I.M. Kolthoff and P.J. Elving), John Wiley & Sons, Inc., New York, Chapter 19A; (b) Kolthoff, I.M. (1974) *Anal. Chem.*, **46**, 1992.

33 (a) Ref. [1a], p. 57; (b) Ref. [2a], p. 2; (c) Ref. [5b], p. 1.

34 Brønsted, J.N. (1928) *Chem. Ber.*, **61**, 2049.

2
Solvation and Complex Formation of Ions and Behavior of Electrolytes

Solvation is a process in which solute particles (molecules or ions) in a solution interact with the solvent molecules surrounding them. Solvation in an aqueous solution is called *hydration*. The *solvation energy* is defined as the standard chemical potential of a solute in the solution referred to that in the gaseous state.[1] The solvation of a solute has a significant influence on its dissolution and on the chemical reactions in which it participates. Conversely, the solvent effect on dissolution or on a chemical reaction can be predicted quantitatively from the knowledge of the solvation energies of the relevant solutes. In this chapter, we mainly deal with the energetic aspects of ion solvation and its effects on the behavior of ions and electrolytes in solutions.

During the past two decades, studies on ion solvation and electrolyte solutions have made remarkable progress with the interplay of experiments and theories. Experimentally, X-ray and neutron diffraction methods and sophisticated EXAFS, IR, Raman, NMR and dielectric relaxation spectroscopies have been successfully used to obtain structural and/or dynamic information about ion–solvent and ion–ion interactions. Theoretically, microscopic or molecular approaches to the study of ion solvation and electrolyte solutions were made by Monte Carlo and molecular dynamics calculations/simulations, as well as by improved statistical mechanics treatments. Some topics essential to this book have been included in this chapter. For more details of recent progress, see Refs [1–4].

2.1
Influence of Ion Solvation on Electrolyte Dissolution

Ion solvation is vital to the dissolution of an electrolyte [5–10]. Figure 2.1 shows the Born–Haber cycle for the dissolution of a crystalline electrolyte, MX. In process I, M^+

---

[1] For an electrically neutral molecule, the solvation energy ΔG_{sv}° is equal to the Gibbs energy of transfer ΔG_t° of the molecule from a vacuum into the solvent. However, for an electrically charged ion, the following relationship holds:

$$\Delta G_t^\circ = \Delta G_{sv}^\circ + z_i F\chi$$

where z_i is the ionic charge and χ is the surface potential at the vacuum/solution interface. For an electrically neutral electrolyte, the influence of the surface potential cancels out between the cation and the anion. Thus, the electrolyte can be treated like a neutral molecule.

Electrochemistry in Nonaqueous Solutions, Second, Revised and Enlarged Edition. Kosuke Izutsu
Copyright © 2009 WILEY-VCH Verlag GmbH & Co. KGaA, Weinheim
ISBN: 978-3-527-32390-6

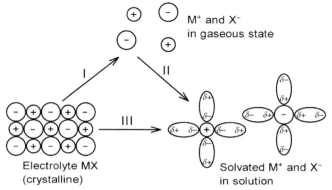

Figure 2.1 Dissolution process of crystalline electrolyte MX into a solvent (see text).

and X^- ions, which are electrostatically strongly bound in the crystal, are dissociated (separated from one another) and are brought into a gaseous state. In process II, the M^+ and X^- ions in the gas phase dissolve into the solvent by being solvated. In process III, the crystal of MX directly dissolves into the solvent, forming the solvated M^+ and X^- ions. The Gibbs energies for the three processes are related as follows:

$$\Delta G_{III}^{\circ} = \Delta G_{I}^{\circ} + \Delta G_{II}^{\circ}$$

Here, the subscripts I, II and III denote the processes I, II and III, respectively. If we denote the lattice Gibbs energy of crystal MX by ΔG_{lat}°, we get $\Delta G_{I}^{\circ} = -\Delta G_{lat}^{\circ}$.[2] ΔG_{II}° is equal to the sum of the solvation energies of M^+ and X^-, and, if MX is completely dissociated into free ions in the solution, it is equal to the solvation energy of MX, ΔG_{sv}°. ΔG_{III}° corresponds to the Gibbs energy of dissolution of electrolyte MX, ΔG_{s}°. Thus, we get Eq. (2.1)

$$\Delta G_{s}^{\circ} = \Delta G_{sv}^{\circ} - \Delta G_{lat}^{\circ} \tag{2.1}$$

The values of the thermodynamic parameters for the dissolution of lithium and sodium halides in water and in propylene carbonate (PC) are given in Table 2.1.

If the solubility product constant of electrolyte MX is expressed by $K_{sp}(MX)$, Eq. (2.2) is obtained as the relation between ΔG_{s}° and K_{sp} (MX):

$$\Delta G_{s}^{\circ} = -RT \ln K_{sp}(MX) \tag{2.2}$$

2) The term 'lattice energy' sometimes means lattice enthalpy, but it does not apply here.

Table 2.1 Thermodynamic parameters for the dissolution of lithium and sodium halides (25°C; kJ mol^{-1}).

Electrolyte	ΔH°_{lat}	$-T\Delta S^\circ_{lat}$	ΔG°_{lat}	Water				Propylene carbonate (PC)			
				ΔH°_{sv}	$-T\Delta S^\circ_{sv}$	ΔG°_{sv}	ΔG°_{s}	ΔH°_{sv}	$-T\Delta S^\circ_{sv}$	ΔG°_{sv}	ΔG°_{s}
LiF	−1040	78	−962	−1036	88	−948	14.2	—	—	—	96.2
LiCl	−861	73	−788	−899	70	−829	−40.6	−869	102	−767	22.1
LiBr	−819	72	−747	−869	66	−803	−55.6	−848	99	−749	−5.4
LiI	−762	69	−693	−825	57	−768	−75.3	−825	101	−724	−31.4
NaF	−923	78	−845	−923	82	−841	4.2	—	—	—	76.1
NaCl	−787	73	−714	−783	60	−723	−8.8	−761	83	−678	43.9
NaBr	−752	72	−680	−753	56	−697	−16.3	−741	89	−652	28.5
NaI	−702	70	−632	−710	49	−661	−28.9	−723	95	−628	4.2

$\Delta H^\circ_{lat}, \Delta S^\circ_{lat}, \Delta G^\circ_{lat}$: lattice enthalpy, entropy and Gibbs energy of the crystalline electrolyte; $\Delta H^\circ_{sv}, \Delta S^\circ_{sv}, \Delta G^\circ_{sv}$: enthalpy, entropy, and Gibbs energy of solvation of the electrolyte; ΔG°_{s}: Gibbs energy of solution of the crystalline electrolyte. Taken from Table 2.1 in Ref. [6], p. 25.

From this equation, the solubility of MX, s, is obtained to be 1, 10^{-2}, 10^{-4} and 10^{-6} M (M = mol dm^{-3}) for ΔG°_{s} of 0, 22.8, 45.7 and 68.5 kJ mol^{-1}, respectively, at 25°C and using $s = K_{sp}^{1/2}$. If ΔG°_{s} has a negative value, the solubility is expected to exceed 1 M. Thus, from Eq. (2.1), the electrolyte is easily soluble if the sum of the solvation energies of the ions constituting the electrolyte is larger than the lattice Gibbs energy (in absolute value) or very near to it. From the ΔG°_{s} values in Table 2.1, it is apparent that all of the lithium and sodium halides are easily soluble in water. In PC, however, the solubilities are much lower than those in water, and LiF, NaF and NaCl are difficult to dissolve.

In general, ΔG°_{sv} and ΔG°_{lat} have large negative values, which are, interestingly, close to each other in magnitude. Thus, ΔG°_{sv}, which is obtained as the difference between the two values (Eq. 2.1), is relatively small. If the values of ΔG°_{sv} in two solvents differ by several percentages, its influence on ΔG°_{sv} may cause a big difference between the solubilities of the electrolyte in the two solvents. This actually happens between water and PC, as shown in Table 2.1.

For reference, the standard Gibbs energies and enthalpies of hydration of some single ions and neutral molecules are given in Table 2.2.

2.2
Some Fundamental Aspects of Ion Solvation

2.2.1
Ion–Solvent Interactions Affecting Ion Solvation

As described above, the role of ion solvation is crucial in the dissolution of electrolytes. Ion solvation also has significant effects on chemical reactions

Table 2.2 Standard Gibbs energies and enthalpies of hydration of single ions and neutral molecules (25°C; kJ mol^{-1}).

Cations	ΔG_{hydr}°	ΔH_{hydr}°	Anions	ΔG_{hydr}°	ΔH_{hydr}°	Neutral molecules	ΔH_{hydr}°
H^+	−1056	−1094	F^-	−472	−519	H_2O	−44.0^b
Li^+	−481	−522	Cl^-	−347	−376	CH_3OH	−44.7^b
Na^+	−375	−407	Br^-	−321	−345	NH_3	−34.6^b
K^+	−304	−324	I^-	−283	−300	H_2S	−19.2^b
Et_4N^+	—	−127^a	ClO_4^-	−214	−232	$CdCl_2$	−194^b
Ph_4As^+	—	−42^a	BF_4^-	−200	−220	$HgCl_2$	−69.1^b
Mg^{2+}	−1838	−1931	BPh_4^-	—	−47^a	CdI_2	−141^b
Al^{3+}	−4531	−4688	SO_4^{2-}	−1090	−1138	HgI_2	−62.3^b

From Table 5.10 in Marcus, Y. (1985) *Ion Solvation*, John Wiley & Sons, Inc., New York, except
^{a}Ref. [6], p.16; and ^{b}Ref. [6], p.25.

and equilibria. Ion–solvent interactions that may participate in ion solvation are shown in Table 2.3 [11]. Their characteristics are outlined below.

Electrostatic Interactions
The electrostatic part of the ionic solvation energy, ΔG_{el} (kJ mol^{-1}), corresponds to the difference between the electrostatic free energy of an ion *in vacuo* and that of the ion in a solution of relative permittivity ε_r. It is roughly given by the Born equation:

$$\Delta G_{el} = -\frac{N_A z^2 e^2}{4\pi\varepsilon_0 \cdot 2r}\left(1 - \frac{1}{\varepsilon_r}\right) = -\frac{69.4 z^2}{r}\left(1 - \frac{1}{\varepsilon_r}\right) \tag{2.3}$$

where ze is the ionic charge, r is the ionic radius (nm) and N_A is the Avogadro constant. Figure 2.2 shows the relationship between ΔG_{el} and ε_r for a univalent ion, obtained by assuming a constant ionic radius. The value of $-\Delta G_{el}$ increases with ε_r very rapidly in the low-permittivity region ($\varepsilon_r < 10$) and rather slowly in the

Table 2.3 Ion–solvent interactions influencing ion solvation.^{a}

1. Electrostatic interactions as expressed by the Born equation	(≥80%)
2. Electron (-pair) donor–acceptor interactions	(≤10%)
3. Interactions of anions with hydrogen bond donor solvents	(≤10%)
4. Interactions based on HSAB concept	(≤20%)
5. Interactions by back-donation from d^{10}-cation to solvent molecules	(≤10%)
6. Interactions related to the structure-making and breaking of solvents	(≤5%)

^{a}The values in parentheses show the rough estimate of the contribution from each factor to the total solvation energies of univalent ions (300–500 kJ mol^{-1}) in a solvent of $\varepsilon_r = 25$–100. From Ref. [11].

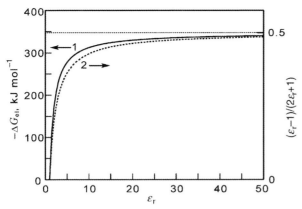

Figure 2.2 The effect of solvent permittivity on the electrostatic solvation energy of an ion (curve 1) and that of a neutral dipolar molecule (curve 2). Curve 1 was obtained from Eq. (2.3) assuming $r = 0.2$ nm. For curve 2, see footnote 4.

high-permittivity region ($\varepsilon_r > 20$). This shows that the difference in ΔG_{el} between two high-permittivity solvents is rather small.[3),4)]

3) The Born equation, proposed in 1920, has been modified in various ways in order to get a single equation that can express the experimental ionic solvation energies. In recent years, the so-called *mean spherical approximation* has often been used in treating ion solvation. In the MSA treatment, the Gibbs energy of ion solvation is expressed by

$$\Delta G_{sv}^{\circ} = \frac{-N_A z^2 e^2}{4\pi\varepsilon_0 \cdot 2(r+\delta_s)}\left(1-\frac{1}{\varepsilon_r}\right)$$

where r is the radius of a spherical ion and $\delta_s = r_s/\lambda_s$, r_s being the radius of a spherical solvent molecule and λ_s the Wertheim polarization parameter, obtained by the relation $\lambda_s^2(1+\lambda_s)^4 = 16\varepsilon_s$ [12]. For water at 25°C, $r_s = 142$ pm and $\lambda_s = 2.65$, and thus $\delta_s = 54$ pm. In the table below, the ΔG_{sv}° (kJ mol^{-1}) values obtained experimentally in water are compared with those obtained by using the Born and MSA models [12].

The values of δ_s in seven dipolar aprotic solvents have been reported to be 80 ± 5 pm for cations and 44 ± 4 pm for anions [13]. The MSA is also used in treating ionic activity coefficients; in a recent study [14], the change in solvent permittivity with electrolyte concentration was taken into account in addition to the change in ionic radius, and excellent agreements were obtained between the experimental and theoretical results for 1:1 electrolytes of up to 2.5 M. For more examples of the MSA treatments, see Ref. [1].

4) According to Kirkwood [15], the electrostatic solvation energy of a neutral spherical molecule with a radius r and a dipole moment μ is expressed by $\Delta G_{el} = -(N_A\mu^2/4\pi\varepsilon_0 r^3)\{(\varepsilon_r - 1)/(2\varepsilon_r + 1)\}$. The relationship between $(\varepsilon_r - 1)/(2\varepsilon_r + 1)$ and ε_r, plotted in Figure 2.2, indicates that the influence of ε_r on molecular solvation is somewhat similar to that on ion solvation.

Ion	Li$^+$	Na$^+$	K$^+$	Rb$^+$	Cs$^+$	F$^-$	Cl$^-$	Br$^-$	I$^-$
r (pm)	88	116	152	163	184	119	167	182	206
Experimental	−529	−424	−352	−329	−306	−429	−304	−278	−243
Born	−779	−591	−451	−421	−373	−576	−410	−377	−333
MSA	−483	−403	−333	−316	−288	−396	−310	−291	−264

For univalent ions in high-permittivity solvents, the total solvation energy is roughly in the range of 300–500 kJ mol^{-1} (see footnote 3) and the electrostatic part ΔG_{el} is considered to amount to 80% or more (Table 2.3). However, if we compare the solvation energies of an ion in two high-permittivity solvents, we find that the difference in ΔG_{el} is often less important than the difference in the solvation energies caused by the interactions described below.

Electron Pair Donor–Acceptor Interactions

In ion solvation, the solvent molecules approach a cation with their negative charge and approach an anion with their positive charge (Figure 2.1). Therefore, cation solvation is closely related to the electron pair donor capacity or Lewis basicity of solvents and tends to become stronger with the increase in donor number (DN). On the other hand, the anion solvation is closely related to the electron pair acceptability or Lewis acidity of solvents and tends to become stronger with the increase in acceptor number (AN).

The effects of DN on the solvation energy of the potassium ion and on the standard potential of the hydrogen electrode, which is linearly related to the solvation energy of the hydrogen ion, are shown in Figure 2.3. Near-linear relations can be observed in both cases [16]. There is also a linear relationship between AN and the solvation energies of the chloride ion in aprotic solvents, as in Figure 2.4 [16]. However, the chloride ion in protic solvents such as water and alcohols behaves somewhat differently from the one in aprotic solvents [17], probably because of the influence of hydrogen bonding (see below).

Interactions of Anions with Hydrogen Bond Donor Solvents

Such small anions as F$^-$, Cl$^-$ and OH$^-$ and anions having a small oxygen atom with a localized negative charge (e.g. CH$_3$COO$^-$, C$_6$H$_5$O$^-$)[5] usually have a strong tendency to accept hydrogen bonds. Thus, these anions are strongly solvated by hydrogen bonding in protic solvents such as water and alcohols. On the other hand, they are solvated only weakly and thus are very reactive in aprotic solvents, which are weak hydrogen bond donors (Table 2.4).

Large anions, such as I$^-$ and ClO$_4$$^-$, have a relatively weak tendency to accept hydrogen bonds. However, they are highly polarizable and interact to a fair extent by dispersion forces (London forces) with the molecules of aprotic solvents, which are also considerably polarizable. Thus, for large anions, the solvation energies in protic solvents (water, alcohols) and those in dipolar aprotic solvents (AN, DMF, DMSO) are not as different as in the case of small anions (Table 2.4).

Interactions Based on the HSAB Concept

According to the hard and soft acids and bases (HSAB) concept, hard acids tend to interact strongly with hard bases, while soft acids tend to interact strongly with soft

---

5) The negative charge on the O atom of CH$_3$COO$^-$ is delocalized by replacing CH$_3$ group with electron-withdrawing CF$_3$ group. In the same way, the negative charge of the O atom of C$_6$H$_5$O$^-$ is delocalized by replacing H atom(s) with electron-withdrawing NO$_2$ group (s), such as nitrophenol, dinitrophenol and picric acid.

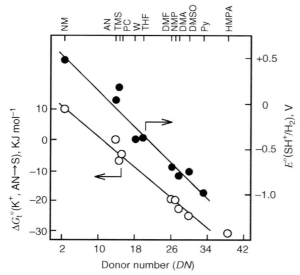

Figure 2.3 Standard Gibbs energies of transfer of the potassium ion from AN to other solvents and standard potentials of the hydrogen electrode, both plotted against the donor number of solvents [16a].

bases. The HSAB concept applies also to solute–solvent interactions. Figure 2.5 shows the polarographic half-wave potentials of metal ions in *N*-methyl-2-pyrrolidi-none (NMP) and *N*-methyl-2-thiopyrrolidinone (NMTP) [16]. Here, we can compare the half-wave potentials in the two solvents because they are referred to the half-wave potential of the bis(biphenyl)chromium(I)/(0) couple, which is considered nearly

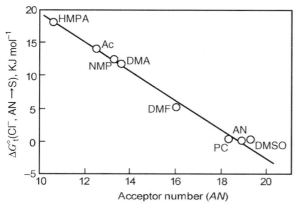

Figure 2.4 Standard Gibbs energies of transfer of the chloride ion from AN to other solvents plotted against the acceptor number of solvents [16a].

Table 2.4 Standard Gibbs energies of transfer of ions from water to nonaqueous solvents ($\Delta G_t^\circ (i, W \rightarrow S)$, kJ mol^{-1}).[a]

Ions	Nonaqueous solvents S															
	MeOH	EtOH	PrOH	Ac	PC	FA	DMF	DMA	DMTF	NMP	AN	NB	DMSO	TMS	HMPA	DCE
H$^+$	10.4	11.1	9		50[b]		-18			-25	46.4	33	-19.4			
Li$^+$	4.4	11	11		23.8	-10	-10		55	-35	25	38	-15			25
Na$^+$	8.2	14	17		14.6	-8	-9.6	-12.1	39	-15	15.1	34	-13.6	-3		26
K$^+$	9.6	16.4	18	4	8.8[c]	-4.3	-10.3	-11.7	27	-11	8.1	23	-13.0	-4	-16	25
Rb$^+$	9.6	16	19	4	7.0[c]	-5	-9.7	-8		-8	6.3	19	-10.4	-9		24
Cs$^+$	8.9	15	17	4	5.4[c]	-6.0	-10.8	(-17)	14	-10	6.0	15	-13.0	-10		
Ag$^+$	6.6	4.9	1	9	18.8	-15.4	-20.8	-29.0	-102	-26	-23.2		-34.8	-4	-44	
Tl$^+$	4.1				11.0		-11.5		-16	-15	8.0		-21.4			
NH$_4^+$	(5)	7								-24	(15)	27				
Me$_4$N$^+$	6	10.9	11	3	-11		-5.3			-3	3	4	-2			16
Et$_4$N$^+$	1	6			-13		-8.0				-7	-5	(-9)			5
Pr$_4$N$^+$		(-6)			-22		-17				-13	-16				
Bu$_4$N$^+$	-21	(-8)			-31		-29				-31	-24				
Ph$_4$As$^+$	-24.1	-21.2	-25	-32	-36.0	-23.9	-38.5	-38.7		-40	-32.8	-36	-37.4	-36	-39	-33
F$^-$	16				56	25	51				71					
Cl$^-$	13.2	20.2	26	57	39.8	13.7	48.3	54.9		51	42.1	35	40.3	47	58	52
Br$^-$	11.1	18.2	22	42	30.0	10.7	36.2	44.0		37	31.3	29	27.4	35	46	38
I$^-$	7.3	12.9	19	25	13.7	7.3	20.4	21		19	16.8	18	10.4	21	30	25
I$_3^-$	-12.6					-7	-27	-30			-15	-23	(-41)			
N$_3^-$	9.1	17.0		43	27	11	36	40		46	37		25.8	41	49	
CN$^-$	8.6	7		48	36	13.3	40				35		35			
SCN$^-$	5.6				7.0	7	18.4	21		18	14.4		9.7	22	20	
NO$_3^-$		14				-12	4				21					7
ClO$_4^-$	6.1	10	17		-3	20	66	70		-12	2				-7	16
CH$_3$COO$^-$	16.0	0.5				-7	-7				61		(50)			
Pic$^-$	-6			-32	-6	-7	-7				-4	-5				
BPh$_4^-$	-24.1	-21.2	-25	-32	-36.0	-23.9	-38.5	-38.7		-40	-32.8	-36	-37.4	-36	-39	-33

[a] Unless otherwise stated, from Marcus, Y. (1983) *Pure Appl. Chem.*, **55**, 977.
[b] Questionable reliability?
[c] Chantooni, M.K., Jr and Kolthoff, I.M. (1980) *J. Chem. Eng. Data*, **25**, 208.

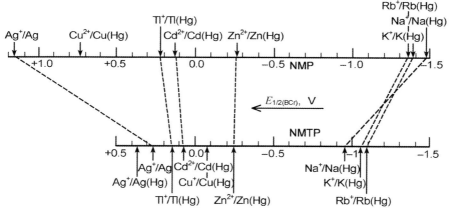

Figure 2.5 Comparison of the half-wave potentials of metal ions in N-methyl-2-pyrrolidinone and those in N-methyl-2-thiopyrrolidinone [16a].

solvent independent (Section 2.3). NMP is a hard base and NMTP is a soft base. NMP coordinates to metal ions (Lewis acids) with its O atom, but NMTP coordinates with its S atom. In Figure 2.5, the half-wave potentials of alkali metal ions are somewhat more negative in NMP than in NMTP, showing that the alkali metal ions (hard acids) solvate more strongly in NMP than in NMTP. On the other hand, the half-wave potential of Ag^+ is much more negative in NMTP than in NMP, indicating a strong solvation of Ag^+ (soft acid) in NMTP. Similar phenomena have been observed in N,N-dimethylformamide (DMF, hard base) and N,N-dimethylthioformamide (DMTF, soft base). However, soft-base solvents are rather exceptional. Most solvents in common use behave as hard bases, although such solvents as AN, BuN and Py are sometimes known to show intermediate characteristics, between hard and soft, as described in Section 4.1.2.

Interactions by Back-Donation from d^{10}-Cation to Solvent Molecules

Acetonitrile (AN) has relatively small DN and usually solvates rather weakly to metal ions. However, it solvates very strongly to Cu^+, Ag^+ and Au^+, which are univalent d^{10}-metal ions. This is because these metal ions have an ability to back-donate their electrons into a π^*-antibonding orbital of the CN group of AN, as shown by $CH_3C \equiv N:\overrightarrow{}Ag^+$. As a result, Cu^+ and Ag^+ in AN are stable and not easily reduced to metal, while the weakly solvated Cu^{2+} is very easily reduced to Cu^+ (see Eq. (4.6)), making Cu^{2+} in AN a strong oxidizing agent.

Interactions Related to the Structure-Making and Structure-Breaking of Solvent

When an ion (or a molecule) is dissolved, a cavity must be formed in the solvent to accommodate it. With an increase in the ionic (or molecular) size and with the strengthening of the interaction between solvent molecules, the energy needed for cavity formation increases. Water molecules are strongly bound to each other by hydrogen bonding and form three-dimensional networks. Thus, the cavity

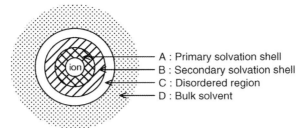

Figure 2.6 Typical model of solvated ions in structured solvents such as water and alcohols.

formation in water needs more energy than in other solvents in which the solvent–solvent interactions are weak. Moreover, if a large hydrophobic ion (tetraalkylammonium ion (R_4N^+), tetraphenylborate ion (Ph_4B^-), etc.) is introduced into water, it rejects the surrounding water molecules. The rejected water molecules are combined to make the structure more rigid (structure making) and decrease the entropy of the system.[6] For these reasons, large hydrophobic ions and molecules are usually energetically unstable in water. They are much more stable in organic solvents, which are free from strong solvent–solvent interactions. On the other hand, if small inorganic ions that are hydrophilic are introduced into water, they interact strongly with water molecules, weaken the structure of the surrounding water and increase the entropy of the system as a whole. Thus, hydrophilic ions in water are energetically stable.

2.2.2
Structure of Solvated Ions

In water, in which hydrogen bonding occurs between water molecules, hydrated (solvated) ions can be depicted by a typical model as shown in Figure 2.6. Region A is the primary solvation shell (sphere), the solvent molecules of which are oriented by interacting directly with the ion; region B is the secondary solvation shell, the solvent molecules of which are still partially oriented by the influence of the ion and by the interaction with the molecules in the primary solvation shell and region D shows the bulk solvent where the influence of the ion is negligible. A structural mismatch between regions B and D is mediated by a disordered region C. A model similar to Figure 2.6 also applies to ions in structured solvents such as alcohols. However, a simpler model, in which regions B and C are not definite, applies to ions in nonstructured polar solvents.

The solvent molecules in the primary solvation shell are constantly renewed by a solvent exchange reaction:

$$MS_m^{n+} + S^* \xrightarrow{k} MS_{m-1}(S^*)^{n+} + S$$

where M^{n+} denotes a metal ion and S and S^* denote solvent molecules. The rate constants (k) of solvent exchange reactions for metal ions have been determined by NMR (fast reactions) or isotope dilution (slow reactions) methods. As shown in

---

6) Hydrophobic molecules of organic compounds, neon, argon, etc. are also structure making.

Table 2.5 Logarithm of solvent exchange rate constant (k (s^{-1})) in various solvents at 25°C.[a]

Solvent	Mg^{2+}	Al^{3+}	Cr^{3+}	Mn^{2+}	Fe^{2+}	Fe^{3+}	Co^{2+}	Ni^{2+}	Cu^{2+}
Water	5.72	0.11	−5.62	7.49	6.64	2.2	6.35	4.58	9.64
MeOH	3.67			5.57	4.70	3.71	4.26	3.00	7.5
AN				7.08	5.82		5.54	3.30	
DMF	1.79	−1.3	−7.26	6.43	6.23	1.79	5.59	3.58	9.0
DMSO		−0.52	−7.49	6.80	6.00	$1._4$	5.65	4.22	

[a]From Funahashi, S. (1998) *Inorganic Reactions in Solutions*, Shokabo, Tokyo, p. 246.

Table 2.5, the k values vary greatly (from $10^{-7.5}$ to $10^{9.6}\,s^{-1}$) by metal ion [18a]. The average lifetimes of solvents in the primary solvation shell also vary widely because they are of the order of ($1/k$). The average lifetimes, determined for Li^+ from the NMR band widths are in the order NM (0.05) < AN (0.6) < THF (1.0) < MeOH (1.6) < water (3.3) < FA (4.0) < NMF (5.6) < DMF (8) ~ DMSO (8) < HMPA (15) [18b]. The values in parentheses show the lifetimes in nanoseconds. This shows that the lifetime increases with the increase in the solvating ability of the solvent.

The *solvation numbers* in the primary solvation shell can be estimated by NMR, IR and Raman spectroscopy or by isotope dilution method. For example, in the NMR peak area method, the solvent molecules in the primary solvation shell of a metal ion (~1 M) give an ^1H-NMR peak separated from that for the bulk solvent if they have lifetimes longer than ~$10^{-4}\,s$. The solvation number of the metal ion is determined directly from the ratio of the two peak areas. Though such metal ions are limited in number at room temperatures (see Table 2.5), they increase at low temperatures (−60 or −100°C). The solvation numbers obtained from NMR peak areas are usually six for such metal ions as Mg^{2+}, Al^{3+}, Ga^{3+}, Zn^{2+}, Mn^{2+}, Fe^{2+}, Co^{2+}, Ni^{2+}, Ti^{3+}, V^{3+}, Cr^{3+}, Fe^{3+} in water, MeOH, AN, DMF, DMSO and NH_3, though they are four for small metal ions such as Be^{2+} and for square planar Pd^{2+} and Pt^{2+}. There is a tendency that bulky solvents, such as HMPA and trimethyl phosphate, give lower solvation numbers than other solvents. If the lifetime is too short to use the NMR method, the IR method can be applied. Figure 2.7 shows

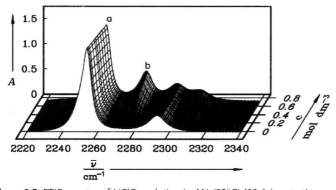

Figure 2.7 FTIR spectra of $LiClO_4$ solution in AN (25°C) [19a] (see text).

the FTIR spectra of the ν(C$-$N) stretching band for a LiClO$_4$ solution in AN (25 °C); (a) is for *free* AN molecules (2253 cm^{-1}) and (b) is for AN molecules in the first solvation shell of Li$^+$ (2276 cm^{-1}) [19a,b]. With increasing LiClO$_4$ concentration, the intensity of band (b) increases. Data analysis yields a solvation number close to four. The difference in ν between bands (a) and (b) increases with the cationic surface charge density in the order NaClO$_4$ < Ba(ClO$_4$)$_2$ < Sr(ClO$_4$)$_2$ < Ca(ClO$_4$)$_2$ < LiClO$_4$, indicating the strengthening of the ion–solvent interaction. IR and Raman spectroscopies are useful to study the strength of ion–solvent interactions. In Ref. [19a], ClO$_4^-$ was considered to be unsolvated in AN. However, a recent study by attenuated total reflectance FTIR spectroscopy showed that the C$-$N stretching band for AN molecules associated with ClO$_4^-$ has a significantly larger molar absorption coefficient than the same band for self-associated AN [19c,d]. Another FTIR study [19d] showed that iodide anion (I$^-$) interacts with the methyl group of AN, which is at the positive end of the molecular dipole. For IR studies of solvated ions in other aprotic solvents, see Ref. [19e,f,g].

X-ray and neutron diffraction methods and EXAFS spectroscopy are very useful in getting structural information of solvated ions. These methods, combined with molecular dynamics and Monte Carlo simulations, have been used extensively to study the structures of hydrated ions in water. Detailed results can be found in the review by Ohtaki [20a] and Ohtaki and Radnai [20b]. The structural study of solvated ions in nonaqueous solvents has not been as extensive, partly because the low solubility of electrolytes in nonaqueous solvents limits the use of X-ray and neutron diffraction methods that need electrolyte of ∼1 M. However, this situation has been improved by EXAFS (applicable at ∼0.1 M), at least for ions of the elements with large atomic numbers, and the amount of data on ion–coordinating atom distances and solvation numbers for ions in nonaqueous solvents are growing [18a,20a,21a,b]. For example, according to the X-ray diffraction method, the lithium ion in formamide (FA) has, on average, 5.4 FA molecules as nearest neighbors with an Li$^+$$-$O distance of 224 pm, while the chloride ion is coordinated by 4.5 FA molecules and the Cl$^-$···N distance is 327 pm; the amino group of FA interacts with the chloride ion in a bifurcated manner through the two hydrogen atoms [21a,b]. The solvation numbers obtained by these methods correspond to the number of solvent molecules in the *first solvation shell* immediately neighboring the ion; here, the solvent molecules may or may not interact strongly with the ion.[7] The structures of solvated halide ions (X$^-$: Cl$^-$, Br$^-$ and I$^-$) in DMSO have been

---

7) Besides spectroscopic (EXAFS, NMR, IR, Raman) and scattering (X-ray and neutron diffraction) methods, transport properties (transference numbers and ionic mobilities) and thermodynamic properties (molar entropies of solvation, compressibilities, etc.) are used to obtain solvation numbers of ions in solution. The results obtained from transport and thermodynamic properties reflect the number of solvent molecules that behave or transport with the ion. Thus, for strongly solvated ions, the results may include the solvent molecules in the primary and secondary solvation shells, or even more and larger values than those obtained by spectroscopic or scattering methods may be obtained. For the methods and the problems associated with the estimation of solvation numbers, see, for example, p. 61 and 139 in Ref. [2]; p. 28 in Ref. [7a]; p. 68 in Ref. [8]; p. 78 and 242 in Ref. [18a].

studied by Raman spectroscopy and X-ray diffraction [21c]. The positively charged S atom in DMSO interacts with X^- and slightly positively charged methyl groups in the coordinating molecules also interact with X^-. The X^- —S, $X^- \cdots$ C and $X^- \cdots$ O distances were determined as follows: Cl^- —S 416, $Cl^- \cdots$ C 363 and $Cl^- \cdots$ O 543, Br^- —S 433, $Br^- \cdots$ C 372 and $Br^- \cdots$ O 544; and I^- —S 437, $I^- \cdots$ C 374 and $I^- \cdots$ O 520 pm. The coordination numbers of DMSO molecules were six, seven and eight for Cl^-, Br^- and I^- ions, respectively, with the uncertainty of ± 1. Rather large uncertainties in the measured solvation numbers are due to the large fluctuations in the solvation structure of the anions.

2.2.3
Ultrafast Ion Solvation Dynamics

Since the end of the 1980s, ultrafast ion solvation dynamics has been studied with great interest by combining femtosecond laser experiments, analytical theories and computer simulations [22]. This is because such ultrafast solvation dynamics is closely related to various chemical processes including electron-transfer reactions and ionic migrations (Sections 4.1.3 and 7.2.1) [23]. In typical studies of ion solvation dynamics, a fluorescent probe molecule is used that is nonpolar in the ground state but is highly polar (possibly ionic) in the electronically excited state. When the probe solute is in the ground state, the solvent dipoles around the solute remain randomly oriented. When the probe solute is excited by an ultrashort laser pulse, the solvent dipoles initially remain randomly oriented around the instantaneously created charge distribution, but, with time, they gradually reorient so that the system is energetically relaxed. Thus, if $E(0)$, $E(t)$ and $E(\infty)$ denote the energies of the fluorescent solute at times 0, t and ∞ ($0 < t < \infty$), respectively, there is a relation $E(0) > E(t) > E(\infty)$. If the emission energy of the excited fluorescent solute is measured as a function of time, it is observed that, with time, the emission maximum shifts to lower energy, i.e. to longer wavelength. This phenomenon is called *time-dependent fluorescence Stokes shift* (TDFSS). The solvation dynamics is monitored by the decay of the *solvation time correlation function* $S(t)$ defined by

$$S(t) = \frac{E(t) - E(\infty)}{E(0) - E(\infty)}$$

As an example, Figure 2.8 shows the experimental $S(t)$–t relation in water [22a,b]. The probe molecule was coumarin 343, which is ionic in the excited state. The experimental results are biphasic: an initial ultrafast Gaussian decay with a time constant of $\sim$55 fs is followed by a slower biexponential decay of time constants 126 and 880 fs. The initial decay constitutes more than 60% of the total solvation. The relationship in Figure 2.8 has been simulated by molecular hydrodynamic theory using the dielectric relaxation parameters of H_2O. It agrees well with the experimental results: the theoretical time constant for the initial Gaussian decay is equal to 52 fs and that for the slower biexponential decay is 134 and 886 fs. The initial decay has been shown to correspond to the intermolecular vibrational band originating from the $O \cdots O$ stretching mode of the $O-H \cdots O$ unit. Similar biphasic relations

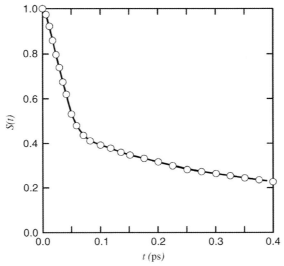

Figure 2.8 Experimentally obtained solvation time correlation function $S(t)$ for the solvation of coumarin 343 in water (taken from Ref. [22a]).

have also been obtained in AN and MeOH [22c,d,e] and, based on the theoretical studies, the initial ultrafast decay has been interpreted by the liberation motion or force free (inertial) motion of solvent molecules. Various factors complicate the study of solvation dynamics: different experimental techniques may give significantly different results, the excited probe solute may behave as a dipole rather than as an ion and the ultrafast decay may be due to the intramolecular relaxation of the excited probe solute [24]. Studies on ultrafast solvation dynamics are still under way and are providing valuable knowledge that helps understand the dynamic solvent effects on various chemical processes.

2.3
Comparison of Ionic Solvation Energies in Different Solvents and Solvent Effects on Ionic Reactions and Equilibria

2.3.1
Gibbs Energies of Transfer and Transfer Activity Coefficients of Ions

The various factors that contribute to ion solvation were discussed in Section 2.2.1. In this section, we deal with the solvent effects on chemical reactions more quantitatively [8,25]. To do this, we introduce two quantities, the Gibbs energy of transfer and the transfer activity coefficient.

If the solvation energy of species i in solvent R (reference solvent) is expressed by $\Delta G_{\mathrm{sv}}^{\circ}(i, \mathrm{R})$ and that in solvent S (a solvent under study) by $\Delta G_{\mathrm{sv}}^{\circ}(i, \mathrm{S})$, the difference between the two is expressed by $\Delta G_t^\circ(i, R \rightarrow S)$ and is called the Gibbs energy of transfer of species i from solvent R to S:

$$\Delta G_t^\circ(i, R \rightarrow S) = G_{sv}^\circ(i, S) - G_{sv}^\circ(i, R)$$

If the species i is electrically neutral, the value of $\Delta G_t^\circ(i, R \rightarrow S)$ can be obtained by a thermodynamic method. For example, if the solubilities of i in solvents R and S are S_R and S_S, respectively, $\Delta G_t^\circ(i, R \rightarrow S)$ can be obtained by Eq. (2.4):

$$\Delta G_t^\circ(i, R \rightarrow S) = RT\ln(S_R/S_S) \tag{2.4}$$

If the species i is an electrolyte MX, which is electrically neutral, it is also possible to obtain the value of $\Delta G_t^\circ(i, R \rightarrow S)$ from the solubilities of MX in the two solvents.

However, if species i is a single ion, the value of $\Delta G_t^\circ(i, R \rightarrow S)$ cannot be obtained by purely thermodynamic means. It is necessary to introduce some extrathermodynamic assumptions. Various extrathermodynamic assumptions have been proposed. Some typical examples are described in the following list. For practical methods of obtaining the Gibbs energies of transfer for ionic species, see[8].

(i) **Reference ion/molecule assumption (assumption of a reference potential system)**
    When a univalent cation I^+ (or anion I^-), which is large in size and symmetrical in structure, is reduced (or oxidized) in solvents R and S to form an electrically neutral I^0, which has essentially the same size and structure as I^+ (or I^-), we assume that $\Delta G_t^\circ(I^+$ or $I^-, R \rightarrow S) = \Delta G_t^\circ(I^0, R \rightarrow S)$ or that the standard potentials of the redox system (I^+/I^0 or I^0/I^-) in R and S are the same. Actually, such redox couples as bis(cyclopentadienyl)iron(III)/bis(cyclopentadienyl)iron(II) (ferrocenium ion/ferrocene, Fc^+/Fc) and bis(biphenyl)chromium(I)/bis(biphenyl)chromium(0) (BCr^+/BCr) seem nearly to meet these requirements

---

8) The followings are the practical procedures for obtaining the Gibbs energies of transfer and transfer activity coefficients of ionic species based on the extrathermodynamic assumptions (i), (ii) and (iii), described in the list:

Assumption (i): When we use the Fc^+/Fc couple as a solvent-independent potential reference, we measure the emfs of the cell Pt|Fc^+ (picrate), Fc, $AgClO_4$(R or S)|Ag. If the emfs in R and S are E_R and E_S, respectively, we get the values of $\Delta G_t^\circ(Ag^+, R \rightarrow S)$ and $\log \gamma_t(Ag^+, R \rightarrow S)$ by the relation $\Delta G_t^\circ(Ag^+, R \rightarrow S) = 2.3RT\log \gamma_t(Ag^+, R \rightarrow S) = F(E_S - E_R)$. Then, we measure the solubilities of a sparingly soluble silver salt (AgX) in R and S to get the values of pK_{sp}(AgX, R) and pK_{sp}(AgX, S). Using these values, we calculate the values of $\Delta G_t^\circ(X^-, R \rightarrow S)$ and $\log \gamma_t(X^-, R \rightarrow S)$. Then, by measuring the solubility of a sparingly soluble salt M^+X^-, we get the values of $\Delta G_t^\circ(M^+, R \rightarrow S)$ and $\log \gamma_t(M^+, R \rightarrow S)$. As an alternative method, if we get the voltammetric half-wave potentials for the

reduction of M^+ in R and S as the values against the half-wave potential of the Fc^+/Fc couple, we can directly get the values of $\Delta G_t^\circ(M^+, R \rightarrow S)$ and $\log \gamma_t(M^+, R \rightarrow S)$ from the relation in Section 8.2.1.

Assumption (ii): We measure the solubilities of the reference electrolyte Ph_4AsPh_4B in R and S (S_R and S_S) and get $\Delta G_t^\circ(Ph_4AsPh_4B, R \rightarrow S)$ from Eq. (2.4). Then, by the relation $\Delta G_t^\circ(Ph_4As^+, R \rightarrow S) = \Delta G_t^\circ(Ph_4B^-, R \rightarrow S) = (1/2)\Delta G_t^\circ(Ph_4AsPh_4B, R \rightarrow S)$, we get $\Delta G_t^\circ(i, R \rightarrow S)$ and $\log \gamma_t(i, R \rightarrow S)$ for $i = Ph_4As^+$ and Ph_4B^-. Then, we measure the solubilities of $AgPh_4B$ in R and S and get $\Delta G_t^\circ(Ag^+, R \rightarrow S)$ and $\log \gamma_t(Ag^+, R \rightarrow S)$. The above procedures are then applicable.

Assumption (iii): We use a cell Ag|0.01 M $AgClO_4$(AN)⫽0.1 M Et_4NPic(AN)⫽0.01 M $AgClO_4$(AN or S)|Ag and measure the emfs in AN and S (E_{AN} and E_S). By neglecting the LJPs on the right side of the salt bridge [0.1 M Et_4NPic (AN)], we get $\Delta G_t^\circ(Ag^+, AN \rightarrow S)$ by the relation $\Delta G_t^\circ(Ag^+, AN \rightarrow S) = F(E_S - E_{AN})$.

in various solvents. Thus, these redox couples are often used as reference systems having solvent-independent potentials.

This assumption, however, has some problems. One is that the relation $\Delta G_t^\circ(I^+$ or $I^-, R \rightarrow S) = \Delta G_t^\circ(I^0, R \rightarrow S)$ does not hold if the relative permittivity of S is much different from that of R. If we divide the solvation energy of I^+ (or I^-) into electrostatic and nonelectrostatic parts, the nonelectrostatic part of $\Delta G_t^\circ(I^+$ or $I^-, R \rightarrow S)$ must be almost equal to $\Delta G_t^\circ(I^0, R \rightarrow S)$. However, as predicted from the Born equation, the electrostatic part of $\Delta G_t^\circ(I^+$ or $I^-, R \rightarrow S)$ is not negligible if there is a large difference between the relative permittivities of S and R.[9]

(ii) **Reference electrolyte assumption** If an electrolyte A^+B^- consists of a cation A^+ and an anion B^- that are large, symmetrical and of very similar size and structure, we can assume the relation $\Delta G_t^\circ(A^+, R \rightarrow S) = \Delta G_t^\circ(B^-, R \rightarrow S)$. As such an electrolyte, tetraphenylarsonium tetraphenylborate (Ph_4AsPh_4B) is the most popular. For Ph_4As^+ and Ph_4B^- (both $\sim$0.43 nm in radius), the relation $\Delta G_t^\circ(A^+, R \rightarrow S) = \Delta G_t^\circ(B^-, R \rightarrow S)$ is almost valid for both nonelectrostatic and electrostatic parts, irrespective of the relative permittivities of R and S. So far, this assumption is considered to be most reliable.

(iii) **Assumption of negligible liquid junction potential** The liquid junction potential (LJP) between electrolyte solutions in different solvents is assumed to be negligible at some particular junctions. The most popular assumption is that the LJP at 0.1 M Et_4NPic (AN)⋮0.01 M $AgClO_4$(S) is negligible (within $\pm$20 mV) [29]. The reliability and the limit of this assumption are discussed in Section 6.4.

Based on these and other extrathermodynamic assumptions, considerable data for the Gibbs energies of ionic transfer have been obtained. The data in Table 2.4 are the Gibbs energies of transfer from water to various nonaqueous solvents [30]. The enthalpies and entropies of ion transfer have also been obtained based on similar extrathermodynamic assumptions. They are important in getting the mechanistic information concerning the ionic transfer between different solvents. The values in Table 2.6 were obtained based on the reference electrolyte assumption [31]. Gritzner [32] determined the Gibbs energies, entropies and enthalpies of transfer of metal ions by polarography using the bis(biphenyl)chromium(I)/(0) couple as solvent-independent potential reference. Marcus compiled in detail these thermodynamic parameters of transfer, including those from water to mixed aqueous organic solvents [33].

9) If the radius of Fc^+ is 0.23 nm (Ref. [8], p. 162), the variation in the *electrostatic* solvation energy of Fc^+ on transfer from water ($\varepsilon_r = 78$) to AN ($\varepsilon_r = 35.9$) is equal to 4.5 kJ mol^{-1}, while that on transfer from water to low-permittivity THF ($\varepsilon_r = 7.6$) is equal to 35.8 kJ mol^{-1}. Here, 1 kJ mol^{-1} corresponds to $\sim$10 mV. The influence of permittivity can be avoided by using the mean of the standard potentials of the I^+/I^0 and I^0/I^- couples as reference because the influ-

ences for both couples cancel out (see Eq. (2.3)). Such I^+/I^0 and I^0/I^- couples can be realized with cobaltocene [26] and some organic molecules [27] but in a limited number of solvents.

For the Fc^+/Fc couple, another reason decreases its reliability as potential reference. In water, we cannot assume that the nonelectrostatic part of the hydration energy of Fc^+ is equal to that of Fc because the hydrated Fc^+ and Fc have different structures [28].

Table 2.6 Standard Gibbs energies, enthalpies, and entropies of transter of ions from water to nonaqueous solvents (25°C).[a]

Ion	$\Delta G_t^\circ, \Delta H_t^\circ, -298\Delta S_t^\circ$ (kJ mol^{-1})				
	MeOH	DMF	DMSO	AN	PC
Na$^+$	8.4, −20.5, 28.9	−10.5, −33.1, 22.6	−13.8, −27.6, 13.8	13.8, −13.0, 26.8	15.1, −6.7, 21.8
Ag$^+$	7.5, −20.9, 28.5	−17.2, −38.5, 21.3	−33.5, −54.8, 21.3	−21.8, −52.7, 31.0	15.9, −12.6, 28.5
Et$_4$N$^+$	0.8, 9.2, −8.4	−8.4, −0.8, −7.5	−12.6, 4.2, −16.7	−8.8, −1.3, −7.5	—, —, —
Bu$_4$N$^+$	−21.8, 21.8, −43.5	−28.5, 15.1, −43.5	—, —, —	−33.1, 18.4, −51.5	—, —, —
Ph$_4$As$^+$, Ph$_4$B$^-$	−23.4, −1.7, −21.8	−38.1, −19.7, −18.4	−36.8, −11.7, −25.1	−32.6, −10.5, −22.2	−35.6, −15.1, −20.5
Cl$^-$	12.6, 8.4, 4.2	46.0, 21.3, 24.7	38.5, 18.8, 19.7	—, —, —	37.7, 28.0, 9.6
I$^-$	6.7, −2.1, 8.8	18.8, −13.8, 32.6	9.2, −13.4, 22.6	18.8, −7.1, 25.9	17.6, −0.8, 18.4
ClO$_4^-$	5.9, −2.5, 8.4	0.4, −22.6, 23.0	−1.3, −19.3, 18.0	—, —, —	—, —, —

[a]Based on the extrathermodynamic assumptions that $\Delta G_t^\circ(Ph_4As^+) = \Delta G_t^\circ(Ph_4B^-)$, $\Delta H_t^\circ(Ph_4As^+) = \Delta H_t^\circ(Ph_4B^-)$ and $\Delta S_t^\circ(Ph_4As^+) = \Delta S_t^\circ(Ph_4B^-)$. Calculated from the data (in kcal mol^{-1}) in Cox, B.G., Hedwig, G.R., Parker, A.J. and Watts, D.W (1974) Aust. J. Chem., 27, 477.

When we consider the solvent effect on the reactivity of a chemical species, it is convenient to use the transfer activity coefficients[10] instead of the Gibbs energies of transfer. The transfer activity coefficient $\gamma_t(i, R \to S)$ is defined by

$$\log\gamma_t(i, R \to S) = \Delta G_t^{\circ}(i, R \to S)/(2.303RT) \tag{2.5}$$

where $2.303RT = 5.71 \text{ kJ mol}^{-1}$ at $25\,^{\circ}\text{C}$. The value of $\log\gamma_t(i, R \to S)$ is negative if the solvation of i is stronger in S than in R, while it is positive if the solvation of i is weaker in S than in R. The value of $\log\gamma_t(i, R \to S)$ is directly related to the reactivity of species i: if $\log\gamma_t(i, R \to S) = 8$, the reactivity of species i in S is 10^8 times that in R, while if $\log\gamma_t(i, R \to S) = -8$, the reactivity of species i in S is only $1/10^8$ that in R. Table 2.7 shows the values of $\log\gamma_t(i, R \to S)$ for the transfer from water to various organic solvents. They were obtained from the values in Table 2.4 using Eq. (2.5). Of course, there is a relationship:

$$\log\gamma_t(i, S_1 \to S_2) = \log\gamma_t(i, R \to S_2) - \log\gamma_t(i, R \to S_1)$$

Various factors that influence ion solvation were discussed in Section 2.2.1. The values in Table 2.7 give quantitative information on the extent of those influences.

2.3.2
Prediction of Solvent Effects by the Use of Transfer Activity Coefficients

We can use the transfer activity coefficients to predict solvent effects on chemical reactions and equilibria [25]. Some examples are shown below.

1. The solubility product constants of precipitate MX in solvents R and S [$K_{sp}(MX, R)$ and $K_{sp}(MX, S)$] are related by the following equations:

$$K_{sp}(MX, S) = \frac{K_{sp}(MX, R)}{\gamma_t(M^+, R \to S)\gamma_t(X^-, R \to S)}$$

---

10) Just as in aqueous solutions, the activity of solute i ($a_{c,i}$) in nonaqueous solutions is related to its (molar) concentration (c_i) by $a_{c,i} = \gamma_{c,i}c_i$, where $\gamma_{c,i}$ is the activity coefficient that is defined unity at infinite dilution. For

(mol l^{-1}), a_0 is the ion size parameter (m) and $A = 1.824 \times 10^6(\varepsilon_r T)^{-3/2} (\text{mol l}^{-1})^{-1/2}$ and $B = 50.29 \times 10^8(\varepsilon_r T)^{-1/2}(\text{mol l}^{-1})^{-1/2}\,\text{m}^{-1}$. The values of A and B at $25\,^{\circ}\text{C}$ in some typical solvents are as follows:

	H₂O	MeOH	AN	PC	DMF	DMSO	THF
ε_r	78.4	32.7	35.9	64.9	36.7	46.5	7.85
$A\,(\text{mol l}^{-1})^{-1/2}$	0.51	1.90	1.65	0.68	1.60	1.12	17.0
$B \times 10^{-9}\,(\text{mol l}^{-1})^{-1/2}\,\text{m}^{-1}$	3.29	5.09	4.86	3.62	4.81	4.27	10.6

nonionic solutes, the activity coefficient remains near-unity up to relatively high concentrations ($\sim$1 M). However, for ionic species, it deviates from unity except in very dilute solutions. The deviation can be estimated from the Debye–Hückel equation, $-\log\gamma_{c,i} = Az_i^2 I^{1/2}/(1 + a_0 B I^{1/2})$. Here, I is the ionic strength and $I = (1/2)\sum c_j z_j^2$

The influence of ionic strength on $\gamma_{c,i}$ is great in solvents of lower permittivities. When we compare ionic activities in different solvents, we have to consider this activity coefficient, in addition to the transfer activity coefficient γ_t. However, in reality, the influence of $\gamma_{c,i}$ is usually negligibly small compared to that of γ_t.

Table 2.7 Transfer activity coefficients of ions from water to nonaqueous solvents $[\log \gamma_t(i, W \rightarrow S)]$.[a]

Ions	Nonaqueous solvents S															
	MeOH	EtOH	PrOH	Ac	PC	FA	DMF	DMA	DMTF	NMP	AN	NB	DMSO	TMS	HMPA	DCE
H^+	1.8	1.9	1.6		8.8[b]		−3.2			−4.4	8.1	5.8	−3.4			
Li^+	0.8	1.9	1.9		4.0	−1.8	−1.8		9.6	−6.1	4.4	6.7	−2.6			4.4
Na^+	1.4	2.5	3.0		2.5	−1.4	−1.7	−2.1	6.8	−2.6	2.6	6.0	−2.4	−0.5		4.6
K^+	1.7	2.9	3.2	0.7	1.5[c]	−0.75	−1.8	−2.0	4.7	−1.9	1.4	4.0	−2.3	−0.7	−2.8	4.4
Rb^+	1.7	2.8	3.3	0.7	1.2[c]	−0.8	−1.7	−1.4		−1.4	1.1	3.3	−1.8	−1.6		4.4
Cs^+	1.6	2.6	3.0	0.7	0.9[c]	−1.1	−1.9	(−3)	2.5	−1.8	1.1	2.6	−2.3	−1.8		4.2
Ag^+	1.2	0.9	0.2	1.6	3.2		−3.0	−5.1	−17.9	−4.6	−4.1		−6.1	−0.7	−7.7	
Tl^+	0.7	1.2			1.9	−2.7	−2.0		−2.8	−2.6	1.4	(2.6)	−3.8			
NH_4^+	(1)	1.2								−4.2		4.7				
Me_4N^+	1.0	1.9		0.5	−2.6		−0.9			−0.5	0.5	0.7	−0.4			2.8
Et_4N^+	0.2	1.0			−3.0		−1.4				−1.2	−0.9	(−1.6)			0.9
Pr_4N^+	(−1.0)				−4.6		−3.0				−2.3	−2.8				
Bu_4N^+	(−1.4)				−6.1		−5.1				−5.4	−4.2				
Ph_4As^+	−3.7	−4.4		−5.6	−6.4	−4.2	−6.8	−6.8		−7.0	−5.8	−6.3	−6.6	−6.3	−6.8	−5.8
F^-	2.8				4.4	4.4	8.9				12.5					
Cl^-	2.3	3.5	4.6	10.0	6.8	2.4	8.5	9.6		8.9	7.4	6.1	7.1	8.2	10.2	9.1
Br^-	1.9	3.2	3.9	7.4	5.1	1.9	6.4	7.7		6.5	5.5	5.1	4.8	6.1	8.1	6.7
I^-	1.3	2.7	3.3	4.4	2.3	1.3	3.6	3.7		3.3	2.9	3.2	1.8	3.7	5.3	4.4
I_3^-	−2.2					−1.2	−4.7	−5.3			−2.6	−4.0	(−7.2)			
N_3^-	1.6	3.0		7.5	4.7	1.9	6.3	7.0		8.1	6.5	6.1	4.5	7.2	8.6	
CN^-	1.5	1.2		8.4	6.3	2.3	7.0				6.1		6.1			
SCN^-	1.0				1.2	1.2	3.2	3.7		3.2	2.5		1.7	3.9	3.5	
NO_3^-		2.5									3.7					1.2
ClO_4^-	1.1	1.8	3.0		−0.9	−2.1	0.7			−2.1	0.4	1.2			−1.2	2.8
CH_3COO^-	2.8					3.5	11.6	12.3			10.7		(8.8)			
Pic^-	−1.1	0.1			−1.1	−1.2	−1.2				−0.7	−0.9				
BPh_4^-	−4.2	−3.7	−4.4	−5.6	−6.4	−4.2	−6.8	−6.8		−7.0	−5.8	−6.3	−6.6	−6.3	−6.8	−5.8

[a] Unless otherwise stated, calculated from the values of $\Delta G_t^\circ(i, W \rightarrow S)$ in Marcus, Y. (1983) Pure Appl. Chem., 55, 977.

[b] Questionable reliability?

[c] Chantooni, M.K., Jr and Kolthoff, I.M. (1980) J. Chem. Eng. Data, 25, 208.

$$p K_{sp}(MX, S) = p K_{sp}(MX, R) + \log \gamma_t(M^+, R \rightarrow S) + \log \gamma_t(X^-, R \rightarrow S) \quad (2.6)$$

For example, using $p K_{sp}(AgCl, W) = 10.0$, $\log \gamma_t(Ag^+, W \rightarrow AN) = -4.1$, $\log \gamma_t(Cl^-, W \rightarrow AN) = 7.4$ in Eq. (2.6), we get the solubility product constant of AgCl in AN as $p K_{sp}(AgCl, AN) = 13.3$. This value agrees well with the experimental value of 12.4–12.9. In PC, the estimated $p K_{sp}(AgCl, PC)$ is 20.3, while the experimental value is 19.9–20.2.

2. The relation between the dissociation constants of acid HA in solvents R and S [$K_a(HA, R)$ and $K_a(HA, S)$] is expressed by

$$K_a(HA, S) = \frac{K_a(HA, R)\gamma_t(HA, R \rightarrow S)}{\gamma_t(H^+, R \rightarrow S)\gamma_t(A^-, R \rightarrow S)}$$

$$p K_a(HA, S) = p K_a(HA, R) + \log \gamma_t(H^+, R \rightarrow S)$$
$$+ \log \gamma_t(A^-, R \rightarrow S) - \log \gamma_t(HA, R \rightarrow S) \quad (2.7)$$

If we know the transfer activity coefficients for H^+, A^- and HA, we can estimate the dissociation constant of acid HA in S from that in R. For example, for the transfer from water to DMSO, $\log \gamma_t(H^+) = -3.4$, $\log \gamma_t(CH_3COO^-) = 8.8$ and $\log \gamma_t(CH_3COOH) = -0.8$ and, for the transfer from water to AN, $\log \gamma_t(H^+)$ 8.1, $\log \gamma_t(CH_3COO^-) = 10.7$ and $\log \gamma_t(CH_3COOH) = 0.4$. Using $p K_a(CH_3COOH, W) = 4.76$, we get $p K_a(CH_3COOH, DMSO) = 11.0$ and $p K_a(CH_3COOH, AN) = 23.2$, in fair agreement with the experimental values [34] of 12.6 in DMSO and 22.3 in AN, respectively.

3. The standard potential of the M^+/M electrode in solvent S [$E^0(S)$] is related to that in R [$E^0(R)$] by

$$E^0(S) - E^0(R) = 0.059 \log \gamma_t(M^+, R \rightarrow S)$$

We can predict the value of $E^0(S)$ from the known values of $E^0(R)$ and $\log \gamma_t(M^+, R \rightarrow S)$. Here, $E^0(S)$ and $E^0(R)$ should be in a common potential scale. Of course, if the values of $E^0(S)$ and $E^0(R)$ in a common scale are known, we can get the value of $\log \gamma_t(M^+, R \rightarrow S)$. This is used to obtain the values of transfer activity coefficients (Sections 6.3.6 and 8.2.1).

Data on the Gibbs energies of transfer and the transfer activity coefficients are of practical use. The problem, however, is that the data now available are not always sufficiently reliable. It is desirable that reliable data are obtained systematically for as many chemical species as possible. For the compiled data, see Ref. [30].

2.4
Solvent Effects on the Complexation of Metal Ions

The complexation of a solvated metal ion with a ligand (complexing agent) corresponds to the substitution of the solvent molecules existing in the first solvation sphere by the molecules or ions of the ligand in solution. Here, the metal ion–solvent interaction competes with the metal ion–ligand interaction. Solvent molecules also interact with the free ligand and the ligand coordinated to the metal ion, affecting the reactivity of the ligand. Thus, the stability of a metal complex is easily influenced by solvent.

The complexation of metal ions with halide and thiocyanate ions has been studied in a variety of solvents, from both thermodynamic and structural aspects [6, 21a, 35]. An example is the stepwise complexation of a silver ion with halide ions X^-:

$$Ag^+ + X^- \rightleftharpoons AgX \quad K_{sp} = [Ag^+][X^-]$$

$$Ag^+ + 2X^- \rightleftharpoons AgX_2^- \quad \beta_2 = [AgX_2^-]/([Ag^+][X^-]^2)$$

Table 2.8 shows the values of pK_{sp} ($= -\log K_{sp}$) and $\log \beta_2$. As described in Section 2.3, the solvent effect on pK_{sp} can be estimated approximately from the values of the transfer activity coefficients of Ag^+ and X^-. In Table 2.8, the values of $K_{sp}\beta_2$ ($= [AgX_2^-]/[X^-]$) are between $10^{-5.4}$ and $10^{-3.4}$ in protic solvents (H_2O, MeOH) and between $10^{0.6}$ and $10^{1.3}$ in aprotic solvents (AN, PC, DMSO). The much larger $K_{sp}\beta_2$ values in aprotic solvents mean that AgX_2^- is much more stable in aprotic solvents than in protic solvents. Thus, if an excess amount of the AgX precipitate is added to a solution containing X^-, a considerable part of the X^- ion is consumed to form soluble AgX_2^-. In contrast, if the AgX precipitate is added to a solution containing an excess amount of X^-, the precipitate is dissolved almost completely,

Table 2.8 Solvent effects on the formation of precipitates and complexes of the silver ion with halide ions (X^-).[a]

Solvent	Indifferent electrolyte		Cl^-	Br^-	I^-
H_2O	5 M NaClO$_4$	pK_{sp}	10.10	12.62	16.35
		$\log \beta_2$	5.40	7.23	10.95
		$K_{sp}\beta_2$	$10^{-4.70}$	$10^{-5.39}$	$10^{-5.40}$
MeOH	1 M LiClO$_4$	pK_{sp}	13.0	15.2	18.2
		$\log \beta_2$	7.9	10.6	14.8
		$K_{sp}\beta_2$	$10^{-5.1}$	$10^{-4.6}$	$10^{-3.4}$
AN	0.1 M Et$_4$NClO$_4$	pK_{sp}	12.4	13.2	14.2
		$\log \beta_2$	13.1	13.8	15.2
		$K_{sp}\beta_2$	$10^{0.7}$	$10^{0.6}$	$10^{1.0}$
PC	0.1 M Et$_4$NClO$_4$	pK_{sp}	20.0	20.5	21.8
		$\log \beta_2$	20.9	21.2	22.8
		$K_{sp}\beta_2$	$10^{0.9}$	$10^{0.7}$	$10^{1.0}$
DMSO	0.1 M Et$_4$NClO$_4$	pK_{sp}	10.4	10.9	12.1
		$\log \beta_2$	11.7	12.0	13.0
		$K_{sp}\beta_2$	$10^{1.3}$	$10^{1.1}$	$10^{0.9}$

[a]Taken from Table X in Ref. [6]. Somewhat different values are found in Table 4 of Ahrland, S. (1990) *Pure Appl. Chem.*, **62**, 2077. $K_{sp}\beta_2 = [AgX_2^-]/[X^-]$.

forming AgX_2^-. This is why an AgX/Ag electrode in a high concentration of X^- cannot be prepared in aprotic solvents (see Section 6.1.2). In protic solvents, however, the formation of AgX_2^- is not appreciable, and it is possible to prepare an AgCl/Ag electrode using a saturated KCl solution.

The complexation of metal ions with neutral molecules has also been studied extensively. We consider here the complexes of metal ions with macrocyclic ligands such as crown ethers and cryptands [36]. The 1 : 1 complexation of a metal ion M^{n+} with macrocyclic ligand L is expressed by Eqs. (2.8) and (2.9), where K is the complex formation constant:

$$M^{n+} + L \rightleftarrows ML^{n+} \tag{2.8}$$

$$K = [ML^{n+}]/\{[M^{n+}][L]\} \tag{2.9}$$

The formation constants ($\log K$) for the complexes of dibenzo-18-crown-6 and cryptand[222] are, as shown in Table 2.9, markedly influenced by solvent.

If we express the complex formation constants in solvents R and S by K_S and K_R, respectively, we get Eq. (2.10) from Eq. (2.9):

$$\log K_S - \log K_R = \log \gamma_t(ML^{n+}, R \rightarrow S) - \log \gamma_t(M^{n+}, R \rightarrow S) - \log \gamma_t(L, R \rightarrow S) \tag{2.10}$$

However, the relationship in Eq. (2.11) is often obtained for complexes between univalent metal ions and cryptands if R and S are both dipolar aprotic solvents:

$$\log K_S - \log K_R \approx \log \gamma_t(M^+, R \rightarrow S) \tag{2.11}$$

This relation tends to occur if the size of the univalent metal ion fits the size of the cavity of the cryptand or is smaller. Then, we get Eq. (2.12), and therefore Eq. (2.13), from Eq. (2.10):

$$\log \gamma_t(ML^+, R \rightarrow S) \approx \log \gamma_t(L, R \rightarrow S) \tag{2.12}$$

$$\Delta G_t^\circ(ML^+, R \rightarrow S) \approx \Delta G_t^\circ(L, R \rightarrow S) \tag{2.13}$$

These results show that the metal ion in the complex ML^+ is almost shielded from outside solvent, by being captured in the cavity of the cryptand. Thus, the solvent–complex interaction is nearly the same as the solvent–ligand interaction (see Table 4.3). However, there are cases when Eqs. (2.12) and (2.13) are not valid. Then, the shielding is not complete and some influence of the metal ion–solvent interaction is observed. Equations (2.12) and (2.13) do not hold also when the solvent is protic, e.g. water. The oxygen and nitrogen atoms of the cryptand interact with the protic solvent and the interaction is stronger with the free cryptand than with the cryptand combined with the metal ion.

Although solvents have significant effects on the stabilities of macrocyclic ligand complexes, the selectivity of the ligand to metal ions is not much influenced by solvents. For example, in Table 2.9, both ligands are most selective to K^+ in all the solvents tested.

Table 2.9 Formation constants (log K) of metal ion complexes with dibenzo-18-crown-6 and cryptand[222].[a]

Ligands	Metal ions	Solvents						
		H_2O	MeOH	EtOH	AN	PC	DMF	DMSO
Dibenzo-18-crown-6	Na^+	0.8_0	4.4	—	4.9_5	5.2_0	2.8_0	1.9_3
	K^+	2.0_5	5.0_5	—	4.7_7	5.1_3	—	2.4_6
	Rb^+	1.5_6	4.2_3	—	3.7_0	3.9_1	—	(1.9)
	Cs^+	1.0	3.5_5	—	3.5_5	3.3_1	1.4_8	1.3_4
	Ag^+	1.5_0	4.0_4	—	—	5.8_2	—	—
Cryptand[222]	Li^+	0.9_8	2.6	≤ 2.3	6.9_7	6.9_4	—	<1.0
	Na^+	3.9_8	7.9_8	8.5_7	9.6_3	10.5_4	6.1_7	5.3_5
	K^+	5.4_7	10.4_1	10.5_0	11.3_1	11.1_9	7.9_5	6.9_9
	Rb^+	4.2_4	8.9_8	9.2_8	9.5_0	9.0_2	6.7_3	5.8_0
	Cs^+	1.4_7	4.4	4.1_7	4.5_7	4.1	2.1_6	1.4_3
	Ag^+	9.6	12.2_0	11.5_1	8.9_9	16.3_1	10.0_5	7.2_2

[a]Cox, B.G. (1981) *Ann. Rep. Prog. Chem. Sect. C*, **78**, 3.

2.5
Selective Solvation of Ions in Mixed Solvents

In order to achieve appropriate solvent properties, we often use mixtures of two or more solvents (mixed solvents). Although mixtures of water and organic solvents are most frequently used, mixtures of two organic solvents are also popular. A book that solely deals with mixed solvents has been published [37]. Mixed solvents play important roles in modern electrochemical technologies, as discussed in Chapter 12.

The compositions of mixed solvents are expressed by mole fraction (x), mass fraction (w) or volume fraction (ϕ). If a mixed solvent consists of n_A moles of solvent A and n_B moles of solvent B, x_A, w_A and ϕ_A for component A can be expressed by $x_A = n_A/(n_A + n_B)$, $w_A = n_A/(n_A + n_B M_B/M_A)$ and $\phi_A = n_A/(n_A + n_B V_B/V_A)$. Here, M_i is the molecular weight of component i and V_i the molar volume of i before mixing. It is assumed that the volume of the mixed solvent is equal to the sum of the volumes of the two constituent solvents, although some volume contraction may occur by mixing.

The problem of relative permittivity of mixed solvents (denoted by ε for simplicity) was theoretically studied by Debye and Onsager [38]. Onsager derived the following equation:

$$\frac{(\varepsilon-1)(2\varepsilon+1)}{9\varepsilon} = \frac{\phi_A(\varepsilon_A-1)(2\varepsilon_A+1)}{9\varepsilon_A} + \frac{\phi_B(\varepsilon_B-1)(2\varepsilon_B+1)}{9\varepsilon_B}$$

This equation can be written with good approximation as $\varepsilon = \phi_A\varepsilon_A + \phi_B\varepsilon_B$, indicating that the permittivity of the mixed solvent changes linearly between ε_A and ε_B if it is

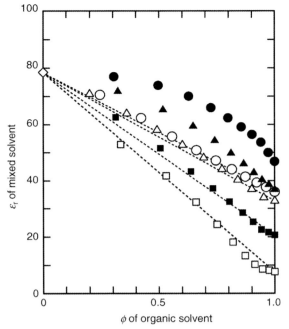

Figure 2.9 Relative permittivities of water–organic solvent mixtures plotted against their volume fractions. Solvent: open circles AN; open triangles MeOH; open squares THF; filled circles DMSO; filled triangles DMF; filled squares Ac. (From the data in Table 7.1, Ref. [8].)

plotted against the volume fraction of the constituent solvents. When the interaction between the constituent solvents is weak, this linear relation holds well. However, when the interaction is strong, a large deviation from linearity is observed. Figure 2.9 shows the relationship between the relative permittivity and the volume fraction of mixtures of water and organic solvents. Because acetonitrile, acetone and tetrahydrofuran interact only weakly with water, near-linear relations are observed for their mixtures with water. However, for DMSO and DMF, which readily interact with water, large deviations from linearity are observed.

The acid–base properties of a mixed solvent is also an important factor influencing the behavior of solutes. Thus, the parameters of the acidity and basicity of mixed solvents have been studied to some extent [39]. Figure 2.10 shows the donor numbers of mixtures of nitromethane and other organic solvents. Because nitromethane has very weak basicity ($DN = 2.7$), the addition of small amounts of basic solvents (HMPA, DMSO, pyridine) increases the donor number remarkably. Figure 2.11, on the other hand, shows the acceptor numbers of mixtures of water and aprotic solvents. Because water is protic and selectively interacts with $Et_3P=O$ (strong Lewis base), many of the relations curve upward. However, with HMPA, the relation curves downward because HMPA is a strong base and easily interacts with

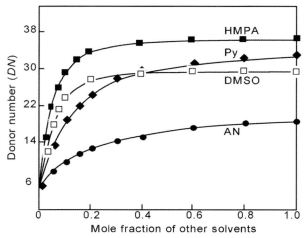

Figure 2.10 Donor numbers of mixtures of nitromethane with other solvents [16a].

water to weaken the interaction between water and $Et_3P=O$. The acidity and basicity of mixed solvents are influenced not only by the acidity and basicity of the constituent solvents but also by the mutual interactions between the molecules of constituent solvents. At present, however, this cannot be treated theoretically.

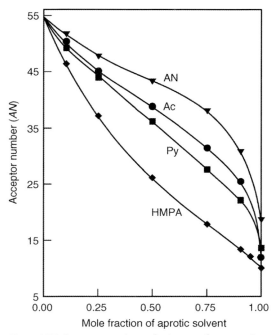

Figure 2.11 Acceptor numbers of mixtures of water and aprotic solvents [16a]. Aprotic solvents are shown on the curves.

For the properties and structures of binary solvent mixtures and the preferential solvation between the binary solvents, see Ref. [37].

In mixed solvents, ions are solvated competitively by the constituent solvents. If one solvent solvates more easily than the other, the so-called *selective solvation* occurs. In general, cations are selectively solvated by strongly basic solvents, anions by strongly acidic solvents. The ion solvation in mixed solvents has been studied extensively by measuring transport behavior and spectroscopic properties [40]. NMR is particularly useful in studying the solvated ionic species in mixed solvents. If the exchange between the solvent molecules in the primary solvation shell and those in the bulk solution is slow enough, isolated NMR peaks are observed for each of the solvated species, providing quantitative information. For example, if we measure the ^{27}Al-NMR spectrum in DMF–DMSO mixtures of different compositions, we get seven peaks corresponding to $i = 0$–6 in $[Al(DMSO)_i(DMF)_{6-i}]^{3+}$. If the solvent exchange is rapid enough, as in the case of Na^+ and Li^+, we get a single NMR signal. Then, from the chemical shift and the peak width, we can determine the mean strength of ion solvation and the mean time of stay of solvent molecules in the solvation shell.

Attempts have been made to treat selective ion solvation in mixed solvents as ligand exchange reactions [41–43]. We express ion X (cation or anion) existing in solvent A by XA_n and in solvent B by $XB_{n'}$. Here, $n \neq n'$ if the molecular size of A is very different from that of B or if A is unidentate and B is bidentate (e.g. PC and DME in Li^+ solvation). Otherwise, it is usual that $n = n'$. Then, the ligand exchange reaction in the mixture of A and B will proceed as follows with the increase in the concentration of B:

$$XA_n + B \rightleftarrows XA_{n-1}B + A(K_1), \quad XA_{n-1}B + B \rightleftarrows XA_{n-2}B_2 + A(K_2), \ldots,$$
$$XA_2B_{n-2} + B \rightleftarrows XAB_{n-1} + A(K_{n-1}), \quad XAB_{n-1} + B \rightleftarrows XB_n + A(K_n)$$

where K_i is the equilibrium constant. If X is selectively (preferentially) solvated by B, all of the above reactions will proceed within small concentration of B. However, if the selectivity is small or the exchange process can proceed ideally following the statistical probability, the relation $K_i = (n-i+1)i^{-1}\beta_n^{1/n}$, where $\beta_n = (K_1 K_2 \ldots K_n)$, is expected. The Gibbs energy of transfer of X from solvent A to solvent mixture (A + B) is given as follows [41]:.

$$\Delta G_t^\circ(X, A \rightarrow A + B)$$

$$= \sum \phi_i \mu_i^{\circ\,elec} - \mu_0^{\circ\,elec} - RT \ln\left[1 + \sum_{i=1}^{n}(x_B/x_A)^i \prod_{j=1}^{i} K_j\right] - nRT \ln x_A \qquad (2.14)$$

where ϕ_i is the fraction of $XA_{n-i}B_i$ and $\sum \phi_i = 1$, $\mu_i^{\circ\,elec}$ and $\mu_0^{\circ\,elec}$ are the chemical potentials of long-range interactions (electrostatic in origin and treated by Born theory) of $XA_{n-i}B_i$ and XA_n, respectively. If solvents A and B are isodielectric, the first two terms on the right side in Eq. (2.14) is zero. Ionic preferential solvation has been handled variously, usually as the deviation from the simple solvation process as in Eq. (2.14) [42].

Cox *et al.* [43] expressed the Gibbs energies of transfer of ion X from solvent A to solvent B and to the mixture (A + B) by Eqs. (2.15a) and (2.15b), respectively:

$$\Delta G_t^\circ (X, A \to B) = -RT \ln \beta'_n \qquad (2.15a)$$

$$\Delta G_t^\circ (X, A \to A + B) = -nRT \ln \phi_A - RT \ln \left[1 + \sum_{i=1}^{n} \beta'_i (\phi_B/\phi_A)^i \right] \qquad (2.15b)$$

Here, ϕ_A and ϕ_B are the volume fractions of A and B, respectively, β'_i is the complex formation constant of $XA_n + iB \rightleftharpoons XA_{n-i}B_i + iA$, that is obtained in the solvent almost pure in A (i.e. very small in the concentration of B). The solvation of B should be stronger than that of A. These equations are important in showing that the data for the complexation between X and B in solvent A are applicable to predicting the solvation of X in solvent B and in mixed solvent (A + B). As discussed in detail in Section 6.3.6, these equations are valid as long as the permittivities of A and B do not differ significantly.

The Gibbs energy of ionic transfer to mixed solvents can be obtained thermodynamically (by solubility measurement or potentiometry) or spectroscopically (by NMR, for example), and various data are available. The Gibbs energies of transfer from water to mixed aqueous organic solvents have been compiled in [44a] for cations and [44b] for anions. The enthalpies and entropies of transfer of electrolytes and ions from water to mixed aqueous organic solvents have also been compiled in [44c].

2.6
Ion Association and Solvent Permittivities

In the above sections, we considered electrolytes that are ionophores.[11] Ionophores, such as sodium chloride, are ionic in the crystalline state and are expected to dissociate into free ions in dilute solutions. In fact, in high-permittivity solvents ($\varepsilon_r > 40$), ionophores dissociate almost completely into ions unless the solutions are of high concentration. When an ionophore is completely dissociated in the solution, its molar conductivity Λ decreases linearly with the square root of the concentration c ($<10^{-2}$ M):

$$\Lambda = \Lambda^\infty - Sc^{1/2}$$

where Λ^∞ is the molar conductivity at infinite dilution and S is the Onsager slope (Section 7.1.1). This is used to confirm the complete dissociation of an electrolyte.

With the decrease in permittivity, however, complete dissociation becomes difficult. Some part of the dissolved electrolyte remains undissociated and forms ion pairs. In low-permittivity solvents, most of the ionic species exist as ion pairs.

---

11) Electrolytes are classified into *ionophores* and *ionogens*. Ionogens (such as hydrogen halides) exist as neutral molecules in their pure state but form ions in solutions through chemical interactions either with solvent molecules or with some added species.

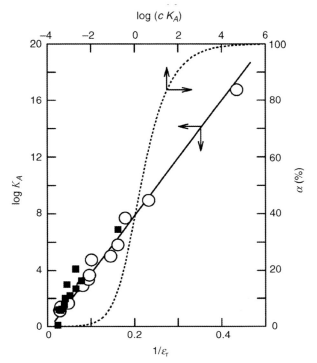

Figure 2.12 Relations between the ion association constants (log K_A) and the reciprocal of solvent permittivity (1/ε_r) (solid line) and between the degree of ion association (α) and log (cK_A) (dotted curve) (open circles: Bu$_4$NPic in AN, NB, MeOH, Ac, Py, DCE, o-dichlorobenzene, acetic acid, chlorobenzene and benzene; filled squares: KCl in ethanolamine, MeOH, EtOH, acetic acid and H$_2$O–dioxane mixtures). The solid line was obtained using Eq. (2.19) for $a = 0.6$ nm.

Ion pairs contribute neither ionic strength nor electric conductivity to the solution. Thus, we can detect the formation of ion pairs by the decrease in molar conductivity Λ. In Figure 2.12, the logarithmic values of ion association constants (log K_A) for tetrabutylammonium picrate (Bu$_4$NPic) and potassium chloride (KCl) are plotted against (1/ε_r) [45].

$$A^+ + B^- \rightleftarrows A^+ B^- \quad (\text{ion pair})$$

$$K_A = [A^+B^-]/([A^+][B^-])$$

Figure 2.12 also shows a curve representing the relation between the values of log (cK_A) and the degree of ion association (α). This curve shows that, for 10^{-2} M KCl in EtOH (log $K_A = 2$), log (cK_A) = 0 and 38% of KCl exists as ion pairs. For 0.1 M KCl in acetic acid (log $K_A = 6.2$), ~99.9% of KCl exists as ion pairs; the fraction dissociated into free ions is only 0.1%. Usually, in solvents with $\varepsilon_r < 10$, the fraction of electrolyte that is dissociated into ions is small except in very dilute solutions.

Conversely, in solvents with $\varepsilon_r > 40$, the fraction of electrolyte that forms ion pairs is small except in fairly concentrated solutions.[12]

The theories of Bjerrum (1926) and Fuoss (1958) on ion association are described in detail in a recent review [46] or books [47]. According to the Debye–Hückel theory, the activity of an ion in the solution deviates from its concentration because of the electrostatic interaction between ions, and the deviation increases with increasing ionic strength (see footnote 10). In the theory, it was postulated that the energy of electrostatic ionic interaction is much less than the kinetic energy of thermal agitation. This is true if the distance between ions is large enough. However, with decreasing distance between ions, the energy of electrostatic interaction increases and approaches the energy of thermal agitation. According to Bjerrum's theory, the energy of electrostatic interaction between two ions ($|z_+z_-|e^2/4\pi\varepsilon_0\varepsilon_r r$) becomes equal to $2k_B T$ (k_B: Boltzmann constant) at a critical distance q and, if two ions of opposite charges are at distance r that satisfies $r \le q$, they are associated; if $r > q$, however, they remain unassociated. Moreover, the distance r cannot become smaller than the 'distance of closest approach' a, which is equal to the sum of the radii of the associated cation and anion. Therefore, ion association occurs if $a \le r \le q$. If $a > q$, ion association does not occur and all ions remain free.

Here, the distance q is given by Eq. (2.16):

$$q = \frac{|z_+z_-|N_A e^2}{8\pi\varepsilon_0\varepsilon_r RT} \tag{2.16}$$

where N_A is the Avogadro constant. In aqueous solutions at $25\,°C$, the value of q is 0.35 nm for univalent ions and 1.4 nm for divalent ions. For many of the univalent hydrated ions in water, the relation $a > q$ holds and their association is difficult. However, in solvents of lower permittivities, the relation $q > a$ holds for most electrolytes, and the ion association is more pronounced.

The probability of two ions of opposite charges being located at a distance r is shown in Figure 2.13. The shaded region in the figure is where ion association is expected to occur: thus, the larger the difference between a and q, the easier is the ion association. According to Bjerrum, the ion association constant K_A (in molar concentration) is expressed by Eq. (2.17):

$$K_A = \frac{4\pi N_A}{1000} \int_a^q r^2 \exp\left(\frac{2q}{r}\right) dr \tag{2.17}$$

An alternative approach to ion association was proposed by Fuoss. He defined the ion pair as two oppositely charged ions that are in contact, i.e. at a distance of $r = a$, and derived the following equation for the ion association constant:

$$K_A = (4\pi N_A/3000)(a/m)^3 \exp(2q/a) \tag{2.18}$$

12) Considerable ion association may occur even in high-permittivity solvents, e.g. $K_A = 2.97 \times 10^4$ for $CF_2HCOOLi$ and 0.62×10^4 for CFH_2COOLi in PC ($\varepsilon_r \sim 64$). A linear relation has been found between the mean charge density of oxygen atoms, $q(O)$, for $CF_xH_yCOO^-$ ($x + y = 3$), obtained by MNDO calculation, and log K_A for the association of $CF_xH_yCOO^-Li^+$ in DMSO (Ref. [53b]).

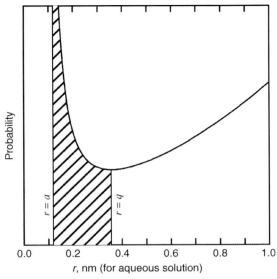

Figure 2.13 The probability for two ions of opposite signs existing at distance r. The scale on the abscissa is for an aqueous solution.

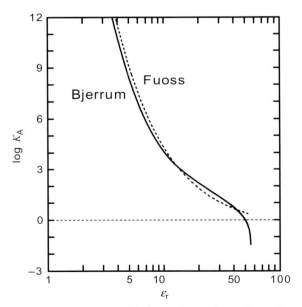

Figure 2.14 Comparison of the $\log K_A - \log \varepsilon_r$ relation obtained by Bjerrum's theory (solid curve) and that obtained by Fuoss' theory (dotted curve). The case of a 1:1 electrolyte with $a = 0.5$ nm.

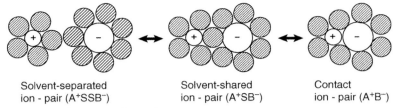

Solvent-separated
ion - pair (A⁺SSB⁻)

Solvent-shared
ion - pair (A⁺SB⁻)

Contact
ion - pair (A⁺B⁻)

Figure 2.15 Three types of ion pairs.

Equation (2.19) is obtained from Eq. (2.18).

$$\log K_A = \log(4\pi N_A a^3/3000) + 0.434|z_+ z_-|N_A e^2/(4\pi\varepsilon_0\varepsilon_r RT) \tag{2.19}$$

From this equation, a linear relation is expected between $\log K_A$ and $(1/\varepsilon_r)$. The solid line in Figure 2.12, obtained by assuming $a = 0.6$ nm in Eq. (2.19), agrees well with the experimental results.

The two curves in Figure 2.14 are the relationships between $\log K_A$ and $\log \varepsilon_r$ for a 1:1 electrolyte. The solid curve was obtained by Bjerrum's theory (Eq. (2.17)) and the dotted curve by Fuoss' theory (Eq. (2.19)), both assuming $a = 0.5$ nm. The big difference between the two theories is that, according to Bjerrum's theory, ion association does not occur if ε_r exceeds a certain value (~50 in Figure 2.14), although this value depends on the value of a. Both theories are not perfect and could be improved. In recent treatments of ion association, noncoulombic short-range interactions between ions are also taken into account [48]. By introducing noncoulombic interactions, $W^*(r)$, Eq. (2.17) is modified to a form as in Eq. (2.20):

$$K_A = \frac{4\pi N_A}{1000} \int_a^q r^2 \exp\left[\frac{2q}{r} - \frac{W^*(r)}{k_B T}\right] dr \tag{2.20}$$

For example, in methylformate, the $W^*(r)$ values are negative for $LiClO_4$, $NaClO_4$ and Bu_4NClO_4 and positive for $LiAsF_6$ and $NaBPh_4$ [49]. Negative $W^*(r)$ values suggest structure-making effects that tend to stabilize the ion pair, while positive $W^*(r)$ values indicate structure-breaking effects that destabilize the ion pair. Thus, K_A values for perchlorates are larger than those expected from Eq. (2.17). Beside this treatment, the ion association has been dealt with by such treatments as the mean spherical approximation (MSA) [50] and the consideration of activity coefficients of charged symmetrical ion pairs [51].

Moreover, as schematically shown in Figure 2.15, three different types of ion pairs, i.e. solvent-separated (A^+SSB^-), solvent-shared (A^+SB^-) and contact (A^+B^-) types, may be formed depending on the strength of ion–solvent interactions, complicating the treatment of ion association. If we consider the following multistep equilibrium, the total ion association constant (K_A) is expressed by Eq. (2.21):

$$A^+_{solv} + B^-_{solv} \overset{K_1}{\rightleftharpoons} A^+ SSB^- \overset{K_2}{\rightleftharpoons} A^+ SB^- \overset{K_3}{\rightleftharpoons} A^+ B^-$$

$$K_A = \frac{[A^+ SSB^-] + [A^+ SB^-] + [A^+ B^-]}{[A^+][B^-]} = K_1(1 + K_2 + K_2 K_3) \tag{2.21}$$

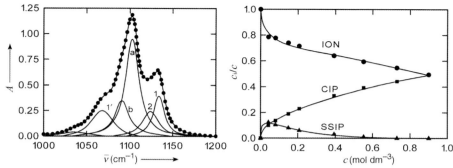

Figure 2.16 Left: Measured (points) and fitted (lines) IR spectrum in the region of the v_3^{ClO} vibration of 0.73 M LiClO₄ in AN at 25°C. Bands a and b are from the free ClO₄⁻, bands 1, 1′ and 2 are from ClO₄⁻ in the contact ion pair. Right: Fractions of free ions (ION), solvent-shared ion pairs (SSIP) and contact ion pairs (CIP) in LiClO₄/AN (25°C) [53a].

Different types of ion association can be identified by using such methods as infrared, Raman and dielectric relaxation spectroscopies [52,53]. For example, the relative concentrations of free ions (c_{ION}), solvent-shared ion pairs (c_{SSIP}) and contact ion pairs (c_{CIP}) in LiClO₄–AN (25°C) were obtained by combining infrared and dielectric relaxation spectroscopic measurements [53]. Because ClO₄⁻ was practically unsolvated in AN, a solvent-separated ion pair was not formed. In the IR spectrum of Figure 2.16 (left), the integrated intensity of bands 1, 1′ and 2 gave c_{CIP}, while that of bands a and b gave ($c_{ION} + c_{SSIP}$). The total ion pair concentration ($c_{CIP} + c_{SSIP}$) was obtained from the dispersion amplitude of the dielectric relaxation spectrum.[13] The results are shown in Figure 2.16 (right). The association constant, $K_A = 19.8$ dm³ mol⁻¹, agreed well with the conductimetric result, $K_A = 21.8$ dm³ mol⁻¹.

In low-permittivity solvents, ions of opposite charges easily form ion pairs. However, if the electrolyte concentrations are increased in these solvents, the formation of triple ions and quadrupoles also occurs as follows:

$$A^+ + A^+ B^- \rightleftarrows A^+ B^- A^+$$
$$B^- + A^+ B^- \rightleftarrows B^- A^+ B^-$$
$$A^+ B^- + A^+ B^- \rightleftarrows (A^+ B^-)_2$$

The positive and negative triple ions can remain without being associated even in low-permittivity solvents, mainly because the triple ions are large. Because the triple ions are conductive, their formation can be detected by the increase in molar conductivity that occurs with increasing electrolyte concentrations (Figure 7.2).

13) Static solution permittivity $\varepsilon(c)$ and static solvent permittivity $\varepsilon_s(c)$ for solutions of various electrolytes at various concentrations (c) have been obtained by dielectric relaxation spectroscopy [54]. Ion pairs contribute to permittivity if their lifetime is longer than their relaxation time. However, free ions do not contribute to permittivity. Thus, $\varepsilon(c) = \varepsilon_s(c)$ for completely dissociated electrolyte solutions and $\varepsilon(c) \geq \varepsilon_s(c)$ for associated electrolyte solutions. The relaxation times of the different ion pair types in Eq. (2.21) are generally so close to one another that only the total ion pair concentration is obtained by dielectric relaxation methods [53].

Highly concentrated electrolyte solutions in low-permittivity solvents are sometimes very conductive and used as electrolyte solutions for lithium batteries. Thus, the conductimetric behavior of triple ions has been studied in detail (Section 7.1.3) [55]. Triple ions are also formed in high-permittivity protophobic aprotic solvents: for example, they are formed in acetonitrile between Li^+ and anions such as Cl^-, CH_3COO^- and $C_6H_5COO^-$ [56]. In AN, these simple ions are only weakly solvated, are highly reactive (Table 2.7) and are stabilized by triple ion formation. In such cases, noncoulombic interactions seem to play an important role.

References

1 Fawcett, W.R. (2004) *Liquids, Solutions, and Interfaces, from Classic Macroscopic Descriptions to Modern Microscopic Details*, Oxford University Press, New York.

2 Bockris, J.O'M. and Reddy, A.K.N. (1998) *Modern Electrochemistry:* vol. 1, *Ionics*, 2nd edn, Plenum Press, New York.

3 Barthel, J.M.G., Krienke, H. and Kunz, W. (1998) *Physical Chemistry of Electrolyte Solutions, Modern Aspects*, Springer, Darmstadt.

4 Mamantov, G. and Popov, A.I. (eds) (1994) *Chemistry of Nonaqueous Solutions, Current Progress*, Wiley-VCH Verlag GmbH, Weinheim, Chapters 1–3.

5 Covington, A.K. and Dickinson, T. (eds) (1973) *Physical Chemistry in Organic Solvent Systems*, Plenum Press, London.

6 Ahrland, S. (1978) Solvation and complex formation in protic and aprotic solvents, in *The Chemistry of Nonaqueous Solvents*, vol. VA (ed. J.J. Lagowski), Academic Press, New York, Chapter 1.

7 (a) Burgess, J. (1999) *Ions in Solution*, Ellis Horwood, Chichester; (b) Burgess, J. (1978) *Metal Ions in Solution*, Ellis Horwood, Chichester.

8 Marcus, Y. (1985) *Ion Solvation*, John Wiley & Sons, Inc., New York.

9 Marcus, Y. (1997) *Ion Properties*, Marcel Dekker, New York.

10 Krestov, G.A. (1991) *Thermodynamics of Solvation. Solutions and Dissolution; Ions and Solvents; Structure and Energetics*, Ellis Horwood, Chichester.

11 Parker, A.J. (1976) *Electrochim. Acta*, **21**, 671.

12 (a) Fawcett, W.R. (1999) *J. Phys. Chem. B*, **103**, 11181; (b) Ref. [1], p. 106.

13 Fawcett, W.R. and Opallo, M. (1992) *J. Phys. Chem.*, **96**, 2920.

14 (a) Fawcett, W.R. and Tikanen, A.C. (1996) *J. Phys. Chem.*, **100**, 4251; (b) Ref. [1], p. 130.

15 (a) Kirkwood, G. (1934) *J. Chem. Phys.*, **2**, 351; (b) Reichardt, C. (2003) *Solvents and Solvent Effects in Organic Chemistry*, Wiley-VCH Verlag GmbH, Weinheim, p. 225.

16 (a) Gutmann, V. (1978) *The Donor–Acceptor Approach to Molecular Interactions*, Plenum Press, New York; (b) Gutmann, V. and Resch, G. (1995) *Lecture Notes on Solution Chemistry*, World Scientific, Singapore.

17 (a) Mayer, U. (1977) *Monatsh. Chem.*, **108**, 1479; (b) Okazaki, S. and Sakamoto, I. (1990) *Solvents and Ions*, San-ei Publishing, Kyoto, p. 315 (in Japanese).

18 (a) Funahashi, S. (1998) *Inorganic Reactions in Solution*, Shokabo, Tokyo, pp. 242, 246, Chapter 4 (in Japanese); (b) Mishustin, A.I. (1981) *Russ. J. Phys. Chem.*, **55**, 844.

19 (a) Barthel, J. and Seser, R. (1994) *J. Solution Chem.*, **23**, 1133; (b) Ref. [1b], p. 47; (c) Loring, J.S. and Fawcett, W.R. (1999) *J. Phys. Chem. A*, **103**, 3608; (d) Fawcett, W.R., Brooksby, P., Verbovy, D., Bakó, I. and Pálinkás, G. (2005) *J. Mol. Liq.*, **118**, 171; (e) Toth, J.P., Ritzhaupt, G. and Paul, J. (1981) *J. Phys. Chem.*, **85**, 1387; (f) Rouviere, J., Dimon, B., Brun, B. and Salvinien, J. (1972) *C. R. Acad. Sci. (Paris) C*, **274**, 458; (g) Perelygin, I.S. and

Klimchuk, M.A. (1975) *Russ. J. Phys. Chem.*, **49**, 76.

20 (a) Ohtaki, H. (2001) *Monatsh. Chem.*, **132**, 1237; (b) Ohtaki, H. and Radnai, T. (1993) *Chem. Rev.*, **93**, 1157.

21 (a) Ohtaki, H. and Ishiguro, S.,Ref. [1c], Chapter 3; (b)Ohtaki, H. and Wada, H. (1985) *J. Solution Chem.*, **14**, 209; (c) Wakabayashi, K., Maeda, Y., Ozutsumi, K. and Ohtaki, H. (2004) *J. Mol. Liq.*, **110**, 43.

22 For example, (water) (a) Jimenez, R., Fleming, G.R., Kumar, P.V. and Maroncelli, M. (1994) *Nature*, **369**, 471; (b) Nandi, N., Roy, S. and Bagchi, B. (1995) *J. Chem. Phys.*, **102**, 1390; (alcohols) (c) Biswas, R., Nandi, N. and Bagchi, B. (1997) *J. Phys. Chem. B*, **101**, 2968; (AN) (d) Rosenthal, S.J., Xie, X., Du, M. and Fleming, G.R. (1991) *J. Chem. Phys.*, **95**, 4715; (e) Cho, M., Rosenthal, S.J., Scherer, N.F., Ziegler, L.D. and Fleming, G.R. (1992) *J. Chem. Phys.*, **96**, 5033; (various solvents) (f) Maroncelli, M. (1993) *J. Mol. Liq.*, **57**, 1.

23 For example, (a) Rossky, P.J. and Simon, J.D. (1994) *Nature*, **370**, 263; (b) Weaver, M.J. and McManis, G.E. III (1990) *Acc. Chem. Res.*, **23**, 294; (c) Bagchi, B. and Biswas, R. (1998) *Acc. Chem. Res.*, **31**, 181; (d) Nandi, N., Bhattacharyya, K. and Bagchi, B. (2000) *Chem. Rev.*, **100**, 2013.

24 (a) Kovalenko, S.A., Ernsting, N.P. and Ruthmann, J. (1997) *J. Chem. Phys.*, **105**, 3504; (b) Biswas, R., Bhattacharyya, S. and Bagchi, B. (1998) *J. Chem. Phys.*, **108**, 4963.

25 For example, (a) Tremillon, B. (1974) *Chemistry in Non-Aqueous Solvents*, D. Reidel Publishing, Dordrecht, the Netherlands, Chapter 5; (b) Tremillon, B. (1997) *Reactions in Solution: An Applied Analytical Approach*, John Wiley & Sons Ltd., Chichester, Chapter 6; (c) Bauer, D. and Breant, M., (1975) in *Electroanalytical Chemistry*, vol. 8 (ed. A.J. Bard), Marcel Dekker, New York, p. 281; (d) Popovych, O. and Tomkins, R.P.T. (1981) *Nonaqueous Solution Chemistry*, John Wiley & Sons, Inc., New York, Chapter 5. This problem has also been discussed in many other publications.

26 Krishtalik, L.I., Alpatova, N.M. and Ovsyannikova, E.V. (1991) *Electrochim. Acta*, **36**, 435.

27 (a) Senda, M. and Takahashi, R. (1974) *Rev. Polarogr. (Jpn.)*, **20**, 56; (b) Madec, C. and Courtot-Coupez, J. (1977) *J. Electroanal. Chem.*, **84**, 177.

28 For example Diggle, J.W. and Parker, A. J. (1973) *Electrochim. Acta*, **18**, 975.

29 Alexander, R., Parker, A.J., Sharp, J.H. and Waghorne, W.E. (1972) *J. Am. Chem. Soc.*, **94**, 1148.

30 (a) Marcus, Y. (1983) *Pure Appl. Chem.*, **55**, 977;   (b) Ref. [8], pp. 166–169; (c) Ref. [9], Chapter 15.

31 (a) Cox, B.G., Hedwig, G.R., Parker, A.J. and Watts, D.W. (1974) *Aust. J. Chem.*, **27**, 477; (b) Cox, B.G., Waghorne, W.E. and Pigott, C.K. (1979) *J. Chem. Soc., Faraday Trans. I*, **75**, 227.

32 Gritzner, G. and Horzenberger, F. (1992) *J. Chem. Soc., Faraday Trans. I*, **88**, 3013; 1995, **91**, 3843; 1996, **92**, 1083.

33 (a) Marcus, Y. (1985) *Pure Appl. Chem.*, **57**, 1103; (b) Ref. [8], pp. 172–173; (c) Ref. [9], Chapter 16.

34 Chantooni, M.K., Jr and Kolthoff, I.M. (1975) *J. Phys. Chem.*, **79**, 1176.

35 (a) Ahrland, S. (1979) *Pure Appl. Chem.*, **51**, 2019; 1990, **62**, 2077; (b) Ref. [18a].

36 (a) Cox, B.G. and Schneider, H. (1989) *Pure Appl. Chem.*, **61**, 171; (b) Chantooni, M.K., Jr and Kolthoff, I.M. (1985) *J. Solution Chem.*, **14**, 1; (c) Chantooni, M.K., Jr, Roland, G. and Kolthoff, I.M. (1988) *J. Solution Chem.*, **17**, 175; (d) Katsuta, S., Kudo, T. and Takeda, Y. (1997) *Curr. Top. Solution Chem.*, **2**, 219.

37 Marcus, Y., (2002) *Solvent Mixtures, Properties and Selective Solvation*, Marcel Dekker, New York.

38 (a) Onsager, L. (1936) *J. Am. Chem. Soc.*, **58**, 1486; (b) Weaver, J.R. and Parry, R.W. (1966) *Inorg. Chem.*, **5**, 703; (c) Ref. [8], Chapter 7.

39 For example, (a) Ref. [16a], Chapter 9; (b) Ref. [16b], Chapter 14; (c) Marcus, Y.

(1994) *J. Chem. Soc., Perkin Trans. 2*, 1015, 1751.

40 For example, (a) Conway, B.E., Bockris, J.O'M. and Yeager, E.(eds) (1983) *Comprehensive Treatise of Electrochemistry*, vol. 5, Plenum Press, New York, Chapter 1; (b) Coetzee, J.F. and Ritchie, C.D.(eds) (1976) *Solute–Solvent Interactions*, vol. **2**, Marcel Dekker, New York, Chapter 11; (c) Ref. [5], Chapter 4; (d) Ref. [8], Chapter 7; (e) Ref. [20a].

41 (a) Covington, A.K. and Newman, K.E. (1976) *Adv. Chem. Ser.*, **155**, 153; (b) Covington, A.K. and Newman, K.E. (1979) *Pure Appl. Chem.*, **51**, 2041.

42 (a) Covington, A.K. and Dunn, M. (1989) *J. Chem. Soc., Faraday Trans. 1*, **85**, 2827, 2835 (Coordinated cluster theory); (b) Covington, A.K. and Newman, K.E. (1988) *J. Chem. Soc., Faraday Trans. 1*, **84**, 1393, (Kirkwood–Buff theory); (c) Marcus, Y. (1983) *Aust. J. Chem.*, **36**, 1719; Marcus, Y. (1988) *J. Chem. Soc., Faraday Trans. 1*, **84**, 1465 (Quasi-lattice quasi-chemical theory); (d) Ref. [37], Chapter 5.

43 (a) Cox, B.G., Parker, A.J. and Waghorne, W.E. (1974) *J. Phys. Chem.*, **78**, 1731; (b) Cox, B.G. and Waghorne, W.E. (1984) *J. Chem. Soc., Faraday Trans. 1*, **80**, 1267.

44 (a) Kalidas, C., Hefter, G. and Marcus, Y. (2000) *Chem. Rev.*, **100**, 819; (b) Marcus, Y. (2007) *Chem. Rev.*, **107**, 3880; (c) Hefter, G., Marcus, Y. and Waghorne, W.E. (2002) *Chem. Rev.*, **102**, 2773.

45 Inami, Y.H., Bodenseh, H.K. and Ramsey, J.B. (1961) *J. Am. Chem. Soc.*, **83**, 4745.

46 Marcus, Y. and Hefter, G. (2006) *Chem. Rev.*, **106**, 4585.

47 (a) Ref. [2], Section 3.8; (b) Ref. [3], Chapters 1 and 3; (c) Barthel, J. and Gores, H.-J.,Ref. [4], Chapter 1; (d) Ref. [5], Chapter 5; (e) Ref. [40a], Chapter 2; (f) Ref. [41].

48 (a) Barthel, J. (1979) *Pure Appl. Chem.*, **51**, 2093; (b) Ref. [3], pp. 7, 33; (c) Ref. [4], p. 56.

49 Salomon, M. (1987) *Pure Appl. Chem.*, **59**, 1165.

50 (a) Krienke, H. and Barthel, J. (1998) *Z. Phys. Chem.*, **204**, 713; (b) Krienke, H. and Barthel, J. (1998) *J. Mol. Liq.*, **78**, 123.

51 Marcus, Y. (2006) *J. Mol. Liq.*, **123**, 8.

52 Ref. [4], Chapter 2.

53 (a) Barthel, J., Buchner, R., Eberspacher, P.N., Munsterer, M., Stauber, J. and Wurm, B., (1998) *J. Mol. Liq.*, **78**, 83; (b) Barthel, J., Gores, H.-J., Neueder, R. and Schmid, A. (1999) *Pure Appl. Chem.*, **71**, 1705.

54 Buchner, R. and Barthel, J. (1994) *Annu. Rep. Prog. Chem., Sect. C, Phys. Chem.*, **91**, 71.

55 (a) Salomon, M. (1989) *J. Power Sources*, **26**, 9; (b) Ref. [49]; (c) Ref. [4], p. 61.

56 (a) Hojo, M., Tanino, A., Miyauchi, Y. and Imai, Y. (1991) *Chem. Lett.*, 1827; (b) Miyauchi, Y., Hojo, M., Moriyama, H. and Imai, Y. (1992) *J. Chem. Soc., Faraday Trans.*, **88**, 3175; (c) Barthel, J., Gores, H.J. and Kraml, L. (1996) *J. Phys. Chem.*, **100**, 3671.

3
Acid–Base Reactions in Nonaqueous Solvents

This chapter discusses acid–base reactions in nonaqueous solvents, with particular emphasis on how they differ from those in aqueous solutions. The problem of pH in nonaqueous solutions is also discussed. The Brønsted acid–base concept that an acid is a proton donor and a base is a proton acceptor is adopted. The relation between an acid **A** and its conjugate base **B** is expressed by

$$\mathbf{A} \rightleftarrows \mathrm{H}^+ + \mathbf{B}$$

The reaction between an acid $\mathbf{A}_1$ and a base $\mathbf{B}_2$ is a proton transfer from $\mathbf{A}_1$ to $\mathbf{B}_2$, $\mathbf{A}_1$ becoming its conjugate base $\mathbf{B}_1$ and $\mathbf{B}_2$ its conjugate acid $\mathbf{A}_2$:

$$\mathbf{A}_1 + \mathbf{B}_2 \rightleftarrows \mathbf{B}_1 + \mathbf{A}_2$$

Acid–base reactions in nonaqueous solvents have been extensively studied. Many books and reviews on acid–base equilibria and acid–base titrations in nonaqueous solvents are available. References [1–3] are particularly useful.

3.1
Solvent Effects on Acid–base Reactions

We consider a process in which an HA-type acid, such as hydrogen chloride or acetic acid, dissolves in a solvent and dissociates into H^+ and A^-. This process occurs in two steps:

$$\text{Ionization:} \quad \mathrm{HA} \rightleftarrows (\mathrm{H}^+, \mathrm{A}^-)_{\text{solv}} \qquad K_{\mathrm{I}} = \frac{[(\mathrm{H}^+, \mathrm{A}^-)]}{[\mathrm{HA}]}$$

$$\text{Dissociation:} \quad (\mathrm{H}^+, \mathrm{A}^-)_{\text{solv}} \rightleftarrows \mathrm{H}^+_{\text{solv}} + \mathrm{A}^-_{\text{solv}} \qquad K_{\mathrm{D}} = \frac{\gamma^2 [\mathrm{H}^+][\mathrm{A}^-]}{[(\mathrm{H}^+, \mathrm{A}^-)]}$$

where γ represents the activity coefficients of H^+ and A^-, and the activity coefficients of electrically neutral species are assumed equal to unity. In the first step, solvent molecules S act as a Lewis base (electron pair donor) to the H atom and as a Lewis

Electrochemistry in Nonaqueous Solutions, Second, Revised and Enlarged Edition. Kosuke Izutsu
Copyright © 2009 WILEY-VCH Verlag GmbH & Co. KGaA, Weinheim
ISBN: 978-3-527-32390-6

Table 3.1 Effect of solvents on the difficulties of ionization of acid HA and the dissociation of ion pair $(H^+, A^-)_{solv}$.

Solvent classification	Ionization of HA	Dissociation of $(H^+, A^-)_{solv}$
High-permittivity amphiprotic solvents	Easy	Easy
High-permittivity aprotic solvents	Difficult or fairly easy[a]	Easy
Low-permittivity amphiprotic solvents	A little difficult or easy	Difficult
Low-permittivity aprotic solvents	Difficult	Difficult

[a]Difficult in protophobic solvents, but fairly easy in protophilic solvents.

acid (electron pair acceptor) to the A atom, as shown by $m(S) \rightarrow \overset{\delta+ \frown \delta-}{H-A} \rightarrow n(S)$, and ionize the H and A atoms, resulting in the breaking of the H–A covalent bond and the formation of the ion pair $(H^+, A^-)_{solv}$. If the H–A bond is intrinsically weak, the ionization occurs easily. For a given acid, however, the ionization becomes easy, if solvent molecules interact strongly with the H and A atoms. In the second step, the ion pair $(H^+, A^-)_{solv}$ dissociates into free ions. The dissociation process is easy if the solvent has a high permittivity and the distance of closest approach between H^+ and A^- is large enough. Thus, two important solvent properties, i.e. relative permittivity and donor–acceptor properties, significantly influence both ionization and dissociation processes. In Table 3.1, solvents are classified into four groups according to values of relative permittivity and strengths of donor–acceptor properties; the degrees of difficulty of the ionization and dissociation processes are shown for each class.

The acid dissociation constants, K_a, which are determined by potentiometry or conductimetry, usually correspond to[1]

$$K_a = \frac{\gamma^2 [H^+][A^-]}{[HA] + [(H^+, A^-)]} = \frac{K_I K_D}{1 + K_I}.$$

In order that the acid HA dissociates almost completely into H^+ and A^-, i.e. to get a large K_a value, it is essential that both the ionization and the dissociation processes occur easily. This means that both K_I and K_D should be large enough, as is often the case in amphiprotic solvents of high permittivities. In the following sections, we discuss the characteristics of acid–base reactions in each of the four classes of solvents.

3.1.1
Acid–Base Reactions in Amphiprotic Solvents of High Permittivity

The solvents of this class are usually of high permittivity ($\varepsilon_r > 20$) and of fair or strong acidity (proton donor capacity) and basicity (proton acceptability) [4]. Water is the

---

1) For example, K_I, K_D and K_a in acetic acid are 1.0, 2.7×10^{-5} and 1.3×10^{-5}, respectively, for $HClO_4$, and 5.37, 9.4×10^{-7} and 7.9×10^{-7}, respectively, for pyridinium ion (PyH^+) (see Ref. [13c]).

most important solvent in this class. In water, the HA- and BH^+-type acids dissociate as in Eqs. (3.1) and (3.2):

$$HA + H_2O \rightleftharpoons H_3O^+ + A^- \tag{3.1}$$

$$BH^+ + H_2O \rightleftharpoons H_3O^+ + B \tag{3.2}$$

Here, NH_4^+, RNH_3^+, $R_2NH_2^+$, R_3NH^+ and PyH^+ (R: alkyl; Py: pyridine) are examples of BH^+-type acids. An acid (HA or BH^+), with an intrinsically strong proton donor capacity, completely dissociates in water and forms its conjugate base (A^- or B) and H_3O^+. Practically, HA or BH^+ do not remain in the solution. This acid is called a *strong acid*. Examples of strong acids in water are hydrochloric, nitric and perchloric acids. The hydronium ion (H_3O^+) is the strongest acid that can remain in water. The strengths of the acids, which are stronger than H_3O^+, cannot be differentiated and are equivalent to the strength of H_3O^+. This phenomenon is called the *leveling effect*. For acids that are weaker than H_3O^+, the reactions in Eq. (3.1) or (3.2) proceed only partially to the right. These acids are called *weak acids* and their acid strengths are measured by the values of the acid dissociation constant defined by Eq. (3.3) or (3.4):

$$K_a = \frac{a(H_3O^+)a(A^-)}{a(HA)} = \frac{\gamma^2[H_3O^+][A^-]}{[HA]} \tag{3.3}$$

$$K_a = \frac{a(H_3O^+)a(B)}{a(BH^+)} = \frac{[H_3O^+][B]}{[BH^+]} \tag{3.4}$$

A base (A^- or B) in water reacts as in Eq. (3.5) or (3.6) and forms its conjugate acid (HA or BH^+) and a hydroxide ion (OH^-):

$$A^- + H_2O \rightleftharpoons HA + OH^- \tag{3.5}$$

$$B + H_2O \rightleftharpoons BH^+ + OH^- \tag{3.6}$$

The strongest base in water is OH^-. For bases that are much stronger than OH^-, the reactions in Eq. (3.5) or (3.6) proceed completely to the right and form OH^-. These bases are called *strong bases*; their base strengths are equivalent to that of OH^- and cannot be differentiated. For bases that are weaker than OH^-, the reactions in Eq. (3.5) or (3.6) proceed only partially to the right. These bases are called *weak bases*; their base strengths are compared by the values of the base dissociation constant K_b:

$$K_b = \frac{a(HA)a(OH^-)}{a(A^-)} = \frac{[HA][OH^-]}{[A^-]}$$

$$K_b = \frac{a(BH^+)a(OH^-)}{a(B)} = \frac{\gamma^2[BH^+][OH^-]}{[B]}$$

Table 3.2 Temperature effects on the ion product constant of water ($pK_W = -\log K_W$).

Temperature, °C	0	15	25	35	50	100
pK_W	14.943	14.346	13.996	13.680	13.262	12.31

If water molecules interact with each other as an acid and a base, autoprotolysis (autoionization) occurs, as in Eq. (3.7)

$$H_2O + H_2O \rightleftharpoons H_3O^+ + OH^- \tag{3.7}$$

The autoprotolysis constant (autoionization constant) of water, $K_W = a(H_3O^+)a(OH^-) \approx [H_3O^+][OH^-]$, is often called the ion product constant of water. The value of $-\log K_W$ varies with temperature as shown in Table 3.2. In water, there exists the following relation between K_a of an acid and K_b of its conjugate base:

$$K_a \times K_b = K_W$$

Among amphiprotic solvents of high permittivities, there are water-like neutral solvents (e.g. methanol and ethanol), more acidic protogenic solvents (e.g. formic acid) and more basic protophilic solvents (e.g. 2-aminoethanol). There are also amphiprotic mixed solvents, such as mixtures of water and alcohols and water and 1,4-dioxane. The acid–base equilibria in amphiprotic solvents of high permittivity can be treated by methods similar to those in aqueous solutions. If the solvent is expressed by SH, the acid HA or BH^+ will dissociate as follows:

$$HA + SH \rightleftharpoons SH_2^+ + A^-$$

$$BH^+ + SH \rightleftharpoons SH_2^+ + B$$

The acid dissociation constant is expressed by Eq. (3.8) or (3.9):

$$K_a = \frac{a(SH_2^+)a(A^-)}{a(HA)} = \frac{\gamma^2[SH_2^+][A^-]}{[HA]} \tag{3.8}$$

$$K_a = \frac{a(SH_2^+)a(B)}{a(BH^+)} = \frac{[SH_2^+][B]}{[BH^+]} \tag{3.9}$$

For the base A^- or B, the reactions in Eq. (3.10) or (3.11) occur with solvent SH:

$$A^- + SH \rightleftharpoons HA + S^- \tag{3.10}$$

$$B + SH \rightleftharpoons BH^+ + S^- \tag{3.11}$$

The base dissociation constant K_b is expressed by Eq. (3.12) or (3.13):

$$K_b = \frac{a(HA)a(S^-)}{a(A^-)} = \frac{[HA][S^-]}{[A^-]} \tag{3.12}$$

$$K_b = \frac{a(BH^+)a(S^-)}{a(B)} = \frac{\gamma^2[BH^+][S^-]}{[B]} \tag{3.13}$$

The autoprotolysis (autoionization) of an amphiprotic solvent SH occurs as in Eq. (3.14), and the autoprotolysis (autoionization) constant, K_{SH} for pure SH, is defined by Eq. (3.15):

$$SH + SH \rightleftarrows SH_2{}^+ + S^-$$ (3.14)

$$K_{SH} = a(SH_2{}^+)a(S^-) = \gamma^2[SH_2{}^+][S^-]$$ (3.15)

The strongest acid in solvent SH is $SH_2{}^+$ (lyonium ion), while the strongest base in SH is S^- (lyate ion). The strengths of the acids that are much stronger than $SH_2{}^+$ are made equivalent to that of $SH_2{}^+$, while the strengths of the bases that are much stronger than S^- are made equivalent to that of S^-. Here, the acid strength of $SH_2{}^+$ differs from one solvent to another. For example, in formic acid as solvent, $SH_2{}^+$ is very strongly acidic and some strong acids in water behave as weak acids. On the other hand, in 2-aminoethanol as solvent, $SH_2{}^+$ is only weakly acidic, and some weak acids in water behave as strong acids.

It is possible to compare the strengths of weak acids by the values of their acid dissociation constants K_a. Figure 3.1 shows the titration curves for acids (HA or BH^+) of different K_a values. The ordinate shows pa_H, which is defined by $pa_H = -\log a(SH_2{}^+)$. The pa_H corresponds to the pH in aqueous solutions

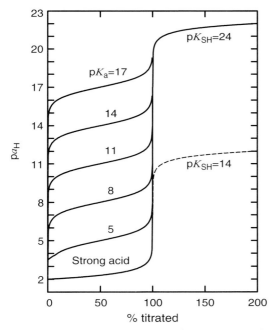

Figure 3.1 Calculated titration curves of a strong acid and weak acids of various pK_a values with a strong base. In the solvent of $pK_{SH} = 24$ and at the acid concentration of 10^{-2} M. The effect of activity coefficient and that of dilution was neglected. (The dashed curve is for the case of $pK_{SH} = 14$ (water)).

(see Section 3.2). The pa_H of nonaqueous solutions can be measured with a glass pH electrode or some other pH sensors (see Sections 3.2.1 and 6.2). For the mixture of a weak acid **A** and its conjugate base **B**, pa_H can be expressed by the Henderson–Hasselbalch equation:

$$pa_H = pK_a + \log \frac{a(\mathbf{B})}{a(\mathbf{A})} \left[= pK_a + \log \frac{a(A^-)}{a(HA)} \quad \text{or} \quad pK_a + \log \frac{a(\mathbf{B})}{a(BH^+)} \right]$$

In Figure 3.1, if the effect of activity coefficient is considered, the pa_H at the half-neutralization point ($pa_{H1/2}$) can be expressed by $pa_{H1/2} = pK_a + \log \gamma$ for $\mathbf{A} = HA$ and by $pa_{H1/2} = pK_a - \log \gamma$ for $\mathbf{A} = BH^+$. γ shows the activity coefficient of A^- or BH^+. In the solvents of this class, just as in water, the solution containing a weak acid and its conjugate base (or a weak base and its conjugate acid) has a buffer action against the pH variation. The buffer action is maximum at $c(\mathbf{A}) = c(\mathbf{B})$ (c: analytical concentration) if $[c(\mathbf{A}) + c(\mathbf{B})]$ is kept constant.

If we consider two solvents of different permittivities (I and II), the pK_a value of an acid **A** in the two solvents should differ by the amount expressed by Eq. (3.16), according to the electrostatic Born theory:

$$pK_{a,II} - pK_{a,I} = \frac{0.217 Ne^2}{RT} \left(\frac{1}{r_H} + \frac{z_B^2}{r_B} - \frac{z_A^2}{r_A} \right) \left(\frac{1}{\varepsilon_{r,II}} - \frac{1}{\varepsilon_{r,I}} \right) \quad (3.16)$$

Here, **A** denotes an acid of type HA^-, HA or BH^+, and z_i and r_i denote the charge and the radius, respectively, of species i. The influence of permittivity on pK_a depends on the charges, radii and the charge locations of the acid and its conjugate base. Table 3.3 shows the pK_a values of some acids and acid–base indicators in water, methanol and ethanol [3]. The solvent effects on pK_a are smaller for BH^+-type acids than for HA^-- or HA-type acids. For the BH^+-type acids, $z_A = 1$ and $z_B = 0$ in Eq. (3.16), and the influence of solvent permittivity is expected to be small.

Table 3.3 Relation between the charges of acids and their pK_a values in water, methanol and ethanol.

Acids	Charges	pK_a (in H_2O)	pK_a (in MeOH)	pK_a (in EtOH)
Acetic acid	0	4.75	9.7	10.4
Benzoic acid	0	4.20	9.4	10.1
Salicylic acid	0	2.98	7.9	8.6
Phenol	0	9.97	14.2	15.3
Picric acid	0	0.23	3.8	3.9
Anilinium	+1	4.60	6.0	5.7
Methyl orange	+1	3.45	3.8	3.4
Neutral red	+1	7.4	8.2	8.2
Bromophenol blue	−1	4.1	8.9	9.5
Bromothymol blue	−1	7.3	12.4	13.2

Relative permittivities, ε_r, are 78.5 in water, 32.6 in MeOH and 24.6 in EtOH. From p. 130 of Ref. [3].

3.1.2
Acid–Base Reactions in Aprotic Solvents of High Permittivity

The solvents of this class are often called dipolar aprotic solvents. They are polar and of very weak acidity (proton donor capacity, hydrogen bond donor capacity and electron pair acceptability). However, with regard to basicity (proton acceptability, hydrogen bond acceptability and electron pair donor capacity), some are stronger than water (protophilic) and some are much weaker than water (protophobic). Examples of protophilic aprotic solvents are DMSO, DMF, DMA, NMP and HMPA; protophobic aprotic solvents include AN, PC, NM and TMS (see Table 1.1 for abbreviations).

The characteristic of acid–base reactions in dipolar aprotic solvents, compared to that in dipolar amphiprotic solvents, is the easy occurrence of homo- and hetero-conjugation reactions [2, 3, 5]. However, before discussing the homo- and hetero-conjugations, we first discuss the solvent effects on the acid dissociation constants in dipolar aprotic solvents.

Solvent Effects on Acid Dissociation Constants

Table 3.4 shows the pK_a values of various acids in some popular organic solvents and in water [6].[2] Solvent effects on pK_a values are significant. For example, the pK_a of acetic acid is 4.76 in water, 12.6 in DMSO and 22.3 in AN. The solvent effects can be predicted by using the transfer activity coefficients for the species participating in the dissociation equilibrium (Section 2.3). The relation between the pK_a values of acid HA in solvents R (water in this case) and S is given by Eq. (2.7):

$$pK_a(HA, S) - pK_a(HA, R) \qquad\qquad (2.7)$$
$$= \log \gamma_t(H^+, R \to S) + \log \gamma_t(A^-, R \to S) - \log \gamma_t(HA, R \to S)$$

For acetic acid, the transfer activity coefficients of H^+, A^- and HA are as follows:

$$H_2O \to DMSO: \quad \log \gamma_t(H^+) = -3.4,$$
$$\log \gamma_t(CH_3COO^-) = 8.8, \quad \log \gamma_t(CH_3COOH) = -0.8$$
$$H_2O \to AN: \quad \log \gamma_t(H^+) = 8.1,$$
$$\log \gamma_t(CH_3COO^-) = 10.7, \quad \log \gamma_t(CH_3COOH) = 0.4$$

The solvation of H^+ is stronger in protophilic DMSO but much weaker in protophobic AN than in amphiprotic water. The solvation of CH_3COO^- is much weaker in aprotic AN and DMSO than in water (for the strong solvation of CH_3COO^- in water, see Section 2.2.1). As for the solvation of CH_3COOH, there is not much difference among the three solvents. From Eq. (2.7) and $pK_a = 4.76$ in water, we get $pK_a = 11.0$

---

2) In Ref. [6], pK_a values have been compiled for many acids and conjugate acids of bases in 12 popular dipolar aprotic solvents (Ac, AN, DMF, DMA, DMSO, HMPA, MIBK, NMP, NM, PC, Py, TMS). The data described for each acid are the conditions and the method for measurement, pK_a and homoconjugation constants, and the reference.

Table 3.4 Dissociation constants of acids and conjugate acids of bases in various solvents (pK_a values, 25 °C).[a]

Solvents	AN	PC	γ-BL	MN	DMSO	DMF	NMP	Py	MeOH	HOAc	H$_2$O
Relative permittivities	35.94	64.92	39	35.87[b]	46.45	36.71	32.2	12.91	32.66	6.17[c]	78.36
Autoprotolysis constants (pK_{SH})	≥33.2	≥29.2	~29	≥24	≥33.3	≥31.6	≥24.2		16.9	14.5	14.0
Acids											
HClO$_4$	(2.1)	(1.3)		2.1	0.4	(Strong)	(Strong)	3.3		4.9	(Strong)
HCl	8.9*	(10.9*)		8.1*	2.1	3.3	3.9	5.7	1.2	8.6	−3.7
HBr	5.5*				1.1	1.8		4.4	0.8	6.1~6.7	−4.1
HNO$_3$	8.9*	(10.4*)		8.8*	1.4	(Decomp.)	2	4.1	3.2	9.4	−1.8
H$_2$SO$_4$ (pK_{a1})	7.8*	(8.4*)		5.1*	1.4	3.1				7.2	
H$_2$SO$_4$ (pK_{a2})	25.9*			21.4	14.7	17.2					1.96
Acetic acid	22.3*	22.5*		20.5*	12.6	13.5	13.3	10.1	9.7		4.76
Benzoic acid	20.7*	19.7*		19.5*	11.0	12.3	12.3	9.8	9.4		4.19
Chloroacetic acid	15.3*			17.0*	8.9	10.1	10.9		7.8		2.88
Dichloroacetic acid	13.2*			14.1*	6.4	7	8.3		6.3		1.48
2,4-Dinitrophenol	16.0*	14.9*		15.9*	5.4	6.4	6.8	4.4	7.9		3.96
2,6-Dinitrophenol	16.5		13.7	16.0	4.9	6.1		4.8	7.7		3.70
Malonic acid (pK_{a1})	15.3*				7.2	8.0			7.5		2.88
Malonic acid (pK_{a2})	30.5				18.6	20.8			12.4		5.68
Methanesulfonic acid	10.0*	8.3*	8.1	(6.0)	1.6	3.0	2				
o-Nitrobenzoic acid	18.2*			16.7*	8.2	9.9	10.2		7.6		2.20
m-Nitrobenzoic acid	19.3*			17.6*	9.2	10.8	10.7		8.2		3.46
p-Nitrobenzoic acid	18.5*			17.6*	9.0	10.6	10.5		8.2		3.40
o-Nitrophenol	22.0*	20.8*		21.4*	11.0	12.2	12.6				7.17
m-Nitrophenol	23.9*	21.2*		22.2*	13.9	13.9	14.3	12.5			8.28
p-Nitrophenol	20.7*	19.7*		20.1*	11.3	12.2	12.5	9.2	11.2		7.15
Oxalic acid (pK_{a1})	14.5*		12.4*		6.2	8.6	(9.2)		6.1		1.27
Oxalic acid (pK_{a2})	27.7				14.9	16.6			10.7		4.29
Phenol	27.2*			(25.7)	16.5	>16	17.6		14.2		9.89

	1	2	3	4	5	6	7	8	9	10
(o)Phthalic acid (pK_{a1})	14.3*		12.4	6.2	6.7			7.4		3.0_0
(o)Phthalic acid (pK_{a2})	29.8			16.0	~16.5			12.1		5.4_0
m-Phthalic acid (pK_{a1})	19.3			9.8				8.6_3		3.5_4
m-Phthalic acid (pK_{a2})	23.0			12.3				10.6		4.6_2
Picric acid	11.0	9.3	10.5	−1	(3.6)	3.5	(Strong)	3.8		0.38
Salicylic acid	16.8*	15.2*	8.4	6.7	8.2		8.6	7.9		2.97
p-Toluenesulfonic acid	8.7*	(6.4*)	14.5*	0.9	2.6		(Strong)		8.4	
Trichloroacetic acid	10.6*	(7.3)	(5.8)	3.5	3.5			4.9	11.5	0.70
Trifluoromethanesulfonic acid	(2.6)	(2.2)	(3.0)	(Strong)	(Strong)					
Bases										
Ammonia	16.5	15.9	15.2	10.5	9.5		10.3	10.8	6.4	9.25
Aniline	10.7	10.1	9.1	3.6	4.4			6.0		4.63
Butylamine	18.3	16.9	11.1	10.5	10.5	5.5	10.3	11.8	8.6	10.64
Diethylamine	18.8	18.0	10.5	10.5	10.4		9.2			10.93
Diphenylguanidine	17.9	17.2	8.6	8.6	9.0		8.9			10.1
Ethylamine	18.4	17.1	11.0	11.0			10.2			10.63
Piperidine	18.9	18.2	10.6	10.6	10.4		10.4			11.12
Pyridine	12.3	11.9	12.0	3.4	3.3				6.1	5.25
Tetramethylguanidine	23.3			13.2	13.7	9.6	12.8			13.6
Tributylamine	18.1	17.5		8.4	8.6					9.9
Triethanol amine	15.9	(15.9)		7.5	7.6		7.5			7.76
Triethylamine	18.7	17.9	9.0	9.0	9.3	3.8	8.7		9.5	10.72
Bromocresol green[d]	18.5			7.4				9.8		4.9
Bromothymol blue[d]	22.3	17.5		11.3			13.2	12.4		7.3
Phenol red[d]	25.1			13.7			13.2	12.8		8.0
Thymol blue[d]	27.2	(20.6)		~15.3			14.5	14.0		9.2

The asterisk (*) shows that homoconjugation occurs appreciably. From Izutsu, K. (1990) Acid–base Dissociation Constants in Dipolar Aprotic Solvents, Blackwell Science, Oxford, and others.

[a]For the abbreviated symbols, see Table 1.1.
[b]At 30 °C.
[c]At 20 °C.
[d]Values of pK_{a2}.

and 23.2 for DMSO and AN, respectively, in fair agreement with the experimental values of 12.6 and 22.3 [7].

In the case of picric acid (HPic), the solvation of Pic$^-$ in aprotic solvents is comparable to or slightly stronger than the solvation in water (Table 2.7). Polarizable Pic$^-$ easily interacts with polarizable aprotic solvents by dispersion forces, while water does not interact as strongly with Pic$^-$ by hydrogen bonding because the negative charge of Pic$^-$ is delocalized. The pK_a of picric acid is 0.38 in water, -1 in DMSO, and 11.0 in AN (Table 3.4). We can attribute most of the difference in pK_a to the difference in the solvation of H$^+$.

pK_a values of BH$^+$ acids are also included in Table 3.4. The following relation can predict the solvent effect of those acids on pK_a:

$$pK_a(BH^+, S) - pK_a(BH^+, R) = \log \gamma_t(H^+, R \to S) + \log \gamma_t(B, R \to S)$$
$$- \log \gamma_t(BH^+, R \to S) \qquad (3.17)$$

Usually, of the three terms on the right-hand side of Eq. (3.17), the first is most important and the influence of the second and third terms is relatively small. However, there are cases when the influence of the second and third terms has an important role. For example, the pK_a of the anilinium ion is 10.7 in AN and 3.6 in DMSO. Because $\log \gamma_t$ (H$^+$, AN $\to$ DMSO) $= -11.5$, we get $\log \gamma_t$(B, AN $\to$ DMSO) $- \log \gamma_t$ (BH$^+$,AN $\to$ DMSO) $= +4.4$ from Eq. (3.17). This value is not negligible. Generally, BH$^+$ is stabilized in basic solvents and B is stabilized in acidic solvents. Thus, we can attribute the above value ($+4.4$) to the stabilization of the anilinium ion (BH$^+$) in strongly basic DMSO.

Now, we consider the solvent effects on the mutual relationship between the pK_a values of different acids. Figure 3.2 shows the relations between pK_a of nonortho-substituted phenols and the Hammett σ values of the substituents [8]. Good linear relations are observed in four solvents.[3] It is of special interest that the slopes in AN, DMF and DMSO are almost the same and are nearly 2.0 times the slope in water. Similar linear relations have also been obtained for nonortho-substituted benzoic acids: the slopes in AN, DMSO and DMF are the same and 2.4 times that in water [8]. The effects of substituents on the pK_a of phenols and benzoic acids are much larger in aprotic solvents than in water. The reason for this is that the anions (conjugate bases) of the acids are hydrated in water by hydrogen bonding: in a series of substituted acids, the acids with higher pK_a values give anions with higher basicities. Thus, the hydrogen bond between the anions and water is stronger for the acids with higher pK_a values. The anions are stabilized by the increased strength of the hydrogen bonding,

---

[3] If we plot pK_a values in one solvent against those in another, we often get a linear relation if the acids are of the same type (HA or BH$^+$ type). In particular, between two aprotic solvents, the slope of the linear relation is nearly unity. Here, for BH$^+$-type acids in a given aprotic solvent, the difference in pK_a values is closely related to the difference in the enthalpies of dissociation, i.e. the difference in the B–H$^+$ bond energies. On the other hand, the pK_a values of a given BH$^+$-type acid in different solvents are related to the difference in the entropies of dissociation [9].

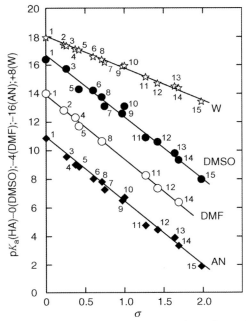

Figure 3.2 Relations between the pK_a values of nonortho-substituted phenols and the Hammett σ values of the substituents [8]. Substituents: 1, none; 2, 4-chloro; 3, 4-bromo; 4, 3-chloro; 5, 3-trifluoromethyl; 6, 3,4-dichloro; 7, 3,5-dichloro; 8, 3-nitro; 9, 3,4,5-trichloro; 10, 4-cyano; 11, 4-nitro; 12, 3,5-dinitro; 13, 3-chloro-4-nitro; 14, 3-trifluoromethyl-4-nitro; 15, 3,4-dinitro.

and the conjugate acids dissociate more easily. Therefore, in water, the hydrogen bonding of water to anions tends to suppress the substituent effect on pK_a. In aprotic solvents with poor hydrogen bond donor capacities, such a suppression effect is not observed. This fact is of analytical importance: when two or more acids are to be titrated simultaneously, the resolution is better in aprotic solvents than in water. The resolutions in MeOH and EtOH are between those of water and aprotic solvents.

In Table 3.4, the pK_{a1} and pK_{a2} values of some diprotic acids are also shown. For sulfuric acid, o-phthalic acid and malonic acid, the values of (pK_{a2} − pK_{a1}) in dipolar aprotic solvents are much higher than those in water. The large (pK_{a2} − pK_{a1}) values are mainly due to the fact that, for those acids in aprotic solvents, monoanions HA$^-$ are highly stabilized by strong intramolecular hydrogen bonding, while dianions A^{2-} are unstable because two adjacent negative charges, which repel each other, cannot be stabilized by the solvation of aprotic solvents. For dicarboxylic acids, the intramolecular hydrogen bonding of HA$^-$ is possible only if the distances between −COOH and −COO$^-$ are small enough. Thus, the values of (pK_{a2} − pK_{a1}) in AN and DMSO are 13.2 and 8.7, respectively, for o-phthalic acid, but 3.7 and 2.5, respectively, for m-phthalic acid. For more details, see Refs [2, 5, 10].

Homo- and Heteroconjugation Reactions

In dipolar aprotic solvents, the dissociation process of some acids (HA) is complicated by homo- and heteroconjugation reactions. In the homoconjugation reaction, some part of the A^- ions formed by the dissociation of HA (Eq. (3.18)) reacts with undissociated HA to form HA_2^-, as shown by Eq. (3.19):

$$HA \rightleftharpoons H^+ + A^- \tag{3.18}$$

$$A^- + HA \rightleftharpoons HA_2^- \tag{3.19}$$

In the heteroconjugation reaction, A^- reacts with an acid other than HA. If we express the acid by HR, the heteroconjugation reaction can be shown by Eq. (3.20) and AHR^- is formed:

$$A^- + HR \rightleftharpoons AHR^- \tag{3.20}$$

Here, HR may be water or an acidic impurity in the solvent. Sometimes A^- reacts with more than one molecule of HA or HR to form $A(HA)_n^-$ or $A(HR)_n^-$ ($n = 2, 3, \ldots$). The homo- and heteroconjugations are due to the hydrogen bonding of HA or HR to A^-.[4]

Homo- and heteroconjugation reactions occur when the A^- ion is solvated only weakly and is very reactive. This occurs in aprotic solvents when A^- is small in size or when the negative charge of A^- is localized on a small oxygen atom, as in the case of carboxylate ions. The homo- and heteroconjugation reactions are also influenced by the solvations of HA and HR and are easier in protophobic solvents, in which HA and HR are solvated only weakly. Thus, the homo- and heteroconjugation reactions tend to occur most easily in protophobic aprotic solvents, such as NM, AN and PC. Some examples of the homoconjugation constants, defined by Eq. (3.21), are shown in Table 3.5.

$$K^f(HA_2^-) = \frac{a(HA_2^-)}{a(A^-)a(HA)} \approx \frac{[HA_2^-]}{[A^-][HA]} \tag{3.21}$$

It is apparent from the log $K^f(HA_2^-)$ values that the homoconjugation occurs more easily in protophobic AN and PC than in protophilic DMF and DMSO.[5] In protophilic DMF and DMSO, fairly strong solvation of HA tends to suppress the homoconjugation. In the cases of 2,6-dinitrophenol and picric acid, the homoconjugation is almost negligible even in protophobic aprotic solvents (see Table 3.5 for the difference from 3,5-dinitrophenol). In these cases, the anion A^- has a delocalized

4) Homoconjugation also occurs with BH^+-type acids ($BH^+ + B \rightleftharpoons B_2H^+$). However, the extent of homoconjugation is small or negligible for BH^+ acids of most amines, although for BH^+ of some tertiary phosphine oxides ($R_3P{=}O$), homoconjugation occurs to a considerable extent [6].

5) Homoconjugation occurs slightly even in amphiprotic solvents such as water and alcohols. The increase in solubility of benzoic acid in concentrated aqueous solutions of alkali metal benzoate (salting-in) has been explained by a

mechanism involving homoconjugation. In general, homoconjugation increases the solubility of sparingly soluble salts. If HA is added to the saturated solution of sparingly soluble salt MA, the species HA_2^- is formed by homoconjugation. Because the radius of HA_2^- is larger than that of A^-, the solubility of MHA_2 is higher than that of MA, for electrostatic reasons. By measuring the influence of HA on the solubility of MA, it is possible to determine the homoconjugation constant (see Section 7.3.2).

Table 3.5 Homoconjugation constants of acids [log $K^f(HA_2)$]

Acids	AN	PC	DMF	DMSO
Hydrogen chloride	2.2	4.2	2.2	—
Acetic acid	3.7	4	2.6	1.5
Benzoic acid	3.6	4.0	2.4	1.8
Salicylic acid	3.3	3.2	1.7	1.5
p-Nitrophenol	3.7	3.6	2.2	1.8
3,5-Dinitrophenol	4.6	—	3.2	1.5
2,6-Dinitrophenol	$K^f \sim 0$	—	$K^f \sim 0$	$K^f \sim 0$
Picric acid	0.3	$K^f \sim 0$	$K^f = 0$	$K^f = 0$

negative charge and is polarizable, and it interacts, by dispersion forces, with an aprotic solvent that is also polarizable. In these cases, the acid HA is also polarizable and is stabilized by interacting with solvent.

The titration curve of a weak acid HA with a strong base (usually R_4NOH) is influenced significantly by homoconjugation (Figure 3.3) [11]. When the homoconjugation is significant, the titration curve is divided into two parts on either side of the half-neutralization point. Before the half-neutralization point, the pa_H is lower than in the absence of homoconjugation. After the half-neutralization point, however, the pa_H is higher than in the absence of homoconjugation. This phenomenon is explained as follows: the pa_H of the solution is determined by Eq. (3.22), irrespective

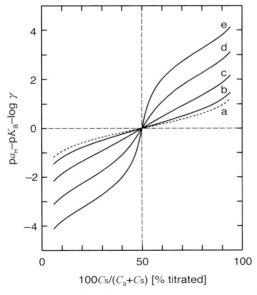

Figure 3.3 The effect of homoconjugation on the neutralization titration curve of a weak acid HA. Values of $K^f(HA_2{}^-)(C_a + C_s)$: 0 for a, 10 for b, 10^2 for c, 10^3 for d and 10^4 for e. The original figure in Ref. [11] was modified.

of the occurrence of homoconjugation:

$$pa_H = pK_a + \log \frac{a(A^-)}{a(HA)} = pK_a + \log \frac{\gamma[A^-]}{[HA]} \tag{3.22}$$

Before the half-neutralization point, A^- formed by the addition of titrant (strong base) reacts with the excess HA to form HA_2^-, thus the value of the $[A^-]/[HA]$ ratio is kept lower than in the absence of homoconjugation. On the other hand, after the half-neutralization point, the remaining HA reacts with excess A^- to form HA_2^-, thus the value of the $[A^-]/[HA]$ ratio is higher than in the absence of homoconjugation. At the half-neutralization point, $pa_{H1/2} = pK_a + \log \gamma$ irrespective of the occurrence of homoconjugation. From the influence of homoconjugation on the pa_H titration curve, we can determine the homoconjugation constant.[6] A practical procedure is given in Section 6.3.1. Conductimetry can also be applied to determine the homoconjugation constant, as described in Section 7.3.2.

In the absence of homoconjugation, as is usually the case in aqueous solutions, the mixture of a weak acid and its conjugate base or a weak base and its conjugate acid works as a pa_H (pH) buffer. The buffer capacity is maximum when $c(A) = c(B)$ (see p. 67). In the presence of homoconjugation, however, the problem of the pH buffer is much more complicated.[7] Figure 3.4 shows the effect of homoconjugation

---

6) According to Kolthoff and Chantooni [11], the following relation is obtained, in a good approximation, among $a(H^+)$, K_a, $K^f(HA_2^-)$, the analytical concentration of HA $(C_a \equiv [HA] + [HA_2^-])$, and that of R_4NA $(C_s \equiv [A^-] + [HA_2^-])$:

$$a(H^+)^2 \gamma^2 C_s - a(H^+) \gamma K_a \{(C_a + C_s) + K^f(HA_2^-)(C_s - C_a)^2\} + K_a^2 C_a = 0$$

Here, we assume that $[H^+] \ll C_s$, Et_4NA and Et_4NHA_2 are completely dissociated and the solution is free from acidic or basic impurities that may influence the dissociation equilibrium. If the values of K_a, $K^f(HA_2^-)$, C_a and C_s are known, we can get $a(H^+)$ by solving the qua-

where r is defined by $r = a(H^+)\gamma/K_a$ and is equal to the ratio of $a(H^+)\gamma$ at a given point on the titration curve and that at the half-neutralization point because $K_a = a(H^+)_{1/2}\gamma_{1/2}$.

7) The buffer capacity (β) of a solution is defined as the number of moles of strong base needed to increase the pa_H of 1 l of the solution by one unit and can be expressed by

$$\beta = dC_s/dpa_H = -2.3[H^+](dC_s/d[H^+])$$

Thus, the buffer capacity β at each point of the titration curve is inversely proportional to the slope of the tangential line to the curve. When homoconjugation occurs, β is expressed approximately by the following relation [11]:

$$\beta = \frac{-2.3\{K_a K^f(HA_2^-) - (C_a - C_s)^2 + K_a(C_a + C_s) - 2C_s[H^+]\}}{[H^+] + 4K_a K^f(HA_2^-)(C_a - C_s)^2 - K_a^2/[H^+]}$$

dratic equation. Figure 3.3 shows the simulated titration curves for various values of $K^f(HA_2^-) \times (C_a + C_s)$. If we compare the experimental titration curve with the simulated ones (curve-fitting method), we can get the value of $K^f(HA_2^-)$ (see Figure 6.6). The value of $K^f(HA_2^-)$ can also be obtained by the following relation:

$$K^f(HA_2^-) = \{r^2 C_s - r(C_a + C_s) + C_a\}/\{r(C_a - C_s)^2\}$$

At the half-neutralization point, however, this relation does not hold and β is expressed as follows, if $K^f(HA_2^-)(C_a + C_s) \gg 1$:

$$\beta_{1/2} = (2.3/2)\{(C_a + C_s)/2K^f(HA_2^-)\}^{1/2}$$

The buffer capacities for various $K^f(HA_2^-) \times (C_a + C_s)$ values are shown in Figure 3.4 against the values of $(pa_H - pK_a - \log \gamma)$, which is shown on the ordinate in Figure 3.3.

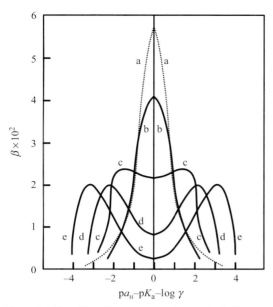

Figure 3.4 The effect of homoconjugation on the buffer capacity
(β) of a solution containing HA and Et$_4$NA [11]. Values of
$K^f(HA_2^-)(C_a + C_s)$: 0 for a, 1 for b, 10 for c, 10^2 for d and 10^3 for e.

on the buffer capacity (β) of a buffer consisting of a weak acid HA and its conjugate base A$^-$. On the abscissa of the figure, zero corresponds to the solution of $C_a = C_s$ (see footnote 6 for C_a and C_s), and the positive and negative sides are for $C_a < C_s$ and $C_a > C_s$, respectively. By the increase in the homoconjugation constant, the buffer capacity at $C_a = C_s$ decreases and the β versus pa_H curve shows a minimum there. Therefore, to get a good pH buffer in an aprotic solvent, we have to select an acid/ conjugate base couple that is free from homoconjugation. In practice, to calibrate a glass pH electrode in aprotic solvents, a 1 : 1 mixture of picric acid and picrate (usually an R$_4$N salt) or a weak base (e.g. aniline, imidazole, guanidine) and its conjugate acid (see footnote 4) is used.

The term heteroconjugation is justified only when the acid HR is much weaker than HA, otherwise the reaction HR + A$^- \rightleftarrows$ R$^-$ + HA more or less occurs. The most important HR acid is water. The heteroconjugation constants of water to some anions have been determined in AN: for example, $K^f_{A(HR)^-} \approx 0.5$ for picrate, $K^f_{A(HR)^-} = 2.3$, $K^f_{A(HR)_2^-} = 15.2$ and $K^f_{A(HR)_3^-} = 10.2$ for benzoate, $K^f_{A(HR)^-} = 3.6$ and $K^f_{A(HR)_2^-} = 8.0$ for methanesulfonate [12].

If water is added to a pH buffer in a dipolar aprotic solvent, the pa_H of the buffer is affected. Generally, if the buffer is in a protophobic aprotic solvent and consists of an acid/conjugate base couple that easily homoconjugates, the pa_H of the buffer is influenced considerably by the addition of hydrogen bond donors or acceptors. Here again, the effects of homo- and heteroconjugations are important. Figure 3.5a shows

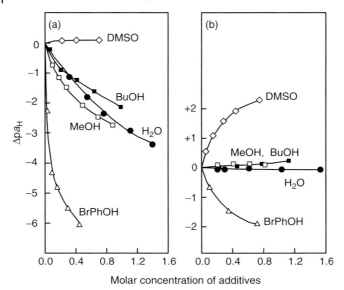

Figure 3.5 Effects of the addition of proton donors and acceptors on the pa_H of the benzoic acid–benzoate buffers in AN [12]. Solutions: (a) 3.6 mM HBz + 30.5 mM Et$_4$NBz and (b) 65 mM HBz + 2.18 mM Et$_4$NBz (see text).

the effects of hydrogen bond donors (water, MeOH, BuOH, p-bromophenol) and acceptor (DMSO) on the pa_H of the [3.6 mM benzoic acid (HBz) + 30.5 mM Et$_4$NBz] solution in AN ($pa_H = 23.4$) [12]. The addition of hydrogen bond donors (HR) decreases the pa_H of the solution. The decrease in pa_H is due to the decrease in the concentration of free Bz$^-$ by the heteroconjugation reaction Bz$^-$ + HR $\rightleftarrows$ Bz (HR)$^-$. The influence of DMSO, which is a poor hydrogen bond donor, is very small. Figure 3.5b shows the influence on the pa_H of the (65 mM HBz + 2.18 mM Et$_4$NBz) solution in AN ($pa_H = 16.8$). Because Bz$^-$ exists as Bz(HBz)$^-$, by the homoconjugation with HBz, weak hydrogen bond donors such as water and alcohols cannot influence pa_H. Only p-bromophenol, a strong hydrogen bond donor, competes with HBz in reacting with Bz$^-$ and decreases pa_H (for A$^-$ of benzoate and HR of p-bromophenol, $K^f_{A(HR)^-} = 3.6 \times 10^2$ and $K^f_{A(HR)_2^-} = 3.9 \times 10^3$, while $K^f_{HA_2^-} = 4.0 \times 10^3$). DMSO, a hydrogen bond acceptor, increases paH, because it reacts with the free HBz and decreases its activity.

On the other hand, the pa_H of picric acid/picrate buffers in AN are not influenced much by water and alcohols of up to 0.5 M, because in buffers, homo- and hetero-conjugation reactions do not occur appreciably [12]. In protophilic DMSO, the influences of hydrogen bond donors and acceptors on the pa_H of buffers are smaller than in protophobic AN. DMSO solvates to hydrogen bond donor HR, and suppresses the heteroconjugation, while DMSO suppresses the influence of a hydrogen bond acceptor by competing with it.

3.1.3
Acid–Base Reactions in Amphiprotic Solvents of Low Permittivity

t-Butanol ($\varepsilon_r = 12$, neutral), acetic acid ($\varepsilon_r = 6.2$, protogenic) and ethylenediamine ($\varepsilon_r = 12.9$, protophilic) are examples of amphiprotic solvents of low permittivity [3, 4]. In these solvents, an acid HA is ionized rather easily and forms an ion pair because the solvents can work both as an electron pair donor and acceptor:

$$HA \rightleftarrows (H^+, A^-)_{solv}$$

However, because of the low solvent permittivities, the dissociation of the ion pair to free ions is difficult. Thus, even a strong acid in water behaves as a weak acid, e.g. the pK_a of $HClO_4$ is 3.9 in t-butanol, 4.9 in acetic acid, and 3.1 in ethylenediamine.

In these solvents, ion pairs are also formed by the interactions between a base B and the solvent SH (Eq. (3.23)) and by the acid–base reaction between an acid HA and a base B (Eq. (3.24)):

$$B + SH \rightleftarrows (BH^+, S^-)_{solv} \tag{3.23}$$

$$HA + B \rightleftarrows (BH^+, A^-)_{solv} \tag{3.24}$$

Here again, the dissociation of ion pairs to free ions does not occur appreciably because of the low permittivities. In these solvents, even tetraalkylammonium salts can dissociate only slightly, as shown by the low dissociation constant (1.6×10^{-6} M) of Et_4NPic in acetic acid (see also Section 2.6).

For acid–base equilibria in acetic acid, the original reports by Kolthoff and Bruckenstein are important [13]. Reference [4] provides a comprehensive review of acid–base equilibria in amphiprotic solvents.

3.1.4
Acid–Base Reactions in Aprotic Solvents of Low Permittivity

In Section 1.2, some aprotic solvents of low permittivity (Py, THF, diethyl ether, MIBK) were classified as dipolar aprotic solvents because they have some basicity and behave like polar solvents. Therefore, the solvents considered here are aliphatic or aromatic hydrocarbons and halogenated hydrocarbons that are classified as inert solvents [3, 14]. The solvents of this class interact only very weakly with acids and bases, and the ionization and dissociation of acids do not occur appreciably in them. However, if both an acid (HA) and a base (B) are added to the solvent, a salt BHA is formed by the neutralization reaction

$$HA + B \rightleftarrows BHA \tag{3.25}$$

The salt is formed by hydrogen bonding and is usually in a polymer state $(BHA)_n$. The bond between BH and A is ionized to various degrees, from nonionized (BH . . . A)$_n$ to completely ionized $(BH^+ . . . A^-)_n$. The ionization becomes more pronounced with increasing acid–base interactions. When acid HA forms a dimer $(HA)_2$ in the inert solvent, as in the case of carboxylic acids in benzene, the salts denoted by $B(HA)_2$ [(BH$^+$ . . . A$^-$ . . . HA), (AH . . . B . . . HA), etc.] are formed.

The formation of the salt BHA can be detected quantitatively by such methods as potentiometry, conductimetry, UV/vis and IR spectroscopies, dielectric polarization measurement, and differential vapor pressure measurement. If the equilibrium constant of Eq. (3.25) is expressed by $K = [BHA]/([HA][B])$, we get

$$\log([BHA]/[HA]) = \log K + \log[B] \text{ or } \log([BHA]/[B]) = \log K + \log[HA]$$

By plotting the relation of log [BHA]/[HA] versus log [B] or log [BHA]/[B] versus log [HA], a linear relation of unit slope can be obtained, and the intercept corresponds to the value of log K. In fact, the values of log K have been obtained for various acid–base couples in inert solvents [3, 14] and they are usually between 2 and 7. If the values of log K are obtained for various bases using a given acid as reference, it is possible to compare the strengths of the bases. On the other hand, if the values of log K are obtained for various acids using a given base as reference, it is possible to compare the strengths of the acids.

3.2
pH Scales in Nonaqueous Solutions

3.2.1
Definition of pH in Nonaqueous Solutions

pH, defined by $pH = -\log a(H^+)$, is widely used as a measure of acid–base properties of aqueous solutions. However, the term 'pH' is also used as a measure of acid–base properties of nonaqueous solutions.[8]

In 1985, the IUPAC Commission of Electroanalytical Chemistry defined the pH for solutions in organic solvents of high permittivity and in water–organic solvent mixtures [15]. According to them, the pH is notionally defined by Eq. (3.26), where m shows the molality and γ the activity coefficient:

$$pH = -\log a(H^+) = -\log\{m(H^+)\gamma(H^+)/m^0\} \tag{3.26}$$

However, because $a(H^+)$ is a single ion activity, which is thermodynamically indeterminable, an operational pH definition is given as outlined in (1) and (2) below:

1. The pH values of the solutions X and S in solvent s and the emfs of cells (I) and (II), $E(S)$ and $E(R)$, are related by Eq. (3.27).

   Reference electrode|salt bridge in solvent s‖solution X in solvent s|H$_2$|Pt (I)

   Reference electrode|salt bridge in solvent s‖solution S in solvent s|H$_2$|Pt (II)

---

8) In Section 3.1, we dealt with the pa$_H$ value, defined by pa$_H = -\log a(H^+)$, for solutions in amphiprotic and aprotic solvents of high permittivity. Hereafter, the symbol pH is used instead of pa$_H$.

$$pH(X) = pH(S) + \frac{E(S) - E(X)}{\ln 10 \times (RT/F)} \tag{3.27}$$

Here, X is the test solution and S is the standard solution. A pH glass electrode may be used instead of the hydrogen electrode in cells (I) and (II).

2. 0.05 mol kg^{-1} potassium hydrogen phthalate (KHPh) in solvent s is used as the most fundamental pH buffer (Reference Value pH Standard, RVS) and the reference pH values (pH$_{RVS}$) are assigned to it. Other pH buffers can also be used as primary standards if pH values are appropriately assigned to them. In assigning pH values to RVS and the primary standards, an absolute method, which uses the Harned cell and the Bates–Guggenheim convention, is employed.

The above notional and operational pH definitions for solutions in nonaqueous and mixed solvents are the copy of the IUPAC Recommendations 1985 for aqueous solutions [16a]. These recommendations for aqueous solutions were, however, criticized and were replaced by the new one (Recommendations 2002) [16b]. The differences between Recommendations 1985 and 2002 are explained in Section 6.2.1; one of them is that the Reference Value pH Standard (0.05 mol kg^{-1} KHPh) in Recommendation 1985 is not special but is treated as one of the primary standards in Recommendations 2002. Here, the pH values for the primary standards in Recommendations 2002 are the same as those for the RVS and the primary standards in Recommendations 1985.

Application of Recommendations 2002 to nonaqueous and mixed solvents has been tried, as described in Section 6.2.2 [15c]. However, in this case also, the pH values for the standard solutions do not differ between the two recommendations. Thus, we can use the standard solutions for Recommendations 1985. At present, pH values for 0.05 mol kg^{-1} KHPh and some other primary standards have been reported for 100% formamide and the mixtures between water and 12 organic solvents (see footnote 6 in Section 6.2) [17]. When these standard solutions are available, pH can be determined accurately with a pH meter and a glass electrode, just as in aqueous solutions.

However, difficult problems arise there when the IUPAC method is applied to the solutions in neat organic solvents or water–poor mixed solvents. For example, in these solvents often the KHPh is not soluble enough and the buffer action of KHPh is too low in solutions of aprotic nature [18].[9] Another problem is that the response of the glass electrode is often very slow in nonaqueous solvents,[10] although this has been considerably improved by the use of pH-ISFETs [19]. These difficulties for pH measurements in nonaqueous and mixed solvents are dealt with in detail in Section 6.2, with the practical methods and their applications.

---

9) The low buffer capacities of the KHPh solution in solvents of aprotic nature is caused by the increase in the $(pK_{a2} - pK_{a1})$ value of phthalic acid. In H$_2$O–DMF and H$_2$O–DMSO mixtures, the buffer capacity of 0.05 mol kg^{-1} KHPh is not enough if the water content is less than $\sim$30 v/v% [18].

10) The glass electrode responds especially slowly in protophilic aprotic solvents such as DMSO and DMF, sometimes taking 1 h to reach a steady potential. In such cases, the use of Si$_3$N$_4$- and Ta$_2$O$_5$-type pH-ISFETs is very promising because they almost respond instantaneously and with Nernstian or near-Nernstian slopes [19].

3.2.2
pH Windows in Nonaqueous Solvents and pH Scales Common to Multisolvents

pH Windows in Nonaqueous Solvents
In aqueous solutions, the pH of an acid solution in which $a(H^+) = 1\, mol\, kg^{-1}$ or $1\, mol\, l^{-1}\, (= M)$, depending on the definition [16], is equal to zero. Because the ion product constant of water, $K_W\, [= a(H^+)a(OH^-)]$, is equal to $10^{-14}\, M^2$ at 25 °C, the pH of the solution of $a(OH^-) = 1\, M$ is equal to 14. Of course, there exist aqueous solutions of pH < 0 or pH > 14. But it is reasonable to assume that the pH window of water is approximately between 0 and 14 and its width is $\sim$14 ($= pK_W$).

In an amphiprotic solvent SH, the autoprotolysis occurs as in Eq. (3.28), and the autoprotolysis constant $K_{SH}\, [= a(SH_2^+)a(S^-)]$ is constant at a given temperature:

$$SH + SH \rightleftarrows SH_2^+ + S^- \tag{3.28}$$

If the activity of SH_2^+ or the pH is known, the activity of the lyate ion S^- is obtained by the relation $a(S^-) = K_{SH}/a(SH_2^+)$. Some pK_{SH} values in pure and mixed solvents are listed in Tables 3.4 and 6.6. Most of the pK_{SH} values in amphiprotic solvents are less than 20. Therefore, the widths of the pH windows are usually less than 20 in amphiprotic solvents.

In most aprotic solvents, which have weak proton donor capacities, it is difficult to get stable lyate ions. The lyate ion in DMSO (dimsyl ion, $CH_3SOCH_2^-$) is somewhat stable. Alkali metal dimsyls in DMSO are strongly basic and have been used as titrants (Ref. [5], p. 356), but this is rather exceptional. Generally, aprotic solvents do not undergo appreciable autoprotolysis. It is especially true for protophobic aprotic solvents, which have very weak proton donating and accepting abilities. Efforts to determine K_{SH} values have been made even in such solvents. However, if a small amount of water is present as an impurity in such solvents, the H_3O^+ and OH^- ions formed from water may complicate the autoprotolysis process. The K_{SH} values determined may not be the product $a(SH_2^+)a(S^-)$ but the product $a(SH_2^+)a(OH^-)$ or even $a(H_3O^+)a(OH^-)$ (see Section 6.2.3). Moreover, OH^- ion added as R_4NOH cannot remain stable because of Hoffmann degradation (R_4NOH forms alkene, R_3N and water). Impurities other than water may also affect the K_{SH} values. Thus, it is very difficult to get true K_{SH} values in aprotic solvents and the literature data highly vary [6]. The pK_{SH} values for aprotic solvents are usually higher than 20 (Tables 3.4 and 6.6) and, especially for some protophobic aprotic solvents, they may approach $\sim$40. Thus, the pH windows are much wider in aprotic solvents than in water (see below).

In aprotic solvents, in which pK_{SH} values are not definitive, it is difficult to estimate the activity of the lyate ion from the pH of the solution. It is different from the case in water, in which $a(OH^-)$ can be estimated from the pH value. Although the pH scale in aprotic solvents has such a disadvantage, it is still useful to quantitatively understand the acid–base aspects of chemical phenomena. Wider use of the pH concept in nonaqueous solutions is desirable.

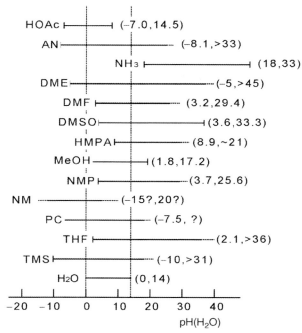

Figure 3.6 pH windows in various solvents shown by a common pH scale. The pH scale in water is used as reference. The values in parentheses correspond to $-\log \gamma_t(H^+, W \rightarrow S)$ and pK_{SH}, respectively.

pH Scales Common to Multisolvents

In the previous section, the pH of the solution of $a(H^+) = 1 \, \text{mol kg}^{-1}$ was defined equal to zero in each solvent.[11] However, the solvation of H^+ differs from one solvent to another and, even in a solution of pH $= 0$, the reactivity of H^+ differs drastically by solvent. In order to compare the acid–base properties in different solvents, it is convenient to define a pH scale that is common to various solvents [20]. Figure 3.6 shows the pH windows in various solvents in such a common pH scale, using the pH in water as reference. In the figure, the left margin of the pH window corresponds to the value of $-\log \gamma_t(H^+, W \rightarrow S)$ (see Table 2.7) and the width of the window corresponds to the value of pK_{SH}. If the solvent is of weaker basicity than water, the pH window expands to the left (more acidic) side than water. On the other hand, if the solvent is of weaker acidity than water, the pH window expands to the right (more basic) side than water.

These expanded pH windows open up various chemical possibilities:

11) If the density of the solvent is between 0.8 and 1.2 g cm^{-3}, the difference between the pH on molality and that on molarity is within 0.1.

(i) In solvents with a wide pH region on the left, the solvated protons (SH_2^+) have a very strong acidity and some acids, which behave as strong acids in water, tend to behave as weak acids of different strengths. (The term *differentiating solvent* is used for a solvent that can differentiate the strengths of acids (or bases) that are equivalent in water.) Thus, they can be determined separately by titration. Moreover, in such a solvent, some bases, which are too weak to titrate in water, can be titrated and their strengths can be determined.

(ii) In solvents with a wide pH region on the right, the lyate ion (S^-) has very strong basicity and some bases, which behave as strong bases in water, tend to behave as weak bases of different strengths. They can be determined separately by titration. Moreover, in such a solvent, some acids, which are too weak to titrate in water, can be titrated and their strengths can be determined.

In protophobic aprotic solvents, the pH window expands on both sides and all of the advantages in (i) and (ii) can be realized. Figure 3.7 shows the titration curve for a mixture of several acids in 4-methyl-2-pentanone (MIBK). In the figure, $HClO_4$ and HCl, which are strong acids in water, are differentiated, and phenol, which is a very weak acid, can accurately be determined [21]. Acid–base titrations in nonaqueous solvents have wide applicability [1, 3, 22]. The titrations are especially useful in determining medicinal substances because many of them are weak acids or bases. The indicator method can be used to detect end points. However, potentiometric

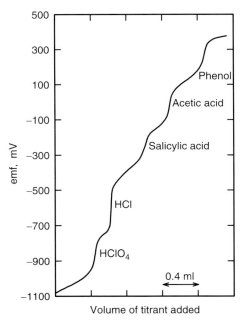

Figure 3.7 Potentiometric titration curve of a mixture of acids in MIBK [21]. Titrated with 0.2 M Bu_4NOH using a glass electrode–Pt electrode system.

Table 3.6 Formation constants of complexes of H^+ in AN with solvents B.[a]

Solvents B	log β_1	log β_2
DMF	5.5 (6.1[b])	7.0 (7.8[c])
DMSO	5.4 (5.8[b])	8.2 (9.3[c])
NMP	5.7	9.1
DMA	7.3	10.0
HMPA	8.0	11.0

$\beta_1 = [BH^+]/([B][H^+])$, $\beta_2 = [B_2H^+]/([B]^2[H^+])$.
[a]From Ref. [23a] (Izutsu and Hiraoka) except [b] and [c].
[b]By Kolthoff et al. [23b]
[c]By Ugo et al. [23c]

titrations that use a pH meter, a pH sensor and an automatic burette are most useful. Valuable physicochemical information, including acid dissociation constants, can be obtained from potentiometric titration curves (Chapter 6).

The hydrogen ion in protophobic aprotic solvents is very reactive. For example, judging from the values of transfer activity coefficient, H^+ is 10^8 times more reactive in AN than in water. Thus, if basic substances are added to the solution in AN, they easily combine with H^+. Table 3.6 shows the complex formation constants of H^+ with some basic (donor) solvents [23]. Complexation of up to two solvent molecules occurs by the addition of small amounts of donor solvents.[12] If water is added as donor solvent to the H^+ solution in AN, up to four water molecules are combined with H^+. The formation constants for the reactions $H^+ + W \rightleftarrows Hw^+$, $H^+ + 2W \rightleftarrows H_{2w}^+$, $H^+ + 3W \rightleftarrows H_{3w}^+$ and $H^+ + 4W \rightleftarrows H_{4w}^+$ have been determined spectrophotometrically using Hammett indicators [24]. They are log $\beta(Hw^+) = 2.2$, log $\beta(H_{2w}^+) = 3.9$, log $\beta(H_{3w}^+) = 4.8$ and log $\beta(H_{4w}^+) = 5.3$, respectively. Figure 3.8 shows the distribution of each species as a function of free water concentration (i.e. total H_2O minus H_2O combined to H^+). It is apparent that even when the concentration of free water is as low as 0.01 M, the fraction of free H^+ is only about 30% and, if the concentration of free water is 1 M, more than half of the H^+ exists as H_{4w}^+. Moreover, H^+ in AN gradually reacts with the AN molecules and loses its reactivity. The H^+ in other protophobic aprotic solvents also tends to react with water, with other basic impurities and with the solvent itself. Sulfolane (TMS) is said to be the only protophobic solvent that does not react appreciably with H^+. To get a wide pH window on the acid side (the left side in Figure 3.6), the solvent must be free of water and other basic impurities, and the acid solution must be prepared fresh, just before measurement. Electrolytic (or coulometric) generation of H^+ is convenient for this purpose: dry H^+ can be obtained, for example, by anodic oxidation of hydrogen absorbed into a palladium electrode [25].

To get a wide pH window on the basic side (the right side in Figure 3.6), the solvent must be aprotic and free of water and other acidic impurities. Because stable lyate

12) When H^+ in AN is titrated with a protophilic
solvent as in Table 3.6, a clear pH jump occurs
on the titration curve at the equivalence point.

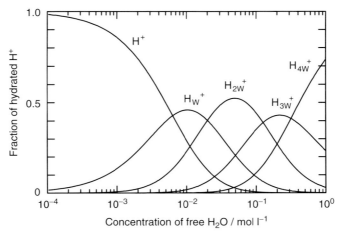

Figure 3.8 Step-wise complexation of H^+ in AN with water, indicated as a function of the free water concentration [24].

ions cannot be obtained in most aprotic solvents, R_4NOH ($R = Et$ or Bu) is often used as a strong base for titration. Alkali metal hydroxides are insoluble and are not applicable in aprotic solvents. However, R_4NOH in aprotic solvents is also unstable because it is decomposed (Hoffmann degradation). Therefore, while titrating the acids in aprotic solvents, the standard solution of R_4NOH in methanol, (toluene + methanol) or 2-propanol is used as a titrant. In that case, some amount of alcohol is introduced with the titrant and may decrease the width of the pH window. There is also gradual decomposition of the added R_4NOH. These factors must be kept in mind when carrying out pH titrations in aprotic solvents.

References

1 Kolthoff, I.M. and Elving, P.J. (1979) *Treatise on Analytical Chemistry*, 2nd edn, Part I, vol. 2, Interscience Publishers, New York, Chapter 19.
2 Kolthoff, I.M. (1974) *Anal. Chem.*, **46**, 1992.
3 Safrik, L. and Stransky, Z. (1986) *Titrimetric Analysis in Organic Solvents: Comprehensive Analytical Chemistry*, vol. 22, Elsevier, Amsterdam.
4 Popov, A.I. and Caruso, H., Acid–base equilibria and titrations in non-aqueous solvents. B. Amphiprotic solvents, in Ref. [1], pp. 303–348.
5 Kolthoff, I.M. and Chantooni, M.K., Jr, Acid–base equilibria and titrations in non-

aqueous solvents. A. General introduction to acid–base equilibria in nonaqueous organic solvents; C. Dipolar aprotic solvents, in Ref. [1], pp. 239–302, 349–384.
6 Izutsu, K. (1990) *Acid–Base Dissociation Constants in Dipolar Aprotic Solvents*, IUPAC Chemical Data Series No. 35, Blackwell Science, Oxford.
7 Chantooni, M.K., Jr and Kolthoff, I.M. (1975) *J. Phys. Chem.*, **79**, 1176.
8 (a) Chantooni, M.K., Jr and Kolthoff, I.M. (1976) *J. Phys. Chem.*, **80**, 1306; (b) Kolthoff, I.M. and Chantooni, M.K., Jr (1971) *J. Am. Chem. Soc.*, **93**, 3843.

9 Izutsu, K., Nakamura, T., Takizawa, K. and Takeda, A. (1985) *Bull. Chem. Soc. Jpn.*, **58**, 455.

10 Kolthoff, I.M. and Chantooni, M.K., Jr (1975) *J. Am. Chem. Soc.*, **97**, 1376; 1976, **98**, 5063, 7465.

11 Kolthoff, I.M. and Chantooni, M.K., Jr (1965) *J. Am. Chem. Soc.*, **87**, 4428.

12 (a) Kolthoff, I.M. and Chantooni, M.K., Jr (1967) *Anal. Chem.*, **39**, 1080; (b) Kolthoff, I.M. and Chantooni, M.K., Jr (1969) *J. Am. Chem. Soc.*, **91**, 6907.

13 (a) Kolthoff, I.M. and Bruckenstein, S. (1956) *J. Am. Chem. Soc.*, **78**, 1; 1957, **79**, 1; (b) Bruckenstein, S. and Kolthoff, I.M. (1956) *J. Am. Chem. Soc.*, **78**, 10, 2974; 1957, **79**, 5915; (c) Kolthoff, I.M. and Bruckenstein, S. (1959) *Treatise on Analytical Chemistry*, vol. 1 (eds I.M. Kolthoff and P.J. Elving), Part I, Interscience Publishers, New York, pp. 475–542.

14 Steigman, J., Acid–base equilibria and titrations in non-aqueous solvents. D. Inert solvents, in Ref. [1], pp. 385–423.

15 (a) Mussini, T., Covington, A.K., Longhi, P. and Rondinini, S. (1985) *Pure Appl. Chem.*, **57**, 865; (b) Rondinini, S., Mussini, P.R. and Mussini, T. (1987) *Pure Appl. Chem.*, **59**, 1549; (c) Rondinini, S. (2002) *Anal. Bioanal. Chem.*, **374**, 813.

16 (a) Covington, A.K., Bates, R.G. and Durst, R.A. (1985) *Pure Appl. Chem.*, **57**, 531; (b) Buck, R.P. *et al.* (2002) *Pure Appl. Chem.*, **74**, 2169.

17 Mussini, P.R., Mussini, T. and Rondinini, S. (1997) *Pure Appl. Chem.*, **69**, 1007.

18 Izutsu, K. and Yamamoto, H. (1998) *Talanta*, **47**, 1157.

19 (a) Izutsu, K., Nakamura, T. and Hiraoka, S. (1993) *Chem. Lett.*, 1843; (b) Izutsu, K. and Ohmaki, M. (1996) *Talanta*, **43**, 643; (c) Izutsu, K. and Yamamoto, H. (1996) *Anal. Sci.*, **12**, 905.

20 For example, (a) Tremillon, B. (1974) *Chemistry in Non-Aqueous Solvents*, D. Reidel, Dordrecht, the Netherlands, Chapter 4; (b) Tremillon, B. (1997) *Reactions in Solution: An Applied Analytical Approach*, John Wiley & Sons, Inc., Chichester, Chapter 6; (c) Bauer, D. and Breant, M. (1975) *Electroanalytical Chemistry*, vol. 8 (ed. A.J. Bard), Marcel Dekker, New York, p. 281.

21 Bruss, D.B. and Wyld, G.E.A. (1957) *Anal. Chem.*, **29**, 232.

22 For example, (a) Charlot, G. and Tremillon, B. (1969) *Chemical Reactions in Solvents and Melts*, Pergamon Press, Oxford; (b) Gyenes, I. (1970) *Titrationen in Nichtwasserigen Medien*, F. Enke, Stuttgart; (c) Fritz, J.S. (1973) *Acid–Base Titrations in Nonaqueous Media*, Allyn & Bacon, Needham Heights, MA; (d) Kratochvil, B. (1978) *Anal. Chem.*, **50**, 153R; 1980, **52**, 151R; (e) Izutsu, K. and Nakamura, T. (1987) *Bunseki*, 392; 1992, 366.

23 (a) Izutsu, K. and Hiraoka, S.,unpublished results; (b) Kolthoff, I.M., Chantooni, M.K., Jr and Bhowmik, S. (1967) *Anal. Chem.*, **39**, 1627; (c) Ugo, P., Daniele, S. and Mazzocchin, G.-A. (1985) *Anal. Chim. Acta*, **173**, 149.

24 Chantooni, M.K., Jr and Kolthoff, I.M. (1970) *J. Am. Chem. Soc.*, **92**, 2236.

25 Mihajlovic, R.P., Vajgand, V.J. and Dzudovic, R.M. (1991) *Talanta*, **38**, 673.

4
Redox Reactions in Nonaqueous Solvents

This chapter deals with the fundamental aspects of redox reactions in nonaqueous solutions. In Section 4.1, we discuss solvent effects on the potentials of various types of redox couples and on reaction mechanisms. Solvent effects on redox potentials are important in connection with the electrochemical studies of such basic problems as ion solvation and electronic properties of chemical species. We then consider solvent effects on reaction kinetics, paying attention to the role of dynamic solvent properties in electron-transfer (ET) processes. In Section 4.2, we deal with the potential windows in various solvents to show the advantages of nonaqueous solvents as media for redox reactions. In Section 4.3, we describe some examples of practical redox titrations in nonaqueous solvents. Because many of the redox reactions are realized as electrode reactions, the subjects covered in this chapter will also appear in Part Two in connection with electrochemical measurements.

4.1
Solvent Effects on Various Types of Redox Reactions

4.1.1
Fundamentals of Redox Reactions

A redox reaction includes an electron-transfer process. For example, if Ce^{4+} is added to an aqueous solution of Fe^{2+}, a redox reaction, $Fe^{2+} + Ce^{4+} \rightleftarrows Fe^{3+} + Ce^{3+}$, occurs and Fe^{2+} (reducing agent) is oxidized to Fe^{3+}, donating an electron to Ce^{4+}. On the other hand, Ce^{4+} (oxidizing agent) is reduced to Ce^{3+}, gaining an electron from Fe^{2+}. The oxidation of Fe^{2+} to Fe^{3+} also occurs at an electrode kept at appropriate potentials. In this case, the electrode gains an electron from Fe^{2+}. In redox reactions, electron transfer occurs with the aid of either oxidizing or reducing agents or with the aid of electrodes.

The redox properties of a solution are expressed quantitatively by the redox potential, which is measured as the potential of an inert redox electrode (e.g. platinum electrode) immersed in the solution under study. Thus, for an aqueous solution containing the oxidized and reduced forms (Ox, Red) of the reaction $Ox + ne^- \rightleftarrows Red$, the redox potential is equal to the emf of cell (I):

$$Pt|H_2(p = p^0)|H^+ (a = 1)(w)\|Ox(a_{Ox}), Red(a_{Red})(w)|Pt \qquad (I)$$

Electrochemistry in Nonaqueous Solutions, Second, Revised and Enlarged Edition. Kosuke Izutsu
Copyright © 2009 WILEY-VCH Verlag GmbH & Co. KGaA, Weinheim
ISBN: 978-3-527-32390-6

The electrode on the left of the junction ($\|$) is the standard hydrogen electrode (SHE), the potential of which is defined as zero at all temperatures. P^0 is the standard pressure and is equal to 10^5 Pa (1 bar) by the IUPAC Recommendation (1982).[1] The liquid junction potential at $\|$ is kept negligible by an appropriate salt bridge. The potential of the electrode on the right of the junction is expressed by the Nernst equation:

$$E = E^0 + \frac{RT}{nF} \ln \frac{a_{\mathrm{Ox}}}{a_{\mathrm{Red}}} \tag{4.1}$$

where a_{Ox} and a_{Red} are the activities of Ox and Red, respectively, and E^0 is the standard redox potential, which is equal to the redox potential at $a_{\mathrm{Ox}} = a_{\mathrm{Red}} = 1$. E^0 is related to the equilibrium constant K of reaction (4.2) by $E^0 = (RT/F)\ln K$.

$$(1/n)\mathrm{Ox} + (1/2)\mathrm{H}_2 \rightleftharpoons (1/n)\mathrm{Red} + \mathrm{H}^+ \qquad K = \frac{a_{\mathrm{Red}}^{1/n} a_{\mathrm{H}^+}}{a_{\mathrm{Ox}}^{1/n} a_{\mathrm{H}_2}^{1/2}} \tag{4.2}$$

Each redox couple has its own E^0 value. If we rewrite Eq. (4.1) using the concentrations of Ox and Red, we get

$$E = E^{0'} + \frac{RT}{nF} \ln \frac{[\mathrm{Ox}]}{[\mathrm{Red}]} \tag{4.3}$$

$E^{0'}$ is the potential at $[\mathrm{Ox}] = [\mathrm{Red}]$ and is often called the *formal potential*. The relation between E^0 and $E^{0'}$ is expressed by $E^{0'} = E^0 + (RT/nF)\ln(\gamma_{\mathrm{Ox}}/\gamma_{\mathrm{Red}})$, where γ denotes the activity coefficient.

We now consider the titration of Red_1 with an oxidizing agent Ox_2, where the oxidation of Red_1 proceeds by $\mathrm{Red}_1 \rightleftharpoons \mathrm{Ox}_1 + n_1\mathrm{e}^-$ and the reduction of Ox_2 by $\mathrm{Ox}_2 + n_2\mathrm{e}^- \rightleftharpoons \mathrm{Red}_2$. The titration reaction is expressed by Eq. (4.4).

$$n_2\mathrm{Red}_1 + n_1\mathrm{Ox}_2 \rightleftharpoons n_2\mathrm{Ox}_1 + n_1\mathrm{Red}_2 \tag{4.4}$$

In Figure 4.1, the titration curve passes I (before the 100% titration) and I′ (after the 100% titration) if $n_1 = n_2 = 1$, II and II′ if $n_1 = n_2 = 2$, I and II′ if $n_1 = 1$ and $n_2 = 2$ and II and I′ if $n_1 = 2$ and $n_2 = 1$. In all cases, the potential is equal to the formal potential of the $\mathrm{Ox}_1/\mathrm{Red}_1$ couple ($E_1^{0'}$) at 50% titration, while it is equal to the formal potential of the $\mathrm{Ox}_2/\mathrm{Red}_2$ couple ($E_2^{0'}$) at 200% titration. If we consider, for simplicity, the case of $n_1 = n_2 = n$, the equilibrium constant of reaction (4.4), given by $K = a_{\mathrm{Ox}_1} a_{\mathrm{Red}_2}/(a_{\mathrm{Red}_1} a_{\mathrm{Ox}_2})$, is related to E_1^0 and E_2^0 by $K = \exp[(nF/RT)(E_2^0 - E_1^0)]$. The increase in the value of $(E_2^0 - E_1^0)$ increases the value of K, and the redox reaction (4.4) proceeds more completely to the right.

Some typical redox couples and metal ion–metal couples are shown below, with the corresponding Nernst equations. The effects of solvents on these reactions will be discussed in the next section.

---

[1] Before 1982, $p^0 = 101\,325$ Pa (1 atm) was used. The difference in the standard pressure has a negligible influence on the value of standard potential, as shown by E^0 (101 325 Pa) $= E^0$ (10^5 Pa) $+ 0.17$ mV.

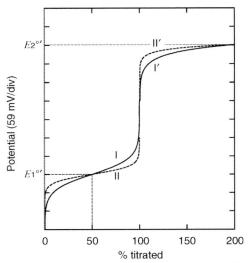

Figure 4.1 Curves for the potentiometric titration of Red_1 with oxidizing agent Ox_2. Before the 100% titration, curve I is for $n_1 = 1$ and II for $n_1 = 2$, where n_1 is for $Ox_1 + n_1e^- \rightleftarrows Red_1$. After 100% titration, curve I' is for $n_2 = 1$ and II' for $n_2 = 2$, where n_2 is for $Ox_2 + n_2e^- \rightleftarrows Red_2$ (see text).

1. $M^{n+} + ne^- \rightleftarrows M$ (M : metal)

$$E = E_a^0 + \frac{RT}{nF} \ln a(M^{n+})$$

2. $M^{n+} + ne^- + Hg \rightleftarrows M(Hg)$ (M(Hg) : metal amalgam)

$$E = E_b^0 + \frac{RT}{nF} \ln \frac{a(M^{n+})a(Hg)}{a(M(Hg))}$$

3. $Ox + ne^- \rightleftarrows Red$

$$E = E_c^0 + \frac{RT}{nF} \ln \frac{a_{Ox}}{a_{Red}}$$

4. $Ox + mH^+ + ne^- \rightleftarrows Red$

$$E = E_d^0 + \frac{RT}{nF} \ln \frac{a_{Ox}}{a_{Red}} + \frac{mRT}{nF} \ln a(H^+)$$

(Hydrogen ions participate in the reduction, and the potential is influenced by the pH of the solution; the potential shift being $-59.16(m/n)$ mV/pH at 25 °C.)

5. $ML_p^{(n+pb)+} + ne^- \rightleftarrows M + pL^b$ ($ML_p^{(n+pb)+}$:metal comples; b: zero or negative)

$$E = E_e^0 + \frac{RT}{nF} \ln \frac{a(ML_p^{(n+pb)+})}{a(L^b)^p}, \qquad E_e^0 = E_a^0 - \frac{RT}{nF} \ln K$$

(K is the complex formation constant and $K = a(ML_p^{(n+pb)+})/\{a(M^{n+})a(L^b)^p\}$. With the increase in K, E_e^0 shifts to the negative direction.)

Table 4.1 Standard potentials of M^{n+}/M and $M^{n+}/M(Hg)$ electrodes in water.[a]

Metals	$E^0(M^{n+}/M)$	$E^0(M^{n+}/M(Hg))$
Na	−2.717 (V versus SHE)	−1.959 (V versus SHE)
K	−2.928	−1.975
Zn	−0.763	−0.801
Cd	−0.402	−0.380
Tl	−0.327	−0.294
Pb	−0.126	−0.120

[a]Mainly from Ref. [2].

Data of the standard potentials for inorganic redox systems in aqueous solutions have been compiled by IUPAC [1]. The standard potentials for some M^{n+}/M and $M^{n+}/M(Hg)$ couples are shown in Table 4.1 [2]. For alkali metals, the standard potentials of $M^+/M(Hg)$ are about 1 V more positive than those of M^+/M. This is because alkali metals have strong affinities to mercury and are stable in the amalgams. It is impossible to measure the potentials of alkali metal electrodes directly in aqueous solutions because alkali metals react with water. To determine the potential of an alkali metal electrode in an aqueous solution, we measure the potential of the corresponding amalgam electrode in an aqueous solution and then the difference between the potentials of alkali metal and alkali metal amalgam electrodes using an appropriate nonaqueous solution [2].[2)]

2) To get the potential of the sodium electrode in aqueous 0.100 M NaCl against a calomel electrode in the same solution, one measures the emfs of cells (1) and (2). Cell (1) is to measure the potential of the amalgam electrode (Na(Hg)) and cell (2) is to measure the potential of the Na electrode against the Na (Hg) electrode. The sum of the emfs of the two cells (−3.113 V) corresponds to the emf of the hypothetical cell (3) and is equal to the potential of the Na electrode in the aqueous solution.

Na(Hg) but dissolves and dissociates electrolyte NaI. The composition of the electrolyte solution in cell (2) does not influence its emf. Instead of measuring the emfs of cells (1) and (2), one can measure the emf (−3.113 V) of the double cell (4). The advantage of the double cell is that the emf is not influenced by the concentration of Na(Hg).

$$(+)Hg|Hg_2Cl_2|0.100 \text{ M NaCl}_{aq}|Na-Hg|NaI$$
$$(ethylamine)|Na(-) \qquad\qquad (4)$$

$$(+)Hg|Hg_2Cl_2|0.100 \text{ M NaCl}_{aq}|Na-Hg(0.206\%)(-) \ [E = -2.268 \text{ V}, 25\,°C] \qquad (1)$$

$$(+)Na-Hg(0.206\%)|NaI(ethylamine)|Na(-) \ [E = -0.845 \text{ V}] \qquad (2)$$

$$(+)Hg|Hg_2Cl_2|0.100 \text{ M NaCl}_{aq}|Na(-) \ [E = -3.113 \text{ V}] \qquad (3)$$

Because Na(Hg) slowly reacts with water, the Na(Hg) electrode in cell (1) is preferably a flow type, in which the electrode surface is continuously renewed. Ethylamine is used in cell (2) because it does not react with Na and

The standard potential of the Na^+/Na couple can be obtained from these emf values. However, it can also be obtained by calculating from thermodynamic data [2], and the result agrees well with the result by emf measurement.

4.1.2
Solvent Effects on Redox Potentials and Redox Reaction Mechanisms

In this section, solvent effects are considered for each of the above reactions, focusing on the standard redox potentials and the reaction mechanisms. It should be noted that the potentials here are based on a scale common to all solvents, so that the potentials in different solvents can be compared with one another (see Section 2.3).

1. $M^{n+} + ne^- \rightleftharpoons M$

The variation in the standard potential of this type of reaction is directly related to the variation in the solvation energy of metal ion M^{n+}. If the standard potentials in solvents R and S are expressed by $E^0(R)$ and $E^0(S)$, respectively,

$$E^0(S) - E^0(R) = \left(\frac{RT}{nF}\right) \ln \gamma_t (M^{n+}, R \to S)$$

$$= \left(\frac{0.0592}{n}\right) \log \gamma_t (M^{n+}, R \to S) \quad [V, 25°C] \quad (4.5)$$

where $\gamma_t(M^{n+}, R \to S)$ is the transfer activity coefficient of M^{n+} from solvent R to S. The standard potential is more negative in the solvent in which M^{n+} is solvated more strongly. The relation in Eq. (4.5) is used in obtaining the transfer activity coefficients or the Gibbs energies of transfer of metal ions by measuring electrode potentials. In Table 4.2, the standard potentials of M^{n+}/M couples in various solvents are indicated by using a common scale, which is referred to the potential of the standard hydrogen electrode in water. The values in the table were obtained on the basis of extrathermodynamic assumptions, similar to those employed in obtaining the ionic Gibbs energies of transfer and the ionic transfer activity coefficients in Tables 2.4 and 2.7. The values in Table 4.2 give approximate information on the effects of ion solvation on the standard potentials.

Sometimes, the reaction mechanisms themselves change with the solvent. For example, the reduction of Cu^{2+} to Cu^0 occurs in one step in water (in the absence of a complexing agent that stabilizes Cu(I)), but it occurs in two steps in AN, $Cu^{2+} \to Cu^+ \to Cu^0$. In AN, Cu^+ is extremely stable because of its strong solvation (Section 2.2) and the step $Cu^{2+} \to Cu^+$ easily occurs at a very positive potential (see Eq. (4.6)), while the step $Cu^+ \to Cu^0$ occurs at a very negative potential.

2. $M^{n+} + ne^- + Hg \rightleftharpoons M(Hg)$

Equation (4.5) is also valid in this case. Reactions of this type are realized in polarography at a dropping mercury electrode, and the standard potentials can be obtained from the polarographic half-wave potentials ($E_{1/2}$). Polarographic studies of metal ion solvation are dealt with in Section 8.2.1. Here, only the results obtained by Gritzner [3] are outlined. He was interested in the role of the HSAB concept in metal ion solvation (Section 2.2.1) and measured half-wave potentials

Table 4.2 Standard potentials of M^{n+}/M electrodes in various solvents (values referred to SHE in water (V, 25 °C)).

Solvents[a]	Electrode systems								
	$H^+/(1/2)H_2$	Li^+/Li	Na^+/Na	K^+/K	Rb^+/Rb	Cs^+/Cs	Ag^+/Ag	Tl^+/Tl	Cu^{2+}/Cu
H₂O	0.000	−3.040	−2.714	−2.936	−2.943	−3.027	0.799	−0.336	0.339
MeOH	0.10	−2.99	−2.63	−2.84	−2.84	−2.94	0.87	−0.29	0.47
EtOH	0.12	−2.93	−2.57	−2.77	−2.78	−2.87	0.85	−0.26	0.58
PrOH	0.09	−2.93	−2.54	−2.75	−2.75	−2.85	0.81		0.56
TFE				−2.53			1.32		
En(OH)₂	0.05	−3.04	−2.74	−2.96			0.81		
Ac				−2.90	−2.90	−2.99	0.89		
PC	0.52	−2.79	−2.56	−2.88	−2.95	−3.10	0.99	−0.22	0.73
FA		−3.14	−2.80	−2.98	−3.00	−3.09	0.64		
DMF	−0.19	−3.14	−2.81	−3.04	−3.04	−3.14	0.58	−0.46	0.25[b]
DMA			−2.84	−3.06	−3.03	−3.20	0.50		
DMTF		−2.47	−2.31	−2.66		−2.88	−0.26	−0.50	
NMP	−0.26	−3.40	−2.87	−3.05	−3.03	−3.13	0.53	−0.49	
AN	0.48	−2.73	−2.56	−2.88	−2.88	−2.97	0.56	−0.25	0.65[b]
NM		−2.54	−2.45	−2.74	−2.92	−3.02	1.02		
NB	0.34	−2.65	−2.54	−2.70	−2.75	−2.87		−0.18	
DMSO	−0.20	−3.20	−2.85	−3.07	−3.05	−3.16	0.44	−0.56	0.09
TMS			−2.75	−2.98	−3.04	−3.13	0.76		0.71
HMPA				−3.10			0.32		
1,1-DCE			−2.41	−2.63	−2.64	−2.74			
1,2-DCE			−2.46	−2.67	−2.68	−2.78			

From Marcus, Y. (1985) *Pure Appl. Chem.*, **57**, 1129.
[a]TFE = 2,2,2-trifluoroethane, En(OH)₂ = 1,2-ethanediol. For other solvents, see Table 1.1.
[b]Mean value of the standard potentials for Cu^{2+}/Cu^+ and Cu^+/Cu^0.

for the reductions of alkali and alkaline earth metal ions, Tl^+, Cu^+, Ag^+, Zn^{2+}, Cd^{2+}, Cu^{2+} and Pb^{2+}, in 22 different solvents. He used the half-wave potential of the BCr^+/BCr couple as a solvent-independent potential reference. As typical examples of the hard and soft acids, he chose K^+ and Ag^+, respectively, and plotted the half-wave potentials of metal ions against the half-wave potentials of K^+ or against the potentials of the 0.01 M Ag^+/Ag electrode. The results were as follows:

(i) Figure 4.2 shows that the half-wave potentials of Rb^+ in various solvents are linearly related to those of K^+. This is also true for other alkali metal ions, showing that they are hard acids.

(ii) In hard-base solvents, the half-wave potentials for Tl^+, Zn^{2+}, Cd^{2+}, Cu^{2+} and Pb^{2+} are linearly related to those of K^+. However, in NMTP, DMTF, HMPTA and 2,2'-thiodiethanol, which are soft bases, and in AN and BuN, which are between hard and soft bases, the half-wave potentials of these metal ions are not correlated with those of K^+.

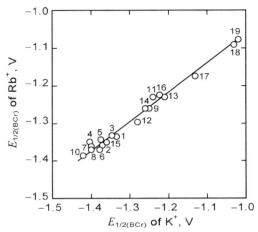

Figure 4.2 Relation between the half-wave potentials
(V versus BCr^+/BCr) of Rb^+ and those of K^+ in various
organic solvents [3]. Solvents are 1, NMF; 2, DMF; 3,
N,N-diethylformamide; 4, DMA; 5, N,N-diethylacetamide;
6, NMP; 7, TMU; 8, DMSO; 9, TMS; 10, HMPA; 11, MeOH; 12, Ac;
13, PC; 14, γ-BL; 15, trimethyl phosphate; 16, AN; 17, BuN; 18,
NMTP; 19, DMTF.

(iii) If the half-wave potentials for Tl^+, Zn^{2+}, Cd^{2+}, Cu^{2+} and Pb^{2+} are plotted
against the potentials of the Ag^+/Ag electrode, two linear relations are
obtained, one for the hard-base solvents and the other for the soft base or
intermediate solvents (Figure 4.3). These relations can be applied to
determine whether a given solvent solvates to a given metal ion as a

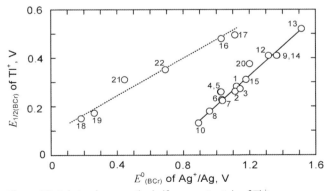

Figure 4.3 Relation between the half-wave potentials of Tl^+
(V versus BCr^+/BCr) and the potentials of Ag^+/Ag electrode
(V versus BCr^+/BCr) in various solvents [3]. Solvents are 20, FA;
21, HMPTA; 22, thiodiethanol. For other solvents, see Figure 4.2.

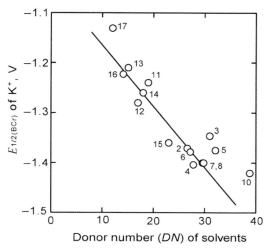

Figure 4.4 Relation between the half-wave potentials of K^+ (V versus BCr^+/BCr) in various organic solvents and the donor number of solvents [3]. For solvents, see Figure 4.2.

hard base or as a soft base. 2,2'-Thiodiethanol ($(HOCH_2CH_2)_2S$, an amphidentate ligand in coordination chemistry) interacts with Tl^+, Zn^{2+}, Cd^{2+} and Pb^{2+} with its sulfur atom (as a soft base), while with Cu^{2+} with its oxygen atom (as a hard base). Dimethyl sulfoxide (DMSO) also has sulfur and oxygen atoms, but it solvates to metal ions with its oxygen atom as a hard base.

(iv) For hard-acid metal ions in hard-base solvents, almost linear relations are observed between the half-wave potentials and the donor number (DN) of the solvents (Figure 4.4), supporting the fact that the DN is the scale for hard bases.

3. $Ox + ne^- \rightleftarrows Red$

Many redox reactions are included in this class. Examples with $n=1$ are $Cu^{2+} + e^- \rightleftarrows Cu^+$, $Fe^{3+} + e^- \rightleftarrows Fe^{2+}$, $Fe(bpy)_3{}^{3+} + e^- \rightleftarrows Fe(bpy)_3{}^{2+}$, $Fc^+ + e^- \rightleftarrows Fc^0$, $BCr^+ + e^- \rightleftarrows BCr^0$, $Q^{\bullet+} + e^- \rightleftarrows Q^0$ and $Q^0 + e^- \rightleftarrows Q^{\bullet-}$, where Fc^0, BCr^+ and Q^0 show ferrocene, bis(biphenyl)chromium(I) and organic compounds, respectively.

If the standard potentials of a redox couple in solvents R and S are expressed by $E^0(R)$ and $E^0(S)$, respectively, then

$$E^0(S) - E^0(R) = \left(\frac{RT}{nF}\right) \ln \frac{\gamma_t(Ox, R \to S)}{\gamma_t(Red, R \to S)}$$

$$= \left(\frac{0.059}{n}\right) \{\log \gamma_t(Ox, R \to S) - \log \gamma_t(Red, R \to S)\}$$

$$[V, 25\,^\circ C] \tag{4.6}$$

When log $\gamma_t(\text{Ox}, \text{R} \rightarrow \text{S}) = \log \gamma_t(\text{Red}, \text{R} \rightarrow \text{S})$, $E^0(\text{S}) = E^0(\text{R})$, i.e. the standard potential does not vary with the solvent. This is considered to be nearly the case for the Fc^+/Fc^0 and $\text{BCr}^+/\text{BCr}^0$ couples, which are used as solvent-independent potential references. However, in general, the standard potential shifts to negative or positive direction with the solvent: $E^0(\text{S}) > E^0(\text{R})$ if $\log \gamma_t(\text{Ox}, \text{R} \rightarrow \text{S}) > \log \gamma_t(\text{Red}, \text{R} \rightarrow \text{S})$ and $E^0(\text{S}) < E^0(\text{R})$ if $\log \gamma_t(\text{Ox}, \text{R} \rightarrow \text{S}) < \log \gamma_t (\text{Red}, \text{R} \rightarrow \text{S})$.

For redox couples of metal ions ($\text{M}^{n+}/\text{M}^{n'+}$ where $n > n'$), the standard potentials usually shift to negative direction with the increase in permittivity or in Lewis basicity of solvents. This is because M^{n+} is usually more stabilized than $\text{M}^{n'+}$ by its transfers from solvent of lower permittivity to that of higher permittivity or from solvent of lower basicity to that of higher basicity. However, there are cases in which the metal ion of lower valency is very strongly solvated. For example, because of the strong solvation of Cu^+ and Fe^{2+} in AN, the standard potentials of $\text{Cu}^{2+}/\text{Cu}^+$ and $\text{Fe}^{3+}/\text{Fe}^{2+}$ couples in AN are much more positive than those expected from the donor number of AN. Thus, Cu^{2+} and Fe^{3+} in AN are strong oxidizing agents and are used in redox titrations (see Section 4.3.1).

Many of the electrophilic (Lewis acid) organic compounds are reduced electrochemically at the cathode or chemically by reducing agents; especially in aprotic solvents that are of very weak acidity, the reaction of the first reduction step is often $Q^0 - e^- \rightleftarrows Q^{\bullet-}$. On the other hand, many of the nucleophilic (Lewis base) organic compounds are oxidized electrochemically at the anode or chemically by oxidizing agents; in solvents of low basicity, the reaction of the first oxidation step is often $Q^0 \rightleftarrows Q^{\bullet+} + e^-$.[3] Here, $Q^{\bullet-}$ and $Q^{\bullet+}$ are the radical anion and cation, respectively, and are extremely reactive if their charges are localized. However, if the charges are more or less delocalized, the radical ions are less reactive. Some radical ions are fairly stable and we can study their characteristics by ESR and UV/vis spectroscopies (Sections 8.3.1, 8.3.2, 9.2.1 and 9.2.2). The potentials of the $Q^0/Q^{\bullet-}$ and $Q^{\bullet+}/Q^0$ couples are usually measured by cyclic voltammetry, employing high-voltage scan rates and low-temperature techniques if necessary (Sections 8.3 and 8.4). The standard potential of the $Q^0/Q^{\bullet-}$ couple tends to shift to positive direction by the strengthening of $Q^{\bullet-}$-solvation, i.e. with increase in solvent permittivity or solvent acidity, while the standard potential of the $Q^{\bullet+}/Q^0$ couple tends to shift to negative direction by the strengthening of $Q^{\bullet+}$-solvation, i.e. with increase in solvent permittivity or solvent basicity (Section 8.3.1).[4] If the effect of solvent permittivity alone is considered, the shift of the standard potential of $Q^0/Q^{\bullet-}$ and

[3] When an organic compound Q^0 is reduced to form its radical anion $Q^{\bullet-}$, it accepts an electron to its lowest unoccupied molecular orbital (LUMO). On the other hand, when Q^0 is oxidized to form a radical cation $Q^{\bullet+}$, it gives an electron from its highest occupied molecular orbital (HOMO). If the reduction and oxidation potentials and the difference between them are measured by cyclic voltammetry, the information on LUMO, HOMO and their differences can be obtained, though the influences of solvation energies must be taken into account (Section 8.3).

[4] These tendencies are understandable from Eq. (4.6). For the $Q^{\bullet+}/Q^0$ couple, the nonelectrostatic part of $\log \gamma_t$ for $Q^{\bullet+}$ is nearly equal to $\log \gamma_t$ of Q^0. For the $Q^0/Q^{\bullet-}$ couple, on the other hand, the nonelectrostatic part of $\log \gamma_t$ for $Q^{\bullet-}$ is nearly equal to $\log \gamma_t$ of Q^0. Thus, the solvent effects on the standard potentials of the two couples are mainly determined by the electrostatic parts of $\log \gamma_t$ for $Q^{\bullet+}$ and $Q^{\bullet-}$.

that of $Q^{\bullet+}/Q^0$ should occur in opposite directions but by equal magnitudes, as expected from the Born equation (Eq. (2.3)). For this reason, the mean of the standard potentials of $Q^0/Q^{\bullet-}$ and $Q^{\bullet+}/Q^0$ is considered to be solvent independent. Actually, such an extrathermodynamic assumption has been proposed (see footnote 9 in Chapter 2).

The second reduction step in aprotic solvents is the formation of dianions, Q^{2-}:

$$Q^{\bullet-} + e^- \rightleftarrows Q^{2-}$$

For aromatic hydrocarbons, the experimental standard potential of the second step is about 0.5–0.8 V more negative than that of the first step. In the gas phase, the standard potential of the second step is expected to be 4–5 V more negative than that of the first step. This difference in the solution and in the gas phase is caused by the fact that the solvation energy of ions depends approximately on the square of the charge number; the solvation energy of the dianion Q^{2-} is four times that of the anion radical $Q^{\bullet-}$, while the solvation energy of the neutral Q^0 is nearly negligible [4]. Moreover, the dianions Q^{2-} are more protophilic (basic) than $Q^{\bullet-}$ and are easily converted to QH_2 (or QH^-), withdrawing protons from solvent or solvent impurities (possibly water), although Q^{2-} with delocalized charges can remain somewhat stable.

With compounds such as 9,10-diphenylanthracene and in protophobic solvents such as AN, the formation of dications has been confirmed from the second oxidation step in cyclic voltammetry (Section 8.3.2).

$$Q^{\bullet+} \rightleftarrows Q^{2+} + e^-$$

Some organic compounds undergo multistep one-electron reduction processes. For example, buckminsterfullerene (C_{60}) in appropriate aprotic solvents is reduced in six steps, each corresponding to a reversible one-electron process: $Q^0 \rightleftarrows Q^{\bullet-} \rightleftarrows Q^{2-} \rightleftarrows Q^{3-} \rightleftarrows Q^{4-} \rightleftarrows Q^{5-} \rightleftarrows Q^{6-}$ (see Figure 8.14) [5]. It is interesting to note that these multicharged anions are stable at least in voltammetric timescale (see footnote 7 in Chapter 8). Here again, the delocalization of the negative charges suppresses the reactivity of the anions.

4. $Ox + mH^+ + ne^- \rightleftarrows Red$

The redox reactions of organic compounds are influenced by the Brønsted acid–base property of the solvents. In protic solvents such as water, many organic compounds (Q^0) are reduced by a one-step two-electron process, $Q^0 + 2H^+ + 2e^- \rightarrow QH_2$, although processes such as $Q^0 + H^+ + 2e^- \rightarrow QH^-$ and $Q^0 + 2e^- \rightarrow Q^{2-}$ also occur occasionally.[5] For the reversible processes, the number of hydrogen ions (m) participating in the reduction can be determined from the pH dependence of the standard potential or the voltammetric half-wave potential. If we add a proton donor (water or weak Brønsted acid) step-wise to the solution of

5) In aprotic solvents, the first reduction step of dissolved oxygen is also a reversible one-electron process to form superoxide ions ($O_2 + e^- \rightleftarrows O_2^{\bullet-}$). However, in aqueous solutions, hydrogen ions participate in the reduction and two- or four-electron reaction occurs as shown by $O_2 + 2H^+ + 2e^- \rightarrow H_2O_2$ or $O_2 + 4H^+ + 4e^- \rightarrow 2H_2O$.

Q^0 in an aprotic solvent and run voltammetric measurements, the reduction mechanism of Q^0 gradually changes from the two one-electron steps in an aprotic environment to the one two-electron step in a protic environment. This gradual change is important in elucidating the role of proton donors in the reduction of organic compounds and to determine the basicity of radical anions (Section 8.3.1). Detailed electrochemical and nonelectrochemical studies on redox reactions of organic compounds and solvent effects on them are discussed in Chapters 8 and 9.

5. Reduction of Metal Complexes

When a metal complex $ML_p^{(n-pb)+}$ is reduced in solvents R and S by $ML_p^{(n-pb)+} + ne^- \rightleftarrows M + pL^b$, the standard potentials in the two solvents can be correlated by Eq. (4.7).

$$E^0(R) - E^0(S) = (RT/nF)\ln \gamma_t(ML_p^{(n+pb)+}, R \rightarrow S) - (pRT/nF)\ln \gamma_t(L^b, R \rightarrow S)$$

(4.7)

Table 4.3 shows the standard potentials of the Ag^+/Ag and AgL^+/Ag electrodes in various aprotic solvents, where L stands for cryptands [6]. The standard potential of the Ag^+/Ag electrode varies with the solvent by more than 500 mV. However, the standard potentials of the AgL^+/Ag electrodes are not much influenced by solvents, the variations being ~80 mV at most. For silver cryptates, AgL^+, the relation log $\gamma_t(AgL^+, R \rightarrow S) \approx \log \gamma_t(L, R \rightarrow S)$ holds, as described in Section 2.4. From this and $p = 1$ in Eq. (4.7), the relation $E^0(R) \approx E^0(S)$ can be expected.

In some cases, the reduction mechanism of a metal complex varies with the solvent. For example, the complex $[Fe(bpy)_3]^{2+}$ (bpy = 2,2'-bipyridine) in aqueous solutions is reduced at a dropping mercury electrode directly to metal iron by a two-electron process $[Fe(II) \rightarrow Fe(0)]$. In aprotic solvents, however, it is reduced in three steps, each corresponding to a reversible one-electron process, and the final product

Table 4.3 Standard potentials of silver cryptates–silver electrodes (AgL^+/Ag) [6].[a]

Solvents S	E^0 (mV versus Ag/0.01 M Ag$^+$(AN))			
	Ag$^+$/Ag	AgL$^+$/Ag; L =		
		cryp(222)	cryp(221)	cryp(211)
DMSO	5	−428	−560	−364
DMA	95	−399	−536	−321
NMP	115	−405	−534	−328
AN	117	−407	−542	−332
PrN	171	−364	−495	−294
DMF	176	−409	−554	−345
BuN	221	−367	−500	−299
Ac	422	−383	−514	−287
TMS	429	−433	−562	−362
PC	532	−428	−567	−350

[a] The potential values contain the liquid junction potentials at 0.1 M Et$_4$NClO$_4$(AN)/0.1 M Et$_4$NClO$_4$(S).

is $[Fe(bpy)_3]^-$, which is relatively stable [7]:

$$[Fe(bpy)_3]^{2+} \overset{+e^-}{\rightleftharpoons} [Fe(bpy)_3]^+ \overset{+e^-}{\rightleftharpoons} [Fe(bpy)_3]^0 \overset{+e^-}{\rightleftharpoons} [Fe(bpy)_3]^- \qquad (4.8)$$

Similar processes also occur with 2,2'-bipyridine and 1,10-phenanthroline complexes of metals such as Co, Cr, Ni and Ru. It is also known from the ESR study that, in the second step of Eq. (4.8), the electron is accepted not by the central metal ion but by the ligand, giving a radical anion of the ligand (see Section 8.2.2). The low-valency complexes are stabilized in aprotic solvents because aprotic solvents are of such weak acidity that they cannot liberate the coordinating ligand and its radical anion from the central metal ion. Aprotic solvents are suitable to study the chemistry of low-valency metal complexes.

4.1.3
Dynamical Solvent Effects on the Kinetics of Redox Reactions

In the past two decades, studies on the kinetics of electron-transfer (ET) processes have made a considerable progress in many chemical and biological fields. Of special interest to us is that the dynamical properties of solvents have remarkable influences on the ET processes that occur either heterogeneously at the electrode or homogeneously in the solution. The theoretical and experimental details of the dynamic solvent effects on ET processes have been reviewed in the literature [8]. The following is an outline of the important role of dynamical solvent properties in ET processes.

Equation (4.9) shows the ET process at the electrode and Eq. (4.10) shows this process in the solution.

$$Ox_1 + electrode(e^-) \rightarrow Red_1 \qquad (4.9)$$

$$Ox_1 + Red_2 \rightarrow Red_1 + Ox_2 \qquad (4.10)$$

The homogeneous self-exchange ET process is the case where $Red_2 = Red_1$ and $Ox_2 = Ox_1$ in Eq. (4.10).

According to the Marcus model, the standard rate constant, k_s, for reaction (4.9) can be expressed as follows if the reaction is an adiabatic outer-sphere process[6]:

$$k_s = \kappa K_p \nu_n \exp\left(\frac{-\Delta G^*}{RT}\right); \qquad \nu_n = \tau_L^{-1}\left(\frac{\Delta G^*}{4\pi RT}\right)^{1/2} \qquad (4.11)$$

or

$$\ln\frac{k_s}{(\Delta G^*)^{1/2}} + \frac{\Delta G^*}{RT} = \ln\frac{\kappa K_p}{(4\pi RT)^{1/2}} + \ln\tau_L^{-1} \qquad (4.11')$$

---

6) In the outer-sphere electrode reaction, the reactant and the product do not interact strongly with the electrode surface, and usually they are at a distance of at least a solvent layer from the electrode; the original configuration of the reactant is nearly maintained in the activated complex. In the inner-sphere electrode reaction, there is a strong interaction (specific adsorption) of the reactant, intermediates or products with the electrode. See Ref. [8d], p. 116.

where κ is the transmission coefficient ($\sim$1); K_p is the pre-equilibrium constant that describes the statistical probability of the formation of an electrode–reactant configuration that is appropriate to the electrode reaction; ν_n is the nuclear frequency factor (in s^{-1}), which represents the frequency of attempts on the barrier; τ_L is the longitudinal solvent relaxation time (Table 1.3); ΔG^* is the activation energy and $\Delta G^* = \Delta G_{is}^* + \Delta G_{os}^*$. ΔG_{is}^* is the inner shell activation energy due to the reorganization of reactant itself, ΔG_{os}^* is the outer shell activation energy due to the reorganization of the solvent around the reactant, and here we assume that $\Delta G_{os}^* \gg \Delta G_{is}^*$. By using the Born model, ΔG_{os}^* can be expressed by Eq. (4.12):

$$\Delta G_{os}^* = \frac{N_A e^2}{32\pi\varepsilon_0}\left(\frac{1}{r} - \frac{1}{2R}\right)\left(\frac{1}{\varepsilon_{op}} - \frac{1}{\varepsilon_s}\right) \tag{4.12}$$

where r is the radius of the reactant, $2R$ is the distance between the reactant and its image charge in the electrode, ε_{op} and ε_s are, respectively, the optical and static permittivities and $(\varepsilon_{op}^{-1} - \varepsilon_s^{-1})$ is the solvent Pekar factor (Table 1.3). The adiabatic ET process is obtained when the interaction between Ox and the electrode is considerable and, as the solid curves in Figure 4.5 show, the splitting in the energy curves at the point of intersection is larger than the kinetic energy of thermal agitation (k_BT). Then, the process from (Ox + e$^-$) to Red occurs with unit probability through the lower solid curve (denoted by $\curvearrowright$). In contrast, the nonadiabatic ET process occurs, as shown by the dashed curve in Figure 4.5, when the splitting at the intersection is smaller than k_BT and Ox may jump to its excited state (denoted by $\nearrow$). For a nonadiabatic process, τ_L^{-1} in Eq. (4.11) should be replaced by $\tau_L^{-\alpha}$ where $0 < \alpha < 1$.

It is apparent from Eq. (4.11) that the solvent effect on k_s has two components, i.e. an energetic component and a dynamic component. Here, the relation between

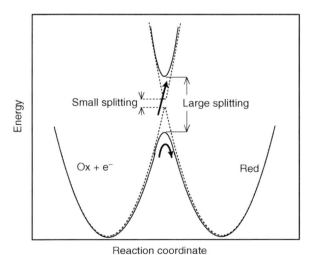

Figure 4.5 Schematic energy-reaction coordinate profiles for symmetrical ET processes having small and large energy splittings at the intersection point.

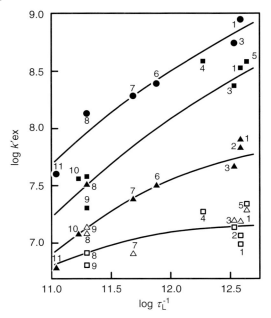

Figure 4.6 Log k'_{ex} versus log (τ_L^{-1}) relations for self-exchange reactions of 5 metallocenes in 11 solvents. τ_L in s and k'_{ex} in $M^{-1}s^{-1}$. Solvents: 1, AN; 2, propionitrile; 3, Ac; 4, D$_2$O; 5, NM; 6, DMF; 7, DMSO; 8, benzonitrile; 9, NB; 10, TMU, 11, HMPA. Filled circles, $Cp'_2Co^{+/0}$ (Cp' = pentamethylcyclopentadienyl); filled squares, $Cp^e_2Co^{+/0}$ (Cp^e = (carboxymethyl) cyclopentadienyl); filled triangles, $Cp_2Co^{+/0}$ (Cp = cyclopentadienyl); open triangles, $Cp_2Fe^{+/0}$; open squares, (hydroxylmethyl) ferrocenium/ferrocene [8a].

$\{\ln[k_s/(\Delta G^*_{os})^{1/2}] + \Delta G^*_{os}/RT\}$ and $\ln \tau_L^{-1}$ (or $-\ln \tau_L$) is expected to be linear, with a slope of unity for an adiabatic process. Moreover, if the dynamic component predominates, a near-linear relation can be expected between log k_s and log τ_L^{-1}. Both of these relations have been confirmed experimentally using the electrode processes of large organic and organometallic compounds, as shown in Sections 8.2.2 and 8.3.1 and in Figure 8.9.

The rate constant, k_{ex}, for a homogeneous self-exchange ET process can also be expressed by an equation similar to Eq. (4.11). In this case, in Eq. (4.12), 32 is replaced by 16 and 2R shows the internuclear distance in the homogeneous-phase precursor state. Here again, the influence of solvent on k_{ex} contains both energetic and dynamic components. The dependence of k_{ex} on solvent dynamics has been studied using k_{ex} values, which are often obtained by ESR line-broadening techniques (Section 9.2.2).[7] Figure 4.6 shows the relation between log k'_{ex} and log τ_L^{-1} for various metallocene couples in 11 Debye solvents, where k'_{ex} was obtained by correcting the measured k_{ex}

---

7) For the relation between the rate constant for homogeneous self-exchange ET process (k_{ex}) and the standard rate constant of the corresponding electrode reaction (k_s), see footnote 4 in Chapter 9.

values for the solvent-dependent ΔG_{os}^*. Although the data are scattered, the dependence of log k'_{ex} on log τ_L^{-1} is apparent for the redox couples having higher log k'_{ex} values.

It is fascinating that the solvent relaxation time from picoseconds to femtoseconds, which is related to solvent reorganization around the reactant (Sections 1.1.1 and 2.2.3), plays an important role in determining the rates of both heterogeneous and homogeneous ET processes.[8] It should be noted, however, that various complicating factors exist in these studies: for example, some correlation is occasionally found between ΔG_{os}^* and $(\tau_L)^{-1}$, making it difficult to discriminate the dynamic contribution from the energetic one [8a]; the ultrafast component of the solvation dynamics may lead to a significant enhancement of the ET rate and to a weakening of the dependence on τ_L [8e].

4.2
Redox Properties of Solvents and Potential Windows

When a small piece of sodium metal is put into water, it reduces water and generates sodium hydroxide and molecular hydrogen that burns over the surface of water. On the other hand, when sodium metal is put into oxygen-free acetonitrile (AN), it remains without reducing AN. The third case is when sodium metal is put into oxygen-free hexamethylphosphoric triamide (HMPA). It gradually dissolves in HMPA, forming bluish-violet solvated electrons and solvated sodium ions. Solvated electrons in water (hydrated electrons) are very unstable, but solvated electrons in HMPA are fairly stable even at room temperatures. It is interesting to note that sodium metal, which is a strong reducing agent, reacts in very different ways in different solvents. This is a result of the difference in acid–base and redox properties of the solvents.

Just as the pH windows are useful in discussing the applicability of solvents as media for acid–base reactions, the potential windows (sometimes called electrochemical windows) are convenient to predict the usefulness of solvents as media for redox reactions. It is desirable that the potential windows are expressed on the basis of a common (solvent-independent) potential scale, like the pH windows based on a common (solvent-independent) pH scale (Figure 3.6).

For water, the reactions that determine the potential window are simple and well defined. The negative end of the potential window is determined by the reduction of water, which generates hydrogen gas and hydroxide ions (Eq. (4.13)). The positive end, on the other hand, is determined by the oxidation of water, which generates oxygen gas and hydrogen ions (Eq. (4.14)).

$$2H_2O + 2e^- \rightleftarrows H_2 + 2OH^- \quad E^0 = -0.81 \text{ V} \tag{4.13}$$

$$2H_2O \rightleftarrows O_2 + 4H^+ + 4e^- \quad E^0 = +1.23 \text{ V} \tag{4.14}$$

8) See footnote 17 of Chapter 8 for the method of determining the rate constant of homogeneous ET process as expressed by Eq. (4.10) by cyclic voltammetry.

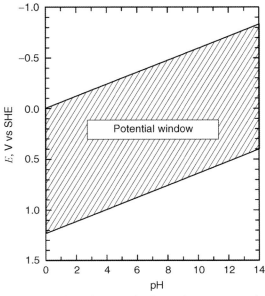

Figure 4.7 Relation between the thermodynamic potential window and pH in aqueous solutions.

As shown in Figure 4.7, the thermodynamic potential window of water depends on pH, but its width is ~1.23 V, irrespective of the pH of the solution. If the entire width of the potential window of water is estimated by the positive potential limit at pH $= 0$ and the negative potential limit at pH $= 14$, it is ~2.06 ($= 1.23 + 14 \times 0.059$) V.

For nonaqueous solvents, it is rare that the thermodynamic potential windows are definitely determined. For most amphiprotic solvents (SH), the negative ends of the potential windows are determined by their reductions, which generate molecular hydrogen and the corresponding lyate ions (S$^-$).

$$2SH + 2e^- \rightleftarrows H_2 + 2S^-$$

For alcohols (ROH) and acetic acid, they are as follows:

$$2ROH + 2e^- \rightleftarrows H_2 + 2RO^-$$

$$2CH_3COOH + 2e^- \rightleftarrows H_2 + 2CH_3COO^-$$

However, their oxidation processes are often complicated, and it is not easy to define the positive ends of the potential windows thermodynamically. In aprotic solvents, both the reduction and the oxidation processes of solvents are complicated, and the definite estimation of thermodynamic potential windows is almost impossible.

For these reasons, it is usual to determine practical potential windows voltammetrically, using appropriate indicator electrodes. A voltammogram is measured in a solvent under study, in the presence of an electroinactive supporting electrolyte. The negative end of the potential window is where the reduction current begins to flow, while the positive end is where the oxidation current begins to flow. However, for the reduction and oxidation of solvents to occur at the electrode, considerable overpotentials are usually needed. Moreover, the magnitudes of overpotentials are seriously influenced by the materials and the surface conditions (activity, roughness, etc.) of the electrode used (Section 8.1). Therefore, the potential windows thus determined are wider than the thermodynamically expected widths and are somewhat indeterminate, unless the conditions of measurements are clearly defined. Sometimes, the supporting electrolyte or the impurity in the solvent is more electroactive than the solvent under study and makes the potential window narrower than that of the pure solvent (Section 11.1.2). Data on potential windows in various solvents have been compiled in various books and review articles [9]. However, the data are fairly scattered and in using such data, the conditions of measurements should be checked carefully.

Figure 4.8 shows the potential windows obtained at a bright platinum electrode based on the Fc^+/Fc (solvent-independent) potential scale. Because of the overpotentials, the window in water is $\sim$3.9 V, which is much wider than the thermodynamic value (2.06 V). The windows for other solvents also contain some overpotentials for the reduction and the oxidation of solvents. However, the general tendency is that the negative potential limit expands to more negative values with the decrease in solvent acidity, while the positive potential limit expands to more

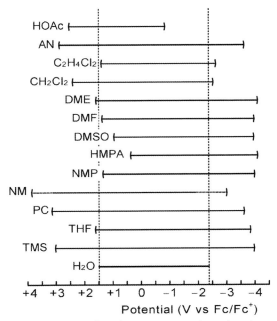

Figure 4.8 Potential windows in various solvents based on a common potential scale (versus Fc^+/Fc). Obtained by voltammetry at a smooth Pt electrode at $10\,\mu A\,mm^{-2}$.

Table 4.4 Merits of solvents with wide potential windows

Solvent with wide negative potential region (weak in acidity)	• Solvent is difficult to reduce • Strong reducing agent can remain in the solvent • Difficult-to-reduce substances can be reduced
Solvent with wide positive potential region (weak in basicity)	• Solvent is difficult to oxidize • Strong oxidizing agent can remain in the solvent • Difficult-to-oxidize substances can be oxidized

positive values with the decrease in solvent basicity. This means that solvents of weak acidity are difficult to reduce, while those of weak basicity are difficult to oxidize. This is in accordance with the fact that the LUMO and HOMO of solvent molecules are linearly related to AN and DN, respectively, of solvents [10].

The potential windows in Figure 4.8 are voltammetric results and include overpotentials. The potential windows applicable in the bulk of solutions may be somewhat narrower. However, the above tendency must also be true in solutions. Thus, in solvents with wide potential windows on the two sides, many redox reactions, which are impossible in water, become possible. They are summarized in Table 4.4.

4.3
Redox Titrations in Nonaqueous Solutions

Redox titrations in nonaqueous solvents are applicable in many fields in the same manner as acid–base titrations in nonaqueous solvents. Various inorganic and organic substances, which cannot be titrated in aqueous solutions, can be titrated in nonaqueous solvents. Examples of redox titrations in nonaqueous solvents have been reviewed by Barek and Berka [11].[9] In this section, redox titrations are briefly outlined, focusing on the oxidizing and reducing agents for nonaqueous solvents. Unlike acid–base reactions, the mechanisms of redox reactions are usually complicated; the electron-transfer processes are often followed by breaking or formation of chemical bonds. Therefore, the following is somewhat descriptive.

4.3.1
Titrations with Oxidizing Agents

Examples of oxidizing agents used in organic solvents are Cu(II), Fe(III), Co(III), Mn(III), Ce(IV), Pb(IV), Cr(VI), Mn(VII), halogens (Cl_2, Br_2 and I_2) and halogen compounds.

9) There are many books and review articles dealing with acid–base titrations in nonaqueous solvents. However, there are only a few review articles covering the redox titrations in nonaqueous solvents, of which Ref. [11] seems to be the most detailed.

Table 4.5 Standard potentials in acetonitrile (25 °C) [12,13].

Reactions	E^0 (V)^a
$Ag^+ + e^- \rightleftarrows Ag$	0.000
$Ag^+ + e^- \rightleftarrows Ag(Hg)$	0.087
$[Ag^+(0.01\ M\ AgNO_3) + e^- \rightleftarrows Ag]$	(−0.131)
$Cu^{2+} + e^- \rightleftarrows Cu^+$	0.679
$Cu^+ + e^- \rightleftarrows Cu$	−0.604
$Cu^+ + e^- \rightleftarrows Cu(Hg)$	−0.594
$Fe^{3+} + e^- \rightleftarrows Fe^{2+}$	1.44

^{a}Values referred to the standard potential of Ag^+/Ag in AN.

In AN, Cu(I) is solvated very strongly but Cu(II) only moderately, resulting in the standard potentials of Cu(II)/Cu(I) and Cu(I)/Cu(0) couples shown in Table 4.5. Cu(II) is a strong oxidizing agent and is reduced to Cu(I) by oxidizing other substances. The standard solution of Cu(II) is prepared either by dissolving $Cu(AN)_4(ClO_4)_2$ in dry AN [12] or by coulometric anodic oxidation of the Cu electrode. The titrations are carried out using redox indicators or by potentiometry, using a platinum indicator electrode and an Ag/Ag^+ reference electrode. Cu(II) is also used in DMF and Py as oxidizing agent. Fe(II), Cr(II) and Ti(III) in DMF can be determined by titration with Cu(II).

When I^- is titrated with $Cu(ClO_4)_2$ in AN, it is oxidized in two steps, $I^- \rightarrow I_3^-$ and $I_3^- \rightarrow I_2$. The formal potentials of the two steps are $+0.396$ and -0.248 V versus Ag/Ag^+, respectively. Many organic compounds, such as hydroquinone, ascorbic acid, ferrocene and its derivatives, allylamine, hydroxylamine, phenylhydrazine, thiourea and SH compounds, can also be titrated with Cu(II) in AN. Figure 4.9 shows the titration curves of tetramethylbenzidine (TMB) in AN [14]. In dry AN, TMB is oxidized in two steps as follows:

$$TMB + Cu(II) \rightarrow TMB\ Green^+ + Cu(I)$$

$$TMB\ Green^+ + Cu(II) \rightarrow TMB\ Orange^{2+} + Cu(I)$$

In AN containing $\sim$0.1 M water, however, the oxidation occurs in three steps. The reactions of the first two steps are the same as those above, and the third step corresponds to the oxidation of $TMB\cdot H^+$, which is produced in the presence of water. There are cases in which the titration curves are improved by the presence of water. Figure 4.10 shows the titration curves of hydroquinone (HQ). Only in the presence of $\sim$0.1 M water can the end point be detected [15]. Water acts as the acceptor of the proton, which is generated as a result of the oxidation of HQ:

$$Hydroquinone + 2Cu(II) + 2H_2O \rightleftarrows Quinone + 2Cu(I) + 2H_3O^+$$

The titration with Cu(II) is also influenced by the anions of electrolytes because the potential of the Cu(II)/Cu(I) couple changes by the complexation of anions with Cu(II) and Cu(I).

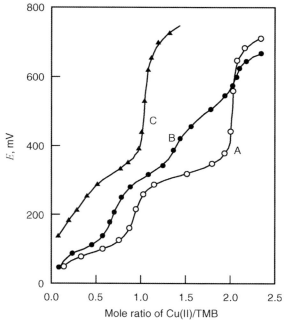

Figure 4.9 Potentiometric titration of TMB with Cu(II) in acetonitrile [14]. Curve A in anhydrous AN, B in the presence of ~0.1 M H_2O and C in the presence of $HClO_4$.

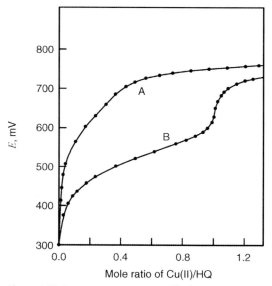

Figure 4.10 Potentiometric titration of hydroquinone with Cu(II) in acetonitrile [15]. Curve A in anhydrous AN and B in the presence of ~0.1 M H_2O.

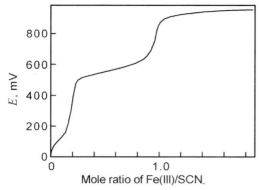

Figure 4.11 Potentiometric titration of KSCN with 0.1 M solution of hydrated Fe(III) in acetonitrile [13].

The formal potential of the Fe(III)/Fe(II) couple in AN is $+1.57$ V versus Ag/0.01 M Ag$^+$(AN), about 1.3 V more positive than in aqueous solutions on a common (solvent-independent) potential scale. This is because Fe(II) in AN is solvated strongly, Fe(III) only moderately. Fe(III) in AN is a strong oxidizing agent. Both Fe(III) and Fe(II) react with moisture in the air and are easily hydrated. The potential of the hydrated Fe(III)/Fe(II) couple in AN is $+1.1$ V versus Ag/0.01 M Ag$^+$(AN). It is possible to prepare an anhydrous Fe(ClO$_4$)$_3$ solution by the anodic oxidation of the absolutely anhydrous Fe(ClO$_4$)$_2$ solution. However, in normal titrations, the standard solution of hydrated Fe(III) is prepared by dissolving Fe(ClO$_4$)$_3$·6H$_2$O in AN.

Fe(III) has been used to titrate SCN$^-$, I$^-$ and various organic compounds in AN. Figure 4.11 shows the curve obtained by titrating SCN$^-$ in AN with hydrated Fe(III) [13]. The first step is for the reaction between Fe(III) and SCN$^-$, forming complexes [Fe(SCN)$_6$]$^{3-}$ and [Fe(SCN)$_5$OH]$^{3-}$. The second step is for the reaction $2SCN^- + 2Fe(III) \rightarrow (SCN)_2 + 2Fe(II)$.

4.3.2
Titrations with Reducing Agents

Examples of reducing agents in nonaqueous solvents are Cr(II), Sn(II), Fe(II), Ti(III), ascorbic acid and organic radical anions. In DMF containing 0.05 M HCl, the potential of Cr(III)/Cr(II) is -0.51 V, that of Ti(IV)/Ti(III) is 0.00 V and that of Fe(III)/Fe(II) is 0.405 V (versus aqueous SCE). Although Cr(II) is a strong reducing agent, the titration with it must be carried out under nitrogen atmosphere. Ti(III) is used in DMF to titrate inorganic substances such as Cu(II), Fe(III), Sb(V), I$_2$ and Br$_2$ and organic substances such as azo dyes, quinones, nitro and nitroso compounds and oximes. Coulometric titrations, in which Ti(II) is generated by electroreduction of TiCl$_4$, have also been carried out [16].

The radical anions of organic compounds are strong reducing agents and applicable as titrants. The standard solutions of radical anions are prepared by reducing the organic compounds in aprotic solvents, either with sodium metal or by cathodic reductions. However, because the radical anions usually react with oxygen, the titrations must be carried out under oxygen-free atmosphere. Coulometric titrations are convenient. For example, coulometric titrations of such organic compounds as anthracene, nitromethane, nitrobenzene, benzophenone and azobenzene have been carried out using biphenyl radical anions, which were coulometrically generated in DMF, THF and DME. The titration reactions are rapid and the end point can be detected either potentiometrically or visually using the color of the radical anions.

References

1 Bard, A.J., Parsons, R. and Jordan, J. (eds) (1985) *Standard Potentials in Aqueous Solutions*, Marcel Dekker, New York.

2 Mussini, T., Longhi, P. and Rondinini, S. (1985) *Pure Appl. Chem.*, **57**, 169.

3 Gritzner, G. (1986) *J. Phys. Chem.*, **90**, 5478.

4 Evans, D.H. (2008) *Chem. Rev.*, **108**, 2113.

5 Echegoyen, L.E., Herranz, M.Á. and Echegoyen, L. (2006) *Bard-Stratmann Encyclopedia of Electrochemistry*, vol. 7a *Inorganic Chemistry*, (eds F. Scholz and C.J. Pickett), Wiley-VCH Verlag GmbH, Weinheim, p. 145.

6 Lewandowski, A. (1989) *J. Chem. Soc., Faraday Trans. 1*, **85**, 4139.

7 (a) Tanaka, N. and Sato, Y. (1966) *Inorg. Nucl. Chem. Lett.*, **2**, 359; (b) Tanaka, N. and Sato, Y. (1968) *Electrochim. Acta*, **13**, 335; (c) Tanaka, N. and Sato, Y. (1968) *Bull. Chem. Soc. Jpn.*, **41**, 2059, 2064; 1969, **42**, 1021.

8 For example (a) Weaver, M.J. and McManis, G.E. III (1990) *Acc. Chem. Res.*, **23**, 294; Weaver, M.J. (1992) *Chem. Rev.*, **92**, 463; (b) Galus, Z. (1995) *Advances in Electrochemical Science and Engineering*, vol. 4 (eds H. Gerischer and C.W. Tobias) Wiley-VCH Verlag GmbH, Weinheim, pp. 217–295; (c) Miller, C.J. (1995) *Physical Electrochemistry, Principles, Methods, and Applications* (ed. I. Rubinstein), Marcel Dekker, New York, Chapter 2; (d) Bard, A.J. and Faulkner, L.R. (2001) *Electrochemical Methods: Fundamentals and Applications*, John Wiley & Sons, Inc., New York, pp. 115–132; (e) Bagchi, B. and Gayathri, N. (1999) *Electron Transfer: From Isolated Molecules to Biomolecules, Part 2, Advances in Chemical Physics*, vol. 107 (eds J. Jortner and E. Bixon), John Wiley & Sons, Inc., New York, pp. 1–80.

9 (a) Marcus, Y. (1998) *The Properties of Solvents*, John Wiley & Sons, Inc., New York, pp. 188–197; (b) Gritner, G. (1990) *Pure Appl. Chem.*, **62**, 1839; (c) Badoz-Lambling, J. and Cauquis, G. (1974) *Electroanalytical Chemistry*, 2nd edn (ed. H.W. Nürnberg), John Wiley & Sons, Inc., New York, Chapter 5; (d) Ue, M., Ida, K. and Mori, S. (1994) *J. Electrochem. Soc.*, **141**, 2989 (data at glassy carbon electrode).

10 Sabatino, A., LaManna, G. and Paolini, I. (1980) *J. Phys. Chem.*, **84**, 2641.

11 Barek, J. and Berka, A. (1984) *CRC Crit. Rev. Anal. Chem.*, **15**, 163.

12 Senne, I.K. and Kratochvil, B. (1972) *Anal. Chem.*, **44**, 585.

13 Kratochvil, B. and Long, R. (1970) *Anal. Chem.*, **42**, 43.

14 Kratochvil, B. and Zatko, D.A. (1968) *Anal. Chem.*, **40**, 422.

15 Kratochvil, B., Zatko, D.A. and Markuszewski, R. (1966) *Anal. Chem.*, **38**, 770.

16 (a) Bufatina, M.A., Abdullin, I.F. and Budnikov, G.K. (1991) *Zh. Anal. Khim.*, **46**, 139; (b) Bufatina, M.A., Abdullin, I.F. and Budnikov, G.K. (1991) *J. Anal. Chem. USSR*, **46**, 105.

Part Two
Electrochemical Techniques and Their Applications in Nonaqueous Solutions

Electrochemistry in Nonaqueous Solutions, Second, Revised and Enlarged Edition. Kosuke Izutsu
Copyright © 2009 WILEY-VCH Verlag GmbH & Co. KGaA, Weinheim
ISBN: 978-3-527-32390-6

5
Overview of Electrochemical Techniques

Electrochemistry deals with the phenomena associated with charge separation and charge transfer, which occur homogeneously in solutions or heterogeneously at electrode/solution interfaces. Electrochemistry has a long history and began more than 200 years ago with the invention of Volta's electric pile in 1799. But progress in recent years has been remarkable. Today, electrochemistry plays important roles in developing new areas of science and technology and makes essential contributions to solving global energy and environmental problems.

Electroanalytical chemistry, on the other hand, deals with methods of measuring electrochemical phenomena and their applications to chemical analyses. The first analytical application of electrochemistry was in 1801 by Cruikshank, who electrolyzed solutions of copper and silver salts and suggested that the electrolytic deposits could serve as a means of identifying those metals. Since then, a great variety of electrochemical techniques have been developed in the field of electroanalytical chemistry. Most of the traditional techniques are based on the measurement of electrolytic currents and/or electrode potentials. During the last two decades, however, a new generation of electroanalytical chemistry has begun, mainly by skillful combinations of electrochemical techniques and various nonelectrochemical techniques of chemistry, physics and biology.

In Part II, we deal with various electrochemical techniques and show how they are applicable in nonaqueous solutions. This chapter gives an overview of electrochemical techniques, from the principles of basic techniques to some recent developments. It will help readers from nonelectrochemical fields to understand the latter chapters of Part II. Many books are available to readers who want to know more about electrochemical techniques [1]. In particular, the excellent book by Bard and Faulkner [1a] provides the latest information on all important aspects of electroanalytical chemistry.

5.1
Classification of Electrochemical Techniques

In order to give an overview of the many electrochemical techniques, it is convenient to classify them. Table 5.1 is an example of such a classification. All electrochemical techniques use some kind of electrode. However, they are divided into two groups:

Electrochemistry in Nonaqueous Solutions, Second, Revised and Enlarged Edition. Kosuke Izutsu
Copyright © 2009 WILEY-VCH Verlag GmbH & Co. KGaA, Weinheim
ISBN: 978-3-527-32390-6

Table 5.1 Classification of electrochemical techniques.

I Methods based on electrode reactions
   A Methods that electrolyze the electroactive species under study completely
     Electrogravimetry; coulometry
   B Methods that electrolyze the electroactive species under study only partially
     Polarography and voltammetry (DC, AC, SW, pulse methods for each); amperometry;
     chronopotentiometry; polarography and voltammetry at the interface between two immiscible
     electrolyte solutions (ITIES)
   C Methods that do not electrolyze the electroactive species under study
     Potentiometry; tensammetry
II Methods not based on electrode reactions
     Conductimetry; high frequency method

---

the methods in group I are based on electrode reactions but those in group II are not. The methods in group I are further divided into three subgroups A, B, and C. The electroactive species under study is completely (quantitatively) electrolyzed in the methods in group A, only partially (~0.1% or less) electrolyzed in the methods in group B and not electrolyzed at all in the methods in group C. It is apparent from Table 5.1 that there are many more techniques based on electrode reactions. Therefore, before outlining each of the electrochemical techniques, we begin with a discussion of the fundamentals of electrode reactions.

5.2
Fundamentals of Electrode Reactions and Current–Potential Relations

In many electrochemical techniques, we measure current–potential curves for electrode reactions and obtain useful information by analyzing them. In other techniques, though we do not actually measure current–potential curves, the current–potential relations at the electrodes are the basis of the techniques. Thus, in this section, we briefly discuss the current–potential relations at the electrode.

We consider an electrode reaction as in Eq. (5.1):

$$Ox + ne^- \rightleftarrows Red \qquad (5.1)$$

Here, both Ox (oxidized form) and Red (reduced form) exist in the solution and their concentrations in the bulk of the solution are C_{Ox} and C_{Red} (mol cm^{-3}), respectively. The potential of the electrode E (V) can be controlled by an external voltage source connected to the electrolytic cell (Figure 5.1). The electrode reaction usually consists of the following three processes:

1. Transport of the electroactive species from the bulk of the solution to the electrode surface.
2. Electron transfer (ET) at the electrode.
3. Transport of the reaction product from the electrode surface to the bulk of the solution.

The current–potential relation for process (2) and that for all three processes are discussed in the following sections.

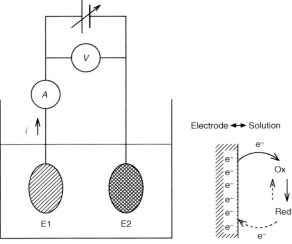

Electrode ◀▶ Solution

Figure 5.1 An electrolytic cell (left) and an electron-transfer reaction at the electrode (right). E1, working or indicator electrode, at which the reaction under study occurs; E2, counter electrode with a constant potential; A, current measuring device; V, voltage measuring device.

5.2.1
Current–Potential Relation for Electron Transfer at the Electrode

Two processes can occur at the electrode surface: one is the process in which Ox is reduced to Red by accepting electron(s) from the electrode and the other is the process in which Red is oxidized to Ox by donating electron(s) to the electrode. If we express the rate of reduction, i.e. the amount of Ox reduced per unit surface area and per unit time, by v_f (mol cm^{-2} s^{-1}) and the rate of oxidation by v_b (mol cm^{-2} s^{-1}), they are given by Eqs. (5.2) and (5.3):

$$v_f = k_f C_{Ox}^0 = k_{f0} C_{Ox}^0 \exp\left(-\alpha \frac{nFE}{RT}\right) \tag{5.2}$$

$$v_b = k_b C_{Red}^0 = k_{b0} C_{Red}^0 \exp\left\{(1-\alpha)\frac{nFE}{RT}\right\} \tag{5.3}$$

where k_f and k_b are the rate constants (cm s^{-1}) at potential E, k_{f0} and k_{b0} are the rate constants (cm s^{-1}) at $E = 0$, α is the transfer coefficient $(0 < \alpha < 1)$[1] and C_{Ox}^0 and C_{Red}^0 are the concentrations (mol cm^{-3}) of Ox and Red at the electrode surface.

1) The transfer coefficient α is a measure of the symmetry of the energy barrier in the reaction coordinate and is often between 0.3 and 0.7. Here, we simply consider that when the electrode potential changes by ΔE to negative (positive) direction, the fraction $\alpha \Delta E$ works to lower (heighten) the energy barrier for the forward process, while the fraction $(1 - \alpha)\Delta E$ works to heighten (lower) that for the backward process.

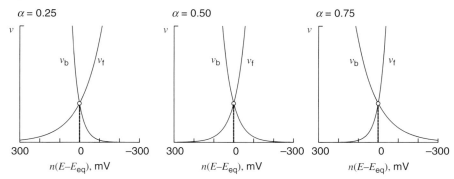

Figure 5.2 Dependence of v_f and v_b on electrode potential E and transfer coefficient α. E_{eq} is the equilibrium potential.

Equations (5.2) and (5.3) show that k_f and k_b depend on the electrode potential and the value of α. Thus, v_f and v_b also depend on the electrode potential and α as shown in Figure 5.2. It is noted that, in some potential regions, the oxidation and reduction processes occur simultaneously. However, the net reaction is a reduction if $v_f > v_b$ and an oxidation if $v_f < v_b$.

When $v_f = v_b$, the electrode reaction is in equilibrium and no change occurs overall. The potential at which the reaction is in equilibrium is the *equilibrium potential*. If we express it by E_{eq} (V), we get the following Nernst equation from Eqs. (5.2) and (5.3):

$$E_{eq} = E^{0\prime} + \frac{RT}{nF} \ln \frac{C_{Ox}}{C_{Red}}$$

where bulk concentrations, C_{Ox} and C_{Red}, are used because the surface concentrations are equal to the bulk concentrations at the equilibrium potential. $E^{0\prime}$ (V) is the formal potential and is expressed by

$$E^{0\prime} = \frac{RT}{nF} \ln \frac{k_{f0}}{k_{b0}}$$

$E^{0\prime}$ is equal to the equilibrium potential at $C_{Ox} = C_{Red}$ and is related to the standard potential E^0 by $E^{0\prime} = E^0 + (RT/nF)\ln(\gamma_{Ox}/\gamma_{Red})$. The rate constants for reduction (k_f) and oxidation (k_b) are equal to each other at $E^{0\prime}$. The rate constant at $E^{0\prime}$ is often called the *standard rate constant* and denoted by k_s in this book.

If we convert v_f and v_b to the corresponding electric currents, i_f and i_b (A):

$$i_f = nFAv_f \quad \text{and} \quad i_b = -nFAv_b$$

A is the electrode surface area (cm^2) and, in this book, a positive current value is assigned to the reduction current and a negative current value to the oxidation current.

In practice, i_f and i_b are not measurable separately and the net current i (A), which flows through the external circuit, is measured.

$$i = i_f + i_b$$

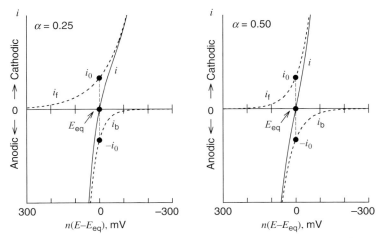

Figure 5.3 Current–potential relations for electrode reactions with $\alpha = 0.25$ (left) and 0.50 (right).

Figure 5.3 shows the current–potential relations for $\alpha = 0.25$ and 0.50.[2] At $E = E_{eq}$, the net current (i) is equal to zero but currents of the same magnitudes (i_0) flow in opposite directions. i_0 is called the *exchange current*. The net current is an oxidation current *(anodic current)* at $E > E_{eq}$ and a reduction current *(cathodic current)* at $E < E_{eq}$.

When the exchange current (i_0) is large enough, the electrode potential is very near to the equilibrium potential (E_{eq}), even when a given current i_c flows at the electrode (see η in Figure 5.4a). When the exchange current is very small, however, the electrode potential must deviate considerably from E_{eq} (to the negative side for reduction and to the positive side for oxidation) for the current i_c to flow (see η in Figure 5.4b). The difference between the equilibrium potential $(i = 0)$ and the potential at a given current value $(i = i_c)$ is called *overpotential* and is denoted by η. As is evident from Figure 5.4, η depends on the value of i_c.

When the overpotential (η) is very small (e.g. <5 mV) for all the current region to be measured, the electrode process is said to be *reversible*. For a reversible electrode process, the Nernst equation almost applies with respect to the surface concentrations of Ox and Red, even under the flow of current (Eq. (5.4)).

$$E = E^{0'} + \frac{RT}{nF} \ln \frac{C_{Ox}^0}{C_{Red}^0} \tag{5.4}$$

On the other hand, if the exchange current is very small and a large overpotential is needed for the current to flow, the electrode process is said to be *irreversible*. In some cases, the electrode process is reversible (the overpotential is small) for small current

2) From the current–potential relation, we can determine the current at a given potential and, inversely, the potential at a given current.

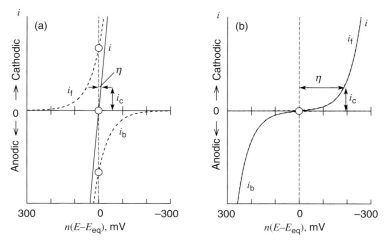

Figure 5.4 Current–potential relations for electrode reactions with (a) a large exchange current (small overpotential) and (b) a small exchange current (large overpotential).

values but irreversible (the overpotential is large) for large current values. Such a process is said to be *quasireversible*.

5.2.2
Current–Potential Relations and Mass Transport

In the above, we assumed that the surface concentrations C_{Ox}^0 and C_{Red}^0 do not depend on the current that flows at the electrode. Then, the reduction current increases exponentially to infinity with the negative shift of the potential, while the oxidation current tends to increase exponentially to infinity with the positive shift of the potential (Figure 5.3). However, in reality, such infinite increases in current do not occur. For example, when a reduction current flows, Ox at the electrode surface is consumed to generate Red and the surface concentration of Ox becomes lower than that in the bulk of the solution. Then, a concentration gradient is formed near the electrode surface and Ox is transported from the bulk of the solution toward the electrode surface. Inversely, the surface concentration of Red becomes higher than that in the bulk of the solution and Red is transported away from the electrode surface. Here, the rate of mass transport is not infinite. If we consider, for simplicity, a Nernst diffusion layer of a thickness δ (Figure 5.5) and express the diffusion coefficients of Ox and Red by D_{Ox} and D_{Red}, respectively, we get Eqs. (5.5) and (5.6):

$$i = \frac{nFAD_{Ox}(C_{Ox} - C_{Ox}^0)}{\delta_{Ox}} = K_f(C_{Ox} - C_{Ox}^0) \tag{5.5}$$

$$i = \frac{nFAD_{Red}(C_{Red}^0 - C_{Red})}{\delta_{Red}} = K_b(C_{Red}^0 - C_{Red}) \tag{5.6}$$

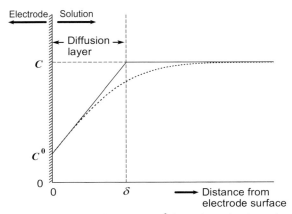

Figure 5.5 Nernst diffusion layer. C^0 depends on the electrode potential (see text).

Here, $K_f = nFAD_{Ox}/\delta_{Ox}$ and $K_b = nFAD_{Red}/\delta_{Red}$, and K_f and K_b are constant if δ is constant.[3] From Eq. (5.4), which applies to a reversible electrode process, C^0_{Ox} approaches zero at $E \ll E_{eq}$, while C^0_{Red} approaches zero at $E \gg E_{eq}$. If we express the currents under these conditions by i_{fl} and i_{bl}, respectively,

$$i_{fl} = K_f C_{Ox} \quad (\text{for} \quad C^0_{Ox} \rightarrow 0) \tag{5.7}$$

$$i_{bl} = -K_b C_{Red} \quad (\text{for} \quad C^0_{Red} \rightarrow 0) \tag{5.8}$$

In the subscripts fl and bl, 'l' means 'limiting' (see below). If we get C^0_{Ox} and C^0_{Red} from Eqs. (5.5)–(5.8) and insert them into Eq. (5.4), we get Eq. (5.9) as the current–potential relation for a reversible process:

$$E = E_{1/2} + \frac{RT}{nF} \ln \frac{i_{fl} - i}{i - i_{bl}} \tag{5.9}$$

where

$$E_{1/2} = E^{0'} + \frac{RT}{nF} \ln \frac{K_b}{K_f} \tag{5.10}$$

The current–potential relation in Eq. (5.9) is shown by curves 1–3 in Figure 5.6. Curve 2 is for $C_{red} = 0$ and curve 3 is for $C_{Ox} = 0$. The curves are S-shaped and the currents at potentials negative enough and positive enough are potential independent, being equal to i_{fl} and i_{bl}, respectively. These currents are called *limiting*

[3] If the thickness of the diffusion layer (δ) is time independent, a steady limiting current is obtained. However, if the electrode reaction occurs at a stationary electrode that is kept at a constant potential, the thickness of the diffusion layer increases with time by the relation $\delta \sim (\pi Dt)^{1/2}$, where t is the time after the start of reaction. Then, the limiting current is proportional to $t^{-1/2}$ and decreases with time. To keep the value of δ time independent, the electrode must be rotated or the solution must be stirred. See, however, the case of a dropping mercury electrode (Section 5.3) and that of an ultramicroelectrode (Section 5.5.4).

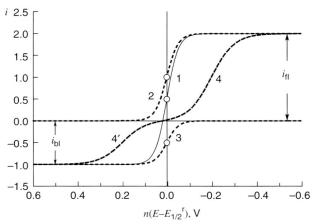

Figure 5.6 Current–potential curves for the processes controlled by the mass transport of electroactive species. Curves 1 to 3 are for reversible processes in solutions containing (1) both Ox and Red, (2) Ox only and (3) Red only. Curves 4 and 4′ are for an irreversible reduction (curve 4) and oxidation (curve 4′) in the solution containing both Ox and Red. $E^r_{1/2}$ is the half-wave potential for the reversible process.

currents and are proportional to the bulk concentrations of Ox and Red, respectively (Eqs. (5.7) and (5.8)). The potential at $i = (i_{fl} + i_{bl})/2$ is equal to $E_{1/2}$ in Eq. (5.10) and is called the *half-wave potential*. Apparently from Eq. (5.10), the half-wave potential is independent of the concentrations of Ox and Red and is almost equal to the standard redox potential E^0, which is specific to each redox system. From the fact that the limiting current is proportional to the concentration of the electroactive species and that the half-wave potential is specific to the redox system under study, the current–potential relation can be used both in quantitative and qualitative analyses.

Curves 4 and 4′ in Figure 5.6 show an example of the current–potential relation obtained for an irreversible electrode process. For a reversible electrode process, the reduction wave appears at the same potential as the oxidation wave, giving an oxidation–reduction wave if both Ox and Red exist in the solution (curves 1, 2 and 3 in Figure 5.6). For an irreversible process, however, the reduction wave (curve 4) is clearly separated from the oxidation wave (curve 4′), although the limiting currents for the two waves are the same as those in the reversible process. The current–potential relation for the irreversible reduction process can be expressed by

$$E = E_{1/2} + \frac{RT}{\alpha n F} \ln \frac{i_{fl} - i}{i} \tag{5.11}$$

$$E_{1/2} = \frac{RT}{\alpha n F} \ln \frac{\delta_{Ox} k_{f0}}{D_{Ox}} = E^{0'} + \frac{RT}{\alpha n F} \ln \frac{\delta_{Ox} k_s}{D_{Ox}} \tag{5.12}$$

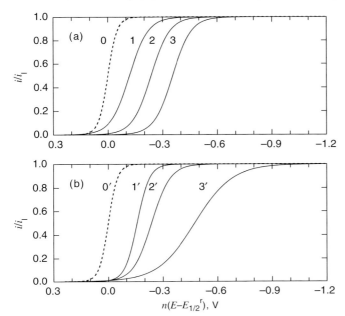

Figure 5.7 Current–potential relations for an irreversible electrode reaction. (a) The influence of reaction rate constant. k_s at $E^{0'}$: curve 1, 10^{-3}; 2, 10^{-4} and 3, 10^{-5} cm s^{-1} ($\alpha = 0.50$). (b) The influence of transfer coefficient. α: curve 1', 0.75; 2', 0.50 and 3', 0.25 ($k_s = 10^{-4}$ cm s^{-1}). (For the case $n = 1$, $\delta = 10^{-3}$ cm and $D_{Ox} = 10^{-5}$ cm^2 s^{-1}. Dashed curves 0 and 0' are for a reversible electrode process.)

where k_s is the rate constant at the formal potential $E^{0'}$. On the other hand, the current–potential relation for the irreversible oxidation process can be expressed by

$$E = E_{1/2} + \frac{RT}{(1-\alpha)nF}\ln\frac{i}{i_{bl}-i} \tag{5.13}$$

$$E_{1/2} = \frac{RT}{(1-\alpha)nF}\ln\frac{D_{Red}}{\delta_{Red}k_{b0}} = E^{0'} + \frac{RT}{(1-\alpha)nF}\ln\frac{D_{Red}}{\delta_{Red}k_s} \tag{5.14}$$

As is apparent from Eqs. (5.12) and (5.14), the half-wave potentials for irreversible processes are independent of the concentrations of Ox and Red, but they are influenced by the values of transfer coefficient (α) and reaction rate constant (k_s). Figure 5.7 illustrates the influences of such parameters.

5.3
DC Polarography – Methods that Electrolyze Electroactive Species Only Partially (1)

Polarography (now called DC polarography) was invented by J. Heyrovský in 1922. In DC polarography, a dropping mercury electrode (DME) is used as an indicator

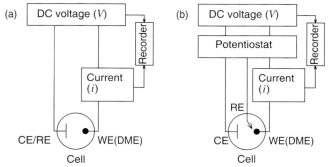

Figure 5.8 Schematic diagram of polarographic (or voltammetric) circuits for two-electrode (a) and three-electrode (b) systems. WE (DME): indicator or working electrode (dropping mercury electrode in the case of polarography); RE: reference electrode; CE: counter electrode; DC voltage (*V*): DC voltage source; current (*i*): current measuring device.

electrode and current–potential curves are measured to obtain various information from them [2]. New types of polarography and voltammetry, which were developed on the basis of DC polarography, are now available as useful electroanalytical techniques (Sections 5.4 and 5.5). However, we outline DC polarography here because it is still the most fundamental among the methods that measure current–potential relations at the indicator electrode. A history of DC polarography can be found in [3].

The apparatus for DC polarography usually consists of three parts, i.e. the circuit to control the potential of the indicator electrode (DME), the circuit to measure the electrolytic current, and the electrolytic cell. Classically, two-electrode devices, as shown in Figure 5.8a, were used, but now three-electrode devices as shown in Figure 5.8b are predominant. In three-electrode devices, the electrolytic cell is equipped with three electrodes: a DME, a reference electrode and a counter electrode (Figure 5.9). The drop time of the modern DME is controlled mechanically and is usually between 0.1 and 3 s. The potential of the DME is accurately controlled by a potentiostatic circuit, even when electric current flows through the cell. The electrolytic solution usually contains a supporting electrolyte and an electroactive species that is reduced or oxidized at the electrode.

Figure 5.10 shows current–potential curves (*polarograms*) obtained in 0.1 M HCl (curve 1) and in 0.1 M HCl + 1.0 mM Cd^{2+} (curve 2). Here, 0.1 M HCl is the supporting electrolyte and 1.0 mM Cd^{2+} is the electroactive species. In Figure 5.10, the polarograms were measured without damping the current oscillation, and the current at the end of drop-life is measured.[4] In curve 1, the anodic current due to the oxidation (anodic dissolution) of the Hg drop begins to flow at +0.1 V (versus

---

4) In *tast polarography*, only the current at the end of each drop is sampled and recorded on the polarogram. Then, the current oscillation as in Figure 5.10 does not occur.

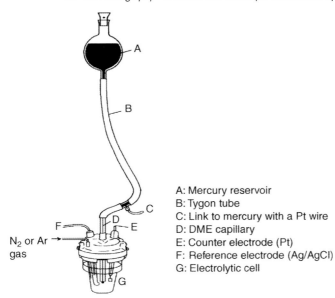

A: Mercury reservoir
B: Tygon tube
C: Link to mercury with a Pt wire
D: DME capillary
E: Counter electrode (Pt)
F: Reference electrode (Ag/AgCl)
G: Electrolytic cell

N₂ or Ar gas

Figure 5.9 A cell for polarographic measurements [1e].

Ag/AgCl electrode) and the cathodic current due to the reduction of hydrogen ion begins to flow at about −1.1 V. Between the two potential limits, only a small current (*residual current*) flows. In curve 2, there is an S-shaped step due to the reduction of Cd^{2+}, i.e. $Cd^{2+} + 2e^- + Hg \rightleftarrows Cd(Hg)$. In DC polarography, the

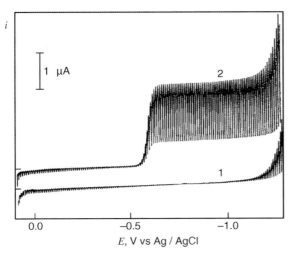

E, V vs Ag / AgCl

Figure 5.10 An example of a DC polarogram recorded without damping. Polarogram 1 in 0.1 M HCl supporting electrolyte and 2 in 1.0 mM Cd^{2+} + 0.1 M HCl.

current–potential curve for the electrode reaction is usually S-shaped and is called a *polarographic wave*.

The limiting current is proportional to the concentration of the electroactive species, whereas the half-wave potential is specific to the electroactive species, being close to the standard potential of the electrode reaction. Thus, by measuring polarographic waves, we can run qualitative and quantitative analyses. In DC polarography, many inorganic and organic substances (ions, complexes and molecules), which are reduced or oxidized at the DME, can be determined at concentration ranges between 5×10^{-6} and 10^{-2} M. Moreover, when several kinds of electroactive species coexist in the solution, they give step-wise waves and can be determined simultaneously (Figure 5.11). The DME has two major advantages: one is the extraordinarily large hydrogen overpotential, which widens the cathodic limit of the potential window (for the anodic potential limit, see p. 129); the other is that, due to the continuous renewal of the Hg drop, the DME needs no pretreatment. With solid electrodes, the surface must be pretreated chemically or by mechanical polishing. However, the DME must be used with great care because of the high toxicity of mercury.

The reversible polarographic waves for the reduction and oxidation processes can be expressed by Eqs. (5.15) and (5.16), respectively:

$$\text{Reduction wave}: \quad E = E_{1/2} + \frac{RT}{nF} \ln \frac{i_d - i}{i} \left(= E_{1/2} + \frac{0.0592}{n} \log \frac{i_d - i}{i}, \ 25\,^\circ\text{C} \right)$$

$$(5.15)$$

$$\text{Oxidation wave}: \quad E = E_{1/2} - \frac{RT}{nF} \ln \frac{i_d - i}{i} \left(= E_{1/2} - \frac{0.0592}{n} \log \frac{i_d - i}{i}, \ 25\,^\circ\text{C} \right)$$

$$(5.16)$$

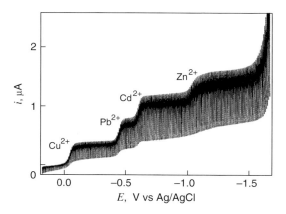

Figure 5.11 An example of DC polarogram for a multicomponent system. Electroactive species: Cu^{2+}, Pb^{2+}, Cd^{2+} and Zn^{2+} in 1.0 mM each; supporting electrolyte: 0.1 M tartaric acid + 0.5 M ammonium acetate.

where i_d is the limiting diffusion current.[5] When the current at the end of a drop-life is measured (Figure 5.10), the limiting diffusion current $(i_d)_{max}$ (μA) is given by the following Ilkovic equation:

$$(i_d)_{max} = (\pm)708nD^{1/2}Cm^{2/3}\tau^{1/6} \tag{5.17}$$

Here, D is the diffusion coefficient in $cm^2\,s^{-1}$, C is the concentration in mM, m is the mercury flow rate in $mg\,s^{-1}$, τ is the lifetime of the mercury drop in s and $(+)$ in $(\pm)$ is for reduction and $(-)$ is for oxidation. Equation (5.17) is the basis of quantitative polarographic analysis, showing that $(i_d)_{max}$ is proportional to the concentration of the electroactive species. Moreover, it is used to obtain the values of diffusion coefficient.

Equations (5.15) and (5.16) show that, for reversible processes, the $\log[(i_d - i)/i]$ versus E relation is linear and its slope is $(59.2/n)$ mV at 25 °C. We can confirm the reversibility of the electrode reaction if these criteria are satisfied experimentally; we can get the value of n from the slope if the reaction is known to be reversible. This procedure is called *wave analysis*.

We can obtain various physicochemical information from the half-wave potentials of reversible polarographic waves, as described below:

1. The half-wave potential $E_{1/2}$ (at 25 °C) for the process $Ox + ne^- \rightleftarrows Red$ is given by

$$E_{1/2} = E^{0\prime} + \frac{0.0592}{n}\log\left(\frac{D_{Red}}{D_{Ox}}\right)^{1/2} (\approx E^{0\prime})$$

From this relation, we can obtain the formal redox potential ($E^{0\prime}$) accurately by taking the second term on the right-hand side into account or approximately by neglecting it. In order to get $E^{0\prime}$ by potentiometry, both Ox and Red must be fairly stable in the solution, even when one of them is generated by redox titration. In the polarographic method, Ox or Red is contained in the solution and the other is generated *in situ* at the DME. A lifetime of several seconds is enough for the generated species.

2. $E_{1/2}$ for the process $Ox + ne^- + mH^+ \rightleftarrows Red$ is given by

$$E_{1/2} = E^{0\prime} + \frac{0.0592}{n}\log\left(\frac{D_{Red}}{D_{Ox}}\right)^{1/2} - \left(\frac{0.0592m}{n}\right)pH$$

The value of (m/n) is determined from the relation between $E_{1/2}$ and pH. If n is known, we can get the value of m. Then, using the values of m and n, we get the value of $E^{0\prime}$.

---

[5] The mass transfer from the bulk solution to the electrode surface occurs in three modes, i.e. (i) diffusion caused by concentration gradient, (ii) migration caused by gradient of electric potential, and (iii) convection caused by stirring or by temperature gradient in the solution. In polarography, the influence of migration is removed by the addition of supporting electrolyte. Thus, the mass transfer to the DME is mainly diffusion controlled.

3. $E_{1/2}$ for the process $M^{n+} + ne^- + Hg \rightleftarrows M(Hg)$ is given by

$$(E_{1/2})_s = E^{0'}_{M^{n+}/M(Hg)} + \frac{0.0592}{n}\log\left(\frac{D_{M(Hg)}}{D_{M^{n+}}}\right)^{1/2} \left(\approx E^{0'}_{M^{n+}/M(Hg)}\right) \quad (5.18)$$

We can get the formal potential for the amalgam formation by Eq. (5.18). Data on the diffusion coefficients of metals in mercury ($D_{M(Hg)}$) are available in the IUPAC report (Galus, Z. (1984) *Pure Appl. Chem.*, **56**, 636).

4. $E_{1/2}$ for the process $ML_p^{(n+bp)+} + ne^- + Hg \rightleftarrows M(Hg) + pL^b$ (L^b: ligand anion or molecule) is given by

$$(E_{1/2})_c = (E_{1/2})_s - \frac{0.0592}{n}\log K - \frac{0.0592p}{n}\log[L^b] \quad (5.19)$$

Here, K is the formation constant for $M^{n+} + pL^b \rightleftarrows ML_p^{(n+pb)+}$. We can determine p and K from Eq. (5.19), i.e. we obtain the value of p from the linear relation between $(E_{1/2})_c$ and $\log[L^b]$ and then the value of K from $[(E_{1/2})_c - (E_{1/2})_s]$ at a given $[L^b]$ value. When a consecutive complex formation ($M^{n+} \rightleftarrows ML^{(n+b)} \rightleftarrows ML_2^{(n+2b)} \rightleftarrows \cdots \rightleftarrows ML_p^{(n+pb)}$) occurs, we can determine the consecutive formation constants by analyzing the relation between $(E_{1/2})_c$ and $\log[L^b]$ more closely, based on the principle similar to those described in Section 6.3.3. Thus, polarography has been widely used in coordination chemistry.

The wave for the totally irreversible one-electron reduction is expressed by Eq. (5.20).

$$E = E_{1/2} + \frac{0.916RT}{\alpha F}\ln\frac{i_d - i}{i} \quad \left(= E_{1/2} + \frac{0.0542}{\alpha}\log\frac{i_d - i}{i}, 25\,°C\right) \quad (5.20)$$

$$E_{1/2} = E^{0'} + \frac{0.0592}{\alpha}\log\left(1.349\frac{k_s\tau^{1/2}}{D_{Ox}^{1/2}}\right) \quad (5.21)$$

where α is the transfer coefficient, $E^{0'}$ is the formal potential, k_s is the standard rate constant and τ is the drop time. The α value is obtained by wave analysis that plots the relation of $\log[(i_d - i)/i]$ versus E. If the value of $E^{0'}$ is known, k_s can be obtained from Eq. (5.21). Roughly speaking, the DC polarographic wave appears reversible if $k_s > 2 \times 10^{-2}\,cm\,s^{-1}$ but behaves totally irreversible if $k_s < 3 \times 10^{-5}\,cm\,s^{-1}$.

Several items that need further explanations are described below:

Polarograph The instrument for measuring the current–potential curves in polarography. The first polarograph, invented by Heyrovský and Shikata in 1924, was a two-electrode device (Figure 5.8a) and could record the current–potential curves automatically on photographic paper. It was the first analytical instrument that recorded analytical results automatically. Polarographs with three-electrode circuits became popular in the 1960s. Nowadays, most of the instruments for polarography and voltammetry are three-electrode devices. In the case of two-electrode devices, there is a relation $V = (E_{DME} - E_{REF}) + iR$, where V is the voltage applied to the cell, E_{DME} and E_{REF} are the potentials of the DME and the counter electrode with a constant potential, R is the cell resistance

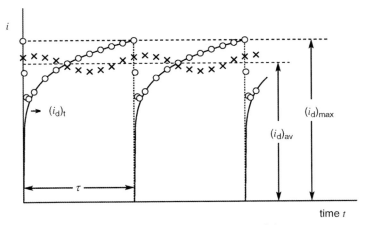

Figure 5.12 Current–time curve at a DME (solid curve) and the oscillation of the pen recorder measured without (○) and with (×) damping.

and i is the current through the cell. If R or $|i|$ is large, the current–potential curve is distorted due to the influence of the ohmic (iR) drop. In the case of three-electrode device, the influence of the iR-drop is eliminated to a considerable extent (see Section 5.9).

Ilkovic equation The equation for the diffusion current at a DME. For each mercury drop, the diffusion current at time t after the beginning of a drop-life $(i_d)_t$ is given by

$$(i_d)_t = 708nD^{1/2}Cm^{2/3}t^{1/6}$$

The $(i_d)_t$–t relation is shown by the solid curve in Figure 5.12. When a pen recorder is used to record the current, the response will be as shown by open circles or by crosses, depending on whether the recorder is used without or with damping. In the case without damping, it is appropriate to measure the current at the end of each drop-life. Then, the diffusion current is given by Eq. (5.17), where τ is a drop time.

$$(i_d)_{max} = 708nD^{1/2}Cm^{2/3}\tau^{1/6} \qquad (5.17)$$

In the case with damping, the mid point of the oscillation corresponds well to the average current during a drop-life. Thus, we get the average diffusion current $(i_d)_{av}$ as shown by Eq. (5.22).

$$(i_d)_{av} = (1/\tau)\int_0^\tau (i_d)_t dt = 607nD^{1/2}Cm^{2/3}\tau^{1/6} \qquad (5.22)$$

Nowadays, it is usual to measure the current without damping and read the values at the end of each drop-life (see Figure 5.10).

Supporting electrolyte The electrolyte that is added to the electrolytic solution to make it electrically conductive as well as to control the reaction conditions. The supporting electrolyte also works to eliminate the *migration current* that flows in its

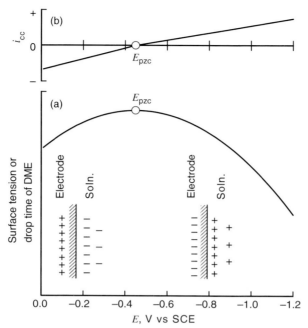

Figure 5.13 (a) Electrocapillary curve at a DME, measured as drop time– or surface tension–potential curve, and a simplified model of the double layer. (b) Charging current–potential curve.

absence. It may be a salt, an acid, a base or a pH buffer, which is difficult to oxidize or to reduce. It is used in concentrations between 0.05 and 1 M, which is much higher than that of electroactive species (usually 10^{-5}–10^{-2} M). The supporting electrolyte sometimes has a great influence on the electrode reaction, changing the potential window of the solution, the double-layer structure, or reacting with the electroactive species (e.g. complex formation with metal ions). It must be selected carefully.

Residual current A small current that flows in the solution free of electroactive species (see curve 1 in Figure 5.10). The residual current in DC polarography is mainly the *charging current*, which is for charging the double layer on the surface of the DME (Figure 5.13).[6]

6) Generally, the charging current (i_{cc}, μA) at an electrode is expressed by

$$i_{cc} = \frac{dq}{dt} = C_d(E_{pzc} - E)\frac{dA}{dt} - AC_d\frac{dE}{dt}$$

where q is the charge at the double layer (μC), C_d is the double-layer capacity (μF cm^{-2}), A is the electrode area (cm^2), E_{pzc} is the potential of zero charge (V) and t is the time (s). Under polarographic conditions, the potential of the DME during a drop-life is practically constant

($dE/dt \approx 0$) and the charging current is due to the change in the electrode area. Thus, the charging current at t (s) after the beginning of a drop-life is $(i_{cc})_t = 5.7 \times 10^{-3}\, m^{2/3}\, t^{-1/3} C_d \times (E_{pzc} - E)$ and that at the end of a drop-life is $(i_{cc})_{max} = 5.7 \times 10^{-3}\, m^{2/3}\tau^{-1/3}C_d(E_{pzc} - E)$. C_d at a DME is on the order of $10 - 20\,\mu F\,cm^{-2}$ but changes appreciably with the electrode potential and with the solution composition. The charging current is *nonfaradaic*, the electrolytic current is *faradaic*.

Polarography and voltammetry Both methods are the same in that current–potential curves are measured. According to the IUPAC recommendation, the term 'polarography' is used when the indicator electrode is a liquid electrode whose surface is periodically or continuously renewed, like a dropping or streaming mercury electrode. When the indicator electrode is some other electrode, the term 'voltammetry' is used. However, there is some confusion in the use of these terms.

Indicator electrode and working electrode In polarography and voltammetry, both terms are used for the microelectrode at which the process under study occurs. The term 'working electrode' is somewhat more popular, but, according to the IUPAC report (*Pure Appl. Chem.*, 1979, **51**, 1159), 'indicator electrode' is more appropriate here. The term 'working electrode' is recommended for the case when the change in the bulk concentration of the substance under study occurs appreciably by the electrolysis at the electrode, while 'indicator electrode' for the case when the change is negligible.

The main analytical limitations of DC polarography and the methods to overcome them are as follows:

Lower limit of analytical determination Because of interference by the charging current, the lower limit of determination is at best $\sim 5 \times 10^{-6}$ M for divalent metal ions. This limit is much reduced by eliminating the charging current appropriately (Section 5.4).

Tolerable amount of the substance electrolyzed before the analyte(s) In DC polarography, several electroactive species can be determined simultaneously because they give step-wise waves. However, if a substance that is easier to reduce or to oxide than the analyte(s) exists in high concentrations (50 times or more), the determination of the analyte(s) is difficult. This problem is overcome by obtaining the derivative curve of the DC polarographic wave (Section 5.4).

Potential limit on the anodic side In DC polarography, the measurable potential limit on the anodic side is much more negative than the oxidation potential of water. It is caused by the anodic dissolution of the DME (Hg $\rightarrow$ Hg^{2+} or Hg$_2^{2+}$). This problem is overcome by using platinum or carbon electrodes (Section 5.5).

5.4
New Types of Polarography – Methods that Electrolyze Electroactive Species Only Partially (2)

In order to overcome the drawbacks of DC polarography, various new types of polarography and voltammetry were developed. Some new polarographic methods are dealt with in this section. They are useful in chemical analyses as well as in studying electrode reactions.

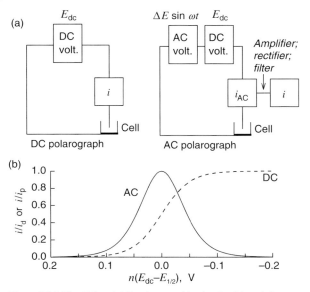

Figure 5.14 The DC and AC polarographic circuits (a) and the current–potential curves for DC and AC polarographies (b).

5.4.1
AC Polarography

In AC polarography [4], as shown in Figure 5.14a, AC voltage ($\Delta E \sin \omega t$) of, for example, 10 mV in peak-to-peak amplitude and 100 Hz in frequency is superimposed on the DC voltage (E_{dc}) and the sum is applied to the DME. Then, from the current that flows through the circuit, the AC component is sampled and recorded against the DC voltage, after being amplified, rectified and filtered. For reversible electrode reactions, the AC current that flows through the circuit ($I(\omega t)$) is correlated with the DC potential of the DME by Eq. (5.23):

$$I(\omega t) = \frac{n^2 F^2 AC(\omega D)^{1/2}\Delta E}{4RT \cosh^2(j/2)} \sin\left(\omega t + \frac{\pi}{4}\right) \quad (5.23)$$

Here, A is the electrode area, C and D are the concentration and the diffusion coefficient of the electroactive species, ΔE and ω ($=2\pi f$) are the amplitude and the angular frequency of the AC applied voltage, t is the time and $j = nF(E_{dc} - E_{1/2})/RT$. For reversible processes, the AC polarographic wave has a symmetrical bell shape and corresponds to the derivative curve of the DC polarographic wave (Figure 5.14b). The peak current i_p, expressed by Eq. (5.24), is proportional to the concentration of electroactive species and the peak potential is almost equal to the half-wave potential in DC polarography.

$$i_p = \frac{n^2 F^2 AC(\omega D)^{1/2}\Delta E}{(32)^{1/2} RT} \quad (5.24)$$

However, the peak current in AC polarography markedly depends on the reversibility of the electrode process, being very small for an irreversible process. We can apply this dependence to study the kinetics of the electrode reactions.

The AC current measured in AC polarography contains, in addition to the faradaic current caused by the electrode reaction, the charging current caused by the charging–discharging of the double layer at the electrode surface. Due to the interference of the charging current, the lower limit of determination in AC polarography is about the same as that in DC polarography. However, there are two AC polarographic methods that eliminate the charging current. One is *phase-sensitive AC polarography* and the other is *second harmonic AC polarography*. The former is based on the fact that the AC component of the faradaic current is 45° out of phase with the applied AC voltage ($E_{ac} = \Delta E \sin \omega t$), while that of the charging current is 90° out of phase with E_{ac}. The charging current is eliminated by measuring the in-phase component of the AC current using a lock-in amplifier. In the latter, the 2ω-component of the AC current is measured using a lock-in amplifier, here the 2ω-component being free from the charging component. Both methods are highly sensitive, the lower limit of determination being $\sim 10^{-7}$ M, and are useful in studying the kinetics of electrode reactions.[7]

5.4.2
SW Polarography

Instead of the sine-wave voltage in AC polarography, square-wave (SW) voltage of small amplitude is superposed on DC voltage (E_{dc}) and the sum is applied to the DME. For each cycle of the applied SW voltage, the faradaic current and the charging current as in Figure 5.15 flow through the circuit. The charging current decreases rapidly (exponentially with time t) and is practically zero near the end of the half cycle. On the other hand, the faradaic current ($\propto t^{-1/2}$) decreases rather slowly and still remains considerably at the end of half cycle. The feature of this method is that the current near the end of each half cycle is sampled to eliminate the charging current [5]. SW polarography gives a bell-shaped wave, as in AC method, and is useful for chemical analyses. Its lower limit of determination is 10^{-7} M or less. Moreover, a readily reducible or oxidizable substance is tolerable up to 10 000 times or more. However, this method is not sensitive to species that undergo irreversible electrode reactions.

5.4.3
Pulse Polarography

This method is classified into normal pulse polarography (NPP) and differential pulse polarography (DPP), based on the modes of applied voltage [6].

---

7) See Section 5.5.6 for an electrochemical impedance spectroscopy or an AC impedance method.

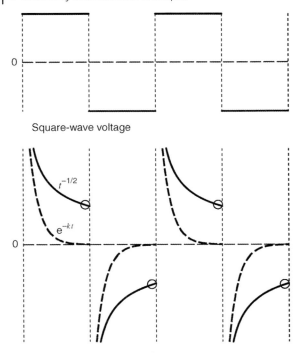

Square-wave voltage

Faradaic current (solid curves)

Capacity current (dashed curves)

Figure 5.15 Square-wave voltage (e.g. 200 Hz, 10 mV) (upper) and the faradaic current (solid curve) and the charging current (dashed curve) (lower) in SW polarography. The currents after attenuation of the charging current (shown by small circles) are sampled.

Normal pulse polarography As shown in Figure 5.16a, for each drop of the DME, the potential is kept for a fixed time (2–4 s) at an initial potential E_i, at which no electrolysis occurs, and then, near the end of the drop-life, a pulse voltage (V_{pulse}) is applied for a duration of 5–100 ms. V_{pulse} is increased from one drop to the next. At the beginning of each voltage pulse, both the faradaic and charging currents are large. However, due to the rapid decrease in the charging current, the current near the end of the pulse contains only the faradaic component. This situation resembles that in SW polarography (see Figure 5.15). In this method, the current near the end of the pulse is sampled and recorded against the potential of the DME during the pulse, i.e. ($E_i + V_{pulse}$). The polarographic wave is S-shaped, as in DC polarography. However, because the influence of charging current has been eliminated, the lower limit of determination is $\sim 10^{-7}$ M. In the inverse NPP method, the initial potential E_i is set at the limiting current region of a substance under study and the pulse potential is varied inversely. This method is used to measure the inverse reaction (reoxidation for reduction and re-reduction for oxidation) to study the electrode kinetics.

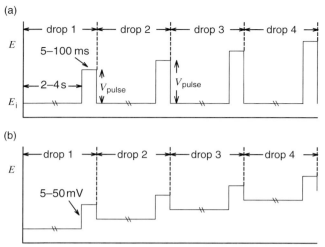

Figure 5.16 Applied voltages in normal pulse (a) and differential pulse (b) polarographies.

Differential pulse polarography As shown in Figure 5.16b, the potential before the pulse changes from one drop to the next. However, the pulse voltage ΔE, which is applied near the end of a drop-life, is the same (5–20 mV) for each drop. The current that flows before the pulse is the same as that in DC polarography. When the pulse voltage is applied, both the faradaic and charging currents suddenly increase. But the charging current decreases rapidly and, near the end of the pulse, only the faradaic component remains. In this method, the current just before the pulse and the current near the end of the pulse (τ_s (s) after the beginning of the pulse) are sampled and the difference is recorded against the potential before the pulse. The polarogram is bell-shaped, with the peak current expressed by Eq. (5.25). The peak potential is similar to the half-wave potential in DC polarography.

$$i_p = \frac{nFAD^{1/2}C}{\pi^{1/2}\tau_s^{1/2}} \cdot \frac{1-\sigma}{1+\sigma} \quad \left[\sigma = \exp\left(\frac{nF\,\Delta E}{2RT}\right)\right] \tag{5.25}$$

This method is the most sensitive of the polarographic methods now available and the lower limit of determination is $\sim 5 \times 10^{-8}$ M. It is fairly sensitive even for substances that undergo irreversible electrode reactions. DPP is very useful in trace analyses.

5.5
Voltammetry and Related New Techniques – Methods that Electrolyze Electroactive Species Only Partially (3)

Voltammetry is a term used to include all the methods that measure current–potential curves (voltammograms) at small indicator electrodes other than the

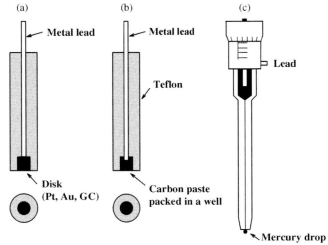

Figure 5.17 Examples of voltammetric indicator electrodes.
(a) Disk electrode (GC, platinum, gold, etc.); (b) carbon paste
electrode; (c) simple hanging mercury drop electrode.

DME [7]. There are various types of voltammetric indicator electrodes, but disk electrodes, as shown in Figure 5.17, are popular. The materials used for disk electrodes are platinum, gold, graphite, glassy carbon (GC), boron-doped diamond,[8] carbon paste, etc. and they can be modified in various ways. For electrode materials other than mercury, the potential windows are much wider on the positive side than for mercury. However, electrodes of stationary mercury drop, mercury film and mercury pool are also applicable in voltammetry. In some cases, voltammograms are recorded in flowing solutions or by rotating the indicator electrodes (see Section 5.5.3). Moreover, all the polarographic methods (DC, AC, SW and pulse methods) are used in voltammetry. Voltammetry is dealt with in Ref. [1] and, in more detail, in the books of Ref. [8].

5.5.1
Linear Sweep Voltammetry

This method is sometimes abbreviated to LSV. In this method, a static indicator electrode ($A\,cm^2$ in area) is used and its potential is scanned at constant rate v ($V\,s^{-1}$)

---

8) The boron-doped diamond-film electrode is a new type of voltammetric indictor electrodes. It is promising because of its high chemical stability, low background currents and wide potential windows (high overpotentials for the oxidation and the reduction of water). See the book by Fujishima, A., Einaga, Y., Rao, T.N. and Tryk, D.A. (2005) *Diamond Electrode*, Elsevier, Amsterdam, and an article by Kulandainathan, M.A., Hall, C., Wolverson, D., Foord, J.S., MacDonald, S.M. and Marken, F. (2007) *J. Electroanal. Chem.*, **606**, 150.

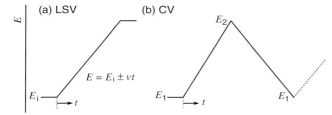

Figure 5.18 Applied voltage in (a) linear sweep voltammetry and (b) cyclic voltammetry.

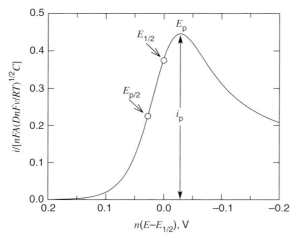

Figure 5.19 Linear sweep voltammogram for a reversible process. E_p, peak potential; $E_{p/2}$, half-peak potential; $E_{1/2}$, half-wave potential; i_p, peak current.

from an initial value (E_i) to the positive or negative direction (Figure 5.18). A typical linear sweep voltammogram is shown in Figure 5.19. In contrast to DC polarography, there is no limiting current region. After reaching a peak, the current decreases again.[9] For a reversible reduction process, the peak current i_p (A) is expressed by Eq. (5.26), where D and C are the diffusion coefficient ($cm^2 s^{-1}$) and the concentration ($mol\, cm^{-3}$) of the electroactive species:

$$i_p = 0.446nFAC\left(\frac{nF}{RT}\right)^{1/2} v^{1/2}D^{1/2} = (2.69 \times 10^5)n^{3/2}AD^{1/2}v^{1/2}C \quad [25\,°C]$$

$$(5.26)$$

9) The current decrease after the peak occurs because the thickness of the diffusion layer increases with time (see footnote 3). Even if the potential scan is stopped after the peak, the current continues to decrease with time in the same way.

The peak potential E_p and the half-peak potential $E_{p/2}$ are related to the half-wave potential $E_{1/2}$ in DC polarography by Eqs. (5.27)–(5.29).

$$E_p = E_{1/2} - 1.11\frac{RT}{nF} = E_{1/2} - \frac{0.0285}{n} \quad \text{[V, 25 °C]} \tag{5.27}$$

$$E_{p/2} = E_{1/2} + 1.09\frac{RT}{nF} = E_{1/2} + \frac{0.0280}{n} \quad \text{[V, 25 °C]} \tag{5.28}$$

$$|E_p - E_{p/2}| = 2.2\frac{RT}{nF} = \frac{0.0565}{n} \quad \text{[V, 25 °C]} \tag{5.29}$$

i_p and E_p (or $E_{p/2}$) give quantitative and qualitative information on the electroactive species.[10] i_p is proportional to $v^{1/2}$ (v = voltage scan rate), while the charging current i_{cc} is proportional to v. Thus, the i_p/i_{cc} ratio decreases with v.

Equations (5.30)–(5.32) are for a totally irreversible reduction process:

$$i_p = (2.99 \times 10^5)n(\alpha n_a)^{1/2}AD^{1/2}v^{1/2}C \quad \text{[25 °C]} \tag{5.30}$$

$$E_p = E^{0'} - \left(\frac{RT}{\alpha n_a F}\right)\left[0.780 + \ln\left(\frac{D^{1/2}}{k_s}\right) + \ln\left(\frac{\alpha n_a Fv}{RT}\right)^{1/2}\right] \tag{5.31}$$

$$|E_p - E_{p/2}| = 1.857\frac{RT}{\alpha n_a F} = \frac{0.0477}{\alpha n_a} \quad \text{[V, 25 °C]} \tag{5.32}$$

where α is the transfer coefficient, n_a is the number of electrons that participate in the rate determining step, $E^{0'}$ is the formal potential, k_s is the standard rate constant and $RT/F = 0.0257$ at 25°C. For irreversible process, the value of i_p depends on α and, if $\alpha = 0.5$, it is about 20% smaller than that for a reversible process. The peak potential E_p depends on the scan rate (v) and becomes ($30/\alpha n_a$) mV more negative for each 10-fold increase in v.

In this method, a wide range of scan rate (v) is possible, i.e. from $1\,\text{mV s}^{-1}$ to $1000\,\text{V s}^{-1}$ with a conventional apparatus, and up to $1\,000\,000\,\text{V s}^{-1}$ or even more with a combination of a sophisticated apparatus and an ultramicroelectrode.

5.5.2
Cyclic Voltammetry

Cyclic voltammetry is often abbreviated CV. In this method, the potential is linearly scanned forward from E_1 to E_2 and then backward from E_2 to E_1, giving a triangular

10) When two or more electroactive species are reduced (or oxidized) or when one electroactive species is reduced (or oxidized) in two or more steps in LSV, the analysis of the voltammogram becomes difficult. For example, in Figure 5.20, it is difficult to accurately measure the current for the second peak ($i'p$). Two methods are applicable in such a case: (i) we stop the potential sweep at point S after the first peak, wait until the current decreases almost to zero and then restart the sweep to measure the second peak (curve 2). (ii) We mathematically transform the linear sweep voltammogram into an S-shaped voltammogram by the *convolution* or *semi-integral method*. Then, two-step S-shaped waves are obtained for a two-component system.

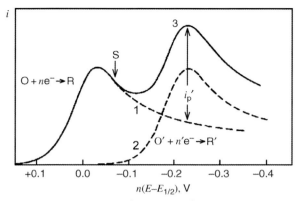

Figure 5.20 Linear sweep voltammogram for a two-component system (see footnote 10).

potential cycle (Figure 5.18). Figure 5.21 shows some examples of cyclic voltammograms for the process $Ox + ne^- \rightleftarrows Red$, where only Ox is in the solution. Curve 1 is when the process is reversible. In the forward scan, a cathodic peak is obtained by the reduction of Ox to Red, as in LSV. In the backward scan, an anodic peak appears due to the reoxidation of the Red, which was generated during the forward scan. For a reversible process, the cathodic and anodic peak currents are equal in magnitudes ($|i_{pc}| = |i_{pa}|$) and the cathodic peak potential (E_{pc}) is $(58/n)$ mV more negative than the anodic peak potential (E_{pa}). These are important criteria for reversibility. Moreover,

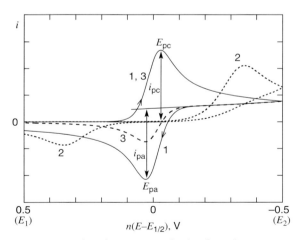

Figure 5.21 Cyclic voltammograms for the electrode reaction $Ox + ne^- \rightleftarrows Red$, which is reversible (curve 1), irreversible but $\alpha = 0.5$ (curve 2) and reversible but accompanied by a conversion of Red to an electroinactive species (curve 3). $E_{1/2}$ is for reversible process.

the half-wave potential, which is used to obtain the formal redox potential, is obtained by $E_{1/2} = (E_{pc} + E_{pa})/2$. By decreasing the reversibility, the difference between the two peak potentials increases. Curve 2 is for a process that is considerably irreversible. Compared with curve 1, the cathodic peak appears at much more negative potential, the anodic peak at much more positive potential. If the process is completely irreversible, the anodic peak does not appear in the measurable potential region. From the irreversible CV curve, we can obtain kinetic parameters (rate constant and transfer coefficient) for the electrode reaction, usually by a simulation method. Curve 3 is for the case in which $Ox \rightleftarrows Red \xrightarrow{k} A$, i.e. Red can be reversibly reoxidized to Ox but, before the reoxidation, some part of the Red is converted to nonelectroactive species A. The cathodic peak appears in the same way as in curve 1, but the anodic peak current is smaller than that in curve 1. From the decrease in the anodic peak current, we can get the rate constant k. In CV, the voltage scan rate can be varied over a wide range, the highest scan rate reaching $10^6\,V\,s^{-1}$ or more. Thus, the CV method is applicable to study electrode processes, even when the products or intermediates are very short-lived ($\leq 10\,\mu s$). The CV method is often used in nonaqueous solutions, as described in Chapter 8.

5.5.3
Voltammetry at Rotating Disk and Rotating Ring-Disk Electrodes

A rotating disk electrode (RDE) and a rotating ring-disk electrode (RRDE) [9] are used to study electrode reactions because the mass transfer at the electrode(s) can be treated theoretically by hydrodynamics. At the RDE, the solution flows toward the electrode surface as shown in Figure 5.22, bringing the substances dissolved in it. The current–potential curve at the RDE is S-shaped and has a potential-independent limiting current region, as in Figure 5.6. The limiting current i_l (A) is expressed by Eq. (5.33) if it is controlled by mass transfer:

$$i_l = 0.62nF\pi r^2 \omega^{1/2} v^{-1/6} D^{2/3} C \tag{5.33}$$

where r (cm) is the radius of the disk electrode, ω (rad s^{-1}) is its angular velocity, v (cm^2 s^{-1}) is the kinematic viscosity of the solution, D (cm^2 s^{-1}) is the diffusion coefficient of electroactive species and C (mol cm^{-3}) is its concentration. The limiting current is proportional to the square root of the rate of rotation. On the other hand, the residual current, which is mainly charging current, does not depend on the rate of rotation. Thus, the increase in the rate of rotation increases the (limiting current)/(residual current) ratio, a favorable condition for analysis.[11] The rates of rotation can be controlled over a wide range (100–10 000 rpm). The

---

11) In order to eliminate the influence of the charging current at an RDE, it is advisable to measure the current–potential curve manu-ally so that the electrode potential for each point of the curve is kept constant (see foot-note 6).

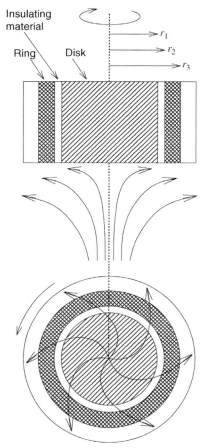

Figure 5.22 Streamlines near the rotating ring-disk electrode.

material of the disk electrode can be platinum, gold, mercury (plated on solid metal), carbon (graphite, glassy carbon and carbon paste), semiconductor or functional polymer.

The advantage of the RDE is doubled by the use of a RRDE. The RRDE consists of a central disk and a concentric ring surrounding it (Figure 5.22). The small gap between ring and disk is insulated by Teflon or epoxy resin. The materials of the ring and disk electrodes can be chosen separately so that best results are obtained at each electrode. The potentials of the two electrodes can be controlled independent of each other by using a bipotentiostat. The rotator and various types of ring–disk electrodes are also commercially available. When the RRDE is rotated, mass transfer occurs, as shown in Figure 5.22. Some fraction of the substance that reaches the disk electrode is then transported to the surface of the ring electrode. This fraction, called the *collection efficiency* and denoted by N, is a function of the electrode geometry but is independent of the rate of rotation.[12] If a reaction occurs at the disk electrode, the fraction N of the product will reach the ring electrode and will be detected. This makes the RRDE a powerful tool in studying electrode reactions. For example, if the product at the disk electrode is unstable and rapidly converted to a substance undetectable at the ring electrode, the experimental N value will decrease. The time needed to transfer from the disk to the ring varies with the rate of rotation. If the experimental N value is obtained as a function of the rate of rotation, the kinetics of the conversion reaction can be analyzed. In this respect, voltammetry at an RRDE somewhat resembles cyclic voltammetry.

5.5.4
Ultramicroelectrodes

The conventional voltammetric indicator electrodes are 0.5–5 mm in diameter. However, *ultramicroelectrodes* (UME) [10] that have dimensions of 1–20 μm are also used as indicator electrodes. The tiny electrodes have some definite advantages over conventional ones:

(i) The small electrodes are suited for measurements in solutions of small volumes or in small spaces (e.g. intracellular measurements).

(ii) Because the electrode surface area is very small, the current that flows through the cell is very small (pA to nA level). This keeps the ohmic drop very small and allows measurements in solutions of high electric resistance, e.g. in solutions without supporting electrolyte or in solvents of low permittivities. Moreover, it allows very rapid voltage scan rates, sometimes at $10^6\,\mathrm{V\,s^{-1}}$ or more. If this electrode is used in CV, unstable products or intermediates can be detected or determined. These are extremely important in voltammetry in nonaqueous solutions, as discussed in Chapter 8.

(iii) Figure 5.23 shows the cyclic voltammograms of ferrocene recorded with a UME at high (a) and low (b) scan rates. The curve recorded at high scan rate has a shape expected for conventional cyclic voltammetry. However, when recorded at low scan rate, the forward and backward curves nearly overlap and the S-shaped waves have limiting current plateaus. A sketch of the diffusion layer at the UME is shown in Figure 5.24. If the duration of electrolysis is very short, the thickness of the diffusion layer is less than the size of the electrode. However, if it is long enough, the thickness of the diffusion layer becomes much larger than the size of the electrode; then, the mass transport to the electrode mainly occurs by radial diffusion, which is time independent. This is the reason why a limiting current plateau is obtained in the voltammogram at the UME. This characteristic is of practical importance when the electrode is used in solutions of small volume. In order to get a stationary

12) If the radius of the disk is r_1 and the inner and outer radii of the ring are r_2 and r_3, respectively, the calculated N value depends on the values of r_2/r_1 and r_3/r_2. For example, $N =$ 0.144 for $r_2/r_1 = 1.04$ and $r_3/r_2 = 1.04$, $N =$ 0.289 for $r_2/r_1 = 1.04$ and $r_3/r_2 = 1.14$, $N =$ 0.414 for $r_2/r_1 = 1.04$ and $r_3/r_2 = 1.30$ and $N = 0.270$ for $r_2/r_1 = 1.10$ and $r_3/r_2 = 1.14$.

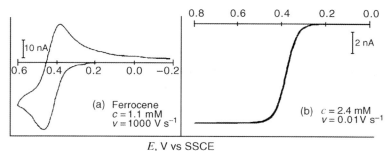

E, V vs SSCE

Figure 5.23 Influence of the scan rate on the cyclic voltammograms of ferrocene at an ultramicroelectrode. Electrode radius: 5 μm; supporting electrolyte: 0.6 M Et_4NClO_4 in AN.

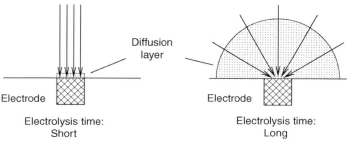

Figure 5.24 Influence of electrolysis duration on diffusion at an ultramicroelectrode.

limiting current with a conventional electrode, the electrode must be rotated to keep the thickness of the diffusion layer constant (see above).

Recently, arrayed electrodes, which consist of several UMEs of the same type or of different types, are prepared and used in sophisticated ways [10e].

5.5.5
Modified Electrodes

If the surface of a metal or carbon electrode is covered with a layer of some functional material, the electrode often shows characteristics that are completely different from those of the bare electrode. Electrodes of this sort are generally called *modified electrodes* [11] and various types have been developed. Some have a monomolecular layer that is prepared by chemical bonding (chemical modification). Some have a polymer coat that is prepared either by dipping the bare electrode in a solution of the polymer, by evaporating the solvent (ethanol, acetone, etc.) of the polymer solution placed on the electrode surface or by electrolytic polymerization of the monomer in solution. The polymers of the polymer-modified electrodes are conducting polymers, redox polymers or ion-exchange polymers and can perform various functions. The applications of modified electrodes are really limitless, as shown by some examples in Table 5.2. The use of modified electrodes in nonaqueous solvents will be discussed in Chapter 12 in connection with modern electrochemical technologies.

Table 5.2 Examples of applications of modified electrodes.

Area	Applications
Energy	Fuel cell; polymer battery; photoelectric cell; capacitor
Information	Storage element; liquid crystal display device; electrochromic display device; electrochemiluminescence device; photoelectric transducer
Analysis	Biosensor; ion sensor; detector in HPLC and FIA; gas sensor; voltammetric indicator electrode; reference electrode
Synthesis	Asymmetric synthesis; selective electrolysis; electrochemical polymerization

5.5.6
Other Items Related to Voltammetry

Electrochemical Impedance Spectroscopy
Electrochemical impedance spectroscopy (EIS) is a method to measure the cell or electrode impedance as a function of frequency (usually $10^{-2}–10^5$ Hz) of the alternating signal of small amplitude (by varying the electrode potential). In order to get the standard rate constant (k_s) of an electron-transfer process, the faradaic impedance (R_s and C_s) is separated from the total impedance (Figure 5.25) and, by using the mathematical equations for R_s and C_s, the value of k_s is calculated. The data in Table 8.6 were obtained by such impedance method. Earlier, impedance was measured with an impedance bridge but now more conveniently with lock-in amplifiers or frequency-response analyzers (FRA) and others. EIS is very useful in studying electrode dynamics and, in addition to the simple heterogeneous ET processes and double-layer structures, it is applicable to a variety of complex electrochemical processes associated with coupled homogeneous reactions, adsorption/desorption, corrosion, electrodeposition, polymer film formation, semiconductor electrode, liquid–liquid interface, etc. Though EIS has a long history, its fundamentals and applications are still making a remarkable progress. But we do not discuss this topic anymore because the theories and the instrumentation are somewhat

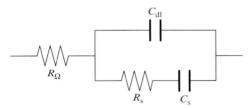

Figure 5.25 Equivalent circuit of the cell for the measurement of electron transfer at the electrode. R_s and C_s are for the faradaic impedance, C_{dl} is for the double-layer capacitance and R_Ω is for the solution resistance.

complicated to treat in this book. There are many books and review articles [12], of which Ref. [12a] is the latest and Ref. [12f] the most concise.

Voltammetry at the Interface Between Two Immiscible Electrolyte Solutions
The electrodes used in conventional polarography and voltammetry are electronic conductors such as metals, carbons or semiconductors, and in an electrode reaction, an electron transfer occurs at the electrode/solution interface. However, polarography and voltammetry can measure the charge transfers (ion transfer, electron transfer and combination of ion transfer and electron transfer) at the interface between two immiscible electrolyte solutions (*ITIES*). Typical such interfaces are water/nitrobenzene and water/1,2-dichloroethane. Moreover, water/hydrophobic ionic liquids are also the target interfaces.

Extensive studies have been carried out [13]. Polarography and voltammetry at liquid–liquid interfaces are of analytical importance because they are applicable to ionic species that are neither reducible nor oxidizable at the conventional electrodes. They are also useful in studying the charge transfer processes at liquid–liquid interfaces or at membranes: solvent extractions, phase transfer catalyses, ion transports at biological membranes, etc. are included among such processes.

This topic is discussed in Chapter 14 because it is closely related to nonaqueous solvents and ionic liquids.

Stripping Voltammetry and Electrochemical Biosensors
These are outlined in footnotes [13] and [14]. They are very popular in aqueous solutions, but their applications in nonaqueous solutions are very rare, being limited to only a few cases.

13) *Stripping voltammetry* [14]: It is used when the analyte is too dilute to determine directly even with a polarographic method of highest sensitivity. In the *anodic stripping voltammetry* (ASV) of a metal ion, it is cathodically deposited (preconcentrated) onto an indicator electrode as metal or metal amalgam and is then determined voltammetrically by scanning the potential to positive direction. In the *cathodic stripping voltammetry* (CSV), which is often applied to anions and organic analytes, the voltammetric measurement after preconcentration is carried out by scanning the potential to negative direction. Various indicator electrodes (hanging mercury drop electrode, mercury film electrode, glassy carbon electrode, modified electrode, etc.) are used. In addition to electrodeposition, such processes as adsorption, ion-exchange and extraction are possible for preconcentration. The method has been applied to many inorganic, organic and biological analytes of down to 10^{-10} M or less and is one of the most sensitive analytical methods. Coetzee *et al.* [14d]

applied the anodic stripping method to determine electropositive alkali and alkaline earth metals using organic solvents such as DMSO.

14) *Electrochemical biosensors* [15]: Here we mean *biomimetic sensors*, which utilize the ability of biological materials (enzymes, antibodies, etc.) to recognize specific compounds and to catalyze their reactions with great specificity. Many of the biosensors are electrochemical sensors, based on amperometric or potentiometric measurements. For example, in the case of an amperometric glucose sensor, glucose oxidase immobilized on a platinum electrode catalyzes the oxidation of glucose:

$$\text{Glucose} + O_2 \xrightarrow{\text{glucoseoxidase}} \text{Gluconic acid} + H_2O_2$$

Glucose is determined amperometrically by the decrease in the reduction current of O_2 or by the increase in the oxidation current of H_2O_2. For an example of the study on electrochemical biosensors in nonaqueous systems, see Ref. [15c].

5.5.7
Combination of Voltammetry and Nonelectrochemical Methods

Voltammetric current–potential curves are important in elucidating electrode processes. However, if the electrode process is complicated, they cannot provide enough information to interpret the process definitely. Moreover, they cannot give direct insight into what is happening on a microscopic or molecular level at the electrode surface. To overcome these problems, many characterization methods that combine voltammetry and nonelectrochemical techniques have appeared in the past 30 years. Many review articles on combined characterization methods [16] are available. Only four examples are described here. For applications of these combined methods in nonaqueous solutions, see Chapter 9.

Measurement of Mass Change at the Electrode Surface by the Use of QCM
The quartz crystal microbalance (QCM) uses the property that makes the resonance frequency of a quartz crystal sensitive to mass changes at the nanogram level at the surface of the electrode attached to the crystal. There is a linear relation between the mass change (Δm) and the frequency change (ΔF) as shown by $\Delta F = -k\,\Delta m$, where k is a constant (Sauerbrey equation). The QCM, originally used only in the gas phase, was found in 1980 applicable also in electrolyte solutions [17a] and soon it was combined with voltammetry to measure the current–potential curves simultaneously with the frequency–potential curves and to run *in situ* microelectrogravimetry [17b]. The electrochemical use of the QCM is now very popular and is referred to as EQCM (*electrochemical quartz crystal microbalance*) [18]. The electrode on the crystal (platinum, gold or silver; bare or modified) is used as the voltammetric indicator or working electrode, as shown in Figure 5.26. Information about the mass change at the electrode is very useful in elucidating the phenomena at or near the electrode surface. The EQCM is now used to study phenomena such as underpotential metal deposition, adsorption of substances on the electrode and surface processes on the modified electrodes.

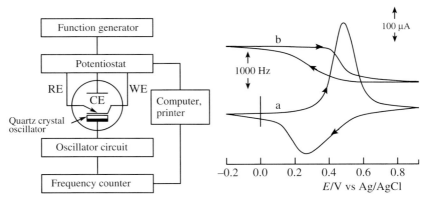

Figure 5.26 Apparatus for combining QCM measurement and cyclic voltammetry (left) and a cyclic voltammogram (a) and a frequency–potential curve (b) obtained simultaneously (right).

Electrogravimetry in the order of microgram is also possible with the EQCM. Moreover, the QCM responds to such properties as permittivity, viscosity, conductivity and density of electrolyte solutions. The apparatus for EQCM are now commercially available. For the use of EQCM in nonaqueous solutions, see Section 9.2.4.

Measurements of UV/vis Absorbances

By using an optically transparent or a gauze indicator electrode, it is possible to measure an absorbance–potential curve, simultaneously with a voltammetric current –potential curve, for the product or intermediate of an electrode reaction that absorbs UV or visible light [19]. Optically transparent electrodes (OTEs) are prepared either by coating a glass or quartz plate or a plastic film with optically transparent oxide (tin oxide, indium tin oxide) or with vapor-deposited metal (platinum, gold and silver). The tin oxide electrode, which is the most popular, is an n-type semiconductor doped with antimony and is $\sim$1 μm in thickness. It has a wider potential range than a platinum electrode on the anodic side, although the cathodic side is rather narrow due to the reduction of the tin oxide. Figure 5.27 shows the absorption spectrum at various OTEs. The fields that combine electrochemistry with spectroscopies of UV/vis, IR, ESR, etc. are called *spectroelectrochemistry*. For the use in nonaqueous solutions, see Section 9.2.1.

ESR Measurements

When a substance with an unpaired electron (such as a radical ion) is generated by an electrode reaction, it can be detected and its reactivity studied by ESR measurement [20]. This method is very important in electrochemistry in nonaqueous solutions and is discussed in Section 9.2.2.

Scanning Electrochemical Microscopy

Recently, scanning probe techniques have played important roles in the field of nanotechnologies; scanning tunneling microscopy (STM), invented by Binning and

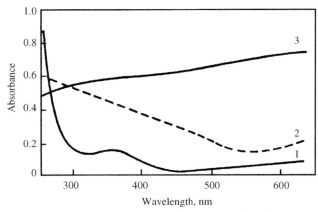

Figure 5.27 UV/vis absorption curves for OTEs coated on the quartz plate. 1: SnO_2 (20 Ω cm^{-2}), 2: Au–PbO$_2$ (11 Ω cm^{-2}), 3: Pt (20 Ω cm^{-2}).

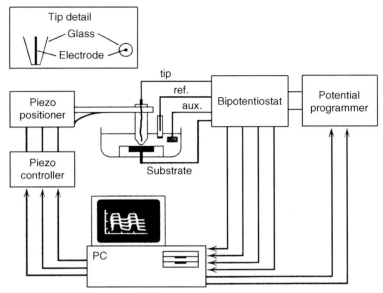

Figure 5.28 SCEM apparatus [21].

Rohrer in 1982, is a high-resolution tool for studying atomic- or molecular-scale structures of surfaces in vacuum and the electrode–electrolyte interfaces in solution; atomic force microscopy (AFM), another scanning probe technique, is used to study surface topography and surface forces. Scanning electrochemical microscopy (SECM), developed by Bard and based on STM, is powerful in measuring currents and current–potential relations caused by electrochemical reactions in microscopic spaces [21]. A schematic diagram of an SECM apparatus is shown in Figure 5.28; the bipotentiostat is used to control the potentials of the tip and the substrate, while the piezo controller is used to move the tip and to control the distance (*d*) between the tip and the substrate. The principles of SECM are shown in Figure 5.29.

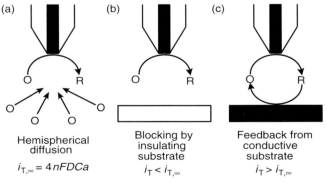

Figure 5.29 Three illustrations of SECM. (a) Hemispherical diffusion to the disk-shaped tip far from the substrate; (b) blocking by insulating substrate; (c) feedback from conductive substrate [21].

In (a), hemispherical diffusion occurs and the limiting current $i_{T,\infty}$ is expressed by $i_{T,\infty} = 4nFDCa$, where a is the radius of the tip disk electrode; in (b), the substrate is a blocking insulator and the limiting current i_T is less than $i_{T,\infty}$; in (c), the substrate is conductive and R formed at the tip is reoxidized there to regenerate O and to give $i_T > i_{T,\infty}$. By analyzing the relation between $(i_T/i_{T,\infty})$ and (d/a) values, it is possible to study the reaction kinetics of both heterogeneous and homogeneous processes. Examples of the applications of SCEM are described in Section 9.2.5 in connection with the electrochemical studies in nonaqueous solutions.

5.6
Electrogravimetry and Coulometry – Methods that Completely Electrolyze Electroactive Species

A method that completely electrolyzes the substances under study is used in electrogravimetry and coulometry. The method is also useful in electrolytic separations and electrolytic syntheses. Electrolysis is carried out either at a controlled potential or at a controlled current.

5.6.1
Controlled-Potential Electrolysis and Controlled-Current Electrolysis

We consider the case in which substance Ox in the solution is completely reduced to Red at the electrode. During the electrolysis, the concentration of Ox gradually decreases and approaches zero. This process is illustrated in Figure 5.30 by the current–potential curves: the decrease in the limiting current corresponds to the decrease in the concentration of Ox [22].

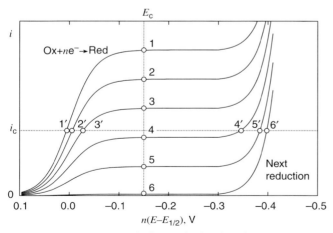

Figure 5.30 Current–potential relations for the electrolytic reduction of Ox, at the instants of 0, 20, 40, 60, 80 and 99% electrolysis.

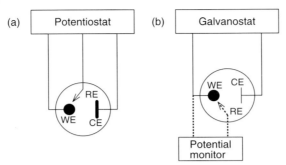

Figure 5.31 The circuits for (a) controlled-potential electrolysis and (b) controlled-current electrolysis (for the circuits based on operational amplifiers, see Figs. 5.42 and 5.43).

Controlled-Potential Electrolysis

Three electrodes are used in controlled-potential electrolysis, i.e. a working electrode, a reference electrode and a counter electrode (Figure 5.31a). During the electrolysis, the potential of the working electrode is kept constant against the reference electrode (e.g. E_c in Figure 5.30), with the aid of a potentiostat. E_c is usually in the limiting current region and the current decreases in the order 1 to 6. Here, if we express the initial concentration and the initial current by C_0 and i_0, respectively, the concentration and the current after t seconds, C_t and i_t, are expressed by

$$C_t = C_0 \exp(-kt) \quad \text{and} \quad i_t = i_0 \exp(-kt) \tag{5.34}$$

k is a constant and is given by $k = Am_{Ox}/V$, where A is the electrode area, V is the solution volume and m_{Ox} is the mass transfer coefficient of Ox, which is equal to D_{Ox}/δ (p. 119). The value of k depends on the particular cell geometry and the rate of stirring. From Eq. (5.34), it is apparent that the concentration and the current decrease exponentially with time. By increasing the k value, we can decrease the time needed for complete electrolysis. On the other hand, by controlling the electrode potentials appropriately, we can perform electrolytic depositions, separations and/or syntheses as desired.

Controlled-Current Electrolysis

Two electrodes are used in controlled-current electrolysis, i.e. a working electrode and a counter electrode (Figure 5.31b). During the electrolysis, the current through the cell is kept constant (e.g. i_c in Figure 5.30) with the aid of a galvanostat. The potential of the working electrode changes in the order 1′–6′. When the potential is at 1′, 2′ or 3′, only the reduction of Ox occurs at the working electrode. However, the concentration of Ox gradually decreases, and it becomes impossible to supply current i_c solely by the reduction of Ox. Then, in order to make up for the deficiency, the working electrode shifts its potential to 4′, 5′ and 6′, and the next reduction process (hydrogen generation in an acidic solution) occurs. This is the disadvantage of the controlled-current method. However, it is sometimes used because the apparatus and the electrolytic cell needed are simpler than in the controlled-potential method.

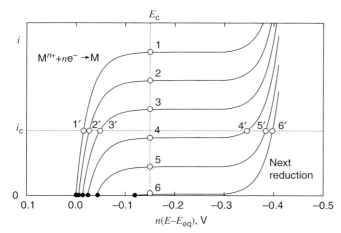

Figure 5.32 Current–potential relations for the electrolytic reduction of metal ion M^{n+}, at the instants of 0, 20, 40, 60, 80, and 99% electrolysis.

5.6.2
Electrogravimetry

In electrogravimetry [22], the analyte, mostly metal ion, is electrolytically deposited quantitatively onto the working electrode and is determined by the difference in the mass of the electrode before and after the electrolysis. A platinum electrode is usually used as a working electrode. The electrolysis is carried out by the controlled-potential or the controlled-current method. The change in the current–potential relation during the process of metal deposition is shown in Figure 5.32. The curves in Figure 5.32 differ from those in Figure 5.30 in that the potentials at $i = 0$ (closed circles) are equal to the equilibrium potential of the M^{n+}/M system at each instant. To apply the curves in Figure 5.32 to the case of a platinum working electrode, the electrode surface must be covered with at least a monolayer of metal M. Then, if the potential of the electrode is kept more positive than the equilibrium potential, the metal (M) on the electrode is oxidized and is dissolved into solution. On the other hand, if the potential of the electrode is kept more negative than the equilibrium potential, the metal ion (M^{n+}) in the solution is reduced and is deposited on the electrode.

In order to deposit more than 99.9% of the metal ion M^{n+} by controlled-potential electrolysis, the potential of the working electrode should be kept more negative than the equilibrium potential at $[M^{n+}] = C_0/1000$ (C_0: initial concentration). Thus, the electrode potential E_c (V) should be

$$E_c < E^{0\prime} + \frac{0.059}{n} \log\left(\frac{C_0}{1000}\right) = E_{eq} - \left(\frac{0.059 \times 3}{n}\right)$$

When the solution contains two metal ions, M_1^{n1+} and M_2^{n2+} (initial concentrations C_{10} and C_{20}, respectively) and when it is necessary to deposit more than 99.9% of

M_1^{n1+} on the electrode, leaving all of M_2^{n2+} in the solution, the potential E_c should be

$$E_1^{0'} + \frac{0.059}{n_1} \log \left(\frac{C_{10}}{1000} \right) > E_c > E_2^{0'} + \frac{0.059}{n_2} \log C_{20} \tag{5.35}$$

where $E_1^{0'}$ and $E_2^{0'}$ are the formal potentials of M_1^{n1+}/M_1 and M_2^{n2+}/M_2, respectively. The condition in Eq. (5.35) can predict the appropriate potential range for separating the two metal ions.

When metal ion M^{n+} is deposited by the controlled-current method, the electrode potential during the electrolysis changes in the order 1', 2', 3', 4', 5', 6' in Figure 5.32, and the next reduction process occurs near the end of the electrolysis. If the solution is acidic and the next reduction process is hydrogen generation, its influence on the metal deposition is not serious. However, if other metal is deposited in the next reduction process, metal M is contaminated with it. To separate the two metal ions M_1^{n1+} and M_2^{n2+} by the controlled-current method, the solution must be acidic and the reduction of hydrogen ion must occur at the potential between the reductions of the two metal ions. An example of such a case is the separation of Cu^{2+} and Zn^{2+} in acidic solutions. If two metal ions are reduced more easily than a hydrogen ion (e.g. Ag^+ and Cu^{2+}), they cannot be separated by the controlled-current method and the controlled-potential method must be used.

5.6.3
Coulometry and Coulometric Titrations

In coulometry, the analyte is quantitatively electrolyzed, and from the quantity of electricity (in coulombs) consumed in the electrolysis, the amount of analyte is calculated using Faraday's law, where the Faraday constant is 9.6485309×10^4 C mol^{-1}. Coulometry is classified into controlled-potential (or potentiostatic) coulometry and controlled-current (or galvanostatic) coulometry, based on the methods of electrolysis [22,23].

In controlled-potential coulometry, the analyte is electrolyzed quantitatively with 100% current efficiency, and the quantity of electricity Q is measured with a coulometer:

$$Q = \int i_t dt = \frac{m}{(M/nF)}$$

Here, m and M are the amount and the molar mass of the analyte. The coulometer is usually an electronic one that integrates the current during the electrolysis, although chemical coulometers, e.g. a silver coulometer and a gas coulometer, can also be used. In this method, the deposition of the analyte is not a necessary process. All substances that are electrolyzed with 100% current efficiency can be determined. They are, for example, metal ions that undergo a change in the oxidation state (e.g. $Fe^{3+} \rightarrow Fe^{2+}$) and organic compounds that are reduced or oxidized at the electrode. Moreover, the working electrode can be a platinum, gold, carbon or mercury electrode. The disadvantages of the controlled-potential coulometry are the long time the electrolysis takes (20–60 min) to complete and the non-negligible

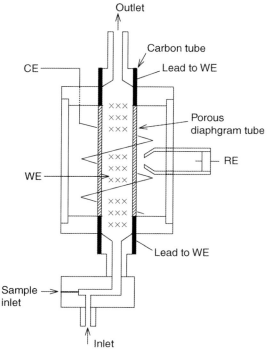

Figure 5.33 A column electrode cell for rapid electrolysis. WE: working electrode of carbon fiber or carbon powder; RE: reference electrode; CE: Pt counter electrode [24a].

effect of the charge consumed by impurities (e.g. oxygen introduced during electrolysis).

Recently *flow coulometry*, which uses a column electrode for rapid electrolysis, has become popular [24]. In this method, as shown in Figure 5.33, the cell has a columnar working electrode that is filled with a carbon fiber or carbon powder and the solution of the supporting electrolyte flows through it. If an analyte is injected from the sample inlet, it enters the column and is quantitatively electrolyzed during its stay in the column. From the peak that appears in the current–time curve, the quantity of electricity is measured to determine the analyte. Because the electrolysis in the column electrode is complete in less than 1 s, this method is convenient for repeated measurements and is often used in coulometric detection in liquid chromatography and flow injection analyses. Besides its use in flow coulometry, the column electrode is really versatile. The versatility can be expanded even more by connecting two (or more) of the column electrodes in series or in parallel. The column electrodes are used in a variety of ways in nonaqueous solutions, as described in Chapter 9.

Controlled-current coulometry is also called *coulometric titration*. An apparatus for controlled-current coulometry is shown in Figure 5.34 for the case of determination of an acid. It consists of a constant current source, a timer, an end point detector (pH

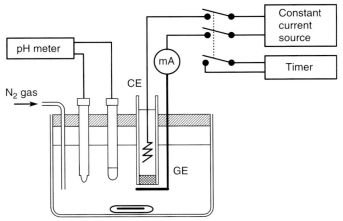

Figure 5.34 Apparatus for controlled-current coulometry. The case for neutralization titration of an acid by *internal* cathodic generation of OH$^-$. GE: generating electrode; CE: counter electrode.

meter) and a titration cell, which contains a generating electrode, a counter electrode in a diaphragm, and two electrodes for pH detection. The timer is actuated when the constant current source is switched on, so that the total electrolysis time can be obtained. OH$^-$ is produced at the generating electrode by the electrolysis of water at constant current i_c (A) and is consumed to neutralize the H$^+$ in the solution. Near the end of the titration, the detector detects the rise in pH and the switch is turn off and on repeatedly until the end point is clearly detected. This procedure is like the manipulation of a burette stopcock in a volumetric titration. If the total duration of the current flow is t (s), the quantity of electricity Q (C) is obtained by $Q = i_c \times t$. The titrant can be generated either inside the titration cell (internal generation) or outside the titration cell (external generation), but it is essential that the current efficiency is 100%. The end point is detected by such methods as potentiometry, amperometry and visual observation of the indicator. This method is applicable to all kinds of titrations, i.e. acid–base, redox, precipitation and complexometric (chelatometric) titrations. The advantages of this method over the classical titrations are that (a) minute amounts of analytes can be determined, (b) unstable titrants can be used by *in situ* generation and (c) the procedure can easily be automated. See Ref. [23] for details.

5.7
Potentiometry – A Method that Does Not Electrolyze Electroactive Species

Potentiometry is a method of obtaining chemical information by measuring the potential of an indicator electrode under zero current flow. It is based on the Nernst equation, which expresses the electrode potential as a function of the activity (or activities) of the chemical species in solutions. The information obtained varies with indicator electrode, from the activity (concentration) of a chemical species to the redox potential in the solution. The potential of the indicator electrode is measured against a reference electrode using a high input impedance mV/pH meter. Potentiometric titration is a method of detecting the titration end point by potentiometry. It also provides various analytical and physicochemical information.

5.7.1
Potentiometric Indicator Electrodes and Reference Electrodes [25]

Potentiometric Indicator Electrodes
Potentiometric indicator electrodes are classified as follows:

Redox electrodes If a platinum electrode is immersed in a solution containing the oxidized and reduced forms (Ox, Red) of a redox reaction $Ox + ne^- \rightleftarrows Red$, its potential is given by the Nernst equation (Section 4.1.1):

$$E = E^0 + \frac{RT}{nF} \ln \frac{a_{ox}}{a_{red}} = E^{0\prime} + \frac{RT}{nF} \ln \frac{[Ox]}{[Red]}$$

where E^0 and $E^{0\prime}$ are, respectively, the standard potential and the formal potential of the redox system. Platinum itself is not involved in the potential-determining reaction. This electrode is called a redox electrode and is used to measure the redox potential of a solution. It works as an indicator electrode for the activity of Ox or Red, if the activity of the other is kept constant. A hydrogen electrode $(H^+|H_2(p = 10^5 \, Pa)|Pt)$, which responds to the pH of the solution with the Nernstian slope, is an example of such cases. Other inert electrode materials such as gold and carbon are also used for redox electrode.

Electrodes of the first kind The electrode of metal M, immersed in a solution of its cation M^{n+}, shows a potential given by

$$E = E^0 + \frac{RT}{nF} \ln a(M^{n+}) = E^{0\prime} + \frac{RT}{nF} \ln[M^{n+}]$$

The potential is determined by reaction $M^{n+} + ne^- \rightleftarrows M$ and the electrode material M is involved in it. The electrodes of this type are called electrodes of the first kind. In principle, these electrodes can indicate the metal ion activities (or concentrations). However, only a few metal ion/metal electrodes work satisfactorily as potentiometric indicator electrodes. An Ag^+/Ag electrode is an example of such electrodes. The potential, determined by reaction $Ag^+ + e^- \rightleftarrows Ag$, is

$$E = E^0 + \frac{RT}{F} \ln a(Ag^+) = E^{0\prime} + \frac{RT}{F} \ln[Ag^+] \tag{5.36}$$

Electrodes of the second kind A typical example is a silver–silver chloride electrode $[Cl^-(aq)|AgCl(s)|Ag]$. To prepare the electrode, a silver wire is coated with AgCl and dipped into a solution containing chloride ions [11g]. Its potential is primarily determined by the reaction $Ag^+ + e^- \rightleftarrows Ag$, and thus by Eq. (5.36). However, because of the relation $K_{sp}(AgCl) = a(Ag^+) \cdot a(Cl^-)$, the potential is determined by the activity of the chloride ion:

$$E = E^0(\mathrm{Ag}^+/\mathrm{Ag}) + 0.0592\log K_{sp} - 0.0592\log a(\mathrm{Cl}^-)$$
$$= E^0(\mathrm{AgCl}/\mathrm{Ag}) - 0.0592\log a(\mathrm{Cl}^-) \quad [25\,^\circ\mathrm{C}]$$

where $E^0(\mathrm{AgCl}/\mathrm{Ag})$ is the standard potential of the silver–silver chloride electrode. The silver–silver chloride electrode is an indicator electrode for chloride ion activity (or concentration). However, more importantly, the silver–silver chloride electrode in saturated (or 3.5 M) KCl is the most popular reference electrode (see below).

Ion-selective electrodes [25a,b] Ion-selective electrodes (ISEs) are usually electrochemical half-cells, consisting of an ion-selective membrane, an internal filling solution and an internal reference electrode (Eq. (5.37)).

| external reference electrode | sample solution | ion-selective membrane | internal solution | internal reference electrode |

ion-selective electrode

(5.37)

They are classified by membrane material into glass membrane electrodes, crystalline (or solid-state) membrane electrodes and liquid membrane electrodes. Liquid membrane electrodes are further classified into liquid ion-exchange membrane electrodes and neutral carrier-based liquid membrane electrodes. Some examples are shown in Figure 5.35 and Table 5.3. If the membrane is

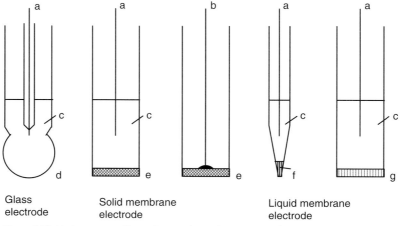

Glass electrode   Solid membrane electrode   Liquid membrane electrode

Figure 5.35 Various types of ion-selective electrodes: (a) internal reference electrode; (b) silver wire for direct contact to the membrane; (c) internal solution; (d) glass membrane; (e) solid-state membrane; (f) ion-exchanger filled at the tip of capillary; (g) ion-exchanger incorporated in PVC membrane.

Table 5.3 Examples of ion-selective electrodes for aqueous solutions.

Type of ISE	Ions	Membrane material
Glass membrane electrodes	H^+	$Na_2O–CaO–SiO_2$
		$Li_2O–Cs_2O–La_2O_3–SiO_2$
	Na^+	$Na_2O–Al_2O_3–SiO_2$
	K^+	$Na_2O–Al_2O_3–SiO_2$
Solid membrane electrodes	F^-	LaF_3
	Cl^-	$AgCl; AgCl–Ag_2S$
	Br^-	$AgBr; AgBr–Ag_2S$
	I^-	$AgI; AgI–Ag_2S$
	CN^-	AgI
	S^{2-}	Ag_2S
	Cu^{2+}	$CuS–Ag_2S$
	Cd^{2+}	$CdS–Ag_2S$
	Pb^{2+}	$PbS–Ag_2S$
Liquid membrane electrode	Cl^-	Dimethyldistearylammonium/Cl^-
	ClO_4^-	1,10-phenanthroline Fe(II)/ClO_4^-
	Ca^{2+}	Didecylphosphate/Ca^{2+}
	Bivalent cation (M^{2+})	Didecylphosphate/M^{2+}
	K^+	Valinomycin/K^+
	Alkali metal ions (M^+)	Crown ether/M^+
Gas-sensing electrode	CO_2	H^+ glass electrode//$NaHCO_3$/GPM[a]
	NH_3	H^+ glass membrane/NH_4Cl/GPM[a]
	SO_2	H^+ glass electrode/$NaHSO_3$/GPM[a]
Enzyme electrode	Urea	NH_4^+ glass electrode/urease membrane

[a] GPM: gas-permeable membrane.

sensitive to ion i of charge z_i and the activities of i in the sample and internal solutions are equal to $a_1(i)$ and $a_2(i)$, respectively, the membrane potential, E_m, which is developed across the membrane is

$$E_m = \frac{RT}{z_i F} \ln \frac{a_1(i)}{a_2(i)}$$

In the ISE, $a_2(i)$ is kept constant. If we express the potential of the ISE by E_{ISE}, that of the external reference electrode by E_{ref}, and the liquid junction potential between the sample solution and the external reference electrode by E_j, the emf of cell (5.37) is given by

$$E = E_{ISE} - E_{ref} + E_j$$

If E_j is negligible or constant, E can be expressed by

$$E = E' + \frac{RT}{z_i F} \ln a_1(i)$$

E' is a constant that depends on the cell configuration.

If the ISE is sensitive to ion i and other ion(s) j, the EMF of the cell is represented by the Nicolsky–Eisenman equation:

$$E = E' + \frac{RT}{z_i F} \ln \left\{ a_1(i) + \sum_{j \neq i} K_{ij}^{pot} a_1(j) \right\}$$

where K_{ij}^{pot} is the selectivity constant [26]. If $K_{ij}^{pot} \ll 1$, the ISE is much more sensitive to ion i than to ion j.

The most popular ISE is a pH glass electrode. However, a great number of ISEs have been developed and widely used in such fields as clinical and environmental analyses. For the use of ISEs in nonaqueous solutions, see Chapter 6.

Ion-selective field effect transistors [25c–e] An ion-selective field effect transistor (ISFET) is a hybrid of an ion-selective electrode and a metal–oxide semiconductor field effect transistor (MOSFET), the metal gate of the MOSFET being replaced by or contacted with a thin film of a solid or liquid ion-sensitive material. The ISFET and a reference electrode are immersed in the solution containing ion i, to which the ISFET is sensitive, and electrically connected as in Figure 5.36. A potential ϕ that varies with the activity of ion i, $a(i)$, as in Eq. (5.38), is developed at the ion-sensitive film:

$$\phi = \text{const.} + \frac{0.059}{z_i} \log a(i) \tag{5.38}$$

In order to keep the drain current (I_D) constant in Figure 5.36, the sum of the gate voltage V_G and ϕ must be kept constant. Thus, if ϕ changes with the ionic

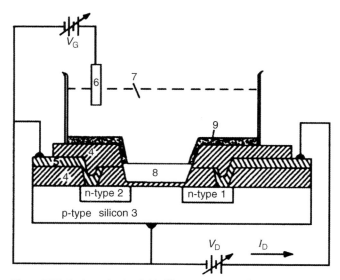

Figure 5.36 An ion-selective field effect transistor. 1, drain; 2, source; 3, substrate; 4, insulator; 5, metal lead; 6, reference electrode; 7, solution; 8, membrane; 9, encapsulant [25c].

Table 5.4 Reference electrodes for aqueous solutions and their potentials (25 °C).[a]

Reference electrode	Half-cell	Condition	E/V versus SHE
Standard hydrogen electrode	$H^+\|H_2(10^5\ Pa)\|Pt$	$a(H^+) = 1$, $p(H_2) = 10^5$ Pa (SHE)	0.0000
Silver–silver chloride electrode	$Cl^-\|AgCl\|Ag$	$a(Cl^-) = 1$	+0.2223
		Saturated KCl	+0.1976
		1 M KCl	+0.2368
		0.1 M KCl	+0.2894
Calomel electrode	$Cl^-\|Hg_2Cl_2\|Hg$	$a(Cl^-) = 1$	+0.2682
		Saturated KCl (SCE)	+0.2415
		Saturated NaCl (SSCE)	+0.2360
		1 M KCl	+0.2807
		0.1 M KCl	+0.3337
Mercury sulfate electrode	$SO_4{}^{2-}\|Hg_2SO_4\|Hg$	$a(SO_4{}^{2-}) = 1$	+0.6158
		Saturated K_2SO_4	+0.650
Mercury oxide electrode	$OH^-\|HgO\|Hg$	$a(OH^-) = 1$	+0.097
		0.1 M NaOH	+0.165

[a] Ref. [27a]. $M = mol\ dm^{-3}$.

activity $a(i)$, we can determine the change by measuring the change in V_G. The advantage of the ISFET is that it is easy to miniaturize. ISFETs have been developed for various ions. Among them, pH-ISFETs with a gate of Si_3N_4, Ta_2O_5 or Al_2O_3 are commercially available, combined with the pH/mV meters for use with them. Recently, pH-ISFETs have been used in routine pH measurements in aqueous solutions. Very promising features of pH-ISFETs in nonaqueous solutions will be described in Chapter 6.

Reference electrodes The primary reference electrode for aqueous solutions is the standard hydrogen electrode (SHE), expressed by $H^+ (a = 1)\|H_2(p = 10^5\ Pa)\|$ Pt (see footnote 1 in Section 4.1). Its potential is defined as zero at all temperatures. In practical measurements, however, other reference electrodes that are easier to handle are used [27]. Examples of such reference electrodes are shown in Table 5.4, with their potentials against the SHE. All of them are electrodes of the second kind. The saturated calomel electrode (SCE) used to be widely used, but today the saturated silver–silver chloride electrode is the most popular.

5.7.2
Potentiometric Titrations

Acid–base, redox, precipitation and chelometric titrations are usually dealt with in textbooks on analytical chemistry. The titration curves in these titrations can be obtained potentiometrically by using appropriate indicator electrodes, i.e. a pH glass electrode or pH-ISFET for acid–base titrations, a platinum electrode for redox titrations, a silver electrode or ISEs for precipitation titrations and ISEs for chelometric titrations. Potentiometric titration curves are often measured with the aid of an automatic burette or an automatic titrator. Measurements of potentiometric titration curves are useful not only for chemical analyses but also for elucidating reaction stoichiometry and for obtaining physicochemical data, such as acid dissociation constants, complex formation constants, solubility products and standard redox potentials. Potentiometric titrations play important roles in chemical studies in nonaqueous solutions, as described in Chapter 6.

5.8
Conductimetry – A Method that Is Not Based on Electrode Reactions

Conductimetry is a method of obtaining analytical and physicochemical information by measuring the conductivities of electrolyte solutions [28]. Conductivity cells have two or four electrodes, but the processes that occur at or near the electrodes are not directly related to the information obtained by conductimetric measurements.

Figure 5.37 shows an electrolyte solution between two plane electrodes. The conductivity of the solution (κ) is expressed by $\kappa = L/AR$, where L is the distance between the two electrodes (cm), A is the electrode area (cm^2) and R is the electrical resistance of the solution (Ω). For dilute electrolyte solutions, the conductivity κ is proportional to the concentrations of the constituent ions, as in Eq. (5.39):

$$\kappa = \sum |z_i| F u_i c_i = \sum \lambda_i c_i \tag{5.39}$$

Here, u_i is the *electric mobility* of ion i and the proportionality constant λ_i ($= |z_i| F u_i$) is the *ionic conductivity* or the *molar conductivity* of ion i.

If a strong electrolyte, consisting of v_+ pieces of cation B^{z+} and v_- pieces of anion B^{z-}, is dissolved to prepare a solution of concentration c, the following relations exist:

$$v_+ z_+ = -v_- z_- = v_- |z_-|, \qquad c = \frac{c_+}{v_+} = \frac{c_-}{v_-}$$

where c_+ and c_- are the concentrations of B^{z+} and B^{z-}, respectively. By substituting these relations in Eq. (5.39), we get

$$\kappa = (u_+ + u_-) z_+ v_+ F c = (u_+ + u_-) |z_-| v_- F c = (v_+ \lambda_+ + v_- \lambda_-) c$$

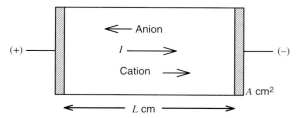

Figure 5.37 Ionic conductivity in electrolyte solutions.

The value Λ, given by Eq. (5.40), is the molar conductivity of an electrolyte:

$$\Lambda = \kappa/c = (u_+ + u_-)z_+ v_+ F = (u_+ + u_-)|z_-|v_- F = v_+ \lambda_+ + v_- \lambda_-$$

$$(5.40)$$

If the conductivity κ and the electrolyte concentration c are expressed in the respective SI units ($S\,m^{-1}$ and $mol\,m^{-3}$), the unit for the molar conductivity Λ is $S\,m^2\,mol^{-1}$. However, κ, c and Λ are often expressed in the respective cgs units ($S\,cm^{-1}$, $mol\,dm^{-3}$ and $S\,cm^2\,mol^{-1}$). In this case, Eq. (5.40) should be replaced by Eq. (5.41):

$$\Lambda(S\,cm^2\,mol^{-1}) = 1000\kappa(S\,cm^{-1})/\{c(mol\,dm^{-3})\} \qquad (5.41)$$

If Λ, λ_+ and λ_- at an infinite dilution are expressed by Λ^∞, λ_+^∞ and λ_-^∞, respectively,

$$\Lambda^\infty = v_+ \lambda_+^\infty + v_- \lambda_-^\infty$$

The values of Λ, λ_+ and λ_- decrease with increasing electrolyte concentration due to the influence of ion–ion interactions. Kohlrausch found the following experimental relation for dilute solutions of strong electrolytes:

$$\Lambda = \Lambda^\infty - kc^{1/2}$$

where k is a constant. A similar relation has been obtained theoretically. According to Debye–Hückel–Onsager limiting law:

$$\Lambda = \Lambda^\infty - (A\Lambda^\infty + B)c^{1/2} \qquad (5.42)$$

where A and B are parameters that depend on ionic charges, viscosity and relative permittivity of solvent and temperature (see footnote 1 in Chapter 7). If we plot the value of Λ against $c^{1/2}$, we get the approximate value of Λ^∞ (for method of obtaining more reliable Λ^∞ value, see Chapter 7). In order to get the values of λ_+^∞ and λ_-^∞, we use transport numbers. The transport number is the fraction of the current carried by a given ion. The transport number of ion i, t_i, is expressed by

$$t_i = \frac{|z_i|u_i c_i}{\sum |z_i|u_i c_i} = \frac{\lambda_i c_i}{\sum \lambda_i c_i}$$

For 1 : 1 electrolyte:

$$t_+ = \frac{u_+}{u_+ + u_-} = \frac{\lambda_+}{\lambda_+ + \lambda_-}, \qquad t_- = 1 - t_+$$

For example, if we know the values of t_K and t_{Cl} for a KCl solution at infinite dilution, we can get λ_K^∞ and λ_{Cl}^∞ from the value of Λ_{KCl}^∞. Once λ_K^∞ and λ_{Cl}^∞ are obtained, we can get the λ^∞ values for other ions by using Kohlrausch's law of independent ionic migration. For example, the value of λ_{Na}^∞ can be obtained by the relation $\lambda_{Na}^\infty = (\Lambda_{NaCl}^\infty - \Lambda_{KCl}^\infty + \lambda_K^\infty)$. Thus, only data on transport numbers for K^+ and Cl^- in the KCl solution are needed.

For the solution of a weak electrolyte, the $\Lambda - c^{1/2}$ relation is quite different from that for the solution of a strong electrolyte. Figure 5.38 shows the $\Lambda - c^{1/2}$ relations for CH_3COOH (weak electrolyte) and HCl (strong electrolyte). The value of Λ for

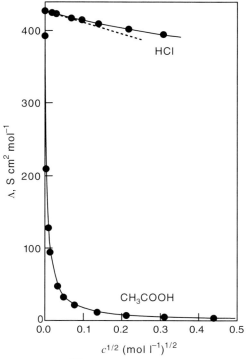

Figure 5.38 Λ–$c^{1/2}$ relationship in aqueous solutions of strong acid HCl and weak acid CH_3COOH.

CH_3COOH decreases abruptly with increase in concentration. If a weak electrolyte consists of v_+ of cation B^{z+} and v_- of anion B^{z-}, the following relations exist:

$$\kappa = \alpha(v_+\lambda_+ + v_-\lambda_-)c, \qquad \Lambda = \alpha(v_+\lambda_+ + v_-\lambda_-)$$

where α is the degree of dissociation of the electrolyte. If α is small enough, the dissociated ions are dilute and behave ideally. In that case

$$\alpha \approx \Lambda/\Lambda^\infty$$

For a 1 : 1 electrolyte, the apparent dissociation constant K' is given by

$$K' = \frac{a^2 c}{1-\alpha} \approx \frac{\Lambda^2 c}{\Lambda^\infty(\Lambda^\infty-\Lambda)} \tag{5.43}$$

We can get an approximate value of K' from Eq. (5.43) by measuring Λ and c. If we consider the effect of concentration on the behavior of dissociated ions, we get Eqs. (5.44) from Eq. (5.42):

$$\alpha = \Lambda/[\Lambda^\infty-(A\Lambda^\infty + B)(\alpha c)^{1/2}] \tag{5.44}$$

The conductivity of a solution is measured using an AC bridge with a two-electrode conductance cell on one arm (Figure 5.39a); a balance is sought, manually or automatically, by adjusting the variable resistance and capacitance in another arm

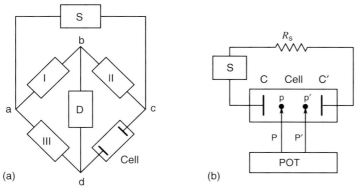

Figure 5.39 Circuits for conductivity measurements with two-electrode cell (a) and four-electrode cell (b). (a) S: AC voltage source; D: detector; I, II, III: bridge elements. (b) S: constant current source; POT: potentiometer; R_s: variable resistor; C and C': electrodes for current flow; P and P': electrodes for voltage measurement.

of the bridge. Usually, AC voltage of a few volts and ~1 kHz is applied to the cell. The impedance caused by the double-layer capacity at the electrodes does not affect the measured values of conductivity. In some cases, the conductance is measured with a four-electrode cell, as shown in Figure 5.39b. For practical methods of measurement, see the reviews in Ref. [28].

Conductimetry is used to detect and determine ionic species in solution. It plays important roles in environmental analyses and in the detection of ion chromatography. It is also used in acid–base, precipitation and chelometric titrations to detect end points. However, the biggest advantage of conductimetry is displayed in the fundamental studies of solution chemistry. Its applications to chemical studies in nonaqueous solutions will be discussed in Chapter 7.

5.9
Electrochemical Instrumentation – Roles of Operational Amplifiers and Microcomputers

The previous sections outlined various electrochemical techniques. For most of these techniques, instruments are commercially available and we can use them conveniently. The common features of modern electrochemical instruments are that operational amplifiers and microcomputers play important roles in them. These are discussed in this section.

5.9.1
Application of Operational Amplifiers in Electrochemical Instrumentation

Operational amplifiers, which are the main components of an analogue computer, were first used in electrochemical instrumentation at the beginning of the 1960s [29].

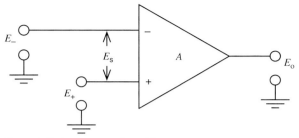

Figure 5.40 An operational amplifier.

Because they are extremely useful in measuring and controlling the electrode potentials and the currents that flow at the electrodes, electrochemical instruments were completely modernized by their introduction. Today, most electrochemical instruments are constructed using operational amplifiers. Knowledge of operational amplifiers will help the reader to understand electrochemical instruments and to construct a simple apparatus for personal use.

1. **Basic operational amplifier circuits** An operational amplifier is generally a differential amplifier, as shown in Figure 5.40, with an inverting input $(-)$, a noninverting input $(+)$ and an output. Ideally, it has the following characteristics: (i) an infinite input impedance (actually $\sim 10^{15}\,\Omega$ maximum), (ii) an almost infinite gain A (actually $A = 10^4 - 10^6$), (iii) an output voltage E_o given by $E_o = A(E_+ - E_-)$ and (iv) a large (but finite) output voltage and a high output current (e.g. ± 10 V and ± 10 mA). From these characteristics, the following can be concluded:

    (1) At each of the two inputs, almost no current flows from the external circuit.

    (2) The value of $(E_+ - E_-)$ is very close to zero (typically $< \pm 0.1$ mV) and, if the noninverting input $(+$ input) is grounded, the inverting input $(-$ input) is held at virtual ground.

Some important circuits constructed with an operational amplifier are shown in Figure 5.41. Their functions are easily understood by use of (1) and (2) above.

Inverting, multiplying or dividing circuit In Figure 5.41a, point s is held at virtual ground, i.e. the potential at point s is actually equal to zero from (2) above. The current i_i, which is equal to (E_i/R_i), flows through the resistor R_i and then, by (1), through the resistor R_f. Thus,

$$E_o = -E_i \times \left(\frac{R_f}{R_i}\right)$$

Circuit (a) works as an inverter of the input voltage because $E_o = -E_i$ for $R_i = R_f$. It also works as a multiplying (or dividing) circuit if the values of R_i and R_f are selected appropriately.

Summing circuit In Figure 5.41b, because point s is held at virtual ground, the currents $i_1 = (E_1/R_1)$, $i_2 = (E_2/R_2)$ and $i_3 = (E_3/R_2)$ flow through the resistors R_1,

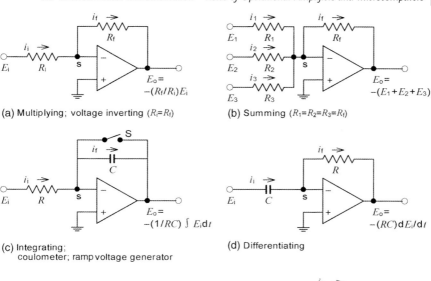

(a) Multiplying; voltage inverting ($R_f=R_t$)

(b) Summing ($R_1=R_2=R_3=R_t$)

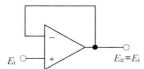

(c) Integrating;
coulometer; ramp voltage generator

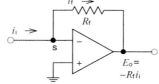

(d) Differentiating

(e) Voltage follower;
high input impedance voltmeter
(pH meter)

(f) Current follower;
current-to-voltage converter

Figure 5.41 Feedback circuits for mathematical operations with an operational amplifier.

R_2 and R_3, respectively, and join at the summing point s to flow through R_f. Thus,

$$E_o = -\left\{ E_1 \left(\frac{R_f}{R_1}\right) + E_2 \left(\frac{R_f}{R_2}\right) + E_3 \left(\frac{R_f}{R_3}\right) \right\}$$

If $R_1 = R_2 = R_3 = R_f$,

$$E_o = -(E_1 + E_2 + E_3)$$

This summing circuit is used in polarography and voltammetry to superimpose AC, SW or pulse voltage to DC applied voltage.

Integrating and differentiating circuits In Figure 5.41c, the capacitor C is charged by current $i_i = (E_i/R)$. Thus, the quantity of charged electricity is expressed by $Q = \int_0^t i_i dt = (1/R) \int_0^t E_i dt$. Because point s is a virtual ground, the output voltage is

$$E_o = -\frac{1}{C} \int_0^t i_i dt = -\frac{1}{RC} \int_0^t E_i dt$$

This integrating circuit is used to give linear and cyclic voltage scans in polarography and voltammetry. It is also used as a coulometer in coulometry.

A simple differentiating circuit is shown in Figure 5.41d. If the capacitor in Figure 5.41c and d is replaced by a diode, logarithmic and exponential circuits can be obtained.

Voltage and current followers Figure 5.41e shows a circuit for a voltage follower. From (1) and (2) above, $E_o = E_i$. Moreover, by using an appropriate operational amplifier, the impedance of the noninverting input can be kept very high ($\sim 10^{13}\ \Omega$ or more). Thus, the voltage E_i of a high-impedance source can be transmitted to a low-impedance circuit, keeping its magnitude and sign unchanged. This circuit is often used in constructing mV/pH meters of high input impedance. It is also used as a voltage follower in a potentiostat. It can prevent a current flowing through the reference electrode, keeping the potential unchanged (VF in Figure 5.42b). The circuit in Figure 5.41f is a current follower and converts the input current to the output voltage $E_o = -R_f i_i$, which can be measured or recorded. Because point s is a virtual ground, the circuit can be inserted in the grounded current-carrying circuit, as CF in Figure 5.42b, without introducing any impedance.

2. **Potential and current controls in electrochemical instrumentation by use of operational amplifiers** It is usual in electrochemical measurements to control the potential of the working (or indicator) electrode or the electrolytic current that flows through the cell. A potentiostat is used to control electrode potential and a galvanostat is used to control electrolytic current. Operational amplifiers play important roles in both of these.

Circuit for potential control and a potentiostat Figure 5.42a shows a potential control circuit. By (1) and (2) above, point s is held at virtual ground and current i ($= E/R_1$) flows through resistor R_1 and then through R_2. If $R_1 = R_2$, the potential at point x is held equal to $-E$, irrespective of the values of impedance Z_1 and Z_2. This relation is used in the circuit of a potentiostat as in Figure 5.42b, where the potential of the RE is equal to $-E$ against ground. Because the working electrode (WE) is grounded, its potential is equal to E against the RE, irrespective of the

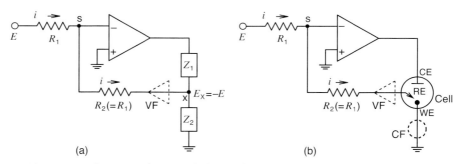

Figure 5.42 Voltage control circuit with the aid of an operational amplifier (a) and a circuit of potentiostat (b). VF: voltage follower; CF: current follower.

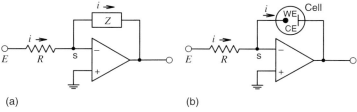

Figure 5.43 Current control circuit with the aid of an operational amplifier (a) and a circuit of galvanostat (b).

impedance of the electrolytic solution and the reaction at the electrode. Here, VF inserted between the resistor R_2 and RE is a voltage follower, which prevents the current flow through RE. On the other hand, CF between WE and the ground is a current follower or a current-to-voltage converter. The current that flows at the WE is measured as a voltage. It should be noted that the performance of the potentiostat is unaffected by the insertion of voltage and current followers.

Circuit for current control and a galvanostat Figure 5.43a shows a current control circuit. The current i $(= E/R)$ flows through the resistor R and then through the impedance Z. This circuit is used in a galvanostat as shown in Figure 5.43b. The current through the cell is controlled at i $(= E/R)$, irrespective of the cell impedance and the reaction at the electrode. When the potential of the working (or indicator) electrode WE is to be measured, an RE is inserted in the cell and the voltage between WE and RE is measured (Figure 5.31b) using a voltage follower and a millivolt meter.

Three-electrode instruments for polarography and voltammetry In Figure 5.44, if E connected to point a is a DC voltage source that generates a triangular voltage cycle, we can use the circuit of Figure 5.44 for measurements in DC polarography as well as in linear sweep or cyclic voltammetry. An integrating circuit as in Figure 5.41c is applicable to generate a triangle voltage cycle. In Figure 5.44, bold lines show positive feedback iR compensation.[15] When a sine-wave or square-wave voltage of small amplitude is to be superimposed on the DC voltage, its voltage source is connected to point b. Then, the potential of the indicator electrode (WE) is controlled at a value equal to the sum of the voltages at points a and b. Of course, appropriate modifications of the current measuring circuit are necessary to get AC or SW polarograms and voltammograms.

15) Positive feedback iR compensation in three-electrode measurements: As described in Section 5.3, the influence of iR drop is significant in two-electrode polarography or voltammetry. The influence is eliminated considerably with three-electrode instruments if the tip of the reference electrode is placed near the surface of the indicator electrode. However, there still remains some iR drop, which occurs by the residual resistance at the indicator electrode or in the solution between the surface of the indicator electrode and the tip of the reference electrode. To eliminate the iR drop, the positive feedback circuit as in Figure 5.44 is used. By adjusting R_c, the voltage that is equal to the uncompensated iR drop is added to the applied voltage. For more detailed method of overcoming the influence of solution resistance, see Ref. [30].

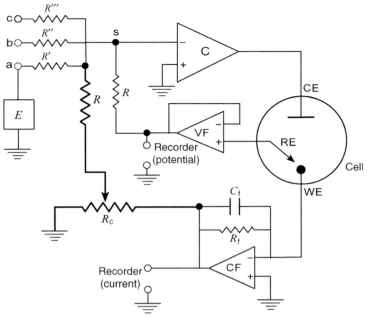

Figure 5.44 Circuit for a three-electrode polarograph. The bold lines are for positive feedback iR drop compensation. C_f is a capacitor to decrease the current fluctuation.

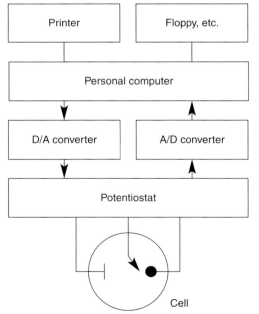

Figure 5.45 A personal computer-aided electrochemical apparatus.

5.9.2
Application of Personal Computers in Electrochemical Instrumentation

In modern instrumentation in polarography and voltammetry, a personal computer (PC) is often connected to a potentiostat, by inserting an analogue-to-digital converter (ADC) and a digital-to-analogue converter (DAC) between them (Figure 5.45) [31]. The DAC converts digital signals from the PC to analogue signals and transfers them to the potentiostat. The ADC, on the other hand, converts analogue signals from the potentiostat to digital signals and transfers them to the PC. This digital instrumentation offers various advantages.

 (i) It is suitable for staircase, square-wave or pulse polarography and voltammetry. Using a simple computer program, the PC generates a staircase, square-wave or pulse voltage to be applied to the cell, and then it reads the current flowing through the cell circuit as digital signals and processes them in a variety of ways. In SW and pulse polarography and voltammetry, the program samples only the current that is solely faradaic (see Figure 5.15).

(ii) The digital instrumentation is useful in rapid-scan cyclic voltammetry. Data of the cyclic voltammogram are first stored in a memory device and then processed or graphically presented. With a sophisticated digital instrument and an ultramicroelectrode, potential sweeps of $1\,000\,000\,\mathrm{V\,s^{-1}}$ or more are possible.

(iii) The digital simulation makes mechanistic studies of electrode processes by cyclic voltammetry much easier. Cyclic voltammograms readily provide qualitative information about electrode processes but, without the simulation technique, it is rather difficult to get quantitative mechanistic information from them.

References

1 For example, (a) Bard, A.J. and Faulkner, L.R. (2001) *Electrochemical Method: Fundamentals and Applications*, 2nd edn, John Wiley & Sons, Inc., New York; (b) Delahay, P. (1954) *New Instrumental Methods in Electrochemistry*, Interscience Publishers, New York; (c) Rossiter, B.W. and Hamilton, J.F. (eds) (1986) *Physical Methods of Chemistry, vol. 2, Electrochemical Methods*, 2nd edn, John Wiley & Sons, Inc., New York; (d) Koryta, J., Dvořák, J. and Kavan, K. (1993) *Principles of Electrochemistry*, 2nd edn, John Wiley & Sons, Inc., New York; (e) Brett, C.M.A. and Brett, A.M.O. (1993) *Electrochemistry, Principles, Methods, and Applications*, Oxford University Press, Oxford; (f) Brett, C.M.A. and Brett, A.M.O. (1998) *Electroanalysis*, Oxford University Press, Oxford; (g) Sawyer, D.T., Sobkowiak, A.J. and Roberts, J.L., Jr (1995) *Electrochemistry for Chemists*, 2nd edn, John Wiley & Sons, Inc., New York; (h) Rubinstein, I. (ed.) (1995) *Physical Electrochemistry*, Marcel Dekker, New York; (i) Kissinger, P.T. and Heinemann, W.R. (eds) (1996) *Laboratory Techniques in Electroanalytical Chemistry*, 2nd edn, Marcel Dekker, New York; (j) Bard, A.J. (ed.) (1966–1998) *Electroanalytical Chemistry*, vols. **1–20**, Marcel Dekker, New York; (k) Bard, A.J., Stratmann, M. and Unwin, P.R. (eds) (2003) *Encyclopedia of Electrochemistry, vol. 3, Instrumentation and Electroanalytical*

Chemistry, Wiley-VCH Verlag GmbH, Weinheim; (l) Wang, J. (2006) *Analytical Electrochemistry,* 3rd edn, John Wiley & Sons, Inc., New York.

2 For example, (a) Kolthoff, I.M. and Lingane, J.J. (1952) *Polarography,* 2nd edn, Interscience Publishers, New York; (b) Heyrovsky, J. and Kuta, J. (1966) *Principles of Polarography,* Academic Press, New York; (c) Bond, A.M. (1980) *Modern Polarographic Methods in Analytical Chemistry,* Marcel Dekker, New York; (d) Ref. [1].

3 (a) Zuman, P. (2001) *Crit. Rev. Anal. Chem.,* **31**, 281; (b) Barek, J. and Zima, J. (2003) *J Electroanalysis,* **15**, 467; (c) Bard, A.J. (2007) *J. Chem. Ed.,* **84**, 644.

4 Breyer, B. and Bauer, H.H. (1963) *Alternating Current Polarography and Tensammetry,* John Wiley & Sons, Inc., New York; Refs [1] and [2].

5 Barker, G.C. and Jenkins, I.L. (1952) *Analyst,* **77**, 685.

6 Barker, G.C. and Gardner, A.W. (1960) *Z. Anal. Chem.,* **173**, 79.

7 Laitinen, H.A. and Kolthoff, I.M. (1941) *J. Phys. Chem.,* **45**, 1079.

8 (a) Bond, A.M. (2002) *Broadening Electrochemical Horizons, Principles and Illustration of Voltammetric and Related Techniques,* Oxford University Press, Oxford; (b) Savéant, J.-M. (2006) *Elements of Molecular and Biomolecular Electrochemistry, An Electrochemical Approach to Electron Transfer Chemistry,* John Wiley & Sons, Inc., NJ; (c) Compton, R.G. and Banks, C.E. (2007) *Understanding Voltammetry,* World Scientific, London.

9 For example, (a) Ref. [1a], Chapter 9; (b) Galus, Z.,Ref. [1c], Chapter 3; (c) Filinovsky, V.Yu. and Pleskov, Yu.V. (1984) *Comprehensive Treatise of Electrochemistry, 9, Electrodics: Experimental Techniques,* (eds E. Yeager, J.O'M. Bockris, B.E. Conway and S. Sarangapani), Plenum Press, New York, 293.

10 (a) Wightman, R.M. and Wipf, D.O. (1989) *Electroanalytical Chemistry,* vol. 15 (ed. J.A. Bard), Marcel Dekker, New York, Chapter

3; (b) Michael, A.C. and Wightman, R.M., Ref. [1i], Chapter 12; (c) Amatore, C.,Ref. [1h], Chapter 4; (d) Štulik, K., Amoatore, C., Holub, K., Mareček, V. and Kutner, W. (2000) *Pure Appl. Chem.,* **72**, 1483; (e) Aoki, K., Morita, M., Horiuchi, T. and Niwa, O. (1998) *Methods of Electrochemical Measurements with Microelectrodes,* IEICE, Tokyo, (in Japanese).

11 (a) Ref. [1a], Chapter 14; (b) Martin, C.R. and Foss, C.A., Jr, Ref. [1i], Chapter 13; (c) Murray, R.W. (1988) *Electroanalytical Chemistry,* vol. 13 (ed. A.J. Bard), Marcel Dekker, New York, Chapter 3; (d) For recommended terminology, Durst, R.A., Bäumner, A.J., Murray, R.W., Buck, R.P. and Andrieux, C.P. (1997); *Pure Appl. Chem.,* **69**, 1317.

12 For example, (a) Orazem, M.E. and Tribollet, B. (2008) *Electrochemical Impedance Spectroscopy,* John Wiley & Sons, Inc., Hoboke, N.J.; (b) Sluyters-Rehbach, M. and Sluyters, J.H. (1970) *Electroanalytical Chemistry,* vol. 4 (ed. A.J. Bard), Marcel Dekker, New York; (c) Sluyters-Rehbach, M. and Sluyters, J.H. (1984) *Comprehensive Treatise of Electrochemistry, 9, Electrodics: Experimental Techniques* (eds E. Yeager, J.O'M. Bockris, B.E. Conway and S. Sarangapani), Plenum Press, New York, p. 177; (d) Lisia, A. (1999) *Modern Aspects of Electrochemistry,* vol. 32 (eds R.E. White, B.E. Conway and J.O'M. Bockris), Plenum Press, New York, Chapter 2; (e) Gabrielli, C.,Ref. [1h], Chapter 6; (f) Ref. [1a], Chapter 10.

13 (a) Girault, H.H.J. and Schiffrin, D.J. (1989) *Electroanalytical Chemistry,* vol. 15 (ed. A.J. Bard), Marcel Dekker, New York, Chapter 1; (b) Samec, Z. and Kakiuchi, T. (1995) *Advances in Electrochemical Science and Engineering,* vol. 4 (eds. H. Gerischer and C.W. Tobias), Wiley-VCH Verlag GmbH, Weinheim, Chapter 5; (c) Volkov, A.G. and Deamer, D.W. (eds) (1996) *Liquid–Liquid Interfaces, Theory and Methods,* CRC Press, New York; (d) Volkov, A.G., Deamer, D.W., Tanelian, D.L. and Markin, V.S. (eds) (1998) *Liquid*

Interfaces in Chemistry and Biology, John Wiley & Sons, Inc., New York; (e) Kihara, S. and Maeda, K. (1994) *Prog. Surf. Sci.*, **47**, 1.

14 (a) Wang, J.,Ref. [1k], Section 2.3; (b) Wang, J. (1988) *Electroanalytical Chemistry*, vol. 16 (ed. A.J. Bard), Marcel Dekker, New York, Chapter 1; (c) Brainina, Kh. and Neyman, E. (1993) *Electroanalytical Stripping Methods*, John Wiley & Sons, Inc., New York; (d) Coetzee, J.F. Hussam, A. and Petrick, T.R. (1983) *Anal. Chem.*, **55**, 120.

15 (a) Kellner, R., Mermet, J.-M., Otto, M.and Widmer, H.M. (eds) (1998) *Analytical Chemistry*, Wiley-VCH Verlag GmbH, Weinheim, pp. 375–429; (b) Thévenot, D.R., Toth, K., Durst, R.A. and Wilson, G.S. (1999) *Pure Appl. Chem.*, **71**, 2333; (c) Nakamura, T. (2007) *Anal. Sci.*, **23**, 253.

16 For example, (a) Ref. [1a], Chapters 16 and 17; (b) Various sections in Ref. [1k]; (c) Abruña, H.D. (ed.) (1991) *Electrochemical Interfaces*, John Wiley & Sons, Inc., New York.

17 (a) Nomura, T. and Minomura, A. (1980) *Nippon Kagaku Kaishi*, 1621; (b) Nomura, T., Nagamine, T., Izutsu, K. and West, T.S. (1981) *Bunseki Kagaku*, **30**, 494, both in Japanese.

18 (a) Bruckenstein, S. and Shay, M. (1985) *Electrochim. Acta*, **30**, 1295; (b) Buttry, D.A. (1991) *Electroanalytical Chemistry*, vol. 17 (ed. A.J. Bard), Marcel Dekker, New York, Chapter 1; (c) Buttry, D.A., Ref. [16c], Chapter 10; (d) Buck, R.P., Lindner, E., Kutner, W. and Inzelt, G. (2004) *Pure Appl. Chem.*, **76**, 1139; (e) Ward, M.D.,Ref. [1h], Chapter 7; (f) Hillman, R.,Ref. [1k], Section 2.7.

19 (a) Ref. [1a], Chapter 17; (b) McCreery, R.L., Ref. [1c], Chapter 7; (c) Crayston, J.A.,Ref. [1k], Section 3.4; (d) Kuwana, T. and Winograd, N. (1974) *Electroanalytical Chemistry*, vol. 7 (ed. A.J. Bard), Marcel Dekker, New York, Chapter 1; (e) Heineman, W.H., Hawkridge, F.M. and Blount, H.N. (1988) *Electroanalytical Chemistry*, vol. 13 (ed. A.J. Bard), Marcel

Dekker, New York, Chapter 1; (f) Kaim, W. and Klein, A. (2008) *Spectroelectrochemistry*, RSC Publishing, Cambridge.

20 (a) McKinney, T.M. (1977) *Electroanalytical Chemistry*, vol. 10 (ed. A.J. Bard), Marcel Dekker, New York, Chapter 2; (b) Goldberg, I.B. and McKinney, T.M.,Ref. [1i], Chapter 29; (c) Wadhawan, J. and Compton, R.G. (2003) *Encyclopedia of Electrochemistry, vol. 2, Interfacial Kinetics and Mass Transport*, (eds A.J. Bard, M. Stratmann and E.J. Calvo), Wiley-VCH Verlag GmbH, Weinheim, Section 3.2.

21 For example, (a) Ref. [1a], Chapter 16; (b) Bard, A.J., Fan, F.-R. and Mirkin, M., Ref. [1h], Chapter 5; (c) Horrocks, B.R., Ref. [1k], Section 3.3; (d) Bard, A.J., Denuault, G., Lee, C., Mandler, D. and Wipf, D.O. (1990) *Acc. Chem. Res.*, **23**, 357; (e) Bard, A.J. and Mirkin, M.V. (2001) *Scanning Electrochemical Microscopy*, Marcel Dekker, New York.

22 Ref. [1a], Chapter 11.

23 Curran, D.J.,Ref. [1i], Chapter 25.

24 (a) Fujinaga, T. and Kihara, S. (1977) *CRC Crit. Rev. Anal. Chem.*, **6**, 223; (b) Sioda, R.E. and Keating, K.B. (1982) *Electroanalytical Chemistry*, vol. 12 (ed. A.J. Bard), Marcel Dekker, New York, Chapter 1.

25 For example, (a) Koryta, J. and Stulik, K. (1983) *Ion-Selective Electrodes*, 2nd edn, Cambridge University Press, Cambridge; (b) Evans, A. (1987) *Potentiometry and Ion-selective Electrodes*, John Wiley & Sons, Inc., New York; (c) Kissel, T.R.,Ref. [1c], Chapter 2; (d) Janta, J. (1990) *Chem. Rev.*, **90**, 691; (e) Covington, A.K. (1994) *Pure Appl. Chem.*, **66**, 565.

26 (a) Umezawa, Y., Umezawa, K. and Sato, H. (1995) *Pure Appl. Chem.*, **67**, 507; (b) Umezawa, Y., Bühlmann, P., Umazawa, K., Tohda, K. and Amemiya, S. (2000) *Pure Appl. Chem.*, **72**, 1851; (c) Umezawa, Y. (ed.) (1990) *Handbook of Ion-Selective Electrodes: Sensitivity Coefficients*, CRC Press, Boca Raton, FL.

27 (a) Ives, D.J.G. and Janz, G.J. (eds) (1961) *Reference Electrodes*, Academic Press, New York; (b) Ref. [1g], Chapter 5.

28 For example, (a) Spiro, M.,Ref. [1c], Chapter 8; (b) Holler, F.J. and Enke, C.G., Ref. [1i], Chapter 8.

29 For example, (a) Ref. [1a], Chapter 15; (b) Kissinger, P.T.,Ref. [1i], Chapter 6; (c) Mattson, J.S., Mark, H.B., Jr and MacDonald, H.C., Jr (eds) (1972) *Electrochemistry: Calculations, Simulations and Instrumentation*, Marcel Dekker, New York; (d) Kalvoda, R. (1975) *Operational Amplifiers in Chemical Instrumentation*, Ellis Horwood, Chichester.

30 Roe, D.K.,Ref. [1i], Chapter 7.

31 (a) Wipf, D., Ref. [1k], Section 1.2; (b) He, P., Avery, J.P. and Faulkner, L.R. (1982) *Anal. Chem.*, **54** 1313A.

6
Potentiometry in Nonaqueous Solutions

Just as in aqueous solutions, potentiometry is the most fundamental and powerful method of measuring pH, ionic activities and redox potentials in nonaqueous solutions. In this chapter, we deal with the basic techniques of potentiometry in nonaqueous solutions and then discuss how potentiometry is applicable to studies of chemistry in nonaqueous solutions. Some topics in this field have been reviewed in Ref. [1a].

6.1
Basic Techniques of Potentiometry in Nonaqueous Solutions

In potentiometry, we measure the emf of a cell consisting of an indicator electrode and a reference electrode. For emf measurements, we generally use a pH/mV meter of high input impedance. The potential of the reference electrode must be stable and reproducible. If there is a liquid junction between the indicator electrode and the reference electrode, we should take the liquid junction potential (LJP) into account.

In aqueous solutions, the method of measuring electrode potentials has been well established. The standard hydrogen electrode (SHE) is the primary reference electrode and its potential is defined as zero at all temperatures. Practical measurements employ reference electrodes that are easy to use, the most popular ones being a silver–silver chloride electrode and a saturated calomel electrode (Table 5.4). The magnitude of the liquid junction potential between two aqueous electrolyte solutions can be estimated by the Henderson equation. However, it is usual to keep the LJP small either by adding the same indifferent electrolyte in the two solutions or by inserting an appropriate salt bridge between the two solutions.

In contrast, in nonaqueous solutions, the method of measuring electrode potential has not been well established. The most serious problem is the reference electrode; there is no primary reference electrode such as the SHE in aqueous solutions and no reference electrode as reliable as the aqueous Ag/AgCl electrode. Thus, various reference electrodes have been employed in practical measurements, making the comparison of potential data difficult. As will be described later, various efforts are being made to improve this situation.

Electrochemistry in Nonaqueous Solutions, Second, Revised and Enlarged Edition. Kosuke Izutsu
Copyright © 2009 WILEY-VCH Verlag GmbH & Co. KGaA, Weinheim
ISBN: 978-3-527-32390-6

Table 6.1 Potentiometric indicator electrodes for use in nonaqueous solutions.

Property measured	Indicator electrodes
Redox potential	Inert redox electrodes (Pt, Au, glassy carbon, etc.)
pH	pH glass electrode; pH-ISFET; iridium oxide pH sensor
Ionic activities	Electrodes of the first kind (M^{n+}/M and $M^{n+}/M(Hg)$ electrodes); univalent cation-sensitive glass electrode (alkali metal ions, NH_4^+); solid-membrane ion-selective electrodes (F^-, halide ions, heavy metal ions); polymer-membrane electrodes (F^-, CN^-, alkali metal ions, alkaline earth metal ions)

6.1.1
Potentiometric Indicator Electrodes for Nonaqueous Solutions

Potentiometric indicator electrodes for aqueous solutions were dealt with in Section 5.7.1. Some of them are also applicable in nonaqueous solutions, if they are durable and can respond appropriately. Examples of such indicator electrodes are given in Table 6.1. A platinum electrode and other inert electrodes are used to measure the redox potential. Metal and metal amalgam electrodes are used to measure the activities of the corresponding metal ions. Univalent cation-sensitive glass electrodes and some solid-membrane ion-selective electrodes (ISEs) are useful to measure the activities of the cations and anions to which they respond. The conventional pH glass electrode is the most popular for pH measurements in nonaqueous solutions, but other pH sensors such as pH-ISFETs are also applicable. Although various types of liquid-membrane ISEs have been developed for aqueous solutions, most of them are not applicable in nonaqueous solutions, because either their sensing membrane or supporting body is destroyed in organic solvents. However, as shown in Table 6.1 and Section 6.3.2, new polymer-membrane ISEs have been developed and used successfully in nonaqueous solutions [2]. Applications of ISEs in nonaqueous solutions have been reviewed [3] and will be discussed in Sections 6.2 and 6.3.

6.1.2
Reference Electrodes for Nonaqueous Solutions

The reference electrodes used in nonaqueous systems can be classified into two types. One type uses, in constructing a reference electrode, the same solvent as that of the solution under study. The other type is an aqueous reference electrode, usually an aqueous Ag/AgCl electrode or SCE. Some reference electrodes are listed in Table 6.2 and are briefly discussed below. For other types of reference electrodes used in nonaqueous solutions, see Ref. [4].

Reference Electrodes that Use the Same Solvent as the Solution Under Study

Silver–silver ion electrode This is the most popular reference electrode used in nonaqueous solutions. Since Pleskov employed it in acetonitrile (AN) in 1948, it has been used in a variety of solvents. It has a structure as shown in Figure 6.1a

Table 6.2 Reference electrodes for use in nonaqueous solutions.

Reference electrodes	MeOH	AN	PC	NM	TMS	DMSO	DMF	NMP
$Ag^+(S)|Ag$	G	G	F	F	F	G	F	F
$AgCryp(22)^+$, $Cryp(22)(S)|Ag$	G	G	G	G	G	G	G	G
$Cl^-(S)|AgCl|Ag$; $AgCl_2^-(S)|Ag$	G	G	G	G	G	G	G	G
I_3^-, $I^-(S)|Pt$	G	G	G	G	G	G	G	G
Aq. SCE; aq. $Cl^-|AgCl|Ag$ + salt bridge(S)	F^a	F^a	F^a	F^a	F^a	F^a	F^a	F^a

G: good for general use and F: applicable under limited conditions (see text).
^{a}Liquid junction potential (in millivolt) at 'satd. $KCl(H_2O)/0.1$ M $Et_4NPic(S)$' estimated in Diggle, J.W., Parker, A.J. (1974) *Aust. J. Chem.*, **27**, 1617: S = MeOH 25, AN 93, PC 135, NM 59, DMSO 172, DMF 174.

and is easy to construct. Its potential is usually reproducible within 5 mV if it is prepared fresh using pure solvent and electrolyte. The stability of the potential, however, is not always good enough. The potential is stable in AN because Ag^+ is strongly solvated in it. In propylene carbonate (PC) and nitromethane (NM), however, Ag^+ is solvated only weakly and the potential is easily influenced by the presence of trace water and other impurities.[1] In dimethylformamide (DMF),

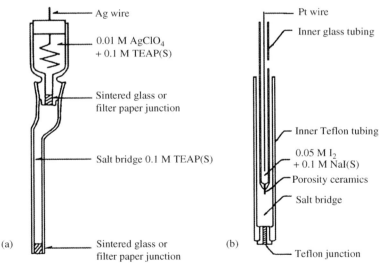

(a)
- Ag wire
- 0.01 M $AgClO_4$ + 0.1 M TEAP(S)
- Sintered glass or filter paper junction
- Salt bridge 0.1 M TEAP(S)
- Sintered glass or filter paper junction

(b)
- Pt wire
- Inner glass tubing
- Inner Teflon tubing
- 0.05 M I_2 + 0.1 M NaI(S)
- Porosity ceramics
- Salt bridge
- Teflon junction

Figure 6.1 Reference electrodes for nonaqueous solutions.
(a) Ag/Ag^+ electrode and (b) Pt/I_3^-, I^- electrode [7]
(TEAP = Et_4NClO_4).

1) If the solvation of Ag^+ is weak, adding 1 or 2(v/v)% AN to the Ag^+ solution is effective in improving the stability of the potential of the Ag^+/Ag electrode (Ref. [24b]).

however, Ag^+ is slowly reduced to Ag^0, causing a gradual potential shift to the negative direction.[2] The shift reaches several tens of millivolts after a few days.

Silver–silver cryptate electrode This type of electrode was developed by this author in order to improve the stability of the silver–silver ion electrode [5]. It has a structure of Ag/5 mM AgCrypClO$_4$, 5 mM Cryp, 50 mM Et$_4$NClO$_4$(S), where Cryp and AgCryp$^+$ denote cryptand(22) and its complex with Ag^+, respectively, and can be prepared in various solvents. Its potential is given by $E = E^0(Ag^+/Ag) - 0.059 \log K + 0.059 \log([AgCryp^+]/[Cryp])$ (K: formation constant of AgCryp$^+$). Because the electrolyte solution is an Ag^+ ion buffer, the potential is stable if the ratio $[AgCryp^+]/[Cryp]$ is kept constant. Thus, the potential is less influenced by solvent impurities than in the case of an Ag^+/Ag electrode. Moreover, as shown in Table 4.3, the potential of this electrode is not so much influenced by the solvent. Recently, Lewandowski et al. used cryptand(222) to develop a similar reference electrode [6].

Silver–silver chloride electrode In protic solvents such as methanol, this type of electrode can be constructed similarly as in aqueous solutions. In aprotic solvents, however, it is difficult to prepare a silver–silver chloride electrode of high Cl^- concentrations. The Ag^+ ion in aprotic solvents strongly interacts with chloride ion to form $AgCl_2^-$ complex (Section 2.4). Thus, the precipitate of AgCl is easily dissolved in solutions of high Cl^- concentrations. When we construct a silver–silver chloride electrode in an aprotic solvent, we keep the free Cl^- concentration low, using a solution of a chloride salt that is either sparingly soluble or of dilute concentration, and saturate AgCl in it. Examples of such electrodes are listed in Table 6.3. Though it takes some time to prepare the AgCl-saturated solution, electrodes of this type work well with a long-term potential stability.

Pt/I$_3^-$, I$^-$ electrode This electrode, proposed by Coetzee and Gardner [7], is of a double-junction type as shown in Figure 6.1b. It is easy to construct in many solvents, and the potential is stable and reproducible. Because the electrode reaction has a high exchange current, the potential shift caused by the current flow

Table 6.3 Silver–silver chloride electrodes in nonaqueous solvents.

Solvents	Electrodes [salt bridge]
AN	AgCl(satd.) + Me$_3$EtNCl(satd.)|AgCl|Ag
NM	AgCl(satd.) + Me$_4$NCl(satd.)|AgCl|Ag, [0.1 M Me$_4$NClO$_4$]
PC	AgCl(satd.) + LiCl(satd.)|AgCl|Ag
TMS	AgCl(satd.) + Et$_4$NCl(satd.)|AgCl|Ag, [0.1 M Et$_4$NClO$_4$]
DMSO	AgCl(satd.) + 0.01 M LiCl + 1 M LiClO$_4$|AgCl|Ag
DMF	AgCl(satd.) + KCl(satd.) + 0.8 M KClO$_4$|AgCl|Ag

2) DMF works as a reducing agent ($2Ag^+ + HCONMe_2 + H_2O \rightarrow 2Ag^0 + Me_2NCOOH + 2H^+$). The reduction process has been used to produce monolayers and stable colloids of silver nanoparticles (Pastoriza-Santos, I. and Liz-Marzán, L.M. (2000) *Pure Appl. Chem.*, **72**, 83).

is relatively small and the potential quickly returns to the equilibrium value when the current is stopped. This is an advantage over a silver–silver ion electrode.

We have previously found that for each of the above reference electrodes, if we construct several pieces at a time and in the same way, their potentials agree to within 1 mV for 1–2 days and their potential shifts are ~0.1 mV/5 h or less [5]. Thus, under well-controlled conditions, all of the above reference electrodes work well as reference electrodes. However, even among the electrodes of the same type, the potentials may differ by 10 mV or more, if they are constructed or used under different conditions. It is difficult to compare such potential data to within 10 mV.[2']

Use of Aqueous Reference Electrodes

Aqueous reference electrodes, such as SCE and Ag/AgCl electrodes, are often used in nonaqueous systems by dipping their tips into nonaqueous solutions of the salt bridge. The tip should not be dipped directly into the solution under study because the solution is contaminated with water and the electrolyte (usually KCl). When we use such aqueous reference electrodes, we must take the liquid junction potential between aqueous and nonaqueous solutions (see the footnote in Table 6.2) into account. If we carefully reproduce the composition of the solutions at the junction, the LJP is usually reproducible within ±10 mV. This is the reason why aqueous reference electrodes are often used in nonaqueous systems. However, the LJP sometimes exceeds 200 mV and it is easily influenced by the electrolytes and the solvents at the junction (Section 6.4). The use of aqueous reference electrodes should be avoided, if possible.

6.1.3
Method of Reporting Electrode Potentials in Nonaqueous Solutions (IUPAC Recommendation)

Because there is no truly reliable reference electrode for use in nonaqueous solutions, various reference electrodes have so far been used. Thus, accurate mutual comparison of the potential data in nonaqueous systems is often difficult. To improve the situation, the IUPAC Commission on Electrochemistry proposed a method for reporting electrode potentials [8]. It can be summarized as follows:

1. Either Fc^+/Fc or BCr^+/BCr couple is selected as reference redox system and the electrode potentials in any solvent and at any temperature are reported as values referred to the (apparent) standard potential of the system. Which of the two couples to select depends on the potential of the system under study; the potential of the reference redox system should not overlap that of the system under study. Because the difference between the standard potentials of the two couples is almost solvent independent (1.132 ± 0.012 V, see Table 6.4), the potential referred to one system can be converted to the potential referred to the other.

2') Various other reference electrodes exist. Among them, the Li^+/Li reference electrode is used very often in studying lithium batteries.

Table 6.4 Potentials of the Ag/Ag^+ and Hg/Hg^{2+} reference electrodes and Fc/Fc^+ reference system in various organic solvents (V versus BCr reference redox system; in 0.1 M Bu_4NClO_4 unless otherwise stated in footnote; at 25 °C).

Solvents	$E(Ag)^a$	$E(Hg)^b$	Fc^c
Alcohols			
Methanol	1.337		1.134^d
Ethanol	1.275	1.349	1.134^d
Ethylene glycol	1.217^e		1.132^e
Ketones			
Acetone	1.315		$1.131^{d,\,f}$
Ethers			
Tetrahydrofuran	1.297	1.367	1.209^f
Esters, lactones			
γ-Butyrolactone	1.364		1.112^d
Propylene carbonate	1.514	1.606	1.114^d
Amides, lactams, ureas			
Formamide	1.200		1.135
N-Methylformamide	1.120^d		1.135
N,N-Dimethylformamide	1.112	1.144	1.127^d
N,N-Diethylformamide	1.143		1.142
N,N-Dimethylacetamide	1.025		1.135
N,N-Diethylacetamide	1.027		1.137
N-Methyl-2-pyrrolidinone	1.032	1.118	1.126^d
1,1,3,3-Tetramethylurea	1.036		1.129^g
Nitriles			
Acetonitrile	1.030	1.336	1.119
Propionitrile	1.026	1.423	1.132
Butyronitrile	1.059	1.427	1.145^g
Isobutylnitrile	1.071		1.131^f
Benzonitrile	1.112	1.448	1.149
Phenylacetonitrile	1.136	1.469	1.150
Nitro compounds			
Nitromethane	1.571	1.686	1.112^d
Nitrobenzene	1.546	1.601	1.140^f
Aromatic heterocyclic compounds			
Pyridine	0.611	0.783	1.149
Halogen compounds			
Dichloromethane	1.562		1.148^f
1,2-Dichloroethane	1.503		1.131
Sulfur compounds			
Dimethyl sulfoxide	0.958	1.022	1.123^d
Sulfolane (tetramethylsulfone)^h	1.349		1.114^d
2,2'-Thiodiethanol	0.691	0.979	1.121
Tetrahydrothiophene	0.700	0.784	—
N,N-Dimethylthioformamide	0.261	0.501	
N-Methyl-2-thiopyrrolidone	0.181	0.452	
Hexamethylthiophosphoric triamide	0.445	0.699	1.153
Phosphorus compounds			

Table 6.4 (Continued)

Solvents	$E(Ag)^a$	$E(Hg)^b$	Fc^c
Trimethylphosphate	1.179	1.292	1.131
Hexamethylphosphoric triamide	0.891	0.929	1.140

From Gritzner, G. (1990) *Pure Appl. Chem.*, **62**, 1839.
[a] Ag/0.01 M Ag^+ (0.1 M Bu_4NClO_4) electrode.
[b] Hg/0.01 M Hg^{2+} (0.1 M Bu_4NClO_4) electrode.
[c] Cyclic voltammetric $(E_{pc} + E_{pa})/2$ of ferrocene.
[d] 0.1 M Et_4NClO_4.
[e] 0.05 M Bu_4NClO_4.
[f] Polarographic $E_{1/2}$.
[g] Pulse-polarographic E_p.
[h] 30 °C.

2. In fact, the potentiometric or voltammetric measurement is carried out using a conventional reference electrode (e.g. Ag^+/Ag electrode).[3] After the measurement in the test solution, Fc or BCr^+ (BPh_4^- salt) is added to the solution and the half-wave potential of the reference system is measured by polarography or voltammetry. Here, the half-wave potential for the reference system is almost equal to its formal potential. Thus, the potential for the test system is converted to the value versus the formal potential of the reference system. The example in Figure 6.2 is for the situation where both the test and the reference systems are measured by cyclic voltammetry, where $E_{1/2} = (E_{pc} + E_{pa})/2$. Curve 1 was obtained before the addition of Fc and curve 2 was obtained after the addition of Fc. It is essential that the half-wave potential of the test system is not affected by the addition of the reference system.

3. In reporting the potential data, the reference redox system used should be indicated by such symbols as $E_{1/2(BCr)}$ and $E^0_{(Fc)}$. Detailed information should also be given concerning the cell construction, solvent purification, impurities in the solution, etc.

For the Fc^+/Fc and BCr^+/BCr couples, the electrode reactions are reversible or nearly reversible in most nonaqueous solvents and the half-wave potentials are not much influenced by solvent impurities such as water. Thus, the half-wave potentials will not vary widely even when they are measured by different persons or at different laboratories. Moreover, the potentials of the Fc^+/Fc and BCr^+/BCr couples are considered almost solvent independent. This justifies the comparison of the potential data in different solvents as far as they are based on this proposal. The data in Table 6.4

---

3) The reference electrode of the Fc^+/Fc couple can be prepared in PC and AN by inserting a platinum wire in the equimolar solution of Fc and Fc^+ (Pic^-). But it is impossible in DMF and DMSO, because Fc^+ is rapidly reduced to Fc in those solvents. The Fc^+/Fc couple is useful as a reference redox system rather than as a reference electrode.

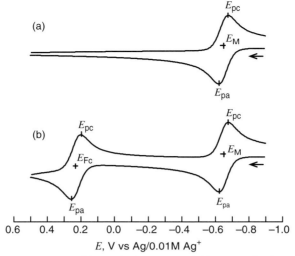

Figure 6.2 Use of the method recommended by IUPAC for the measurement of electrode potentials: the case of cyclic voltammetry. (a) The system under study (M) only; (b) after addition of the reference system (Fc). If E_M and E_{Fc} are the half-wave potentials [$= (E_{pc} + E_{pa})/2$] against the Ag^+/Ag reference electrode; the half-wave potential of M referred to the Fc reference system, $E_{1/2(Fc)}$, is obtained by $E_{1/2(Fc)} = E_M - E_{Fc}$.

are useful in determining the mutual relations between the potentials of the Ag^+/Ag and Hg^{2+}/Hg electrodes and the Fc^+/Fc and BCr^+/BCr couples [9].

6.1.4
Liquid Junction Potential Between Electrolyte Solutions in the Same Solvent

If two electrolyte solutions that are of different concentrations but in the same solvent contact each other at a junction, ion transfers occur across the junction (Figure 6.3). If the rate of transfer of the cation differs from that of the anion, a charge separation occurs at the junction and a potential difference is generated. The potential difference tends to retard the ion of higher rate and accelerate the ion of lower rate. Eventually, the rates of both ions are balanced and the potential difference reaches a constant value. This potential difference is called the *liquid junction potential* [10]. As for the LJP between aqueous solutions, the LJP between nonaqueous solutions can be estimated using the Henderson equation. Generally, the LJP, E_j, at the junction c_1 MX(s)|c_2 NY(s) can be expressed by Eq. (6.1).

$$E_j = -\frac{RT}{F} \cdot \frac{\sum z_i u_i [c_i(2) - c_i(1)]}{\sum z_i^2 u_i [c_i(2) - c_i(1)]} \ln \frac{\sum z_i^2 u_i c_i(2)}{\sum z_i^2 u_i c_i(1)} \tag{6.1}$$

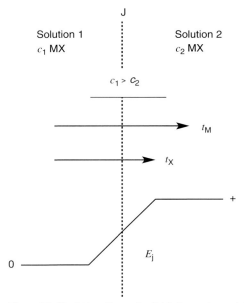

Figure 6.3 Liquid junction potential between two solutions in the same solvent.

where z_i, u_i and c_i denote, respectively, the charge number, mobility and concentration of ion i and Σ denotes the sum over all ionic species. Table 6.5 lists some LJP values calculated from Eq. (6.1).

In electrochemical measurements, the LJP is usually kept small or at a constant value. There are two ways of doing this. One is to add an indifferent electrolyte or a

Table 6.5 Ionic molar conductivities (λ^∞) in some organic solvents and LJPs between solutions in the same solvent (E_j) calculated by the Henderson equation.

Ions and electrolytes at the junction	Concentration ratios	λ^∞ and E_j (mV)					
		AN	TMS	DMF	DMSO	NMA	MeOH
λ^∞ (Et$_4$N$^+$)		84.8	3.94	35.4	17.5	11.5	60.5
λ^∞ (Pic$^-$)		77.7	5.32	37.4	16.8	11.9	47.0
λ^∞ (ClO$_4^-$)		103.7	6.69	52.4	24.1	16.9	70.8
Et$_4$NPic/Et$_4$NPic (E_j)[a]	1/10	−2.5	8.8	1.6	−1.2	1.1	−7.4
	1/100	−5.2	17.6	3.2	−2.4	2.3	−14.8
Et$_4$NClO$_4$/Et$_4$NClO$_4$ (E_j)[a]	1/10	5.9	15.2	11.4	9.4	11.2	4.6
	1/100	11.8	30.6	22.9	18.7	22.5	9.3
Et$_4$NPic/Et$_4$NClO$_4$ (E_j)	1/1	3.8	3.5	4.8	4.9	4.9	5.1
	10/1	3.4	−7.5	−0.3	2.4	0.2	8.3
	100/1	5.2	−16.9	−2.9	2.6	−1.9	14.4
	1/10	7.2	16.9	13.3	11.2	13.1	6.2

[a] $E_j \sim 0$ if the concentration ratio is 1/1.

supporting electrolyte of the same concentration to solutions 1 and 2. The other is to insert, between solutions 1 and 2, a salt bridge of the concentrated solution of an electrolyte whose cationic and anionic mobilities are approximately the same. In aqueous solutions, KCl and KNO_3 are used as such a salt bridge electrolyte. In nonaqueous solutions, tetraethylammonium picrate (Et_4NPic) is often a candidate for such a salt bridge electrolyte because, in various solvents, the molar conductivity of Et_4N^+ is close to that of Pic^- (see Table 6.5). However, there are solvents in which the molar conductivities of Et_4N^+ and Pic^- are considerably different (e.g. TMS and MeOH). In such cases, we use electrolytes such as $(i\text{-}Am)_3BuNBPh_4$ instead (Section 7.2).

In electrochemical measurements in nonaqueous systems, we sometimes use a cell with a junction between electrolyte solutions in different solvents. The problem of the LJP between different solvents is rather complicated and is discussed in Section 6.4.

6.2
pH Measurements in Nonaqueous and Mixed Solvents

pH measurements in nonaqueous and mixed solvents were briefly discussed in Chapter 3. In this section, practical methods of pH measurements in nonaqueous and mixed solvent systems are considered, emphasizing the difference from those in aqueous solutions.

6.2.1
IUPAC Method of pH Measurements in Aqueous Solutions

Before the adoption of the new IUPAC recommendation (Recommendation 2002) [11], the old recommendation (Recommendation 1985) [12] served as a guideline for pH measurement in dilute aqueous solutions. However, the old one was criticized as being metrologically unsatisfactory [13] and was replaced with the new one that is metrologically sound.

According to Recommendation 2002, the pH of dilute aqueous solutions (≤ 0.1 $mol\,kg^{-1}$) can be determined by the following procedure:

The first step of the procedure is to define pH notionally by

$$pH = -\log a_H = -\log m_H \gamma_H / m^0$$

where a is relative activity, m is molality, γ is activity coefficient and m^0 is standard molality. Because pH is a single-ion quantity and is thermodynamically immeasurable, a convention is necessary for its evaluation.

Thus, the second step is to determine primary pH standards or pH(PS) values by means of an absolute method, which requires no reference, by using a Harned cell without transference (I).

$$Pt|H_2|buffer\ S,\ Cl^-|AgCl|Ag \tag{I}$$

The reaction and the Nernst equation for cell I are expressed by

$$(1/2) H_2 + AgCl \rightarrow Ag(s) + H^+ + Cl^-$$

$$E_I = E^0_{Ag/AgCl} - k \log[(m_H \gamma_H/m^0)(m_{Cl} \gamma_{Cl}/m^0)]$$

where E_I is the emf of cell I, $E^0_{Ag/AgCl}$ is the standard potential of the Ag/AgCl electrode, γ_{Cl} is the activity coefficient of the Cl^- ion and $k = (RT/F) \ln 10$. The equation can be rearranged to give the acidity function, $p(a_H \gamma_{Cl})$, since $a_H = m_H \gamma_H/m^0$.

$$p(a_H \gamma_{Cl}) = -\log(a_H \gamma_{Cl}) = (E_I - E^0_{Ag/AgCl})/k + \log(m_{Cl}/m^0) \qquad (6.2)$$

The values of pH for the primary buffer solutions can be obtained in the following way: (1) Fill the Harned cell with HCl at, for example, $m_{HCl} = 0.01$ mol kg^{-1} and determine the value of $E^0_{Ag/AgCl}$; (2) Fill the Harned cell with a buffer at a known ionic strength and measure $p(a_H \gamma_{Cl})$ for at least three molarities of Cl^- to determine $p(a_H \gamma_{Cl})^0$ by extrapolation $m_{Cl} \rightarrow 0$ and (3) Use the Bates–Guggenheim convention $\{\log \gamma^0_{Cl} = -A(I^0_s)^{1/2}/[1 + 1.5((I^0_s)/m^0)^{1/2}]$, where A is the Debye–Hückel limiting slope and (I^0_s) the ionic strength of the buffer with zero added chloride$\}$ to get γ^0_{Cl} and calculate pH.

Seven pH buffers have hitherto been designated as primary standards: they are (i) saturated (at 25 °C) potassium hydrogen tartrate, (ii) 0.1 mol kg^{-1} potassium dihydrogen citrate, (iii) 0.05 mol kg^{-1} potassium hydrogen phthalate, (iv) 0.025 mol kg^{-1} disodium hydrogen phosphate + 0.25 mol kg^{-1} potassium dihydrogen phosphate, (v) 0.03043 mol kg^{-1} disodium hydrogen phosphate + 0.008695 mol kg^{-1} potassium dihydrogen phosphate, (vi) 0.01 mol kg^{-1} disodium tetraborate and (vii) 0.025 mol kg^{-1} sodium hydrogen carbonate + 0.025 mol kg^{-1} sodium carbonate. The values of pH(PS) for these buffers between 0 and 50 °C have been given, though with an uncertainty of 0.004.

The third step is to get secondary pH standards (SS) or pH(SS) values that are obtained by comparing with primary standards or pH(PS). Several cells with liquid junction are used: for example, Pt|H$_2$|buffer S$_2$|KCl (≥ 3.5 mol dm^{-3})|buffer S$_1$|H$_2$|Pt and Ag|AgCl|KCl (≥ 3.5 mol dm^{-3})|buffer S$_1$ or S$_2$|H$_2$|Pt are with uncertainties of pH (SS) ~ 0.015 and Ag|AgCl|KCl (≥ 3.5 mol dm^{-3})|buffer S$_1$ or S$_2$|glass electrode is with an uncertainty of pH(SS) ~ 0.02.

The fourth step is the practical pH measurement. The cell assembly for pH measurement is the cell, Ag|AgCl|KCl (≥ 3.5 mol dm^{-3})|solution pH(S) or pH(X)| glass electrode and a pH meter. The uncertainty of the measurement is different by the calibration method: 0.3 for 1-point calibration, 0.02–0.03 for 2-point calibration and 0.01–0.03 for 5-point calibration.

The pH measurement is thus represented by the traceability chain pH(X) $\rightarrow$ pH (SS) $\rightarrow$ pH(PS) $\rightarrow$ pH as defined, and each step has stated uncertainties.

In Recommendation 1985, the notional pH in aqueous solutions is defined by pH $= -\log a_H$. But, this equation can be rewritten by pH $= -\log m_H \gamma_H/m^0$ or pH $= -\log c_H \gamma_H/c^0$, depending on whether *molal concentration*, m_H [mol kg^{-1}], or *molar concentration*, c_H [mol l^{-1}], is used. The operational pH is defined by the following (1) and (2):

1. Cell (II) contains a sample solution X of pH(X) and its emf is equal to E(X), while cell (III) contains the standard solution S of pH(S) and its emf is equal to E(S):

$$\text{Reference electrode}|\text{KCl(aq., } \geq 3.5 \text{ mol kg}^{-1})||\text{solution X}|\text{H}_2|\text{Pt} \qquad (\text{II})$$

$$\text{Reference electrode}|\text{KCl(aq., } \geq 3.5 \text{ mol kg}^{-1})||\text{solution S}|\text{H}_2|\text{Pt} \qquad (\text{III})$$

The relation among pH(X), pH(S), E(X) and E(S) is given by

$$\text{pH(X)} = \text{pH(S)} + \{E(S) - E(X)\}/\ln 10(RT/F) \qquad (6.3)$$

2. The 0.05 mol kg^{-1} potassium hydrogen phthalate solution is defined as the Reference Value pH Standard (RVS), and the reference pH values, pH(RVS), at various temperatures between 0 and 95 °C are assigned to it. Six other buffer solutions are also used as primary pH standards and the values of pH(PS) are also designated. In addition to RVS and PS, operational pH standards (OS) may be used; the values of OS, pH(OS), are obtained with a cell Pt|H$_2$|OS||KCl $\geq$3.5 mol dm^{-3}||RVS|H$_2$|Pt at 0–95 °C.

Here, the solutions of the RVS plus six PSs in Recommendation 1985 coincide with those of seven PSs in Recommendation 2002. Moreover, the values of pH(RVS) and pH(PS) had been obtained by the same method as the primary method in Recommendation 2002. Therefore, the pH values of the primary standards in Recommendation 2002 and those of pH(RVS) plus pH(PS) in Recommendation 1985 are quite the same.

Practically, the pH measurements are conducted by using pH-sensitive glass electrode or some other pH sensors. The difference in pH between the sample and the standard solutions, pH(X) – pH(S), is given by (1). and the value of pH(S) is given by (2).; thus, we can know the value of pH(X). For dilute aqueous solutions ($\leq$0.1 mol kg^{-1}), pH determined by this method agrees with the notional pH by pH $= -\log a(\text{H}^+) \pm 0.02$. For solutions of ionic strength appreciably higher than 0.1 mol kg^{-1}, as in the case of seawater, the above method needs slight modification [14].

6.2.2
Methods of pH Measurements in Nonaqueous and Mixed Solvents

The following two methods may be considered for measuring pH in nonaqueous and aqueous–organic solvent mixtures.

Method Recommended by IUPAC
Expansion of the applicability of IUPAC Recommendation 2002, which is for dilute aqueous solutions, to nonaqueous and aqueous–organic solvent mixtures has been

discussed [15]. Primary and secondary methods of assigning pH values to primary and secondary standards have been considered.[4] But, the document seems to need some more discussion. Thus, the IUPAC recommendation now available for nonaqueous and aqueous–organic solvent mixtures is described here. It is based on Recommendation 1985 and was proposed by the IUPAC Commission on Electroanalytical Chemistry for pH measurements in organic solvents of high permittivities and in various aqueous–organic solvent mixtures [16, 17].

The pH in nonaqueous solvents or solvent mixtures is notionally defined by

$$pH = -\log a_H = -\log m_H \gamma_H / m^0$$

where a_H is the activity, m_H is the molal concentration (mol kg^{-1}), γ_H is the activity of H$^+$ and m^0 is the standard molality.

In order to get pH values that meet this notional pH definition, the operational pH is defined by (1) and (2) below:

1. The relation among the pH values of the sample and the standard, pH(X) and pH (S), and the emfs of cells (IV) and (V), $E(X)$ and $E(S)$, is given by Eq. (6.4).

   Reference electrode|salt bridge in s‖solution X in s|H$_2$|Pt   (IV)

   Reference electrode|salt bridge in s‖solution S in s|H$_2$|Pt   (V)

---

4) *Primary Standards:* Harned cell is used to develop the primary method.

Pt | H$_2$ | buffer S, MCl in Z | AgCl | Ag

where MCl = KCl or NaCl and Z is the solvent or solvent mixture under investigation. This cell is employed to determine the quantity $p(a_H \gamma_{Cl})$.

$$p(a_H \gamma_{Cl}) = -\log(a_H \gamma_{Cl})$$
$$= (E^Z - E^{0Z})/(RT\ln 10/F)$$
$$+ \log(m_{Cl}/m^0).$$

Here, E^Z and E^{0Z} are the measured emf and the standard emf, respectively, of the cell. E^{0Z} is determined by using a cell Pt|H$_2$|HCl in Z|AgCl|Ag. The procedures for pure aqueous (Z = water) and nonaqueous or mixed (Z = Z) media coincide up to this point; then a substantial difference is introduced for generic solvent Z. pH(PS), where PS means primary standard, is described as follows:

$$pH(PS) = \lim m_{Cl \to 0}[(E^Z - E^{0Z})/(RT\ln 10/F)$$
$$+ \log(m_{Cl}/m^0) - p\gamma_{Cl}]$$

$$p\gamma_{Cl} = (A^Z I^{1/2}/[1 + 1.5(\varepsilon^W \rho^Z/\varepsilon^Z \rho^W)^{1/2}$$
$$\times (I/m^0)^{1/2}]),$$

where ε and ρ are the relative permittivities and densities of the superscripted solvents, $A^Z = A^W(\varepsilon^W/\varepsilon^Z)^{3/2}$ is the relevant Debye–Hückel constant and I is total ionic strength. pH(PS) is obtained by extrapolation to $m_{Cl \to 0}$ of the straight-line plot of pa_H against m_{Cl}.

Thus, in principle, the primary standard, pH (PS), can be obtained by the same procedures as those in Recommendations 2002. But it is under various limiting conditions: for example, (i) the hydrogen electrode must work satisfactorily, (ii) the salt, MCl, should dissociate completely into ions, and for condition (ii), value ε^Z should be larger than 35. Even under satisfactory conditions, however, the uncertainty of the method is larger than in aqueous solutions.

Secondary Standards: The secondary standards can be obtained by comparison methods using various cells with transference (see Section 6.2.1). When a glass electrode is used, the actual slope of response in solvent Z must be used. A cell without transference, Ag|AgCl|buffer S, MCl in Z|glass electrode, may also be used, making it unnecessary to rely on the comparative method that needs some primary pH standard.

$$pH(X) = pH(S) + \frac{E(S) - E(X)}{\ln 10 \times (RT/F)} \qquad (6.4)$$

Here, s shows the solvent and the LJP at ($\vert\vert$) must be equal in cells (IV) and (V).[5] The hydrogen electrode in (IV) and (V) may be replaced by a glass electrode or other appropriate pH sensors.

2. $0.05\,\mathrm{mol\,kg^{-1}}$ potassium hydrogen phthalate (KHPh) in solvent s is used as the RVS and the reference pH values, pH(RVS), are assigned to it. Other buffer solutions are also used as primary pH standards and appropriate pH values are assigned to them.

This pH definition for nonaqueous and mixed solvent systems is practically the same as that for aqueous solutions (Section 6.2.1). Thus, if pH standards are available for the solvent or mixed solvent under study, the glass electrode is calibrated with them and then the pH of the sample solution is measured. The pH(RVS) values for $0.05\,\mathrm{mol\,kg^{-1}}$ KHPh have been assigned to formamide (high permittivity) and aqueous mixtures of 12 organic solvents (see footnote[6] for pH(RVS) at $25\,^{\circ}\mathrm{C}$). Though they are for discrete solvent compositions, the pH(RVS) in between those compositions can be obtained by using a multilinear regression equation [17c].

The IUPAC method is very reliable for pH measurements in water-rich solvent mixtures. However, in water-poor solvent mixtures and neat organic solvents, its applicability is restricted by the following reasons [18]. First, the solubility of KHPh is not sufficient in aqueous mixtures of MeOH and AN with water content less than 10 v/v%. Second, the increase in the aprotic property of the solvent mixtures leads to the loss of the buffer capacity of the KHPh solution. This tendency is apparent from Figure 6.4, which shows the neutralization titration curves of o-phthalic acid (H_2Ph) in various $AN-H_2O$ mixtures. In Figure 6.4, the mixtures

5) The electrolyte in the salt bridge keeps the LJP constant, but its anion should not react with hydrogen ion in the solution under study.

6) The pH values of $0.05\,\mathrm{mol\,kg^{-1}}$ potassium hydrogen phthalate (KHPh) buffers (i.e. RVS in Recommendation 1985) in FA and in water + organic solvent mixtures at $25\,^{\circ}\mathrm{C}$ are shown for wt% of organic solvents.

Pure aqueous solution pH = 4.005;
$MeOH-H_2O$ (10% 4.243, 20% 4.468, 50% 5.125, 64% 5.472, 84.2% 6.232) [17c];
$EtOH-H_2O$ (10% 4.230, 20% 4.488, 40% 4.973, 70% 5.466) [17c];
2-PrOH$-H_2O$ (10% 4.249, 30% 4.850, 50% 5.210, 70% 5.522) [17c];
1,2-Ethanediol$-H_2O$ (10% 4.127, 30% 4.419, 50% 4.790, 70% 5.238) [17c];
2-Methoxyethanol$-H_2O$ (20% 4.505, 50% 5.380, 80% 6.715) [17c];
$AN-H_2O$ (5% 4.166, 15% 4.533, 30% 5.000, 50% 5.461, 70% 6.194) [17c];

1,4-Dioxane$-H_2O$ (10% 4.329, 30% 5.015, 50% 5.782) [17c];
$PC-H_2O$ (8% 4.177, 16% 4.347) [17e];
Ethylene carbonate$-H_2O$ (8% 4.145, 16% 4.301) [17e];
Glycerol$-H_2O$ (20% 4.172, 40% 4.420) [17f];
$THF-H_2O$ (10% 4.26, 20% 4.62, 30% 4.93, 40% 5.31, 50% 5.56, 60% 5.86, 70% 6.17) [17g];
FA (100% 6.43) [17h]
When pH value of KHPh buffer is not available for the solvent composition under study, it is obtained interpolating by multilinear regression [17c, d]. For example, pH for KHPh in EtOH$-H_2O$ can be obtained by pH (KHPh) $= 3.99865 - 0.46452x^{1/2} + 9.5545x - 8.4053x^{2/3}$, where x is the mole fraction of EtOH. In Ref. [17], pH(KHPh) at other temperatures and some pH values for other primary pH standards are also included.

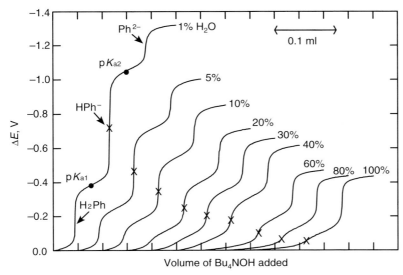

Figure 6.4 Titration curves of 20 ml of (3 mM phthalic acid (H_2Ph) + 2 mM CF_3SO_3H) in various AN–H_2O mixtures with 1.0 M Bu_4NOH (in MeOH). Water content in (v/v)% is shown on each curve. Obtained using Si_3N_4-ISFET (Shindengen Indus. Co.) as pH sensor and titrating at 0.005 ml min^{-1} [18].

of trifluoromethanesulfonic acid (CF_3SO_3H) and H_2Ph were titrated with a strong base, Bu_4NOH (in MeOH), to get an entire picture of the titration curves of H_2Ph. With the decrease in water content, the value of $(pK_{a2} - pK_{a1})$ for H_2Ph markedly increases and the slope at the ×-mark increases, showing a decrease in the buffer action of the HPh^- solution. For AN–H_2O mixtures, the buffer capacity of 0.05 m KHPh is insufficient if the water content is less than 20 v/v%. For DMSO– and DMF–H_2O mixtures, this insufficiency occurs if the water content is less than ~35 v/v%, because protophilic DMSO and DMF interact strongly with water and decrease the protic property of water.

Although the IUPAC method has some drawbacks, it is still reliable and convenient as far as the RVS or other pH standard is available for the solvent under study. Moreover, the method of determining pH(RVS) has been proposed by IUPAC [16, 17]. Thus, if necessary, we can provide ourselves pH(RVS) values for new solvent systems.

Other Methods of pH Measurements in Nonaqueous Solutions
If the IUPAC method is not applicable, one of the following procedures can be used, although the reliability of the data decreases to considerable extent:

1. If there is an acid with a known pK_a value and without any tendency to homoconjugate, the solution of the 1:1 mixture of the acid and its conjugate base is used as the pH standard for calibrating the glass electrode. If the acid is of the HA type, the pH of the solution is calculated by the relation $pH(S) = pK_a + \log \gamma(A^-)$, where $\gamma(A^-)$ is the activity coefficient of A^- calculated by the Debye–Hückel

theory (see footnote 10 in Chapter 2). If the acid is of the BH^+ type, pH $(S) = pK_a - \log \gamma(BH^+)$. The pH of the sample solution is obtained by using Eq. (6.4). For example, the $1:1$ mixture of picric acid and its Et_4N^+ or Bu_4N^+ salt (3–5 mM each) has been used in various organic solvents as pH standard. The $1:1$ mixture of a weak base (e.g. aniline or diphenylguanidine) and its perchlorate or trifluoromethanesulfonate is also applicable as pH standard.[7]

2. If there is a strong acid that completely dissociates, we prepare a solution of the acid of a known concentration and use it to calibrate the glass electrode. The pH of the solution is calculated by estimating $\gamma(H^+)$ using the Debye–Hückel theory.

In the above, we assumed the use of a pH-sensitive glass electrode. The Nernstian response of the glass electrode has been confirmed in many polar organic solvents.[8] However, the response of the glass electrode in nonaqueous solutions is often very slow. In particular, in protophilic aprotic solvents such as DMSO and DMF, it may take over an hour to reach a steady potential. Various efforts have been made to improve the response speed of the glass electrode [19], including preconditioning of the electrode surface, the replacement of the internal solution of the glass electrode (by a nonaqueous solution or by mercury), and searching for the glass composition suitable for nonaqueous solutions. However, these attempts have not been success-ful. The hygroscopic gel layer existing on the surface of the glass electrode is essential for its excellent response in aqueous solutions. However, in nonaqueous solutions, the gel layer seems to retard the response speed of the glass electrode.

Recently, Ta_2O_5- and Si_3N_4-type pH-ISFETs have been used in nonaqueous systems, by making them solvent resistant [20]. In various polar nonaqueous solvents, they responded with Nernstian or near-Nernstian slopes and much faster than the glass electrode. The titration curves in Figure 6.5 demonstrate the fast (almost instantaneous) response of the Si_3N_4-ISFET and the slow response of the glass electrode. Some applications of the pH-ISFETs are discussed in Section 6.3.1.[9]

6.2.3
Determination of Autoprotolysis Constants

The autoprotolysis constant, K_{SH}, for the autoprotolysis of an amphiprotic solvent SH (Eq. (6.5)) is defined by Eq. (6.6):

$$SH + SH \rightleftarrows SH_2^+ + S^- \tag{6.5}$$

---

7) Chemicals such as HPic, $HClO_4$, R_4NPic and $BH^+ClO_4^-$ (B = amine) must be handled with enough care because they are explosive.

8) We can check the Nernstian response of the glass electrode ourselves. The method that uses a weak acid and its conjugate base is described in Section 6.3.1. Another method is to use a strong acid and to get the pH–potential relation of the glass electrode by varying its concentrations. Though the latter method

seems simple, it should be noted that the reaction of the strong acid with basic impu-rities in the solvent or with solvent itself may cause a serious error.

9) The rapid response of the pH-ISFET has also been used in an ionic liquid (IL), ethylam-monium nitrate $[EtNH_3][NO_3]$, to obtain the autoprotolysis constant, K_{SH}, of the IL. See Section 13.3.1.

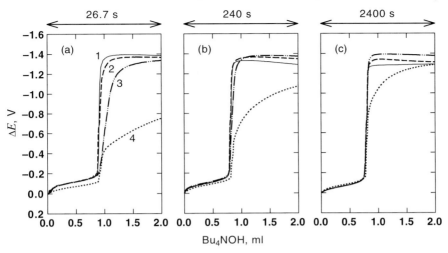

Figure 6.5 Titration curves of 5 mM picric acid in AN with 1 M
Bu$_4$NON (in MeOH), recorded simultaneously with four pH
sensors but at different titration speeds for (a)–(c). In (a), curve 1
is for Si$_3$N$_4$-ISFET, curve 2 for Ta$_2$O$_5$-ISFET, curve 3 for IrO$_2$ pH
sensor and curve 4 for glass electrode. The pH-ISFETs were
obtained from Shindengen Indus. Co. and the IrO$_2$ pH sensor
from TOA Electronics Ltd. [20c].

$$K_{SH} = a(SH_2{}^+)a(S^-)/a(SH)^2$$

or

$$K_{SH} = a(SH_2{}^+)a(S^-) \qquad \text{for } a(SH) = 1 \quad (\text{pure SH}) \tag{6.6}$$

If the solvent is a mixture of water and organic solvent ZH, several equilibria may be
established, more or less simultaneously, when ZH and H$_2$O have similar acid–base
properties:

$$ZH + ZH \rightleftarrows ZH_2{}^+ + Z^-, \qquad H_2O + ZH \rightleftarrows H_3O^+ + Z^-,$$

$$ZH + H_2O \rightleftarrows ZH_2{}^+ + OH^-, \quad H_2O + H_2O \rightleftarrows H_3O^+ + OH^-$$

If solvent ZH is inert in acid–base property, it is not involved in the autoprotolysis but
works only as a medium for the autoprotolysis of H$_2$O. Even in the case of solvent
mixtures, Eqs. (6.5) and (6.6) will be used to express the autoprotolysis processes.

Potentiometry is a useful method to determine pK_{SH} ($= -\log K_{SH}$); in principle, we
get the difference between the potentials of the hydrogen electrode at $a(H_2S^+) = 1$
and $a(S^-) = 1$ and then divide it by the Nernstian slope.[10] The IUPAC Commission
on Electroanalytical Chemistry has proposed a method to determine pK_{SH} in organic
solvents of high permittivities and in water–organic solvent mixtures [21]. In the
method, the standard emf of the cell, Pt|H$_2$($p = 1$ atm)|HX + MX(SH)|AgX|Ag

---

10) Conductometric pK_{SH} determination is discussed in Section 7.3.2.

(M = Li, Na, K, . . .; X = Cl, Br, . . .), is obtained on the one hand and the emf of the cell, $Pt|H_2(p=1 \text{ atm})|MS(m_S) + MX(m_X)(SH)|AgX|Ag$ (S^- = lyate ion), is measured on the other hand by varying m_S and m_X. This method can give reliable pK_{SH} values in amphiprotic organic solvents and water–organic solvent mixtures, if MS is soluble and stable. We can replace the hydrogen electrode by a pH-sensitive glass electrode or other pH sensors. DMSO is an aprotic solvent but alkali metal salts of its lyate ion, dimsyl ion $CH_3SOCH_2^-$, can be obtained fairly easily. Thus, pK_{SH} in DMSO has been obtained by measuring the potentials of a hydrogen electrode in the solution of $HClO_4$ and dimsyl ion (Courtot-Coupez, J., LeDemezet, M. (1969) *Bull. Soc. Chim. Fr.*, 1033). In other neat aprotic solvents, however, it is usually difficult to get stable lyate ions and the IUPAC method is not applicable. One of the methods applied in such a case is to prepare the solution of a strong acid (e.g. $HClO_4$) and that of a strong base (e.g. R_4NOH) and measure the potentials of the pH sensor in the two solutions. Then, pK_{SH} can be obtained approximately from the potential difference between the two solutions (ΔE, mV):

$$pK_{SH} \geq (\Delta E/59.2) - \log\{c(\text{strong acid}) \times c(\text{strong base})\} \tag{6.7}$$

where c denotes the concentration. This method has been used to determine pK_{SH} in some aprotic solvents [22a]. It should be noted, however, that the pK_{SH} values determined by this method have somewhat a indistinct nature (Section 3.2.2). Another method of obtaining pK_{SH} is an indirect method that uses the relation $K_{SH} = K_a \times K_b$, where K_a is the dissociation constant of an acid (HA or BH^+) and K_b is that of a conjugate base (A^- or B). K_a and K_b are determined independently by such methods as spectrophotometry, conductimetry and potentiometry.

Some pK_{SH} values in nonaqueous solvents and water–organic solvent mixtures are listed in Table 6.6 [21, 22]. They are in molal scale but can easily be converted to molar scale by the relation,

$$pK_{SH}(\text{molar}) = pK_{SH}(\text{molal}) - 2\log\rho \tag{6.8}$$

where ρ is the density of solvent SH in kg dm^{-3}.

6.3
Applications of Potentiometry in Nonaqueous Solutions

Potentiometry and potentiometric titrations are widely used in studying various types of reactions and equilibria in nonaqueous systems (Sections 6.3.1–6.3.4). They also provide a convenient method of solvent characterization (Section 6.3.5). Moreover, if the electrode potentials in different solvents can accurately be compared, potentiometry is a powerful method of studying ion solvation (Section 6.3.6).

6.3.1
Acid–Base Reactions in Nonaqueous Solvents

Potentiometry is often used in determining dissociation constants, pK_a, and homoconjugation constants, $K^f(HA_2^-)$, for HA-type acids in aprotic solvents. Therefore, to

Table 6.6 Autoprotolysis constants (pK_{SH}) for pure and mixed solvents.

Pure solvents		Mixed solvents		
Solvents[a]	pK_{SH}	wt%	Mole fraction	pK_{SH}
Sulfuric acid	3.33	Water–methanol[b]		
2-Aminoethanol	5.7	0	0	13.997
Formic acid	6.2	20	0.1232	14.055
N-Methylformamide	10.74	50	0.3599	14.097
Hydrogen fluoride (0 °C)	12.5	70	0.5675	14.218
Water	14.00	90	0.8350	14.845
Acetic acid	14.45	98	0.9650	16.029
1,2-Diaminoethane	15.2	Water–ethanol[b]		
Ethylene glycol	15.84	20	0.0891	14.33
Formamide (20 °C)	16.8	35	0.1739	14.57
Methanol	17.20	50	0.2811	14.88
Ethanol	18.88	80	0.6100	15.91
1-Propanol	19.43	Water–ethylene glycol[b]		
Hexamethylphosphoric triamide*	20.56	10	0.0312	13.85
1-Pentanol	20.65	30	0.1106	13.72
2-Propanol	20.80	50	0.2249	13.66
1-Butanol	21.56	70	0.4038	13.82
Ethyl acetate	22.83	90	0.7232	14.37
Nitromethane*	$\geq$24	Water–1,2-dioxane[b]		
N-Methyl-2-pyrrolidinone*	$\geq$24.2	20	0.0486	14.62
Sulfolane*	24.45	45	0.1433	15.74
Propylene carbonate*	$\geq$29.2	70	0.3230	17.86
N,N-Dimethylacetamide	31.2	Water–acetonitrile[c]		
N,N-Dimethylformamide*	$\geq$31.6	10	0.0465	14.27
Ammonia (-33 °C)	32.5	30	0.1583	14.93
Acetone*	$\geq$32.5	40	0.2264	15.32
Dimethyl sulfoxide*	33.3	50	0.3051	15.71
Acetonitrile*	$\geq$33.3	70	0.5059	16.76

25 °C unless otherwise stated.
[a]Ref. [22a] for solvents with an asterisk* and Ref. [22b] for others.
[b]Ref. [21a].
[c]Ref. [21b].

begin with, we describe the method of obtaining the values of pK_a and $K^f(HA_2^-)$ in an aprotic solvent in which no data of acid–base equilibrium is yet available. Procedures 1–4 seem to be appropriate in such a case [23, 24].

Procedure 1 We select several HA-type acids, for which pK_a values of <7.5 are expected, and accurately determine their pK_a and $K^f(HA_2^-)$ by method(s) other than potentiometry. If the selected acid is a nitro-substituted phenol that has no tendency to homoconjugate (p. 75), we dissolve various amounts of it in the solvent and measure the UV/vis spectrophotometric absorption for the phenolate anion formed by dissociation. For the conductimetric determination of pK_a and $K^f(HA_2^-)$, see Section 7.3.2.

Procedure 2 We construct cell (VI) using the above acids and their tetraalkylammonium salts, R_4NA, and measure the emfs of the cell by varying the ratio C_a/C_s in wide ranges:

$$\text{Ag}|5 \text{ mM AgClO}_4, 50 \text{ mM Et}_4\text{NClO}_4(s)\|C_a \text{ HA} + C_s R_4\text{NA}(s)| \atop \text{glass electrode} \quad \text{(VI)}$$

Then, we get the values of $a(H^+)$ and pH of the solution, C_a HA $+ C_s R_4NA(s)$, using Eq. (6.19) and plot the emf–pH relation.

$$\gamma^2 C_s a(H^+)^2 - \gamma a(H^+) K_a\{(C_a + C_s) + K^f(HA_2^-)(C_s - C_a)^2\} + K_a^2 C_a = 0$$
$$(6.9)$$

If the emf–pH relation is linear and has a slope of ~59.2 mV/pH (at 25 °C), it means that the pH response of the glass electrode is Nernstian. Although the actual confirmation of the Nernstian response is in a limited pH region, we temporarily assume that the response is Nernstian in all pH regions.[11]

Procedure 3 We construct cell (VI) using the HA-type acid under study and its conjugate base and measure the emfs by varying the ratio C_a/C_s. The procedure to vary the ratio C_s/C_a can be replaced by the titration of the acid HA with a strong base (Bu_4NOH in MeOH, toluene–MeOH or 2-PrOH), but the alcohol introduced with the base may have some influence on the results. Because the glass electrode has been calibrated in Procedure 2, the emf values can be converted to pH values.

Procedure 4 We plot the pH versus $[C_s/(C_a + C_s)]$ relation using the above results and get the values of pK_a and $K^f(HA_2^-)$ for the acid under study. The pK_a value is obtained by $pK_a = pH_{1/2} - \log \gamma_{1/2}$, where $pH_{1/2}$ and $\log \gamma_{1/2}$ are the pH and $\log \gamma$ at the half-neutralization ($C_a = C_s$), regardless of the $K^f(HA_2^-)$ value. The $K^f(HA_2^-)$ value, on the other hand, is obtained by

$$K^f(HA_2^-) = \frac{r^2 C_s - r(C_a + C_s) + C_a}{r(C_s - C_a)^2}$$

where $r = a(H^+)\gamma/\{a(H^+)_{1/2}\gamma_{1/2}\}$. More simply, $K^f(HA_2^-)$ can be obtained by curve fitting the experimental relation between $(pH - pK_a - \log \gamma)$ and $[C_s/(C_a + C_s)]$ to the theoretical relation, as in Figure 6.6. See Section 3.1.2 for more details.

Procedures 3 and 4 are for the acids of the HA type, but they can be applied with minor modifications to acids of the BH^+ type, for which homoconjugation is often negligible. From these studies, we find an appropriate pH buffer that is to be used to calibrate the glass electrode in routine pH measurements. Mixtures (1 : 1) of picric

11) If a deviation from the Nernstian response occurs, it can be detected in Procedures 3 and 4, especially for the BH^+-type acids for which homoconjugation is negligible.

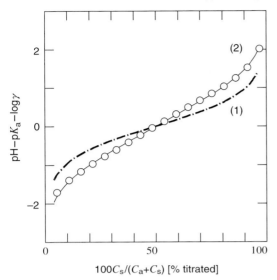

Figure 6.6 Experimental results (circles) and theoretical curves for the relation of $(pH-pK_a-\log \gamma)$ versus % titrated in the titration of 5 mM salicylic acid in γ-BL with 0.09 M Bu$_4$NOH (in toluene–MeOH) [20b]. Obtained with a Ta$_2$O$_5$-ISFET. $K^f(HA_2{}^-)/$ mol^{-1} l is 0 for curve 1 and 600 for curve 2.

acid/tetraalkylammonium picrate and diphenylguanidine/diphenylguanidinium perchlorate are examples of the candidates for such a pH buffer.

Potentiometry with a glass electrode is simple and convenient. We can use it to study the dissociation equilibria of very weak acids, for which spectrophotometry and conductimetry are inapplicable. Many data for acid dissociations in dipolar aprotic solvents have been determined by potentiometry with a glass electrode [22a]. However, the slow response of the glass electrode (Section 6.2.2) is fatal in some nonaqueous solutions, in which the solvent or solute is not stable enough or the moisture introduced during the time-consuming measurement has serious effects on the results. As described above, the response of the pH-ISFET in nonaqueous solutions is much faster than that of the glass electrode [20]. We used the pH-ISFET successfully to study acid–base equilibria in γ-butyrolactone, which is easily hydrolyzed and in which the glass electrode did not give reliable results [20b].

The pH window is very wide in solvents that are weak both in acidity and in basicity. The widths of the pH window are well over 30 in such solvents, compared to about 14 in water (Table 6.6). The usefulness of these expanded pH regions is discussed in Section 3.2.2. In particular, potentiometric acid–base titrations in such solvents are highly useful in practical chemical analyses as well as in physicochemical studies [25]. Acid–base titrations in nonaqueous solvents were popular until 1980s, but now most have been replaced with chromatographic methods. However, the pH-ISFETs are promising to realize simple, rapid and miniature-scale acid–base titrations in nonaqueous solvents. For example, by using an Si$_3$N$_4$-type pH-ISFET, we can get an almost complete titration curve in less than 20 s in the solution containing several different acids [20d].

Some redox processes consume or generate protons. Thus, some electrode reactions are influenced by the pH of the solution, while some conversely influence the pH of the solution. To elucidate the electrode processes, it is desirable to measure both the pH of the solution and the electrode reaction. The relations between pH and the electrode processes have been well studied in aqueous solutions. However, in nonaqueous solutions, such studies have been scarce, except a few cases in recent years. This problem is dealt with in Section 8.3.1.

6.3.2
Precipitation Reactions in Nonaqueous Solutions

By running a potentiometric precipitation titration, we can determine the compositions of the precipitate and its solubility product. Various cation- and anion-selective electrodes as well as metal (or metal amalgam) electrodes work as indicator electrodes. For example, Coetzee and Martin [26] determined the solubility products of metal fluorides in AN, using a fluoride ion-selective LaF_3 single-crystal membrane electrode. Nakamura et al. [2] also determined the solubility product of sodium fluoride in AN and PC, using a fluoride ion-sensitive polymer-membrane electrode, which was prepared by chemically bonding the phtharocyanin cobalt complex to polyacrylamide (PAA). The polymer-membrane electrode was durable and responded in Nernstian ways to F^- and CN^- in solvents such as AN and PC.

6.3.3
Complex Formation Reactions in Nonaqueous Solutions

Many potentiometric studies have been carried out to obtain information about the compositions and the formation constants of metal complexes in nonaqueous solutions. Ion-selective electrodes and metal (or metal amalgam) electrodes are used as indicator electrodes.

For example, the formation constants of $1:1$ complexes of alkali metal ions with crown ethers or cryptands have been determined by Procedure 1 or Procedure 2 below [27].

$$M^+ + L \rightleftharpoons ML^+$$

$$K(ML) = \frac{a(ML^+)}{a(M^+)a(L)} = \frac{\gamma_{ML}[ML^+]}{a(M^+)[L]}$$

Procedure 1 Cell (VII) is constructed using a univalent cation-sensitive glass electrode (Ag electrode for Ag^+) and the emf is measured by titrating metal ion M^+ with ligand L.

$$Ag|5 \text{ mM AgClO}_4, 0.1 \text{ M Et}_4NPic(s)||1 \text{ mM MClO}_4$$
$$+ 10 \text{ mM Et}_4NPic + L(s)|glass \text{ electrode} \qquad\qquad (VII)$$

The emf is converted to pa_M ($= -\log a(M^+)$) and the values of pa_M are plotted against the amount of L added. If M^+ and L react in a $1:1$ ratio, the formation constant can be obtained by the following relation at twice the equivalence point: $\log K(ML) = pa_M + \log \gamma_{ML}$. If the complex is very stable ($\log K(ML) \geq 10$), the concentration of the free M^+ after the equivalence point may be too low for the glass electrode to get the Nernstian response. In that case, an indirect method as in Procedure 2 is used.

Procedure 2 Cell (VIII) is constructed using silver electrodes on the two sides. After the equilibrium of reaction (6.10) is reached in the solution on the left of the junction, the emf of the cell is measured and the equilibrium constant K_{ex} (Eq. (6.11)) is obtained by calculation.[12] If $K(AgL)$ is known, we can get the value of $K(ML)$.

$$Ag|c_1 \, AgClO_4, c_2 \, MClO_4, c_3 \, L(s)\|c_1 AgClO_4, c_2 \, MClO_4(s)|Ag \qquad \text{(VIII)}$$

$$AgL^+ + M^+ \rightleftarrows ML^+ + Ag^+ \qquad \text{(6.10)}$$

$$K_{ex} = \frac{[ML^+][Ag^+]}{[AgL^+][M^+]} = \frac{K(ML)}{K(AgL)} \qquad \text{(6.11)}$$

This method is applicable if $K(AgL)/K(ML)$ is between 10^2 and 10^7.

Potentiometry is also useful to study step-wise complex formations. As an example, we consider here the case in which ion X in inert solvent s is complexed step-wise with other solvents D [28].

$$X + D \rightleftarrows XD \qquad \beta_1 = \frac{a(XD)}{a(X)[D]}$$

$$X + 2D \rightleftarrows XD_2 \qquad \beta_2 = \frac{a(XD_2)}{a(X)[D]^2}$$

Cell (IX) (XY is either X^+Y^- or X^-Y^+) is constructed, D is added step-wise to the sample solution that is initially in pure s and the emf is measured after each addition:

$$Ag|5 \text{ mM AgClO}_4(s)\|0.1 \text{ M Et}_4\text{NPic}(s)\|1 \text{ mM XY,}$$

$$10 \text{ mM Et}_4\text{NPic}(s+D)| \text{ indicator electrode} \qquad \text{(IX)}$$

---

12) We consider that in cell (VIII), s = PC, M^+ = Na^+ and L = dibenzocryptand(2,2,2), and $c_1 = 3.7 \times 10^{-4}$ M, $c_2 = 1.29 \times 10^{-2}$ M and c_3 1.06×10^{-3} M [27b]. At the equilibrium, in the left compartment, $c_1 = [Ag^+] + [AgL^+]$, $c_2 = [M^+] + [ML^+]$ and $c_3 = [L] + [ML^+] + [AgL^+]$. $pa_{Ag} = 8.90$ is obtained from the equilibrium emf and from it, we get $[Ag^+] =$ 1.5×10^{-9} M using the activity coefficient of 0.85. We separately get $K(AgL) = 7.5 \times 10^{15}$ by Procedure 1. Then, $[L] = 3.7 \times 10^{-11}$ M (negligibly small). From these we get $[AgL^+]$ 3.7×10^{-4} M, $[NaL^+] = 6.9 \times 10^{-4}$ M and $[Na^+] = 1.22 \times 10^{-2}$ M. Thus, from Eq. 6.11, we get the final result, $K(NaL^+) = 1.7 \times 10^9$.

If the potential change of the indicator electrode due to the addition of D is expressed by ΔE, we get

$$\exp\left(-\frac{RT\Delta E}{zF}\right) = 1 + \beta_1[D] + \beta_2[D]^2 + \cdots$$

where z is the ionic charge. When D is added in large amount, the decrease in the activity of solvent s must be taken into account (see Eq. (6.18)). The values of $\beta_1, \beta_2, \ldots$ are obtained graphically or by computer calculation.

Using this method, the step-wise complex formation constants of cations with basic solvents and of anions with protic solvents have been determined in relatively inert solvents such as AN [28], PC [29] and acetylacetone (Acac) [30]. The original objective of this study is to determine the step-wise formation constants of complex XD_i $(i = 1, 2, \ldots)$ in inert solvents. However, as described in Section 6.3.6, the results are applicable to evaluate the Gibbs energies of transfer of ion X from solvent s to mixed solvent (s + D) and solvent D [29, 31]. Thus, the results can also be applied to obtain the relative concentrations of the dissolved species of X (i.e. XD_i) as a function of the composition of the mixed solvent.

6.3.4
Redox Reactions in Nonaqueous Solutions

The characteristics of the redox reactions in nonaqueous solutions were discussed in Chapter 4. Potentiometry is a powerful tool for studying redox reactions, although polarography and voltammetry are more popular. The indicator electrode is a platinum wire or other inert electrode. We can accurately determine the standard potential of a redox couple by measuring the electrode potential in the solution containing both the reduced and the oxidized forms of known concentrations. Potentiometric redox titrations are also useful to elucidate redox reaction mechanisms and to obtain standard redox potentials. In some solvents, the measurable potential range is much wider than in aqueous solutions, and various redox reactions that are impossible in aqueous solutions are possible.

In the potentiometric redox titrations shown in Section 4.3, Cu(II) in AN was often used as a strong oxidizing agent. The following is the procedure employed by Senne and Kratochivil [32] for determining the accurate value of the standard potential of the Cu(II)/Cu(I) couple in AN.

The emf of cell (X), E_{cell}, is given by Eq. (6.12),

$$\text{Ag(Hg)}|c_1\,\text{AgClO}_4, c_x\,\text{Et}_4\text{NClO}_4(\text{AN})\|c_2\,\text{CuClO}_4, c_3\,\text{Cu(ClO}_4)_2, \qquad \text{(X)}$$
$$c_x\,\text{Et}_4\text{NClO}_4(\text{AN})|\text{Pt}$$

$$E_{cell} = E_{cell}^0 + 0.05916\log\frac{[\text{Cu(II)}]}{[\text{Cu(I)}][\text{Ag(I)}]} + 0.05916\log\frac{\gamma_{\text{Cu(II)}}}{\gamma_{\text{Cu(I)}}\gamma_{\text{Ag(I)}}} - E_j$$

$$(6.12)$$

where E^0_{cell} is the standard emf of cell (X) and E_j the liquid junction potential at junction ($\vert\vert$). Here, E'_{cell} is defined by

$$E'_{cell} = E_{cell} - 0.05916 \log \frac{[Cu(II)]}{[Cu(I)][Ag(I)]} = E_{cell} - 0.05916 \log \frac{c_3}{c_1 c_2}$$

and Eq. (6.13) is obtained:

$$E'_{cell} = E^0_{cell} + 0.05916 \log \frac{\gamma_{Cu(II)}}{\gamma_{Cu(I)}\gamma_{Ag(I)}} - E_j \qquad (6.13)$$

In order to eliminate the influence of E_j, E'_{cell} is obtained, as shown in Table 6.7, by gradually decreasing c_1, c_2 and c_3 but keeping c_X constant; then, $E^{0'}_{cell}$ is determined as the value of E'_{cell} extrapolated to zero values of c_1, c_2 and c_3. Here, $E^{0'}_{cell}$ can be expressed by

$$E^{0'}_{cell} = E^0_{cell} + 0.05916 \log \frac{\gamma_{Cu(II)}}{\gamma_{Cu(I)}\gamma_{Ag(I)}} \qquad (6.14)$$

If the Debye–Hückel theory is used for the activity coefficients, Eq. (6.14) can be written as

$$E^{0'}_{cell} = E^0_{cell} - \frac{(2)^2(1.65)(0.05916)I^{1/2}}{1 + 4.86 \times 10^{-9} a_j I^{1/2}}$$

Table 6.7 Determination of the standard potential of Cu(II)/Cu(I) couple in acetonitrile.

Et$_4$NClO$_4$ (c_X)	Ag(I) (c_1)	Cu(I) (c_2)	Cu(II) (c_3)	E'_{cell}
10.0 mM	5.26 mM	5.00 mM	5.26 mM	566.5 mV
10.0	3.16	3.00	3.16	566.5
10.0	1.58	1.50	1.58	568.8
10.0	0.53	0.50	0.53	568.4
$E^{0'}_{cell} = 568.8 \pm 0.5$ ($\mu = 0.0100$)				
6.00	4.21	4.00	4.21	568.2
6.00	2.63	2.50	2.63	570.3
6.00	1.05	1.00	1.05	572.8
6.00	0.53	0.50	0.53	573.3
$E^{0'}_{cell} = 574.2 \pm 0.1$ ($\mu = 0.0060$)				
3.00	2.10	2.00	2.10	575.0
3.00	1.26	1.20	1.26	576.4
3.00	0.84	0.80	0.84	578.0
3.00	0.53	0.50	0.53	577.2
$E^{0'}_{cell} = 578.7 \pm 0.5$ ($\mu = 0.0030$)				
1.00	1.05	1.00	1.05	585.4
1.00	0.74	0.70	0.74	583.9
1.00	0.53	0.50	0.53	584.7
1.00	0.21	0.20	0.21	584.6
$E^{0'}_{cell} = 584.2 \pm 0.5$ ($\mu = 0.0010$)				
$E^0_{cell} = 591$ mV ($a_j = 0.0$); E^0(Cu(II)/Cu(I)) $= 679$ mV				

where 1.65 $(\text{mol dm}^{-3})^{-1/2}$ and $4.86 \times 10^{-9} (\text{mol dm}^{-3})^{-1/2} \text{m}^{-1}$ are the constants A and B in AN (see footnote 10 in Chapter 2) and I is the ionic strength, which equals c_X. Here, the relation between $E^{0'}_{\text{cell}}$ and $I^{1/2}/(1 + 4.86 \times 10^{-9} a_i I^{1/2})$ is plotted by adjusting the value of a_i so as to get the best linear relation. Then, $E^{0}_{\text{cell}} = 0.592$ V is obtained from the intercept of the straight line. The standard potential of the $Ag^+/Ag(Hg)$ electrode has been obtained to be 0.087 V versus the standard potential of the Ag^+/Ag electrode. Thus, the standard potential of the $Cu(II)/Cu(I)$ couple in AN is $+0.679$ V versus the standard potential of Ag^+/Ag (Table 4.5).

6.3.5
Potentiometric Characterization of Solvents

We select the most suitable solvent when we run experiments in nonaqueous solutions. However, the solvent usually contains impurities, and if the impurities are reactive, they can have an enormous effect on the reactions and equilibria in the solvent. Thus, the characterization of solvents, including the qualitative and quantitative tests of impurities and the evaluation of their influences, is very important.

Potentiometry and potentiometric titrations are useful for detecting and determining solvent impurities (see Chapter 10). Here, however, we outline the basic principle of the *ion-probe method* proposed by Coetzee *et al.* [1]. In this method, we use a couple of a probe ion and an indicator electrode. The probe ion should interact strongly with the impurity, while the indicator electrode should respond with a Nernstian slope to the probe ion over a wide concentration range. We obtain in the test solvent the response curve of the indicator electrode to the probe ion. If it deviates from the Nernstian response, we can characterize the reactive impurity from the deviation. Figure 6.7 shows some theoretical response curves to the probe ion M^{z+} in the presence of reactive impurity L. In Figure 6.7, the total concentration of L, $c(L)$, is kept constant (10^{-4} M), while the reactivity of L with M^{z+} varies. The response is Nernstian in the absence of L (curve IV). In curves I, II and III, the deviation from the Nernstian response begins at $c(M^{z+}) \approx c(L)$ ($=10^{-4}$ M), and the magnitude of the deviation ΔE (mV) at $c(L) \gg c(M^{z+})$ is expressed by

$$\Delta E = -(59/z)\log(1 + \beta_1[L] + \beta_2[L]^2 + \cdots + \beta_n[L]^n)$$

where β_i is the formation constant of complex ML_i and $[L] \sim c(L)$. From Figure 6.7, we can detect the impurity (L), which reacts with M^{z+}, and can roughly evaluate its concentration and reactivity with M^{z+}. However, we cannot identify the impurity L. It must be identified by other method such as chromatography. Table 6.8 shows some combinations of reactive impurities, probe ions and indicator electrodes for the ion-probe method. For the practical application of this method, see Section 10.3.

6.3.6
Potentiometric Study of Ion Solvation – Applications that Compare Electrode Potentials in Different Solvents

Electrochemical methods play important roles in studying ion solvation. In this section, we deal with the potentiometric study of ion solvation and the difficulties associated with it [31].

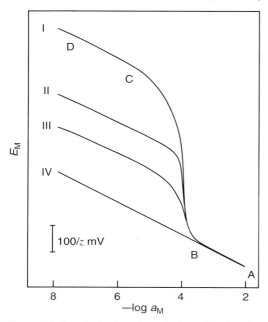

Figure 6.7 Characterization of reactive impurities in solvents by means of ion-probe method [1a]. Theoretical response curves of the indicator electrode to the probe ion (M^{z+}) in the presence of 10^{-4} M reactive ligand (L). (I) $\log \beta_i = 10, 18, 24$ and 18 for $i = 1–4$; (II) $\log \beta_1 = 10$; (III) $\log \beta_i = 6, 11, 15$ and 18 for $i = 1–4$; and (IV) in the absence of L. Slope: $59/z$ mV. M^{z+} is in excess at A to B, and L is in excess at C to D.

Various potentiometric indicator electrodes work as sensors for ion solvation. Metal and metal amalgam electrodes, in principle, respond in a thermodynamic way to the solvation energy of the relevant metal ions. Some ion-selective electrodes can also respond almost thermodynamically to the solvation energies of the ions to which they are sensitive. Thus, the main difficulty in the potentiometric study of ion solvation arises from having to compare the potentials in different solvents, even though there is no thermodynamic way of doing it. In order to overcome this difficulty, we have to employ some method based on an extrathermodynamic assumption. For example, we can use (1) or (2) as follows:

Table 6.8 Reactive impurities and probe ion/indicator electrode.

Reactive impurities	Probe ion/indicator electrode
Proton acceptors (bases such as amines)	H^+/pH glass electrode
Proton donors (acids)	t-BuO$^-$/pH glass electrode
Ligands (with N or S atoms)	Ag^+/Ag electrode; Hg^{2+}/Hg electrode; Cu^{2+}/Cu^{2+} ion-selective electrode
Proton donors; hydrogen bond donors	F^-/F^- ion-selective electrode

1. A method in which the potential of an appropriate reference electrode (or reference redox system) is assumed to be solvent independent.
2. A method in which the LJP between different solvents is assumed to be negligible at appropriate junctions.

In order to study the solvation of metal ion M^+ by method (1), we measure the emf of cell (XI) using solvents S and R.

$$RE|c\,M^+X^-(R\,or\,S)|M \qquad\qquad (XI)$$

where RE shows a reference electrode whose potential is solvent independent. If we denote the emfs in solvents S and R by $E(S)$ and $E(R)$, respectively, then $\Delta G_t^\circ(M^+, R \to S)$ and $\gamma_t(M^+, R \to S)$ can be obtained by

$$\Delta G_t^\circ(M^+, R \to S) = (2.3RT)\log\gamma_t(M^+, R \to S) = F[E(S)-E(R)] \qquad (6.15)$$

When the permittivities of solvents S and R are very different, this affects the Debye–Hückel activity coefficient of M^+ (see footnote 10 in Chapter 2). If necessary, the effect should be estimated (or eliminated) using the Debye–Hückel theory.

In fact, it is not easy to get a reference electrode whose potential is solvent independent. Therefore, we use a reference redox system (Fc^+/Fc or BCr^+/BCr) instead. We measure the potentials of the M^+/M electrode in S and R against a conventional reference electrode (e.g. Ag^+/Ag). At the same time, we measure the half-wave potentials of the reference redox system in S and R using the same reference electrode. Then, the potentials of the M^+/M electrode in S and R can be converted to the values against the reference redox system. In this case, the reliability of the results depends on the reliability of the assumption that the potential of the reference redox system is solvent independent.

In order to study the solvation of metal ion M^+ by method (2), we measure the emf, E, of cell (XII), which contains a junction between solutions in R and S.

$$M|c\,M^+X^-(R)\|c\,M^+X^-(S)|M \qquad\qquad (XII)$$

$\Delta G_t^\circ(M^+, R \to S)$ and $\gamma_t(M^+, R \to S)$ are related to E by

$$\Delta G_t^\circ(M^+, R \to S) = (2.3RT)\log\gamma_t(M^+, R \to S) = F(E-E_j) \qquad (6.16)$$

where E_j is the LJP at the junction ($\|$). If E_j is negligible, we can get $\Delta G_t^\circ(M^+, R \to S)$ and $\gamma_t(M^+, R \to S)$. In practice, however, the LJP between the two solutions is not always small enough, and so we usually insert an appropriate salt bridge (for example, $//0.1\,M\,Et_4NPic(AN)//$ proposed by Parker *et al.* [33]) at the junction to make the LJP negligibly small. The reliability of the results depends on how small the LJP is. For this method to be successful, it is necessary to have some knowledge about the LJP between solutions in different solvents. This problem will be discussed in Section 6.4.

Ion solvation has been studied extensively by potentiometry [31, 34]. Among the potentiometric indicator electrodes used as sensors for ion solvation are metal and metal amalgam electrodes for the relevant metal ions, pH glass electrodes and pH-ISFETs for H^+ (see Figure 6.8), univalent cation-sensitive glass electrodes for alkali

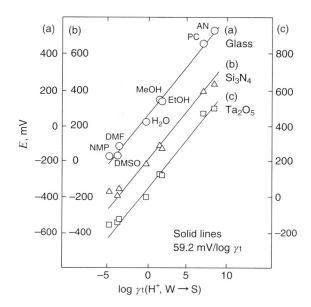

Figure 6.8 Response of (a) glass electrode, (b) Si_3N_4-ISFET and (c) Ta_2O_5-ISFET to the proton transfer activity coefficient [20a].

metal ions, a CuS solid-membrane electrode for Cu^{2+}, an LaF_3-based fluoride electrode for F^- and some other ISEs. So far, method (2) has been employed more often. The advantage of potentiometry is that the number and the variety of target ions increase by the use of ISEs.

The following is an example of the ion solvation study. Cox *et al.* [29] obtained, using cell (XIII), the Gibbs energies of transfer of Ag^+, Li^+ and Na^+ from PC to (PC + DMSO) mixtures as well as from PC to DMSO:

$$Ag|5\ mM\ AgClO_4(PC)\|0.1\ M\ Et_4NPic(PC)\|1\ mM\ MClO_4,$$
$$10\ mM\ Et_4NPic(PC+DMSO)|Ag\ or\ glass\ electrode \tag{XIII}$$

The sensor for ion solvation was a silver wire for Ag^+ and a univalent cation-sensitive glass electrode for Li^+ and Na^+. In determining $\Delta G_t^\circ(M^+, R \to S)$ from Eq. (6.16), they assumed that the LJPs at PC/(PC + DMSO) and PC/DMSO were negligible ($E_j = 0$). At the same time, they determined in PC the step-wise formation constants (β_i) for the complexation of Ag^+, Li^+ and Na^+ with DMSO (see Section 6.3.3), and used them to calculate the Gibbs energies of transfer of those ions from PC to

(PC + DMSO) and to neat DMSO. Equations (6.17) and (6.18) were used in the calculation.

$$\Delta G_t^\circ(X, S \to D) = -RT \ln \beta'_n \qquad (6.17)$$

$$\Delta G_t^\circ(X, S \to S + D) = -nRT \ln \phi_S - RT \ln \left\{ 1 + \sum_{i=1}^{n} \beta'_i \left(\frac{\phi_D}{\phi_S} \right)^i \right\} \qquad (6.18)$$

Here, $X = M^+$, $S = PC$, $D = DMSO$, $\beta'_i = \beta_i(1000\rho_D/M_D)^i$ (ρ_D and M_D: density and molecular weight of D) and ϕ_D and ϕ_S are the volume fractions of D and S. In Figure 6.9, the Gibbs energies of transfer obtained experimentally are compared with those calculated by Eqs. (6.17) and (6.18). Good agreements are observed between the experimental and calculated values. This shows that the Gibbs energies of transfer of an ion from PC to (PC + DMSO) and to DMSO can be estimated using the data obtained in PC for the step-wise complex formation between the ion (Ag$^+$, Li$^+$ or Na$^+$) and the ligand (DMSO). This important conclusion concerning the relation between ion solvation and the ion complexation by solvent is valid, as far as the difference between the permittivities of solvents S and D does not have a significant influence on ion solvation. It should be noted, however, that the simplified relation above does not hold if solvents S and D have markedly different permittivities. This occurs, for example, for the transfer of Li$^+$ from PC ($\varepsilon_r \sim 64$) to DME ($\varepsilon_r \sim 7$) and (PC + DME) mixtures. For the corrections that are necessary in such a case, see Ref. [35].

In studies of ion solvation, electrode potentials in different solvents must be compared. However, if there is a reliable method for it, data on ion solvation can easily be obtained and it becomes possible to estimate solvent effects on various chemical reactions. Thus, establishing a reliable method to compare the potentials in different solvents is vital to solution chemistry.

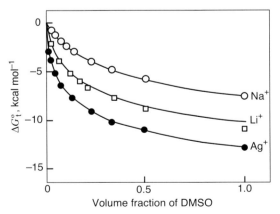

Figure 6.9 Gibbs energies of transfer of Li$^+$, Na$^+$ and Ag$^+$ ions from PC to (PC + DMSO) mixtures and to neat DMSO [29]. Open and closed circles and open squares are experimental results and solid curves are theoretical ones.

6.4
Liquid Junction Potentials between Different Solvents

As described above, establishing a reliable method to compare the potentials in different solvents is very important. One approach to this problem is to develop a method of estimating the LJP between electrolyte solutions in different solvents. However, because of the complicated phenomena at the junction between different solvents, our knowledge on the LJP between different solvents is still rather limited [31, 36, 37]. The following is a brief outline of the LJP between different solvents, mainly based on our recent studies [31, 37].

The LJP between different solvents consists of three components: i.e. (i) a component caused by the differences in electrolyte concentrations on the two sides of the junction and the differences between cationic and anionic mobilities; (ii) a component due to the differences between ion solvation on the two sides of the junction; (iii) a component due to the solvent–solvent interactions at the junction.

The characteristics of the three components are schematically shown in Figure 6.10 for a junction with the same electrolyte on the two sides (c_1 MX(S_1)/c_2 MX(S_2)) [37b]. Component (i) is somewhat similar to the LJP between solutions in the same solvent (Section 6.1.4). Components (ii) and (iii), however, are specific to the junction between different solvents. Fortunately, under appropriate conditions, we can measure the variation in each of the three components separately. Thus, we can study the characteristics of each component.

Component (i) [37d-g] At the junction c_1 MX(S_1)/c_2 MX(S_2), this component is given by Eq. (6.19).[13]

$$E_j(i) = \left(-\frac{RT}{F}\right)\left[(t_{M1}-t_{X1})\ln\frac{a_{MX2}}{a_{MX1}} + (t_{M2}-t_{M1}-t_{X2}+t_{X1})\right.$$
$$\left.\times\left(1-\frac{a_{MX1}}{a_{MX2}-a_{MX1}}\ln\frac{a_{MX2}}{a_{MX1}}\right)\right]$$

(6.19)

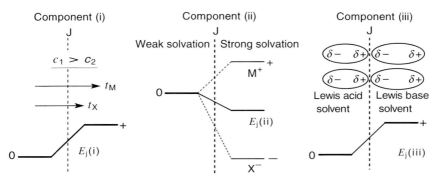

Figure 6.10 Three components for the LJP between electrolyte solutions in different solvents. Junction: c_1 MX(S_1)$\vdots c_2$ MX(S_2).

13) Equation (6.19) is for the case when MX is a 1:1 electrolyte. For the case when MX is $z_1:z_2$ type, see Ref. [37g].

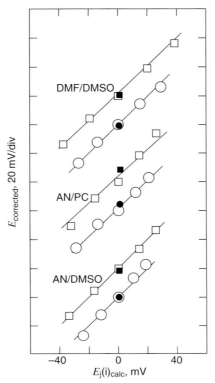

Figure 6.11 Component (i) of the LJP between different solvents: experimental results versus the values calculated by Eq. (6.19). MX = LiClO$_4$ (squares) and NaClO$_4$ (circles). $c_1 : c_2$ = (from left to right) 100 : 1, 10 : 1, 1 : 1, 1 : 10 and 1 : 100 (mM), except for closed circles and squares, where $c_1 : c_2$ = 25 : 25 (mM).

Here, t is the ionic transport number, a is the electrolyte activity and the subscripts 1 and 2 refer to the left and right sides of the junction. For $a_{MX1} = a_{MX2}$, $E_j(i) = 0$. If the experimental (actual) variations in component (i), obtained by changing the ratio c_1/c_2, are plotted against the values calculated by Eq. (6.19), linear relations of unit slopes, as in Figure 6.11, are generally observed. Thus, this component can be estimated by Eq. (6.19).

Component (ii) [37d–g] Equation (6.20) can be derived for this component, under conditions analogous to Eq. (6.19).

$$E_j(ii) = \left(-\frac{1}{2F}\right)\left[(t_{M1} + t_{M2})\,\Delta G_t^\circ(M) - (t_{X1} + t_{X2})\,\Delta G_t^\circ(X)\right] \qquad (6.20)$$

Here, $\Delta G_t^\circ(M)$ and $\Delta G_t^\circ(X)$ are the Gibbs energies of transfer of M$^+$ and X$^-$ from solvent S$_1$ to S$_2$. From this equation, we can predict the characteristics of component (ii): (1) By this component, cation M$^+$ makes the side on which

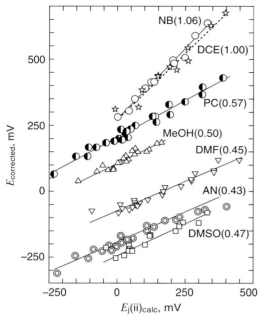

Figure 6.12 Component (ii) of the LJP between different solvents: experimental results versus the values calculated by Eq. (6.20). The case of the junctions between water and organic solvents, 10 mM MX(H_2O)/10 mM MX(S). Solvent S and the slope are shown on each line.

the solvation is stronger more positive, while anion X^- makes the side on which the solvation is stronger more negative (see Figure 6.10); (2) this component is not influenced by electrolyte concentrations. Unfortunately, Eq. (6.20) is valid only when the solvents on the two sides are immiscible. When the solvents are miscible, it holds only qualitatively. Figure 6.12 shows, for the junctions between water and organic solvents, the relations between the experimental (actual) variations in component (ii) and the values calculated by Eq. (6.20). For immiscible junctions (H_2O/NB and H_2O/DCE), linear relations of unit slopes are observed. For other miscible junctions, however, although near-linear relations are observed, the slopes are much smaller than unity.[14] The reasons for the slopes of less than unity have been discussed [38]. The reasons considered are (1) the chemical potentials, μ^0, of M^+ and X^- change gradually through the miscible junctions (thickness: 0.05–1 mm), in contrast to the abrupt change at immiscible junctions (thickness: $\approx$1 nm) and (2) the nature of ionic transfer due to the gradient in μ^0 is different from the nature of migration due to the potential

14) The slopes of near-linear relations at miscible junctions are 0.43–0.50 at H_2O/S, 0.28–0.33 at FA/S, 0.29–0.36 at EG/S and 0.17–0.26 at MeOH/S, where S shows organic solvents.

gradient and that of diffusion due to the concentration gradient. Although the slopes are less than unity, we can use the linear relations to estimate the actual values of component (ii) by multiplying the calculated E_j(ii) and the slope. Component (ii) varies considerably with the electrolyte species, with the variation sometimes reaching more than 300 mV.

Component (iii) [37b, c, 39] This component is due to the solvent–solvent interactions at the junction and, in principle, is electrolyte independent. The magnitude and sign of this component are well explained if we consider that, as in Figure 6.10, the solvent molecules on the two sides interact with each other as a Lewis acid and a Lewis base, and some of the molecules are oriented at the interphase to cause a potential difference. We have no theoretical way of estimating the value of this component, but from the experimental results and under some assumptions, we can make a rough estimate. Component (iii) between solvents that interact strongly, such as H_2O/DMF or H_2O/DMSO, is expected to be over 100 mV (with the H_2O side more negative). At a junction between two aprotic solvents, it is usually within ± 20 mV, because they interact only weakly. It is interesting that at a junction with a mixed solvent on one side (or mixed solvents on the two sides), component (iii) often changes linearly with the volume fraction of the mixed solvent(s). This property can also be explained by the mechanism in Figure 6.10.[15] We can use this fact to estimate component (iii) at such junctions.

Traditionally, the LJPs between different solvents are estimated indirectly. For example, if $\Delta G_t^{\circ}(M^+, R \rightarrow S)$ is known, we can estimate the LJP (E_j) from Eq. (6.16) by measuring the emf (E) of cell (XII). The reliability of the method depends on the reliability of the extrathermodynamic assumption used in determining $\Delta G_t^{\circ}(M^+, R \rightarrow S)$. The author proposed a new method of estimating the LJP between different solvents, in which the three components are estimated separately and then simply added together [37h]. Table 6.9 lists some examples of LJPs estimated by the three-component method and by the traditional indirect method based on the assumption of reference electrolyte (Ph_4AsBPh_4). The results obtained by the two methods agree well. The advantage of the new method is that the influence of electrolytes and solvents on the LJP is predictable.

At a junction between different solvents, the solutions on the two sides are gradually mixed. Thus, the stability and the reproducibility of the LJP are a big concern to us. However, when the electrolyte on the two sides is of the same kind, the

---

15) According to the model of component (iii) (Figure 6.10), the component at a mixed solvent/pure solvent junction should vary linearly with the volume fraction of the mixed solvent. But exceptions, in which the relationships are curves, are observed for mixtures between an amphiprotic solvent (H_2O, FA or *N*-methylformamide) and an aprotic solvent of very low acceptor number (DMF, DMA, NMP, Ac or HMPA). In contrast, for an aprotic solvent of moderate acceptor number (NM, DMSO, AN or PC), linear or near-linear relationships are observed as expected. The curved relationships were explained by the fact that the structure-forming amphiprotic solvent is more likely to and the aprotic solvent less likely to settle at the interphasial region than in the bulk, because the dipolar molecules of the aprotic solvent of very low acceptor number have a vague positive charge center [39a].

Table 6.9 LJPs between different solvents estimated by the three-component method and by a conventional method (mV).[a]

S=	MX=	Three-component method				Conv. method	Difference
		(i)	(ii)	(iii)	$E_j(1)$	$E_j(2)$	$E_j(1) - E_j(2)$
AN	Et$_4$NPic	0	8	37	45	39	6
	Et$_4$NClO$_4$	−2	18	37	53	55	−2
	Et$_4$NI	−2	60	37	95	105	−10
	Et$_4$NCl	−2	131	37	166	169	−3
DMF	Et$_4$NPic	0	2	103	105	111	−6
	Et$_4$NClO$_4$	−2	30	103	131	131	0
	Et$_4$NI	−2	73	103	174	177	−3
	Et$_4$NCl	−2	157	103	258	252	6
MeOH	Et$_4$NPic	1	−18	30	13	6	7
	Et$_4$NClO$_4$	−2	18	30	46	40	6
	Et$_4$NI	−2	22	30	50	49	1
	Et$_4$NCl	−1	40	30	69	81	−12

[a]Liquid junction of the type 25 mM MX(H$_2$O)|25 mM MX(S).

LJP is fairly stable and reproducible. According to our study [37j], at a free diffusion junction, the LJP reaches a steady value within several seconds after the formation of the junction and it is reproducible within ±2 mV. Moreover, the LJP is stable and its drift is less than ±2 mV h^{-1}, even when the estimated LJP value is >200 mV.

In many cases, the electrolytes on the two sides are different. Even in such cases, the LJP is stable and reproducible if the electrolyte concentration on one side is more than 10 times that on the other side.[16] Here, the behavior of the LJP is determined by the electrolyte of higher concentration. As described in Section 6.1.2, aqueous saturated calomel electrode and silver–silver chloride electrode are often used as reference electrodes, with their tips inserted in nonaqueous solutions. This is justified to some extent: if the composition of the nonaqueous solution is fixed, the LJP is fairly reproducible and stable, unless the junction clogs with the solidified electrolyte. The big problem, however, is that the LJP is large in magnitude and varies drastically with the solvent used.

In the study of ion solvation, Parker *et al.* employed the assumption of negligible LJP, considering that the LJP across cell XIV is within ±20 mV [33].

$$Ag|10 \text{ mM AgNO}_3(AN)||100 \text{ mM Et}_4\text{NPic}(AN)||10 \text{ mM AgNO}_3(S)|Ag$$

$$(XIV)$$

The assumption is based on (i) the mobilities of Et$_4$N$^+$ and Pic$^-$ are close to each other, (ii) their ΔG_t° (i, AN → S) values are small and (iii) AN interacts only weakly with

16) For components (a) and (b) at the junctions with different electrolytes on the two sides (c_1 MX(S$_1$)|c_2 NY(S$_2$)), see Ref. [37f].

other aprotic solvents. According to our new estimation method, the LJP at the AN/S junction in (XIV) is within ± 10 mV if S is aprotic but approximately -30 and -50 mV for S of MeOH and H_2O, respectively. Therefore, the assumption is appropriate if S is aprotic, but not appropriate if S is MeOH or H_2O. When the assumption is not applicable, it is recommended to estimate the LJP by the new method and to make a correction for it.

References

1 (a) Coetzee, J.F., Deshmukh, B.K. and Liao, C.-C. (1990) *Chem. Rev.*, **90**, 827; (b) Coetzee, J.F., Chang, T.-H., Deshmukh, B.K. and Fonong, T. (1993) *Electroanalysis*, **5**, 765.

2 (a) Nakamura, T., Tsukamoto, W. and Izutsu, K. (1990) *Bunseki Kagaku*, **39**, 689; (b) Nakamura, T. (1991) *Bunseki*, 642; (c) Nakamura, T. (2009) *Anal. Sci.*, **25**, 33.

3 Pungor, E., Toth, K., Klatsmani, P.G. and Izutsu, K. (1983) *Pure Appl. Chem.*, **55**, 2029; Ref. [1].

4 (a) Lund, H. and Baizer, M.M. (eds) (1991) *Organic Electrochemistry*, 3rd edn, Marcel Dekker, New York, p. 278; (b) Butler, J.N. (1970) *Advances in Electrochemistry and Eectrochemical Engineering*, vol. 7 (eds P. Delahay and C.W. Tobias), Wiley & Sons, New York, p. 77; (c) Mann, C.K. (1969) *Electroanalytical Chemistry*, vol. 3 (ed. J.A. Bard), Marcel Dekker, New York, p. 57.

5 Izutsu, K., Itoh, M. and Sarai, E. (1985) *Anal. Sci.*, **1**, 341.

6 Lewandowski, A., Szukalska, A. and Galinski, M. (1995) *New J. Chem.*, **19**, 1259.

7 Coetzee, J.F. and Gardner, C.W., Jr (1982) *Anal. Chem.*, **54**, 2530, 2625.

8 Gritzner, G. and Kuta, J. (1984) *Pure Appl. Chem.*, **56**, 461.

9 Gritzner, G. (1990) *Pure Appl. Chem.*, **62**, 1839.

10 Koryta, J., Dvorak, J. and Kavan, L. (1993) *Principles of Electrochemistry*, 2nd edn, Wiley & Sons, New York.

11 Buck, R.P. *et al.* (2002) *Pure Appl. Chem.*, **74**, 2169.

12 Covington, A.K., Bates, R.G. and Durst, R.A. (1985) *Pure Appl. Chem.*, **57**, 531.

13 Baucke, F.G.K. (2002) *Anal. Bioanal. Chem.*, **374**, 772.

14 For example (a) Bates, R.G. (1982) *Pure Appl. Chem.*, **54**, 229; (b) Covington, A.K. (1985) *Pure Appl. Chem.*, **57**, 887.

15 Rondinini, S. (2002) *Anal. Bioanal. Chem.*, **374**, 813.

16 Mussini, T., Covington, A.K., Longhi, P. and Rondinini, S. (1985) *Pure Appl. Chem.*, **57**, 865.

17 (a) Rondinini, S., Mussini, P.R. and Mussini, T. (1987) *Pure Appl. Chem.*, **59**, 1549; (b) Rondinini, S., Longhi, P., Mussini, P.R. and Mussini, T. (1987) *Pure Appl. Chem.*, **59**, 1693; (c) Mussini, P.R., Mussini, T. and Rondinini, S. (1997) *Pure Appl. Chem.*, **69**, 1007; (d) Rondinini, S., Mussini, P.R., Mussini, T. and Vertova, A. (1998) *Pure Appl. Chem.*, **70**, 1419; (e) Antonini, D., Falciola, L., Mussini, P.R. and Mussini, T. (2001) *J. Electroanal. Chem.*, **503**, 153; (f) Falciola, L., Mussini, P.R. and Mussini, T. (2000) *J. Solution Chem.*, **29**, 1199; (g) Barbosa, J., Barrón, D., Butí, S. and Marqués, I. (1999) *Polyhedron*, **18**, 3361; (h) Falciola, L., Mussini, P.R., Mussini, T. and Pelle, P. (2002) *Electrochem. Commun.*, **4**, 146.

18 Izutsu, K. and Yamamoto, H. (1998) *Talanta*, **47**, 1157.

19 Bates, R.G. (1973) *Determination of pH, Theory and Practice*, 2nd edn, Wiley & Sons, New York, p. 372.

20 (a) Izutsu, K., Nakamura, T. and Hiraoka, S. (1993) *Chem. Lett.*, 1843; (b) Izutsu, K. and Ohmaki, M. (1995) *Talanta*, **43**, 643; (c) Izutsu, K. and Yamamoto, T. (1996) *Anal. Sci.*, **12**, 905; (d) Izutsu, K., unpublished results.

21 (a) Rondinini, S., Longhi, P., Mussini, P.R. and Mussini, T. (1987) *Pure Appl. Chem.*, **59**, 1693; (b) Barbosa, J. and Sanz-Nebot, V. (1991) *Anal. Chim. Acta*, **244**, 183; Rosés, M., Ràfols, C. and Bosch, E. (1993) *Anal. Chem.*, **65**, 2294; Ràfols, C., Bosch, E., Rosés, M. and Asuero, A.G. (1995) *Anal. Chim. Acta*, **302**, 355.

22 (a) Izutsu, K. (1990) *Acid–Base Dissociation Constants in Dipolar Aprotic Solvents, IUPAC Chemical Data Series No. 35*, Blackwell Scientific Publications, Oxford; (b) Reichardt, C. (2003) *Solvents and Solvent Effects in Organic Chemistry*, 3rd edn, Wiley-VCH Verlag GmbH., Weinheim, p. 497.

23 Kolthoff, I.M. and Chantooni, M.K., Jr (1979) *Treatise on Analytical Chemistry*, 2nd edn, Part I, vol. 2 (eds I.M. Kolthoff and P.J. Elving), Wiley & Sons, New York, pp. 239–302, 349–384.

24 (a) Kolthoff, I.M. and Chantooni, M.K., Jr (1965) *J. Am. Chem. Soc.*, **87**, 4428; (b) Izutsu, K., Kolthoff, I.M., Fujinaga, T., Hattori, M. and Chantooni, M.K., Jr (1977) *Anal. Chem.*, **49**, 503.

25 For example, Safarik, L. and Stransky, Z. (1986) *Titrimetirc Analysis in Organic Solvents*, Elsevier, Amsterdam.

26 Coetzee, J.F. and Martin, M.W. (1980) *Anal. Chem.*, **52**, 2412.

27 (a) Chantooni, M.K., Jr and Kolthoff, I.M. (1985) *J. Solution Chem.*, **14**, 1; (b) Kolthoff, I.M. and Chantooni, M.K., Jr (1980) *Proc. Natl. Acad. Sci. USA*, **77**, 5040; (c) Cox, B.G., Rosas, J. and Schneider, H. (1981) *J. Am. Chem. Soc.*, **103**, 1384.

28 (a) Izutsu, K., Nomura, T., Nakamura, T., Kazama, H. and Nakajima, S. (1974) *Bull. Chem. Soc. Jpn.*, **47**, 1657; (b) Ref. [31] and references therein.

29 Cox, B.G., Waghorn, W.E. and Pigott, C.K. (1979) *J. Chem. Soc., Faraday Trans. I*, **75**, 227.

30 Sakamoto, I., Yamane, N., Sogabe, K. and Okazaki, S. (1989) *Denki Kagaku*, **57**, 253.

31 (a) Izutsu, K. (1991) *Anal. Sci.*, **7**, 1; (b) Izutsu, K. (1998) *Pure Appl. Chem.*, **70**, 24; (c) Ref. [2c].

32 Senne, J.K. and Kratochivil, B. (1972) *Anal. Chem.*, **44**, 585.

33 Alexander, R., Parker, A.J., Sharp, J.H. and Waghorne, W.E. (1972) *J. Am. Chem. Soc.*, **94**, 1148.

34 Izutsu, K. and Nakamura, T. (1990) *Proceedings JSPS/NUS Joint Seminar on Analytical Chemistry* (ed. S.B. Khoo), World Scientific, Singapore, p. 184.

35 Izutsu, K., Nakamura, T., Miyoshi, K. and Kurita, K. (1996) *Elecctrochim. Acta*, **41**, 2523.

36 (a) Alfenaar, M., DeLigny, C.L. and Remijnse, A.G. (1967) *Recl. Trav. Chim. Pays-Bas.*, **86**, 986; (b) Murray, R.C., Jr and Aikens, D.K. (1976) *Electrochim. Acta*, **21**, 1045; (c) Cox, B.G., Parker, A.J. and Waghorne, W.E. (1973) *J. Am. Chem. Soc.*, **95**, 1010; (d) Berne, A. and Popovych, O. (1988) *Aust. J. Chem.*, **41**, 1523; (e) Kahanda, C. and Popovych, O. (1994) *Aust. J. Chem.*, **47**, 921; (f) Senanayake, G. and Muir, D.M. (1987) *J. Electroanal. Chem.*, **237**, 149.

37 (a) Izutsu, K., Nakamura, T., Kitano, T. and Hirasawa, C. (1978) *Bull. Chem. Soc. Jpn.*, **51**, 783; (b) Izutsu, K., Nakamura, T., Takeuchi, I. and Karasawa, N. (1983) *J. Electroanal. Chem.*, **144**, 391; (c) Izutsu, K., Nakamura, T. and Muramatsu, M. (1990) *J. Electroanal. Chem.*, **283**, 435; (d) Izutsu, K., Nakamura, T., Muramatsu, M. and Aoki, Y. (1991) *J. Electroanal. Chem.*, **297**, 41; (e) Izutsu, K., Nakamura, T. and Aoki, Y. (1992) *J. Electroanal. Chem.*, **334**, 213; (f) Izutsu, K., Muramatsu, M. and Aoki, Y. (1992) *J. Electroanal. Chem.*, **338**, 125; (g) Izutsu, K., Arai, T. and Hayashijima, T. (1997)

J. Electroanal. Chem., **426**, 91; (h) Izutsu, K., Nakamura, T., Muramatsu, M. and Aoki, Y. (1991) *Anal. Sci.*, **7** (Suppl.), 1411; (i) Izutsu, K., Nakamura, T., Arai, T. and Ohmaki, M. (1995) *Electroanalysis*, **7**, 884; (j) Izutsu, K., Nakamura, T. and Yamashita, T. (1987) *J. Electroanal. Chem.*, **225**, 255.

38 (a) Izutsu, K. and Kobayashi, N. (2005) *J. Electroanal. Chem.*, **574**, 197; (b) Izutsu, K. (2005) *Rev. Polarogr.*, **51**, 73.

39 (a) Izutsu, K. (2008) *Bull. Chem. Soc. Jpn.*, **81**, 703; (b) Izutsu, K. and Gozawa, N. (1984) *J. Electroanal. Chem.*, **171**, 373; (c) Izutsu, K., Nakamura, T. and Gozawa, N. (1984) *J. Electroanal. Chem.*, **178**, 165, 171.

7
Conductimetry in Nonaqueous Solutions

As described in Section 5.8, the conductivity of electrolyte solutions is a result of the transport of ions. Thus, conductimetry is the most straightforward method to study the behavior of ions and electrolytes in solutions. The problems of electrolytic conductivity and ionic transport number in nonaqueous solutions have been dealt with in several books [1–7]. However, even now, our knowledge of ionic conductivity is increasing, especially in relation to the role of dynamic solvent properties. In this chapter, fundamental aspects of conductimetry in nonaqueous solutions are outlined.

7.1
Dissociation of Electrolytes and Electrolytic Conductivity [1–8]

7.1.1
Molar Conductivity of Dilute Solutions of Symmetrical Strong Electrolytes

Molar conductivity, Λ ($S\,cm^2\,mol^{-1}$), of the solution of a symmetrical (z, z) strong electrolyte is often related to its concentration, c, by

$$\Lambda = \Lambda^\infty - (A\Lambda^\infty + B)c^{1/2} = \Lambda^\infty - Sc^{1/2} \tag{7.1}$$

where Λ^∞ is the molar conductivity at infinite dilution, $A/(mol^{-1/2}\,dm^{-3/2})$ $= 82.04 \times 10^4 z^3/(\varepsilon_r T)^{3/2}$, $B/(S\,cm^2\,mol^{-3/2}dm^{3/2}) = 8.249z^2/\{\eta(\varepsilon_r T)^{1/2}\}$ and ε_r and η are the relative permittivity and the viscosity (P) of the solvent [9].[1] Equation (7.1) shows the Debye–Hückel–Onsager limiting law (1926) and has the same form as the empirical Kohlrausch law (Section 5.8). It shows that the Λ value for a strong electrolyte decreases linearly with the increase in $c^{1/2}$, the intercept and slope being

---

1) (i) For 1:1 electrolytes, the values of A and B (25 °C) are 0.229 and 60.20 for water, 0.923 and 156.1 for MeOH, 1.83 and 89.7 for EtOH, 1.63 and 32.8 for Ac, 0.716 and 22.9 for AN and 0.708 and 125.1 for NM. (ii) For (z_+, z_-)-type electrolytes, Onsager derived $\Lambda = \Lambda^\infty - (A'\Lambda^\infty + B')I^{1/2}$, where $A' = 2.80 \times 10^6 \times$ $|z_+ z_-|q/\{(\varepsilon_r T)^{3/2}(1 + q^{1/2})\}$, $B' = 41.25 \times (|z_+| + |z_-|)/\{\eta(\varepsilon_r T)^{1/2}\}$, $q = \{|z_+ z_-|/ (|z_+| + |z_-|)\} \{(\lambda_+^\infty + \lambda_-^\infty)/(|z_+|\lambda_+^\infty + |z_-|\lambda_-^\infty)\}$, and $I = (\sum c_i z_i^2)/2$. Here, λ_+^∞ and λ_-^∞ are the values for each unit charge of the ions and Λ and Λ^∞ are the so-called equivalent conductivities.

Electrochemistry in Nonaqueous Solutions, Second, Revised and Enlarged Edition. Kosuke Izutsu
Copyright © 2009 WILEY-VCH Verlag GmbH & Co. KGaA, Weinheim
ISBN: 978-3-527-32390-6

Λ^∞ and $-S$ (Onsager slope), respectively.[2] For 1: 1 electrolytes, Eq. (7.1) is accurate up to $\sim$0.003 M in water, but this upper limit may be somewhat lower in nonaqueous solvents. From the linear Λ–$c^{1/2}$ relation, we can get the value of Λ^∞ and can confirm the complete dissociation of the electrolyte. Beyond this concentration range, the experimental Λ values are usually higher than the calculated ones. This occurs because Eq. (7.1) is based on the Debye–Hückel limiting law and the effect of ion size (short-range interaction) has not been taken into account.

Following the Debye–Hückel–Onsager theory, several theories were developed, by Pitts (1953) [10]; Fuoss and Onsager (1957) [11]; Fuoss and Hsia (1967) [12], Justice (1982) [4] and others, taking into account both the long-range and the short-range ionic interactions. These theories can generally be expressed by the equation of the Fuoss–Hsia type.

$$\Lambda = \Lambda^\infty - Sc^{1/2} + Ec \ln c + J_1 c - J_2 c^{3/2} \tag{7.2}$$

where $S = (A\Lambda^\infty + B)$. In the Fuoss–Onsager (1957) theory, $J_2 = 0$. The coefficients S and E are determined by ionic charges and solvent properties, but J_1 and J_2 also depend on the ion parameter, a. For details, see Refs [1, 4–6]. Nowadays the molar conductivity can be measured with high precision (within errors of $\pm$0.05%). Here, the above theories can accurately predict the experimental Λ–$c^{1/2}$ relations for 1: 1 strong electrolytes of up to $\sim$0.1 M. Recently, however, the conductivity of electrolyte solutions at higher concentrations has become important in the fields of electrochemical technologies. This will be discussed in Section 7.1.4.

7.1.2
Molar Conductivity and Association Constants of Symmetrical Weak Electrolytes

When ion association occurs in the solution of a (z, z)-type electrolyte, Eq. (7.3) holds.

$$M^{z+} + A^{z-} \rightleftarrows M^{z+}A^{z-}$$

$$K_A = \frac{1-\alpha}{\alpha^2 c \gamma^2}, \quad \alpha = 1 - \alpha^2 c K_A \gamma^2 \tag{7.3}$$

Here, c is the total concentration of MA, K_A is the association constant, α is the degree of dissociation of the ion pair $M^{z+}A^{z-}$ and γ is the average activity coefficient

---

2) Each ion forms in its neighborhood an *ionic atmosphere* that has a slight excess of ions of opposite sign. The ionic atmosphere reduces the ionic mobility in two ways. First, when an ion moves under an applied electric field in the solution, the center of charge of the atmosphere becomes a short distance behind the moving ion because a finite time ($\sim$10^{-6} s) is necessary to form the atmosphere, and the moving ion is subjected to a retarding force. This effect is called the *relaxation effect*.

Second, in the presence of the ionic atmosphere, a viscous drag is enhanced compared to its absence because the atmosphere moves in an opposite direction to the moving ion. This retarding effect is called the *electrophoretic effect*. In Eq. (7.1), the $A\Lambda^\infty$-term corresponds to the relaxation effect, while the B-term corresponds to the electrophoretic effect. For details, see textbooks of physical chemistry or electrochemistry.

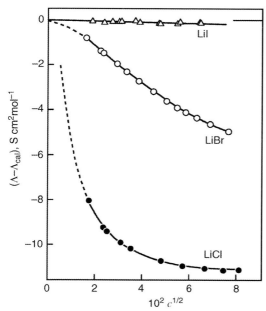

Figure 7.1 $(\Lambda - \Lambda_{cal})$ versus $c^{1/2}$ relations for lithium halide solutions in sulfolane at 30 °C, where Λ is the experimental molar conductivity and Λ_{cal} is the molar conductivity calculated from Eq. (7.1) [1a].

of free ions of concentration αc. Because the ion pairs do not conduct electricity, the molar conductivity Λ in the presence of ion association is less than that in its absence. In Figure 7.1, the difference between the experimental molar conductivity (Λ) and the value calculated from Eq. (7.1), Λ_{cal}, are plotted against $c^{1/2}$ for lithium halides in sulfolane [1a]. For LiI, the difference between Λ and Λ_{cal} is small because ion association is not appreciable ($K_A = 5.6\ mol^{-1}\ l$). For LiBr ($K_A = 278\ mol^{-1}\ l$) and LiCl ($K_A = 13\ 860\ mol^{-1}\ l$), however, Λ is much smaller than Λ_{cal}.

When ion association occurs, the relation $\Lambda = \alpha \Lambda_f$ holds between the molar conductivity for free ions (concentration αc), Λ_f, and that for total electrolyte concentration, Λ. Thus, we get Eq. (7.4) from Eq. (7.3):

$$\Lambda = \Lambda_f - \Lambda \gamma^2 K_A(\alpha c) \tag{7.4}$$

If we use Eq. (7.2) for free ions of concentration αc, we get

$$\Lambda = \Lambda^\infty - S(\alpha c)^{1/2} + E(\alpha c)\ln(\alpha c) + J_1(\alpha c) - J_2(\alpha c)^{3/2} - \Lambda \gamma^2 K_A(\alpha c) \tag{7.5}$$

Because the ions in electrolyte solutions are often more or less associated, Eq. (7.5) is useful in analyzing the conductivity data. The experimental data for Λ and c are subjected to computer analysis, by applying the least-square method, and optimum values of such parameters as Λ^∞, K_A and a (distance of ionic closest approach) are obtained. Sometimes the parameter a is replaced by the Bjerrum's distance q in Section 2.6. In that case, the parameters obtained from Eq. (7.5) are of two kinds, Λ^∞ and K_A.

If the ion association is not extensive ($K_A \leq 20\,\mathrm{mol}^{-1}\,\mathrm{l}$), the value of K_A obtained by the computer analysis is greatly affected by the terms J_1 and J_2 used in Eq. (7.5). Then, the value of K_A does not accurately reflect the extent of ion association.

When the ion association is extensive and the concentration of free ions is low, the influence of the ion–ion interactions can be ignored and the relation $\Lambda = \alpha\Lambda^\infty$ is valid. In that case, we get the following Arrhenius–Ostwald relation:

$$\frac{1}{\Lambda} = \frac{1}{\Lambda^\infty} + \frac{c\Lambda K_A}{(\Lambda^\infty)^2} \tag{7.6}$$

In this equation, the $1/\Lambda$ versus $c\Lambda$ relation is linear, and we get the approximate values of Λ^∞ and K_A from the intercept and the slope, respectively. The values of Λ^∞ and K_A are often used in starting more precise data analyses by means of Eq. (7.5).

Fuoss and Kraus [13] and Shedlovsky [14] improved Eq. (7.6) by taking the effect of ion–ion interactions on molar conductivities into account. Here, Fuoss and Kraus used the Debye–Hückel–Onsager limiting law (Eq. (7.1)), and Shedlovsky used the following semiempirical equation:

$$\Lambda = \alpha\Lambda^\infty - \left(\frac{\Lambda}{\Lambda^\infty}\right) S(\alpha c)^{1/2}$$

In both cases, Eq. (7.6) is modified to Eq. (7.7) by using $Z = S(\Lambda c)^{1/2}(\Lambda^\infty)^{-3/2}$:

$$\frac{T_{(Z)}}{\Lambda} = \frac{1}{\Lambda^\infty} + \frac{K_A}{(\Lambda^\infty)^2} \times \frac{c\gamma^2\Lambda}{T_{(Z)}} \tag{7.7}$$

In the case of Fuoss–Kraus,

$$T_{(Z)} \equiv F_{(Z)} = 1 - Z(1 - Z(1 - \cdots)^{-1/2})^{-1/2}$$

and in the case of Shedlovsky,

$$\frac{1}{T_{(Z)}} \equiv S_{(Z)} = 1 + Z + \frac{Z^2}{2} + \frac{Z^3}{8} + \cdots$$

Values of $F_{(Z)}$ and $S_{(Z)}$ are listed in the literatures [15, 16]. From Eq. (7.7), the $T_{(Z)}/\Lambda$ versus $c\gamma^2\Lambda/T_{(Z)}$ relation is linear, with an intercept of $1/\Lambda^\infty$ and a slope of $K_A/(\Lambda^\infty)^2$, from which we can obtain the values of Λ^∞ and K_A. These Λ^∞ and K_A values are also used in data analyses by Eq. (7.5). The values of K_A obtainable by Eqs. (7.6) and (7.7) range from 10^3 to $10^7\,\mathrm{mol}^{-1}\,\mathrm{l}$.

If $K_A \geq 10^7\,\mathrm{mol}^{-1}\,\mathrm{l}$, the slope of the linear relation is too large to obtain reliable Λ^∞ values (or K_A values). In these cases, we obtain the value of Λ^∞ separately, either by use of Walden's rule or from the known values of ionic molar conductivities (λ_i^∞), and use it to obtain K_A.

7.1.3
Molar Conductivity and the Formation of Triple Ions

In solvents of low permittivities ($\varepsilon_r \leq 10$), most ions are associated as in Eq. (7.3), even at dilute electrolyte concentrations. Moreover, with increasing electrolyte concentrations, triple ions are formed, as in Eq. (7.8), and sometimes even quadrapoles (or dimers) as in Eq. (7.9):

$$2M^+ + A^- \rightleftarrows M_2A^+ \,(K_{t+}) \quad M^+ + 2A^- \rightleftarrows MA_2^- \,(K_{t-}) \tag{7.8}$$

$$2MA \rightleftarrows (MA)_2 \tag{7.9}$$

Ion pairs do not conduct electricity, but triple ions (M_2A^+ and MA_2^-) do. Thus, the formation of triple ions is detected by conductimetric measurement. Figure 7.2 shows the $\log \Lambda - \log c$ relations for $LiAsF_6$ and $LiBF_4$ in 2-methyltetrahydrofuran (2-MeTHF) [17]. After passing a minimum, the value of $\log \Lambda$ increases again with $\log c$, showing that a triple-ion formation occurred in the solution.

In conductimetric studies of triple-ion formation, it is often assumed that M_2A^+ and MA_2^- have the same formation constants ($K_t = K_{t+} = K_{t-}$) and the same molar conductivities at infinite dilutions ($\lambda_t^\infty = \lambda_{t+}^\infty = \lambda_{t-}^\infty$). If we denote the mole fractions of (M^+, A^-) and (M_2A^+, MA_2^-) by α and α_t, respectively, against the total electrolyte concentration c and assume that the concentrations of the ionic species are negligibly small compared to c, we get the relation $\Lambda = \alpha\Lambda^\infty + \alpha_t\lambda_t^\infty$ and, therefore, Eq. (7.10):

$$\Lambda c^{1/2} = K_A^{-1/2}\Lambda^\infty + K_t K_A^{-1/2} c\lambda_t^\infty \tag{7.10}$$

From Eq. (7.10), if we plot the value of $\Lambda c^{1/2}$ against c, we get a linear relation of slope equal to $K_t K_A^{-1/2}\lambda_t^\infty$ and the intercept equal to $K_A^{-1/2}\Lambda^\infty$. If we get the value of Λ^∞

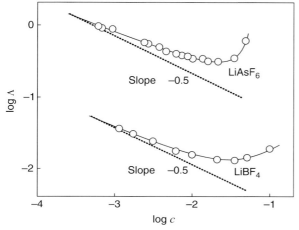

Figure 7.2 Molar conductivities of $LiBF_4$ and $LiAsF_6$ solutions in 2-methyltetrahydrofuran at 25 °C [17].

Table 7.1 Ion pair and triple-ion formation constants (K_A and K_t) at 25 °C[a].

Solvent	ε_r	Electrolyte	K_A, M^{-1}	K_t, M^{-2}
1,2-Dimethoxyethane	7.08	LiAsF$_6$	0.071×10^6	176
		LiBF$_4$	24.0×10^6	30.8
Methyl acetate	6.76	LiClO$_4$	77.8×10^6	38
Tetrahydrofuran	7.58	LiClO$_4$	48.4×10^6	153
2-Me-THF	6.97	LiClO$_4$	182×10^6	33

[a]From Ref. [18c].

from the Walden's rule and assume that $\lambda_t^\infty = \Lambda^\infty/3$ or $2\Lambda^\infty/3$ (λ_t^∞ cannot be determined experimentally), we can obtain the values of K_a and K_t [18]. The values of K_A and K_t in Table 7.1 were obtained in low-permittivity solvents. The cationic and anionic triple ions can remain stable in low-permittivity solvents because triple ions are large in size and are not easily associated (Section 2.6). Thus, some electrolytes are highly soluble in low-permittivity solvents, and they show conductivities comparable to or higher than the solutions in high-permittivity solvents. For example, the solution of 2M LiAsF$_6$ in methyl acetate ($\varepsilon_r = 6.76$, η(viscosity) $= 0.37$ cP) has larger conductivity than 1 M LiClO$_4$ in PC ($\varepsilon_r = 64.9$, $\eta = 2.53$ cP).

As mentioned in Section 2.6, triple-ion formation is not limited to low-perimittivity solvents. It also occurs in high-permittivity solvents if they are of very weak acidity and basicity; for example, K_t for the formation of Li$_2$Cl$^+$ and LiCl$_2^-$ in AN has been determined by polarography to be $\sim 10^5$ M^{-2} [19]. Li$^+$ and Cl$^-$ in AN are only weakly solvated and tend to be stabilized by forming triple ions. For conductimetric studies of triple-ion formation in dipolar protophobic aprotic solvents, see Ref. [20].

The problem of triple-ion formation has been studied in detail because it is related to the lithium battery technologies [18]. In some cases, however, the occurrence of the minimum in the log Λ–log c curve, as observed in Figure 7.2, is not attributed to triple-ion formation but is explained by ion pair formation only. The increase in log Λ at high electrolyte concentrations is attributed either to the increase in the distance of closest approach of ions, the increase in the solution permittivity, or the decrease in the activity coefficient of the ion pairs. Although there is still some controversy, it seems certain that triple ions are actually formed in many cases.

7.1.4
Conductivity of Solutions of Symmetrical Strong Electrolytes at Moderate to High Concentrations

Here, we consider the conductivity of strong electrolyte solutions at moderate to high concentrations in polar nonaqueous solvents. The conductivity of such solutions has been studied extensively because of their importance in the applied fields.

If the conductivity of an electrolyte in a polar solvent is measured up to high concentrations, the conductivity–concentration relation usually shows a maximum as in Figure 7.3. Such a relation is explained by the competition between the increase in the number of charge carriers and the decrease in ionic mobilities, mainly due to

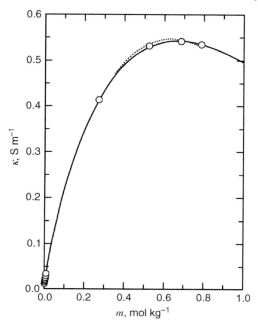

Figure 7.3 Conductivity (κ) versus concentration (m) relation for LiClO$_4$/PC (25 °C). (O) experimental data; (—) Casteel–Amis equation; (- - -) MSA [24b].

the strengthening of ion–ion interactions. Various empirical equations have been reported to express such a relation. The Casteel–Amis equation [21] for the relation between the conductivity κ and the molal concentration m is as follows:

$$\frac{\kappa}{\kappa_{max}} = \left(\frac{m}{\mu}\right)^a \exp\left[b(m-\mu)^2 - \frac{a}{\mu}(m-\mu)\right]$$

where κ_{max} is the maximum value in conductivity, μ is the concentration at which κ_{max} is obtained, and a and b are empirical parameters with no physical meaning. The values of κ_{max} and μ have been listed in Table 1.4 of Ref. [6b]. κ_{max} and μ are mainly determined by solvent viscosity and ionic radii; thus, κ_{max} and μ shift to higher values with increasing temperature at constant solvent compositions and also shift to lower values with increasing viscosity of the solvents at constant temperature.

Recently, Chagnes et al. [22] treated the molar conductivity of LiClO$_4$ in γ-butyrolactone (γ-BL) on the basis of the quasilattice theory. They showed that the molar conductivity can be expressed in the form $\Lambda = (\Lambda^\infty)' - k'c^{1/3}$ and confirmed it experimentally for 0.2–2 M LiClO$_4$ in γ-BL. They also showed, using 0.2–2 M LiClO$_4$ in γ-BL, that the relation $\kappa = \Lambda c = (\Lambda^\infty)'c - k'c^{4/3}$ was valid and that κ_{max} appeared at $c_{max} = [3(\Lambda^\infty)'/4k']^3$, where $d\kappa/dc = 0$.

The concept of mean spherical approximation (MSA, footnote 3 in Chapter 2) has also been used to reproduce the conductivity data of electrolytes of fairly high

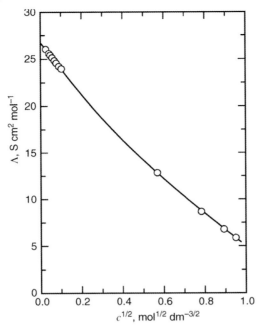

Figure 7.4 Molar conductivity of LiClO$_4$/PC (25 °C). Analysis by Eq. (7.5) ($c < 0.03$ M), $K_A = 5.2$ mol^{-1} dm^3; MSA (0 M $< c < 1$ M), $K_A = 4.2$ mol^{-1} dm^3.

concentrations [23]. The MSA method applies to both associated and nonassociated electrolytes and can give the values of association constant, K_A. Although not described here, details of the method can be found in Ref. [24]. Figure 7.4 shows the results of the MSA calculation. The conductivity of the LiClO$_4$ solution in PC at 25 °C is excellently reproduced up to 1 M, yielding $K_A = 4.2$ dm^3 mol^{-1} [24b]. In contrast, the data analysis as in Section 7.1.2 yields the same curve up to 0.05 M and gives $K_A = 5.2$ dm^3 mol^{-1}. In Figure 7.3, the conductivity obtained by the MSA calculation is also indicated; in this case, only two parameters, i.e. K_A and σ (average ionic diameter), are needed, compared to the four parameters in the Casteel–Amis equation. It is important that the parameters needed in MSA have physical meanings.

7.1.5
Molar Conductivity and Ion Association of Asymmetric Electrolytes

For an asymmetric electrolyte that is completely dissociated, the slope for the Debye–Hückel–Onsager limiting law can be obtained by the method described in footnote 1. For example, for an electrolyte of MA$_2$-type ($z_+ = 2$ and $z_- = 1$), $A' = [5.60 \times 10^6/(\varepsilon_r T)^{3/2}] \times [q/(1 + q^{1/2})]$, $B' = 123.8/[\eta(\varepsilon_r T)^{1/2}]$ and $q = (2/3)(1 + \lambda_-^\infty/\Lambda^\infty)^{-1}$. In the solution of electrolyte MA$_2$, the ion association occurs in two steps, i.e. M^{2+} + A$^-$ $\rightleftarrows$ MA$^+$ and MA$^+$ + A$^-$ $\rightleftarrows$ MA$_2$. If we express the formation constants for the two steps by K_{A1} and K_{A2}, respectively, generally $K_{A1} \gg K_{A2}$. The Fuoss–Edelson equation

(1951) [25] or its modified forms [26] have often been used to analyze the conductivity data of asymmetric electrolytes. However, the Quint–Viallard (1978) [27] and the Lee–Wheaton (1978) [28] equations are used for more sophisticated data analyses. For the analysis of conductivity data of asymmetric electrolytes, see Ref. [29].

7.2
Ionic Conductivities and Solvents

In Section 7.1, molar conductivities of electrolytes of various degrees of dissociation were considered. The limiting molar conductivity of electrolyte M_xA_y (Λ^∞) is given by

$$\Lambda^\infty = x\lambda_M^\infty + y\lambda_A^\infty$$

where λ_M^∞ and λ_A^∞ are the limiting molar conductivities of the component ions. In this section, we will first deal with Stokes' law and Walden's rule, which show the relations between the limiting molar conductivity of ions and the viscosity of solvents. We then discuss methods of determining the limiting molar conductivities of ions.

7.2.1
Stokes' Law and Walden's Rule – Role of Ultrafast Solvent Dynamics

According to the classical Stokes' law, a spherical particle, i, of radius r_i moving with velocity v_i through a static fluid of viscosity η is subjected to a force f, as shown by Eq. (7.11) or (7.12).

$$f = 6\pi\eta r_i v_i \quad \text{(stick)} \tag{7.11}$$

$$f = 4\pi\eta r_i v_i \quad \text{(slip)} \tag{7.12}$$

Equation (7.11) is for the 'stick' condition, i.e. when the solvent immediately adjacent to the spherical particle 'wets' it and so moves along with it. Equation (7.12), on the other hand, is for the 'slip' condition, i.e. when the spherical particle is completely slippery and does not drag along any liquid with it. When the spherical particle is an ion with charge $|z_i|e$ and it is in the liquid under a potential gradient X ($V\,cm^{-1}$), a force f expressed by Eq. (7.13) operates on the particle:

$$f = |z_i|eX \tag{7.13}$$

If the force in Eq. (7.11) or (7.12) balances with that in Eq. (7.13), the particle moves with a constant velocity v_i:

$$v_i = \frac{|z_i|eX}{k'\pi\eta r_i} \quad (k' = 6 \text{ or } 4)$$

The mobility u_i^∞ of an ion i at infinite dilution is the drift velocity for unit potential gradient ($1\,V\,cm^{-1}$) and is expressed by following:

$$u_i^\infty = \frac{v_i}{X} = \frac{|z_i|e}{k'\pi\eta r_i} \quad (k' = 6 \text{ or } 4)$$

Thus, the limiting molar conductivity of ion i (λ_i^∞) is given by Eq. (7.14), where N_A is the Avogadro constant (Section 5.8):

$$\lambda_i^\infty = |z_i| u_i^\infty F = \frac{z_i^2 F^2}{k' \pi N_A \eta r_i} \quad (k' = 6 \text{ or } 4) \tag{7.14}$$

If we express λ_i^∞ in S cm^2 mol^{-1}, η in poise (P) and r_i in centimeter, we get from Eq. (7.14)

$$\lambda_i^\infty \eta = \frac{z_i^2 F^2}{k' \pi N_A r_i} = \frac{4.92 \times 10^{-8} z_i^2}{k' r_i} \quad (k' = 6 \text{ or } 4) \tag{7.15}$$

Equation (7.15) shows that $\lambda_i^\infty \eta$ is proportional to $(1/r_i)$; as shown in Figure 7.5, this is nearly the case for ions much larger than the solvent molecules (e.g. R$_4$N$^+$ larger than Pr$_4$N$^+$ and BPh$_4^-$). However, for small cations such as alkali metal ions, the experimental $\lambda_i^\infty \eta$ values are generally much smaller than those predicted by Eq. (7.15). Traditionally, this has been rationalized by two models [30a]. The first is the solvent-berg model, which postulates that the solvent molecules immediately adjacent to the ion are rigidly bound to it, and thus the ion moves as this kinetic unit; because the effective size of the solvent-berg is larger than that of the bare ion, the mobility is much less than that expected for the bare ion. The second model takes dielectric friction into account; the moving ion orients the solvent dipoles around it,

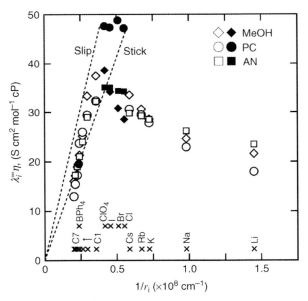

Figure 7.5 $\lambda_i^\infty \eta$-r_i^{-1} relation for cations (open symbols) and anions (solid symbols) in MeOH, PC and AN. C1 → C7 means, from right to left, Me$_4$N$^+$, Et$_4$N$^+$, Pr$_4$N$^+$, Bu$_4$N$^+$, Am$_4$N$^+$, Hex$_4$N$^+$ and Hep$_4$N$^+$. The dashed lines show the theoretical slopes for stick and slip conditions. The data for stick and slip conditions. The data for crystallographic ionic radii (or analogous) and ionic molar conductivities are from Tables 3 and 28 in Marcus, Y. (1997) *Ion Properties*, Marcel Dekker, New York.

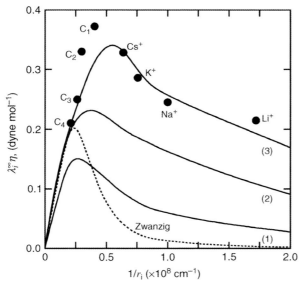

Figure 7.6 Effect of sequential addition of the ultrafast component of the solvent orientational motion on the $\lambda_i^\infty \eta$ values in MeOH at 298 K. Closed circles are experimental results; curves (1), (2) and (3) were obtained by the molecular theory by including the first (slowest) one, the first two and all three Debye relaxations; the dashed line is by the Zwanzig theory [30b].

and these can relax again into a random distribution only after the ion has passed; thus, an electrostatic drag force is exerted on the moving ion. Usually, the friction acting on the moving ion (ζ) is expressed by $\zeta = \zeta_{bare} + \zeta_{DF}$, where ζ_{bare} is the friction predicted by Stokes' law and ζ_{DF} is the dielectric friction. Initially, the dielectric friction was treated by continuum models; the dashed line in Figure 7.6 shows the result obtained by Zwanzig's continuum theory [30e], in which ζ_{DF} is proportional to the Debye relaxation time (τ_D, see Table 1.3). The theory can explain the nonmonotonic $\lambda_i^\infty \eta - r_i^{-1}$ relation, but it overestimates the dielectric friction for small cations. Recently, the dielectric friction is treated by microscopic theories; in particular, Bagchi's group [30a–d] was able to explain the experimental $\lambda_i^\infty \eta - r_i^{-1}$ relations in water, D_2O, MeOH, EtOH, FA and AN almost quantitatively by taking all sequential Debye relaxations into account. The results in MeOH are shown in Figure 7.6: curves 1, 2 and 3 were obtained by considering the first relaxation (48 ps), the first two relaxations (48 and 1.25 ps) and all three relaxations (48, 1.25 and 0.16 ps), respectively.[3] Curve 3 agrees fairly well with the experimental data. The results of Bagchi's group clearly show that the ultrafast solvent relaxation in the

3) For MeOH, the dielectric relaxation data used were obtained by Kindt and Schmuttenmaer (*J. Phys. Chem.* 1996, **100**, 10373) by femtosecond terahertz pulse spectroscopy; τ_1, τ_2 and τ_3 were 48, 1.25 and 0.16 ps, respectively (see Table 1.4). Note that they are considerably different from 51.5, 7.09 and 1.12 ps, respectively, reported by Barthel et al. (Table 2.2 in Ref. [6a]).

Table 7.2 Walden's products of ions in various solvents.[a]

Solvents	η, cP	ε_r	$\lambda_i^\infty \eta$, S cm^2 mol^{-1}P						
			K$^+$ 1.38[b]	Et$_4$N$^+$ 3.37[b]	Bu$_4$N$^+$ 4.13[b]	(i-Am)$_4$N$^+$ 4.43[b]	Cl$^-$ 1.81[b]	Pic$^-$	BPh$_4^-$ 4.21[b]
H$_2$O	0.890	78.3	0.654	0.291	0.173		0.680	0.271	0.187
MeOH	0.545	32.7	0.286	0.333	0.213	0.194	0.285	0.256	0.199
EtOH	1.078	24.6	0.254	0.315	0.213		0.236	0.270	
FA	3.30	109.5	0.421	0.345	0.216		0.565	0.301	0.199
AN	0.341	36.0	0.285	0.292	0.212	0.196	0.337	0.265	0.199
NM	0.627	36.7		0.299	0.214	0.193	0.392		0.204
DMF	0.796	36.7	0.245	0.288	0.212		0.439	0.298	
DMSO	1.99	46.7	0.299	0.322	0.215		0.478	0.334	0.203

[a]The data of $\lambda_i^\infty \eta$ for Et$_4$N$^+$, Bu$_4$N$^+$, (i-Am)$_4$N$^+$ and BPh$_4^-$ are from Ref. [31] and for others from Ref. [1].
[b]Ionic radii ($\times 10^{-8}$ cm), Marcus, Y. (1997) *Ion Properties*, Marcel Dekker, New York, p. 43.

femtosecond region plays a dominant role in determining the ionic mobilities. However, their results also show that, for small cations in solvents with slow rotational and translational relaxation processes (EtOH, PrOH, etc.), the dynamic version of the solvent-berg model also increases its importance to some extent. For the history and the present status of this problem, see Ref. [30].

The Stokes' radius, denoted by r_s, is the ionic radius in the solution and is obtained to satisfy Eq. (7.15). If r_s is not influenced by solvent, the relation of Eq. (7.16) should hold:

$$\lambda_i^\infty \eta = \text{constant} \tag{7.16}$$

This relation is the so-called Walden's rule. Table 7.2 shows the values of $\lambda_i^\infty \eta$ for some cations and anions in various solvents. In the table, the values of $\lambda_i^\infty \eta$ for large ions such as Bu$_4$N$^+$, i-Am$_4$N$^+$ and BPh$_4^-$ are kept constant within $\pm 1\%$ in solvents other than water, suggesting that the Walden's rule is valid [1b, 31]. However, Walden's rule does not hold for small ions such as K$^+$ and Cl$^-$: the values of $\lambda_i^\infty \eta$ are considerably influenced by the solvents. This means that the Stokes' radii, r_s, of these ions vary from one solvent to another, reflecting the variation in ion solvation. Therefore, testing the applicability of Walden's rule is important in studying ion solvation. Moreover, if Walden's rule is applicable for a given ion, the molar conductivity of the ion in the solvent in which it is unknown can be obtained (Section 7.2.2).

For such ions as Et$_4$N$^+$ and Bu$_4$N$^+$, the Stokes radius (r_s) obtained by Eq. (7.15) is usually smaller than the crystal radius (r_x). In order to avoid this contradiction, Robinson and Stokes introduced the concept of effective ionic radius (r_{eff}). They considered that, for very weakly solvated ions such as tetraalkylammonium ions (R$_4$N$^+$), the effective ionic radius r_{eff} must be equal to the crystal ionic radius r_x. They plotted the r_{eff}/r_s versus r_s relation for various R$_4$N$^+$ ions and used it to obtain the values of r_{eff} for other ionic species. If r_{eff} is equal to the real radius of the ion solvated in the solution, we can estimate the solvation number n_i from the following:

$$n_{SV} = \frac{4}{3}\pi\frac{N_A(r_{eff}^3 - r_x^3)}{V_S}$$

Here, V_S is the molar volume of the solvent and N_A is the Avogadro constant. Some of the ionic solvation numbers obtained by this method are listed in Table 7.3.

7.2.2
Method for the Determination of Limiting Molar Conductivities of Ions

The method for the determination of limiting molar conductivities of electrolytes Λ^∞ was discussed in Section 7.1. The limiting molar conductivities of individual ions (λ_i^∞) can be obtained by the following methods.

Direct Method
The limiting molar conductivity of an electrolyte (Λ^∞) and the limiting transference numbers of the ions constituting the electrolyte (t_+^∞, t_-^∞) are determined experimentally, and Λ^∞ is divided into λ_+^∞ and λ_-^∞. This method is direct and does not include any assumption. If the transference numbers can be obtained accurately, this method is the most reliable. Detailed literature is available on the methods of obtaining the limiting transference numbers [1–5, 32].

Indirect Method Based on the Walden's Rule
Walden's rule is applicable to such ions as Bu_4N^+, $(n-Am)_4N^+$, $(i-Am)_4N^+$, Hex_4N^+, $(i-Am)_3BuN^+$, Ph_4As^+, BPh_4^- and $(i-Am)_4B^-$.[4] For these ions, the value of λ_i^∞ in a given solvent can be obtained from the known $\lambda_i^\infty\eta$ value in different solvents. The inaccuracy of this method is ± 1 or 2%.

Indirect Method Using a Reference Electrolyte
The limiting molar conductivity, Λ^∞, for a reference electrolyte $((i-Am)_4N^+B$ $(i-Am)_4^-$, $(i-Am)_3BuN^+BPh_4^-$, etc.) is determined experimentally and $\Lambda^\infty/2$ is considered to be the limiting conductivity of the constituent ions.

When the limiting molar conductivities are to be obtained for a series of ions in a given solvent, the first step is to get the limiting molar conductivity of an ion by one of the above-mentioned methods. Then, the limiting molar conductivities for other ions can be obtained sequentially by applying Kohlrausch's law of independent ionic migration (Section 5.8).

The limiting molar conductivities of ions in various solvents are listed in Table 7.4. The following are some general points about ionic conductivities in nonaqueous solutions:

4) The average values of $(\lambda_i^\infty\eta)$ in organic solvents and at 25 °C have been obtained [31] as follows: Me_4N^+ 0.331±0.044, Et_4N^+ 0.307±0.021, Pr_4N^+ 0.243_1±0.008_5, Bu_4N^+ 0.213_1±0.002_4, $(n-Am)_4N^+$ 0.190_6±0.001_2, $(i-Am)_4N^+$ 0.193_9±0.001_4, Hex_4N^+ 0.173_1±0.001_3, Hep_4N^+ 0.162_0±0.001_4, $(i-Am)_3BuN^+$ 0.200_5±0.001_1, Ph_4As^+ 0.190_3±0.001_4, Ph_4B^- 0.200_8±0.003_1, $(i-Am)_4B^-$ 0.198_6±0.000_4 (S cm² mol⁻¹ P).

Table 7.3 Solvation numbers of ions calculated from the effective ionic radii.

Ions	H$_2$O	MeOH	NB	TMS	AN	PC	FA	DMF	DMSO	HMPA
Li$^+$	7.4	5.0	—	1.4	1.2	3.4	5.4	5.0	3.3	3.4
Na$^+$	6.5	4.6	1.6	2.0	1.4	2.9	4.0	3.0	3.1	1.8
K$^+$	5.1	3.9	1.4	1.5	1.0	2.4	2.5	2.6	2.8	1.6
Rb$^+$	4.7	3.7	—	1.4	—	1.7	2.3	2.5	2.3	1.4
Cs$^+$	4.3	3.3	—	1.3	0.5	1.6	1.9	2.3	2.0	1.2
Ag$^+$	5.9	—	—	—	—	—	—	2.1	—	1.8
NH$_4$$^+$	4.6	3.7	1.4	0.9	—	1.7	1.4	1.8	—	1.6
Me$_4$N$^+$	1.8	—	—	—	—	0.5	—	0.4	0.4	—
Ba^{2+}	9.6	7.9	—	—	—	—	6.3	9.0	—	—
Cl$^-$	3.9	4.1	1.0	0.0	—	1.3	1.0	0.7	1.1	0.3
Br$^-$	3.4	3.8	1.1	0.0	—	1.0	0.8	0.8	1.0	0.5
I$^-$	2.8	3.4	1.2	0.1	—	0.8	0.8	0.7	1.0	—
ClO$_4$$^-$	2.6	2.8	1.1	0.1	—	0.8	0.9	0.6	0.8	0.3

From the table in Marcus, Y. (1985) *Ion Solvation*, John Wiley & Sons, Inc., New York, p. 84.

1. Ionic migration usually slows down in solvents of higher viscosities, although this does not necessarily apply for small inorganic ions.

2. The molar conductivities of alkali metal ions increase, in most solvents, with the increasing crystal ionic radii (Li$^+$ < Na$^+$ < K$^+$ < Rb$^+$ < Cs$^+$).

3. In aprotic solvents, iodide ions migrate most slowly among halide ions. In protic solvents, however, iodide ions migrate the fastest.

4. In dipolar aprotic solvents, the molar conductivity of the fastest anion is larger than that of the fastest cation. However, this does not apply in protic solvents.

5. The molar conductivities of R$_4$N$^+$ decrease with increasing size of the R-group. The molar conductivities of organic anions also decrease with increasing ionic size.

6. In organic solvents with an OH group (MeOH, EtOH, etc.), the conductivity of H$^+$ is abnormally high as it is in water. This is due to the proton jump. Conversely, some lyate ions, such as HCOO$^-$ in formic acid, CH$_3$O$^-$ in MeOH and HSO$_4$$^-$ in sulfuric acid, show abnormally high conductivities similar to that of OH$^-$ in water. In some amines and ethereal solvents, solvated electrons also show high conductivities.

7.3
Applications of Conductimetry in Nonaqueous Solutions

7.3.1
Study of the Behavior of Electrolytes (Ionophores)

Limiting Molar Conductivities of Electrolytes and Individual Ions; Ionic Association Constants; Formation Constants of Triple Ions
As mentioned in Section 7.1, if we determine the molar conductivity of an electrolyte as a function of its concentration and analyze the data, we can get the value of limiting

Table 7.4 Limiting molar conductivities of ions in various solvents ($S\,cm^2\,mol^{-1}$, 25 °C).

Solvents (η/cP)	H^+	Li^+	Na^+	K^+	Rb^+	Cs^+	Ag^+	NH_4^+	$^1/_2Ba^{2+}$	$^1/_2Mg^{2+}$
Water (0.890)	349.81	38.68	50.10	73.50	77.81	77.26	61.90	73.55	63.63	53.05
Methanol (0.545)	146.1	39.59	45.23	52.40	56.23	61.52	50.11	57.60	57.6[a]	60.0[a]
Ethanol (1.096)	62.7	17.05	20.31	23.54	24.90	26.34	20.60	22.05	—	—
1-Propanol (1.947)	45.68	8.21	10.17	11.79	13.22	13.98	—	—	—	—
2-Propanol (2.078)	36.52	—	5.87	7.72	8.41	9.43	—	—	—	—
1-Butanol (2.67)	35.67	7.10	5.79	—	8.88	—	—	—	—	—
2,2,2-Trifluoroethanol (1.78)	—	9.02	—	15.82	—	19.38	—	—	—	—
Ethylene glycol (1.661)	—	1.91	2.88	4.42	4.61	4.53	4.05	—	—	—
Formamide (3.30)	10.41	8.32	9.88	12.39	12.82	13.38	—	14.94	—	—
N-Methylformamide (1.66)	—	10.1	16.01	16.49	17.94	18.52	—	24.59	—	—
N,N-Dimethylformamide (0.793)	35.0	26.1	30.0	31.6	33.2	35.4	35.7	39.4	37.9[b]	39.4[b]
N,N-Dimethylacetamide (0.919)	—	20.29	25.69	25.31	27.01	27.87	27.21	35.91	—	—
Hexamethylphosphoric triamide (3.23)	—	5.64	6.15	6.55	6.94	7.33	6.22	6.45	—	—
1,1,3,3-Tetramethylurea (1.401)	—	14.45	16.06	15.71	16.01	16.91	—	—	—	—
N-Methyl-2-pyrrolidinone (1.666)	—	—	14.69	15.58	—	—	—	—	—	—
Pyridine (0.882)	—	25.9	26.6	31.8	—	—	34.4	46.6	—	—
Dimethyl sulfoxide (1.963)	15.5	11.77	13.94	14.69	14.96	16.19	16.08	—	18.8[b]	18.2[b]
Acetone (0.3116)	—	69.2	70.2	81.1	—	84.1	—	89.5	—	—
Methylethylketone (0.378)	—	59.9	61.2	64.4	—	—	—	79.6	—	—
Propylene carbonate (2.513)	—	7.14	9.13	11.08	11.73	12.16	11.96	—	12.1[b]	12.2[b]
Sulfolane (10.3_{30})	—	4.34[c]	3.61[c]	4.04[c]	4.20[c]	4.34[c]	4.81[c]	4.98[c]	—	—
Acetonitrile (0.344)	—	69.97[d]	77.00[d]	83.87[d]	85.8	87.4	86.0	97.1	102.7[b]	97.1[b]
Nitromethane (0.612–0.627)	64.5	53.89	56.75	58.12	—	—	50.75	62.75	—	—
Nitrobenzene (1.849)	—	—	16.6	17.8	18.6	19.9	—	17.9	—	—
Formic acid (1.610)	79.8	19.5	21.0	24.0	—	—	—	27.1	—	—

(Continued)

Table 7.4 (*Continued*)

Solvents	Me$_4$N$^+$	Et$_4$N$^+$	Bu$_4$N$^+$	Cl$^-$	Br$^-$	I$^-$	NO$_3^-$	ClO$_4^-$	Pic$^-$	CF$_3$SO$_3^-$	BPh$_4^-$
Water	44.9	32.7	19.5	76.35	78.14	76.84	71.46	67.36	30.4	—	—
Methanol	68.79	61.12	39.14	52.38	56.53	62.63	60.95	70.78	46.94	—	36.6^e
Ethanol	29.61	28.73	19.44	21.85	24.50	27.04	24.82	30.5	25.44	—	—
1-Propanol	15.18	15.22	11.00	10.22	11.98	13.74	11.23	16.19	12.34	—	—
2-Propanol	—	—	—	10.38	11.24	12.83	—	15.07	—	—	—
1-Butanol	9.38	10.20	7.80	7.93	8.50	9.52	—	11.26	8.01	—	—
2,2,2-Trifluoroethanol	—	—	—	17.64	19.00	20.92	—	22.77	—	—	—
Ethylene glycol	2.97^e	2.20^e	1.51^e	5.30	5.21	4.79	5.04	4.45	—	—	—
Formamide	12.84	10.44	6.54	17.46	17.51	16.90	17.66	16.63	—	—	6.04^e
N-Methylformamide	—	26.2^e	—	25.61	28.16	28.39	—	27.37	19.75	—	—
N,N-Dimethylformamide	39.53	36.37	26.68	53.8	53.4	51.1	57.1	51.6	37.12	43.6^f	—
N,N-Dimethylacetamide	34.97	32.96	22.91	45.94	43.21	41.80	47.0	42.80	31.50	—	—
Hexamethylphosphoric triamide	8.4^g	9.8^g	6.4^g	19.45	17.48	16.28	19.33	14.97	10.81	—	—
1,1,3,3-Tetramethylurea	22.57	22.07	15.47	—	30.14	—	32.40	28.35	—	—	—
N-Methyl-2-pyrrolidinone	—	—	—	—	—	26.94	—	27.24	20.1	—	—
Pyridine	—	—	—	51.4	51.2	49.0	52.5	47.5	33.60	—	—
Dimethyl sulfoxide	17.93	16.39	10.93	23.41	23.76	23.59	26.84	24.39	17.08	21.7^f	10.2^e

Acetone	96.63	89.49	66.40	109.3	113.9	116.2	127.2	115.8	86.1	—	—
Methylethylketone	—	—	—	—	84.5	86.6	96.8	85.6	—	16.9[f]	—
Propylene carbonate	14.16	13.18	8.98	18.77	19.41	18.82	20.92	18.94	13.12	—	—
Sulfolane	4.28[c]	3.94[c]	2.76[c]	9.29[c]	8.91[c]	7.22[c]	—	6.69[c]	5.32[c]	—	58.02[d]
Acetonitrile	94.52[d]	85.19[d]	61.63[d]	100.4	100.7	102.4[d]	106.2	103.6[d]	77.3	96.3[f]	31.6[e]
Nitromethane	56.1	48.8	34.9	62.50	62.82	63.61	66.65	65.75	45.44	—	—
Nitrobenzene	—	—	—	22.6	21.96	21.28	22.9	21.6	16.5	—	—
Formic acid	—	—	—	26.4	28.2	—	—	29.2	—	—	—

For the values in water, Robinson, R.A. and Stokes, R.H. (1965) *Electrolyte Solutions*, Butterworths, London; for the values in organic solvents, Krumgalz, B.S. (1983) *J. Chem. Soc., Faraday Trans. I*, **79**, 571.

[a]Dobos, D. (1975) *Electrochemical Data*, Elsevier, Amsterdam.
[b]Izutsu, K., Arai, T. and Hayashijima, T. (1997) *J. Electroanal. Chem.*, **426**, 91.
[c]Ref. [1], 30 °C.
[d]Barthel, J. et al. (1990) *J. Solution Chem.*, **19**, 321.
[e]Ref. [1].
[f]Okazaki, S. and Sakamoto, I. (1990) *Solvents and Electrolytes (Japanese)*, Sanei, p. 117.
[g]Fujinaga, T. et al. (1973) *Nippon Kagaku Kaishi*, 493.

molar conductivity Λ^∞ and the quantitative information about ion association and triple-ion formation. If we determine the limiting molar conductivity of an ion (λ_i^∞) by one of the methods described in Section 7.2, we can determine the radius of the solvated ion and calculate the solvation number. It is also possible to judge the applicability of Walden's rule to the ion under study. These are the most basic applications of conductimetry in nonaqueous systems and many studies have been carried out on these problems [1–7].

Various theories have been proposed for analyzing the data of the $\Lambda - c$ relation, by taking into account both the long-range and the short-range interactions between ions (Section 7.1.1). All of them give approximately the same Λ^∞ values. Moreover, they give approximately the same values of ion association constant (K_A), when the K_A values are large enough. However, it is necessary to use a single theory in order to compare the resulting data of Λ^∞ and K_A for a series of electrolytes. For practical information on how to analyze conductivity data, the report by Kay et al. [33], dealing with the conductivity of alkali metal salts in acetonitrile, is useful. Some reports compare the results of various theories [34]. For methods of studying triple-ion formation, some reviews and original papers are available [18, 35].

Determination of the Solubility Products of Sparingly Soluble Salts

When a sparingly soluble 1: 1 electrolyte, saturated in the solution, is completely dissociated, the relation between the solubility product of the electrolyte (K_{sp}) and the conductivity of the solution (κ) can be expressed, as the first approximation, by

$$K_{sp}^{1/2} = c_s = \frac{\kappa}{\Lambda^\infty}$$

When the electrolyte in the saturated solution is partially associated, on the other hand, the relation is modified to

$$K_{sp}^{1/2} = c_s = \frac{\kappa}{\Lambda^\infty} + \frac{\kappa^2 K_A}{(\Lambda^\infty)^2}$$

Here, the association constant K_A can be determined by measuring the conductivity of the unsaturated solutions. Corrections are made by successive approximations for the effect of electrolyte concentrations on molar conductivity Λ and for the effect of activity coefficient on K_A. Here, κ/Λ^∞ is used as the first approximation of the ionic strength.

Determination of Complex Formation Constants

An interesting example is the determination of complex formation constants of alkali metal ions (M^+) with crown ethers (L):

$$M^+ + L \rightleftarrows ML^+, \quad K_{ML} = \frac{[ML^+]}{[M^+][L]}$$

Crown ether L is added step-wise to the 0.1–1.0 mM solution of alkali metal salt MX, until the final concentration of L reaches 5–10 times that of MX, each time by

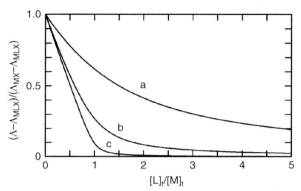

Figure 7.7 Theoretical variation in the molar conductivities of
alkali metal salt (MX) solution by the addition of crown ether (L)
for $[M]_t = 10^{-3}$ M. $K_{ML} = 10^3$ (curve a), 10^4 (curve b) and 10^5
(curve c).

measuring the conductivity and obtaining the Λ value. If the fraction of the free M^+ is
denoted by α, $\Lambda = \alpha\Lambda_{MX} + (1-\alpha)\Lambda_{MLX}$. Λ_{MX} corresponds to Λ before titration,
Λ_{MLX} to Λ after all M^+ are converted to ML^+. If the molar conductivity of M^+ is
different from that of ML^+, K_{ML} can be obtained by the following equation, using the
relation $[L] = [L]_t - [M]_t(\Lambda_{MX} - \Lambda)/(\Lambda_{MX} - \Lambda_{MLX})$ ($[L]_t$ and $[M]_t$ are the total concentrations of L and M^+):

$$K_{ML} = \frac{\Lambda_{MX} - \Lambda}{(\Lambda - \Lambda_{MLX})[L]}$$

Figure 7.7 shows the relation between $(\Lambda - \Lambda_{MLX})/(\Lambda_{MX} - \Lambda_{MLX})$ and $[L]_t/[M]_t$. This
method is usually suitable for determining $\log K_{ML}$ between 2 and 5. However,
if assisted by computer calculations, even $\log K_{ML}$ as low as ~ 0 can be determined [36].
This is because, in this method, the total electrolyte concentration is kept constant
during the titration. Conductimetric studies of crown complexes have been reviewed
by Takeda [37]. For crown complexes with larger $\log K_{ML}$ values, potentiometric
methods are convenient, as described in Section 6.3.3.

7.3.2
Conductimetric Studies of Acid–Base Equilibria

Determination of Autoprotolysis Constants
The autoprotolysis constant $K_{SH} [= a(SH_2^+) \cdot a(S^-)]$ of an amphiprotic solvent SH is
related to the conductivity of the pure solvent κ by

$$\kappa = K_{SH}^{1/2}\Lambda^\infty(SH)$$

where $\Lambda^\infty(SH) = \lambda^\infty(H^+) + \lambda^\infty(S^-)$. Thus, we can determine K_{SH} by measuring κ.
However, the application of this method is limited to the solvents with rather large

K_{SH} values. For the solvents of small K_{SH} values, the impurities in the solvent have an undesirable effect and make the K_{SH} value unreliable.

Determination of Acid Dissociation Constants

A potentiometric method that uses a pH sensor (pH glass electrode, pH-ISFET) as indicator electrode is convenient in studying acid–base equilibria in nonaqueous solvents (Section 6.3). However, in a potentiometric method, the pH sensor must be calibrated beforehand, using a standard buffer of known pH value. The solution is prepared either from a strong acid (completely dissociated) or from a weak acid of a known pK_a value and its conjugate base. When such an acid is not available for the solvent under study, conductimetry is often used to begin studying the acid–base equilibria. By conductimetry, a complete dissociation can be confirmed for a strong acid and the pK_a value and the homoconjugation constant can be determined for a weak acid. Then, we prepare a solution of a known pH value to calibrate the potentiometric pH sensor. Here again, the conductimetric results for very weak acids are often significantly affected by trace amounts of ionic and basic impurities. Thus, the weak acid studied by conductimetry should have $pK_a \leq 7.5$. Many data on acid dissociation constants in dipolar aprotic solvents have been obtained by conductimetry [38].

The procedures typically employed for a weak acid HA are described below [39, 40]:

1. To obtain the limiting molar conductivities of H^+ and A^-, the molar conductivities (Λ) are measured for the tetraalkylammonium salt of A^- (R_4NA) and a strong acid (e.g. CF_3SO_3H), and their limiting molar conductivities (Λ^{∞}) are obtained by calculation. Using the known λ^{∞} values of R_4N^+ and $CF_3SO_3^-$, the λ^{∞} values of H^+ and A^- are obtained.

2. The molar conductivity of HA is measured as a function of its concentration, c_a, and the data are analyzed appropriately according to the mechanism of the acid dissociation.

   (i) If HA is a strong acid and is completely dissociated ($HA \rightarrow H^+ + A^-$), the $\Lambda - c_a^{1/2}$ relation should have the Onsager slope at low concentration regions. From this, we can confirm the complete dissociation of HA and get the value of $\Lambda^{\infty}(HA)$. If more accurate values of $\Lambda^{\infty}(HA)$ are needed, the data are analyzed by one of the theories covered in Section 7.1.1.

   (ii) If HA is a weak acid but dissociates simply by $HA \rightleftarrows H^+ + A^-$, the relation $K_a = (\Lambda/\Lambda^{\infty})^2 c_a$ holds approximately. If the value of $(\Lambda/\Lambda^{\infty})^2 c_a$ is independent of c_a, we can consider that the dissociation reaction is simple and K_a is equal to that value.

   (iii) If HA is a weak acid and the dissociation is accompanied by a homoconjugation reaction:

   $$HA \rightleftarrows H^+ + A^-, \quad A^- + HA \rightleftarrows HA_2^-$$

   we determine the dissociation constant, K_a, and the homoconjugation constant, $K^f(HA_2^-)$, using the French–Roe equation [41], which holds when only a small fraction of HA is dissociated into A^- and HA_2^-.

$$\Lambda\left\{c_a\left[c_a + \frac{1}{K^f(HA_2^-)}\right]\right\}^{1/2} = \Lambda^\infty(HA)\left\{\frac{K_a}{K^f(HA_2^-)}\right\}^{1/2}$$
$$+ c_a\Lambda^\infty(2HA)\{K_aK^f(HA_2^-)\}^{1/2} \qquad (7.17)$$

Here, $\Lambda^\infty(2HA)$ is the sum of λ^∞ for H^+ and HA_2^-. If we plot the relation between $\Lambda\{c_a[c_a + 1/K^f(HA_2^-)]\}^{1/2}$ and c_a, varying the value of $K^f(HA_2^-)$, a straight line is obtained for the most suitable $K^f(HA_2^-)$ value. Then, the slope and the intercept of the line are equal to $\Lambda^\infty(2HA)\{K_aK^f(HA_2^-)\}^{1/2}$ and $\Lambda^\infty(HA)\{K_a/K^f(HA_2^-)\}^{1/2}$, respectively. If HA is not weak enough, Eq. (7.17) should be modified [40].

(iv) If HA is a weak acid and its molar conductivity Λ is independent of concentration c_a, there is a dissociation equilibrium as in Eq. (7.18) and the equilibrium constant is given by Eq. (7.19):

$$2HA \rightleftharpoons H^+ + HA_2^- \qquad (7.18)$$

$$K(2HA) = a(H^+)a(HA_2^-)/[HA]^2 = K_aK^f(HA_2^-) = \alpha^2\gamma^2 \qquad (7.19)$$

Here, $\alpha = \Lambda/\Lambda^\infty(2HA)$ and γ is the activity coefficient of the univalent ion. We can get $K(2HA)$ from Eq. (7.19). If we obtain $K^f(HA_2^-)$ by some other means, we can also get the value of K_a. One method for obtaining $K^f(HA_2^-)$ is to measure the conductivity of the saturated solutions of NaA in the presence of varying concentrations of HA. The conductivity increases with increasing [HA] because of homoconjugation. In the solutions, Eq. (7.20) holds by the electroneutrality principle because $[H^+]$ is negligible:

$$[HA_2^-] = [Na^+] - [A^-] = [Na^+] - \frac{K_{sp}(NaA)}{[Na^+]\gamma^2} \qquad (7.20)$$

$K_{sp}(NaA)$ can be determined from the conductivity of the saturated NaA solution at [HA] = 0. From Eq. (7.20) and approximate values of $\Lambda(NaHA_2)$ and $\Lambda(NaA)$, we get an approximate value of $K^f(HA_2^-)$. Then, after successive approximations, we get final values of $\lambda^\infty(HA_2^-)$ and $K^f(HA_2^-)$.

Conductimetry and conductimetric titration are useful for detecting and determining solvent impurities. The solvent conductivity is a measure of the amount of ionic species in the solvent. The conductivity of well-purified solvent is an important parameter of solvent properties (Table 1.1). An amine as a solvent impurity does not show appreciable conductivity, but if it is titrated with a weak acid, the resulting salt is dissociated and the conductivity increases linearly until the end point (see Figure 10.3). Conductimetric titration is also useful in elucidating reaction mechanisms. The curves in Figure 7.8 are for the titration of 3,5-dinitrobenzoic acid (HA) in AN

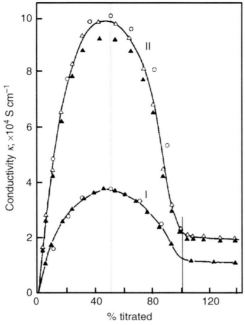

Figure 7.8 Conductimetric titration of 3,5-dinitrobenzoic acid (HA) with triethylamine. (I) 18.7 mM HA, (II) 61.9 mM HA. $-\blacktriangle-$, experimental data; $-\triangle-$, corrected for viscosity; $-\bigcirc-$, calculated values [42].

with triethylamine (B). They show a maximum at the half-neutralization point owing to the fact that the concentration of the conjugate ion HA_2^- is maximum at the half-neutralization point and that the degree of dissociation of the salt $(BH^+HA_2^-)$ is much greater than that of (BH^+A^-). The curves agree well with the theoretical considerations [42]. Conductimetric titrations are also analytically important: various substances can be determined by conductimetric acid–base, precipitation and complexation titrations (see Ref. [39b]).

References

1 (a) Fernandez-Prini, R. (1973) *Physical Chemistry of Organic Solvent Systems* (eds A.K. Covington and T. Dickinson), Plenum Press, New York, Section 5.1; (b) Spiro, M. (1973) *Physical Chemistry of Organic Solvent Systems* (eds A.K. Covington and T. Dickinson), Plenum Press, New York, Section 52, 5.3.

2 Kay, R.L., Evans, D.F. and Matesich, S.M. (1976) *Solute–Solvent Interactions*, vol. 2

(eds J.F. Coetzee and C.D. Ritchie), Marcel Dekker, New York, Chapter 10.

3 Popovych, O. and Tomkins, R.P.T. (1981) *Non-aqueous Solution Chemistry*, John Wiley & Sons, Inc., New York, Chapter 7.

4 Justice, J.-C. (1983) *Comprehensive Treatise of Electrochemistry*, vol. 5 (eds B.E. Conway, J.O'M. Bockris and E. Yeager), Plenum Press, New York, Chapter 3.

5 Spiro, M. (1986) *Physical Methods of Chemistry, vol. 2, Electrochemical Methods*, 2nd edn (eds B.W. Rossiter and J.F. Hamilton), John Wiley & Sons, Inc., New York, Chapter 8.

6 (a) Barthel, J.M.G., Krienke, H. and Kunz, W. (1998) *Physical Chemistry of Electrolyte Solutions*, Springer, Darmstadt; (b) Barthel, J. and Gores, H.-J. (1994) *Chemistry of Non-aqueous Solutions, Current Progress* (eds G. Mamantov and A.I. Popov), Wiley-VCH Verlag GmbH, Weinheim, Chapter 1.

7 Bockris, J.O'M. and Reddy, A.K.N. (1998) *Modern Electrochemistry, vol. 1 Ionics*, 2nd edn, Plenum Press, New York, Chapter 4.

8 Smedley, S.I. (1980) *The Interpretation of Ionic Conductivity in Liquid*, Plenum Press, New York.

9 (a) Onsager, L. (1927) *Physik. Z.*, **28**, 277; (b) Robinson, R.A. and Stokes, R.H. (1970) *Electrolyte Solution*, 2nd edn, Butterworths, London.

10 Pitts, E. (1953) *Proc. Roy. Soc.*, **277A**, 43.

11 Fuoss, R.M. and Onsager, L. (1957) *J. Phys. Chem.*, **61**, 668.

12 Fuoss, R.M. and Hsia, L.L. (1967) *Proc. Nat. Acad. Sci. USA*, **57**, 1550.

13 Fuoss, R.M. and Kraus, C.A. (1933) *J. Am. Chem. Soc.*, **55**, 476.

14 Shedrovsky, T. (1938) *J. Franklin Inst.*, **225**, 739.

15 Fuoss, R.M. (1935) *J. Am. Chem. Soc.*, **57**, 488.

16 Daggett, H.M. (1951) *J. Am. Chem. Soc.*, **73**, 4977.

17 Delsignore, M., Maaser, H.E. and Petrucci, S. (1984) *J. Phys. Chem.*, **88**, 2405.

18 For reviews, see: (a) Salomon, M. (1989) *J. Power Sources*, **26**, 9; (b) Salomon, M. (1987) *Pure Appl. Chem.*, **59**, 1165; (c) Ref. [6b], p. 61–66.

19 Hojo, M., Takiguchi, T., Hagiwara, M., Nagai, H. and Imai, Y. (1989) *J. Phys. Chem.*, **93**, 955.

20 Miyauchi, Y., Hojo, M., Ide, N. and Imai, Y. (1992) *J. Chem. Soc., Faraday Trans.*, **88**, 1425, 3175.

21 Casteel, J.F. and Amis, E.A. (1972) *J. Chem. Eng. Data*, **17**, 55.

22 Chagnes, A., Carré, B., Willmann, P. and Lemordant, D. (2001) *Electrochim. Acta*, **46**, 1783.

23 (a) Ebeling, W. and Rose, J. (1981) *J. Solution Chem.*, **10**, 599; (b) Ebeling, W. and Grigo, M. (1982) *J. Solution Chem.*, **11**, 151; (c) Bernard, O., Kunz, W., Turq, P. and Blum, L. (1992) *J. Phys. Chem.*, **96**, 398, 3833; (d) Turq, P., Blum, L., Bernard, O. and Kunz, W. (1995) *J. Phys. Chem.*, **99**, 822.

24 (a) Chhih, A., Turq, P., Bernard, O., Barthel, J.M.G. and Blum, L. (1994) *Ber. Bunsenges. Phys. Chem.*, **98**, 1516; (b) Barthel, J., Gores, H..-J., Neueder, R. and Schmid, A. (1999) *Pure Appl. Chem.*, **71**, 1705; (c) Ref. [6a], Chapter 6.

25 Fuoss, R.M. and Edelson, D. (1951) *J. Am. Chem. Soc.*, **73**, 269.

26 (a) Doe, H., Kitagawa, T. and Sasabe, K. (1984) *J. Phys. Chem.*, **88**, 3341; (b) Doe, H. and Kitagawa, T. (1985) *Bull. Chem. Soc. Jpn.*, **58**, 2975.

27 Quint, J. and Viallard, A. (1978) *J. Solution Chem.*, **7**, 137, 525, 533.

28 Lee, W.H. and Wheaton, R.J. (1978) *J. Chem. Soc., Faraday Trans II*, **74**, 743, 1456.

29 (a) Pethybridge, A.D. (1986) *Pure Appl. Chem.*, **58**, 1163; (b) Ref. [6a] p. 171.

30 (a) Bagchi, B. and Biswas, R. (1998) *Acc. Chem. Res.*, **31**, 181; (b) Biswas, R., Roy, S. and Bagchi, B. (1995) *Phys. Rev. Lett.*, **75**, 1098; (c) Biswas, R. and Bagchi, B. (1997) *J. Chem. Phys.*, **106**, 5587; (d) Kashyap, H.K., Pradhan, T. and Biswas, R. (2006) *J. Chem. Phys.*, **125**, 174506; (e) Zwanzig, R. (1963) *J. Chem. Phys.*, **38**, 1603; 1970 **52**, 3625; (f) Bagchi, B. (1998) *J. Chem. Phys.*, **98**, 3989.

31 Krumgalz, B.S. (1983) *J. Chem. Soc., Faraday Trans I*, **79**, 571.

32 Kay, R.L. (1973) *Techniques in Electrochemistry*, vol. 2 (eds E. Yeager and A.J. Salkind), Wiley-Interscience, New York, Chapter 2.

33 Kay, R.L., Hales, B.J. and Cunningham, G.P. (1967) *J. Phys. Chem.*, **71**, 3935.

34 (a) Hanna, E.M., Pethybridge, A.D., Prue, J.E. and Spiers, D.J. (1974) *J. Solution Chem.*, **3**, 563, [HMPA]; (b) Rosenfarb, J., Martin, M., Prakash, C. and Caruso, J.A.

(1976) *J. Solution Chem.*, **5**, 311, [TMS];
(c) Hanna, E.M. and Al-Sudani, K. (1987)
J. Solution Chem., **16**, 155, [PC].

35 Salomon, M. and Uchiyama, M.C. (1987)
J. Solution Chem., **16**, 21.

36 Takeda, Y., Mochizuki, Y., Tanaka, M.,
Kudo, Y., Katsuta, S. and Ouchi, M. (1999)
J. Inclusion Phenom., **33**, 217.

37 Takeda, Y. (1991) *Cation Binding by
Macrocycles* (eds Y. Inoue and G.W. Gokel),
Marcel Dekker, New York, Chapter 3.

38 Izutsu, K. (1990) *Acid–Base Dissociation
Constants in Dipolar Aprotic Solvents,
IUPAC Chemical Data Series No. 35,*
Blackwell Scientific Publications, Oxford.

39 (a) Kolthoff, I.M. and Chantooni, M.K.,
Jr (1965) *J. Am. Chem. Soc.*, **87**, 4428.
(b) Kolthoff, I.M. and Chantooni, M.K., Jr
(1979) *Treatise on Analytical Chemistry, Part
I*, 2nd edn, vol. 2 (eds I.M. Kolthoff and P.J.
Elving), John Wiley & Sons, Inc.,
New York, Chapter 19A.

40 Izutsu, K., Kolthoff, I.M., Fujinaga, T.,
Hattori, M. and Chantooni, M.K., Jr (1977)
Anal. Chem., **49**, 503.

41 French, C.M. and Roe, I.G. (1953) *Trans.
Faraday Soc.*, **49**, 314.

42 Kolthoff, I.M. and Chantooni, M.K., Jr
(1963) *J. Am. Chem. Soc.*, **85**, 426; Ref.
[39b], p. 283.

8
Polarography and Voltammetry in Nonaqueous Solutions

Polarography and voltammetry in nonaqueous solutions became popular in the late 1950s.[1] Today they both play important roles in many fields of both pure and applied chemistry. Various substances that are insoluble or unstable in water can dissolve or remain stable in nonaqueous solvents. The measurable windows of temperature and potential are much wider in nonaqueous solvents than in water. Moreover, polarography and voltammetry in nonaqueous solutions are very useful for studying such fundamental problems as ion solvation, electronic properties of inorganic and organic species and reactivities of unstable reaction products and intermediates at the electrodes. In this chapter, our discussion is focused on the applications of these techniques to such fundamental studies. The basic aspects of polarography and voltammetry were outlined in Chapter 5. Combinations of voltammetry and nonelectrochemical techniques are described in Chapter 9. Also, in Chapter 12, the use of nonaqueous solutions in modern electrochemical technologies is considered in connection with the role of voltammetry. Books of Refs [1–5] are concerned with polarography and voltammetry in general and also contain many subjects related to or important in polarography and voltammetry in nonaqueous solutions.

8.1
Basic Experimental Techniques in Nonaqueous Solutions

This section briefly reviews experimental techniques in nonaqueous solutions. Related topics are also dealt with in Section 8.4 and in Chapters 10 and 11. For further information, see Refs [1–12].

8.1.1
Experimental Apparatus for Nonaqueous Systems

The electric resistance of nonaqueous electrolytic solutions is often much higher than that of aqueous ones, and so polarographic and voltammetric measurements in

1) The situation up to the early 1960s is reviewed
   in *Talanta* 1965, **12**, 1211 (Takahashi, R.), 1229
   (Wawzonek, S.).

Electrochemistry in Nonaqueous Solutions, Second, Revised and Enlarged Edition. Kosuke Izutsu
Copyright © 2009 WILEY-VCH Verlag GmbH & Co. KGaA, Weinheim
ISBN: 978-3-527-32390-6

nonaqueous solutions should be made with a three-electrode device. A computer-aided three-electrode instrument, equipped with a circuit for iR-drop compensation, is commercially available for such measurements. However, a simple combination of a potentiostat and a function generator can also be used (Section 5.9).

The cell should be equipped with three electrodes, i.e. an indicator electrode, a counter electrode and a reference electrode. The indicator electrode for polarography is a dropping mercury electrode (DME). However, a variety of indicator electrodes are used in voltammetry. Among the indicator electrodes of conventional sizes are the hanging mercury drop electrode (HMDE), the mercury film electrode and solid electrodes of such materials as platinum, gold and glassy carbon. Nowadays, a boron-doped diamond-film electrode is also applicable in nonaqueous solutions [13]. Solid electrodes for voltammetry are often disk-shaped (0.3–5 mm in diameter) and used either as a stationary or as a rotating electrode. As described in Section 8.4.2, ultramicroelectrodes (UMEs, 1–20 µm in size; Pt, Au, GC, Hg (film or semisphere), etc.) also play important roles in modern voltammetric studies in nonaqueous systems. The potential range of mercury electrodes such as DME and HMDE is wide on the negative side but narrow on the positive side due to the anodic dissolution of mercury. On the other hand, Pt, Au and GC electrodes in well-purified aprotic solvents have potential ranges that are wide on both sides (Table 8.1). However, with these electrodes, the potential limits are easily influenced by trace amounts of impurities, especially water (Section 11.1.2). With the boron-doped diamond-film electrode, the promising features in aqueous solutions (see footnote 8 in Chapter 5) are also valid in nonaqueous solutions, and the potential window in an AN/toluene mixture is ~0.5 V wider than that of the platinum electrode [13].

When three-electrode devices are used, reference electrodes similar to those in potentiometry (Section 6.1.2) are applicable because no appreciable current flows through them. The reference electrodes used in nonaqueous solutions can be classified into two groups [1, 6, 9, 14]. Reference electrodes of the first group are prepared by using the solvent under study and those of the second group are aqueous reference electrodes;

Table 8.1 Measurable potential limits in nonaqueous solvents (V versus Fc$^+$/Fc)^a.

Solvent	Positive limit	Negative limit	Solvent	Positive limit	Negative limit
H$_2$O	+1.5(1)	−2.4	DMSO	+0.9(2)	−3.9(2)
AN	+2.0(4),	−3.1(5),	DMF	+1.3(1)	−3.8(2)
	+2.8(7)	−3.55(2 + LiH)	NMP	+1.35(2)	−3.9(2)
PC	+2.6(2),	−3.6(5)	HMPA	+0.3(2)	−4.05(2)
	+3.1(8)		CH$_2$Cl$_2$	+2.35(5)	−2.5(5)
NM	+2.3(4), +2.9(7),	−1.6(4),	DCE	+1.37(6)	−2.55(6)
	+3.8(7)^b	−3.0(2)	DME	+1.55(5)	−4.05(5)
TMS	+2.0(2),	−3.95(2)	THF	+1.6(5)	−3.85(5)
	+3.0(7)		HOAc	+2.5(3)	−0.8(3)

^{a}From Ref. [7]. The values obtained at a bright Pt electrode at a current density of 10 µA mm^{-2}. The numbers in parentheses show the supporting electrolyte: 1, HClO$_4$; 2, LiClO$_4$; 3, NaClO$_4$; 4, Et$_4$NClO$_4$; 5, Bu$_4$NClO$_4$; 6, Hep$_4$NClO$_4$; 7, Et$_4$NBF$_4$; 8, KPF$_6$.
^{b}Water was removed by anodic pre-electrolysis of the electrolytic solution at positive potentials (see Section 11.1.2).

the first group is the silver–silver ion electrode (e.g. $0.01\,M$ $AgClO_4 + 0.1\,M$ $R_4NClO_4(s)/Ag$, $s =$ solvent under study). It is easy to prepare and is used in a variety of nonaqueous solvents. However, it is generally not suited for long-term use and must be renewed every day before measurement. The reference electrodes of the second group are handy but the liquid junction potential at the aqueous/nonaqueous junction is problematic because its magnitude is sometimes expected to be over 100 mV. Unfortunately, at present, there is no truly reliable reference electrode for use in nonaqueous solutions. Therefore, it has been recommended by IUPAC that electrode potentials in nonaqueous solutions should be reported against the formal potential (or half-wave potential) of the reference redox system, i.e. Fc^+/Fc or BCr^+/BCr [15]. For practical method, see Section 6.1.3. Data on the potential of the $Ag/0.01\,M$ Ag^+ reference electrode and the formal potential of the Fc^+/Fc couple for various solvents are listed in Table 6.4 against the potential of the BCr^+/BCr couple. These data can be applied to convert the potentials measured against the Ag/Ag^+ reference electrode to the values against the formal potential of the BCr^+/BCr or Fc^+/Fc couple.

If the three-electrode instrument is equipped with an iR-drop compensator, most of the iR-drop caused by the solution resistance can be eliminated. However, to minimize the effect of the iR-drop, a Luggin capillary can be attached to the reference electrode with its tip placed close to the indicator electrode. Moreover, for a solution of extremely high resistance, it is effective to use a quasireference electrode of a platinum wire (Figure 8.1a) or a dual-reference electrode (Figure 8.1b), instead of the conventional reference electrode [16].

The design of the electrolytic cell should meet the necessary conditions for the measurement [17]. For example, if a polarographic measurement is to be made under conditions that are strictly free of water and molecular oxygen, a vacuum electrochemical cell as in Figure 8.2 is used by being connected to a vacuum line [18]. When the effect of water is not important, however, general purpose cells (e.g. Figure 5.9) can be used. With such cells, dissolved oxygen is removed before each measurement, usually by bubbling dry inert gas (nitrogen or argon) through the solution. Moreover, during the measurement, the solution in the cell must be kept under inert atmosphere to avoid the recontamination with atmospheric oxygen. The solubility of oxygen is

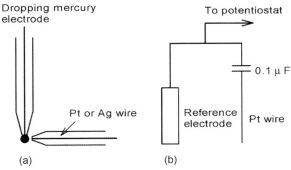

(a) (b)

Figure 8.1 A metal-wire quasireference electrode combined with a dropping mercury electrode (a) and a circuit for dual-reference electrode system (b) [16].

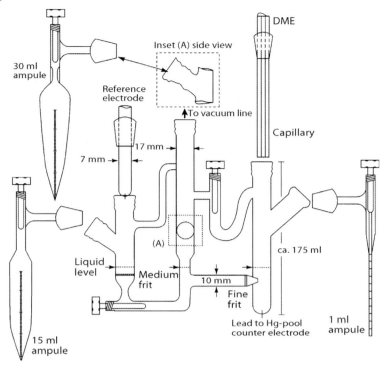

Figure 8.2 Electrochemical cell for use on a vacuum line [17, 18].

often higher in aprotic solvents than in water (p. 253). It should be noted that the vapors of nonaqueous solvents are often harmful (Section 1.1.4). The gas bubbled through the solution must not escape into the laboratory; the gas bubbling should be carried out while the electrolytic cell is in a glove box [19] and the solvent vapor should be trapped.

8.1.2
Solvents and Supporting Electrolytes

The solvent and the supporting electrolyte must be appropriate for each measurement because they have significant effects on electrode processes.

The following criteria should be used to select an appropriate solvent: (i) the electroactive species under study and the reaction products must dissolve and remain sufficiently stable in the solvent; (ii) a polar solvent of weak acidity is suitable for an electrode reaction that occurs at negative potentials or whose measurement is affected by acidic solvents; (iii) a polar solvent of weak basicity is suitable for an electrode reaction that occurs at positive potentials or whose measurement is affected by basic solvents; (iv) the solvent should be easy to purify, low in toxicity, benign to the environment, reasonable in price and should dissolve enough supporting electrolytes. Usually, DMF and DMSO (protophilic aprotic) and AN (protophobic aprotic) are used in case (ii), while AN is used most frequently in case (iii).

It is difficult to keep an electrolyte solution completely free from water, even when the experiment is carried out in a glove box or with a vacuum line. In such cases, procedures such as adding powdered active alumina directly to the electrolyte solution or passing the electrolyte solution through a column packed with active alumina can be used to remove residual water [20]. These procedures are also effective in removing various impurities, which are either acidic, basic, nucleophilic or electrophilic. However, it should be noted that some of the electroactive species may also be adsorbed onto the powder.

The selection of supporting electrolytes is as important as the selection of solvents. Inorganic electrolytes (salts, acids and bases) are widely used as supporting electrolytes in aqueous solutions. However, they are often inappropriate in nonaqueous solutions because of their insufficient solubility, insufficient dissociation or unfavorable effects on electrode reactions. Metal ions are Lewis acids and are not suitable for reactions that reject acidic media. Halide ions, on the other hand, are Lewis bases and not suitable for reactions that reject basic media. Thus, in nonaqueous solutions, tetraalkylammonium salts are often used as supporting electrolytes. For many years, tetraalkylammonium perchlorates, especially Et_4NClO_4 and Bu_4NClO_4, were the most popular supporting electrolytes. However, today their use is avoided because of their explosive nature, and such electrolytes as tetraalkylammonium tetrafluoroborates, hexafluorophosphates and trifluoromethanesulfonates (R_4NBF_4, R_4NPF_6 and $R_4NOSO_2CF_3$) are used instead; these salts are not explosive and have excellent properties as supporting electrolyte (see Section 11.1). As shown by some examples in this chapter and in Chapter 11, supporting electrolytes influence electrode reactions in various ways. It should be noted that even tetraalkylammonium salts sometimes have unfavorable effect on electrode reactions.

8.2
Polarography and Voltammetry of Inorganic Species

In this section, polarography and voltammetry of inorganic species in nonaqueous solutions are dealt with by emphasizing their fundamental aspects.

8.2.1
Polarographic Reductions of Metal Ions

Many of metal ions (M^{n+}) are reduced at a DME to form metal amalgams [M(Hg)]:

$$M^{n+} + ne^- + Hg \rightleftarrows M(Hg)$$

If the reaction is reversible, the S-shaped current–potential curve in DC polarography is expressed by Eq. (8.1) (Section 5.3):

$$E = E_{1/2} + \frac{RT}{nF}\ln\frac{i_d-i}{i} \quad \left(= E_{1/2} + \frac{0.0592}{n}\log\frac{i_d-i}{i}, \quad 25\,^\circ C\right) \tag{8.1}$$

where i_d is the limiting diffusion current and $E_{1/2}$ is the half-wave potential. The limiting diffusion current at the end of a drop-life, $(i_d)_{max}$ (μA), is given by the Ilkovič equation:

$$(i_d)_{max} = 708 n D^{1/2} C m^{2/3} \tau^{1/6} \tag{8.2}$$

where D ($cm^2 s^{-1}$) and C (mM) are the diffusion coefficient and the concentration of the metal ion, m (mg s^{-1}) is the mercury flow rate and τ (s) is the drop time.

We can determine from Eq. (8.2) the values of D for solvated metal ions. The value of D changes with changes in solvent or solvent composition. The viscosity (η) of the solution also changes with solvent or solvent composition. However, the relation between D and η can be expressed by the Stokes–Einstein relation:

$$D = \frac{k_B T}{6 \pi \eta r_s}$$

where k_B is the Boltzmann constant. If the Stokes' law radius (r_s) of the solvated metal ion does not change appreciably with changes in solvent or solvent composition, Walden's rule is valid:

$$D\eta = \text{constant}$$

Conversely, an increase or decrease in the value of $D\eta$ reflects a decrease or increase in the radius of the solvated metal ion. This works as a criterion in determining the effect of solvents on metal ion solvation.

In DC polarography, $E_{1/2}$ (at 25 °C) for the reversible reduction of a simple metal ion is expressed by

$$(E_{1/2})_s = E^{0'}_{M^{n+}/M(Hg)} + \frac{0.0592}{n} \log \left(\frac{D_{M(Hg)}}{D_{M^{n+}}} \right)^{1/2} \approx E^{0'}_{M^{n+}/M(Hg)} \tag{8.3}$$

where $D_{M^{n+}}$ is the diffusion coefficient of the metal ion and $D_{M(Hg)}$ that of the metal atom in mercury. Data for $D_{M(Hg)}$ have been compiled by Galus [21]. Thus, from Eq. (8.3), we can obtain the formal potential for the amalgam formation and, from the formal potential, we can get the value of the standard potential.

If $\Delta G^{\circ}_t (M^{n+}, S_1 \rightarrow S_2)$ is the standard Gibbs energy of transfer of M^{n+} from solvent S_1 to S_2, there is a relationship:

$$\Delta G^{\circ}_t (M^{n+}, S_1 \rightarrow S_2) = nF[(E^0)_{S_2} - (E^0)_{S_1}] \approx nF[(E_{1/2})_{S_2} - (E_{1/2})_{S_1}]$$

The standard potentials and the half-wave potentials in S_1 and S_2 should be expressed on a potential scale that is common to the two solvents. Conversely, if we have such a common potential scale, we can get the value of $\Delta G^{\circ}_t (M^{n+}, S_1 \rightarrow S_2)$ from the difference in the standard potentials or the half-wave potentials in the two solvents. The positive value in $\Delta G^{\circ}_t (M^{n+}, S_1 \rightarrow S_2)$ shows that the solvation of the metal ion is stronger in S_1 than S_2, and vice versa.

Other polarographic methods and cyclic voltammetry are also applicable to the study of the reduction process of a metal ion to its amalgam. In AC polarography, the peak potential for a reversible process agrees with the half-wave potential in DC polarography. Moreover, the peak current in AC polarography provides some infor-

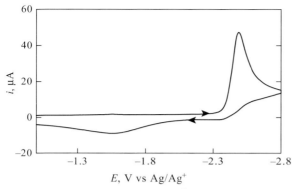

Figure 8.3 Cyclic voltammograms for Ba^{2+} in DMF–50 mM Hex_4NBF_4 obtained at a semispherical Hg electrode at 23 V s^{-1} (Izutsu, Okamura, unpublished results).

mation about the reversibility of the process. In inverse pulse polarography, the metal ion is reduced to its amalgam at the beginning of each mercury drop, and then the reoxidation of the amalgam is observed by applying inverse pulse potentials (Section 5.4). Cyclic voltammetry at a mercury electrode is also convenient to observe both the reduction and reoxidation processes. Figure 8.3 is for Ba^{2+} in DMF. The reoxidation, $Ba(Hg) \rightarrow Ba^{2+}$, occurs at a much more positive potential than the reduction, $Ba^{2+} \rightarrow Ba(Hg)$, suggesting the irreversible nature of the electrode process. For a reversible CV process, the cathodic and anodic peak potentials separate by $(58/n) = 29$ m$\overline{V}$ and their average agrees with the half-wave potential, i.e. $E_{1/2} = (E_{pc} + E_{pa})/2$.

Polarographic reductions have been studied for many kinds of metal ions and in a variety of nonaqueous solvents. Large amounts of data on half-wave potentials are available and have been compiled in some books and reviews [22]. However, many of the old data were obtained using different reference electrodes or aqueous SCE, for which the problem of the liquid junction potential exists (Section 6.1.2). Gritzner [23] compiled half-wave potentials of metal ions as values referred to the BCr$^+$/BCr system, which was recommended by IUPAC. Some of these are listed in Table 8.2. The potential of the BCr$^+$/BCr system is not seriously affected either by the presence of water and other impurities or by differences in experimental conditions. Thus, although the determination of the half-wave potentials is subject to errors of $\sim\pm5$ mV, the reliability of the data in Table 8.2 is fairly high. Moreover, because the BCr$^+$/BCr system is used as a solvent-independent potential reference, the potential data referred to it can be compared to some extent even when the data are obtained in different solvents.

The solvation of metal ions is closely related to the solvent basicity (Section 2.2.1). Gutmann and coworkers measured half-wave potentials of various metal ions by using the BCr$^+$/BCr reference system and plotted them against the solvent donor number (DN), which was proposed by them as the scale of solvent basicity [24]. Figure 8.4 shows some examples of such relations. For all metal ions in the figure, the half-wave potentials shift in the negative direction with increasing DN. These results are reasonable because the DN scale is for the strength of solvents as hard bases and because many of the metal ions act as hard acids and many of the organic solvents act as

Table 8.2 Half-wave potentials of metal ions in various solvents (V versus BCr/BCr^+, 25 °C)[23]

Solvent	Li^+	Na^+	K^+	Rb^+	Cs^+	Cu^+	Ag^+	Tl^+	Mn^{2+}	Co^{2+}	Cu^{2+}	Zn^{2+}	Cd^{2+}	Ba^{2+}	Pb^{2+}
Alcohols															
Methanol	−1.49	−1.22	−1.24	−1.23	−1.20	—	—	0.422	—	—	0.963	−0.397	0.285	−1.06[a]	0.467[h]
Ethanol	−1.46[a]	−1.17[a]	−1.18[b]	−1.18[b]	−1.18[b]	—	—	0.449	—	—	0.936	−0.180	0.224[h]	−0.991	0.525
Ketones															
Acetone	−1.400[a]	−1.224[a]	−1.281[a]	−1.296[a]	−1.270[f]	1.020[a]		0.410[a]	−0.410[a]	0.00[a]	1.23[a]	0.130[d]	0.510[d]	−1.075[a]	0.708
Ethers															
Tetrahydrofuran	−1.435	−1.251	−1.202	−1.204	−1.297			0.408	−0.594	−0.420	0.817	−0.049	0.320	−1.102	0.511
Esters/lactones															
γ-Butyrolactone	−1.34[a]	−1.17[a]	−1.26[a]	−1.26[a]	−1.25[a]			0.41[a]			1.128	0.138	0.511		—[h]
Propylene carbonate	−1.25	−1.068	−1.189	−1.23	−1.22			0.511	−1.08[a]	0.02[a]	1.25[a]	0.21[a]	0.64[a]	−0.941	0.691
Amides/lactams/ureas															
N-Methylformamide	−1.661[a]	−1.321[a]	−1.334[a]	−1.335[a]	−1.294[a]			0.286	—	—	0.723	−0.390	0.132	−1.33[a,i]	0.282
N,N-Dimethylformamide	−1.623	−1.349	−1.371	−1.358	−1.335			0.261	−0.86	−0.55	0.706	−0.291	0.126	−1.305	0.270
N,N-Diethylformamide	−1.618	−1.332	−1.346	−1.332	−1.316			0.272	−0.809	−0.642	0.718	−0.291	0.135		0.270
N,N-Dimethylacetamide	−1.690	−1.380	−1.404	−1.349	−1.344			0.259	−0.88[i]	−0.35[i]	0.725	−0.233	0.129	−1.34	0.259
N,N-Diethylacetamide	−1.765	−1.378	−1.375	−1.343	−1.342			0.260		−0.319	0.736	−0.231	0.109	−1.354	0.272
N-Methyl-2-pyrrolidinone	−1.697	−1.367	−1.405	−1.37[a]	−1.35[a]			0.232	−0.4[a]	−0.13[a]	0.751[h]	−0.26[a]	0.118	−1.39	0.266
1,1,3,3-Tetramethylurea	−1.76	−1.391	−1.401	−1.361		0.765		0.223		−0.164	0.950	0.140	0.249	−1.321	0.220
Nitriles															
Acetonitrile	−1.200[a]	−1.118[a]	−1.223[a]	−1.224[a]	−1.207[a]	0.420	1.155	0.480	−0.34[a]	0.12[a]		0.104	0.460	−0.883	0.686
Propionitrile		−1.026[c]				0.412[c]	1.104[c]	0.485[c]	—	—			0.471		0.669
Butyronitrile		−1.097				0.441	1.189	0.489				0.136	0.469	−0.941	0.686
Isobutyronitrile	−1.211	−1.092	−1.09[d]	−1.11[d]	−1.12[d]	0.440	1.143	0.467		0.08[j]		0.126	0.417		0.658
Benzonitrile	−1.115	−1.044	−1.131[c]	−1.174[c]	−1.183[c]	0.495	1.253	0.495	−0.28[a]	0.21[a,i]		0.243	0.543[a]	−0.88[a]	0.705

Compound															
Nitro compounds															
Nitromethane						1.025	—	0.569	0.43[a]	0.50[a]		0.50[a]		0.824	0.850
Nitrobenzene								0.556	0.42	0.56	1.393	0.59		0.781	0.812
Aromatic heterocycles															
Pyridine	−1.428	−1.201	−1.231	−1.232	−1.215	−0.009	—	0.242	−0.728	−0.277	—	−0.315	0.031	−1.036	0.337
Sulfur compounds															
Dimethyl sulfoxide	−1.86	−1.37[a]	−1.40[a]	−1.37[a]	−1.361	0.595		0.18	−1.00[a]	−0.71[a]	0.724[a]	−0.37[a]	0.02	−1.36[a]	0.179
Sulfolane[f]	−1.26[a]	−1.15[a]	−1.25[a]	−1.26[a]	−1.021[c]	—		0.41[a]	−0.29	0.07[i]	1.234	0.28	0.581	−1.01[a]	0.641
N,N-Dimethylthioformamide	−0.973[a]	−0.906[a]	−1.25[a]	−1.077[e]	−1.147[e]	−0.077[a]	0.396	0.173[a]	−0.410[a]	0.040[a]		−0.243[a]	0.054	−0.778[a]	0.283
N-Methyl-2-thiopyrrolidinone	−1.025[a]	−0.937[e]	−1.031[e]	−1.091[e]	−1.139[e]	−0.119	0.325	0.149	−0.365[a]	−0.05[a]		−0.245[a]	0.059	−0.811	0.272
Hexamethylthiophosphoramide	−1.072	−0.809[e]	—	—	—	0.204	0.58	0.311		−0.434		−0.327	0.252		0.437
Phosphorus compounds															
Trimethylphosphate	−1.721	−1.37[a]	−1.36[a]	−1.350[d]	−1.309			0.31	−1.15[a]	−0.69[a]	0.929[a]	−0.12[a]	0.21[a]	−1.33[a]	0.345
Hexamethylphosphoric triamide	−1.524[g]	−1.421[g]	−1.386[g]	−1.356[g]				0.131			0.551	−0.7	0.055	−1.485	0.158

The data in italics are for reversible electrode processes. Unless otherwise stated, the supporting electrolyte is 0.1 M Bu_4NClO_4. For others:

[a] 0.1 M Et_4NClO_4.
[b] 0.1 M Et_4NI.
[c] 0.05 M Bu_4NClO_4.
[d] 0.05 M Et_4NI.
[e] 0.05 M Bu_4NBPh_4.
[f] 30 °C.
[g] 0.1 M Hep_4NClO_4.
[h] Two-step wave.
[i] With polarographic maximum.
[j] 0.05 M Et_4NClO_4.

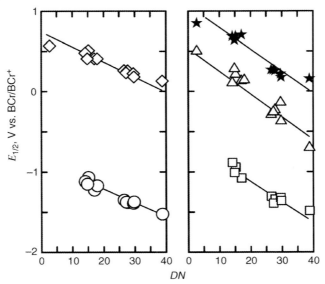

Figure 8.4 Relation between the half-wave potentials of metal ions (V versus BCr^+/BCr) and the donor number of solvents. ○ Na^+; ◇ Tl^+; □ Ba^{2+}; △ Zn^{2+}; ★ Pb^{2+}.

hard bases. As described in Section 4.1.2, Gritzner [25] used polarographic half-wave potentials to discriminate metal ions that behave as soft acid or as acid between hard and soft from those that behave as hard acid. Gritzner [26] also used the polarographic half-wave potentials referred to the BCr^+/BCr to determine the Gibbs energies, enthalpies and entropies of transfer of metal ions. In determining the enthalpies and entropies of transfer, they measured the effects of temperature on the half-wave potentials.

The strength of metal ion solvation affects not only the half-wave potentials but also the rates of electrode reactions of metal ions. For the reduction of a given metal ion, the reaction rate tends to decrease with increasing strength of solvation. The linear relation in Figure 8.5 was obtained for the reduction of a sodium ion; $\Delta G_{sv}^{\circ}(Na^+)$ is the solvation energy of Na^+ and k_s is the standard rate constant at the formal potential [27a].[2] For alkali metal ions in the same solvent, the rate constant tends to increase with increasing ionic radius or with decreasing strength of solvation. In Et_4NClO_4–DMF, for example, the value of k_s is 0.00 011 cm s^{-1} for Li^+, 0.09 cm s^{-1} for Na^+, 0.65 cm s^{-1} for K^+, 0.59 cm s^{-1} for Rb^+ and 1.10 cm s^{-1} for Cs^+ [27b]. The relationship between the kinetic data for metal amalgam formation and solvent properties is discussed in detail in Refs [28, 29].

Polarographic reductions of metal ions are also influenced by the supporting electrolyte:

1. If the anion of the supporting electrolyte forms a complex with the metal ion, the reduction potential of the metal ion shifts to the negative side by complexation

2) Linear log k_s – DN relations have been observed for polarographic reductions of various metal ions [28].

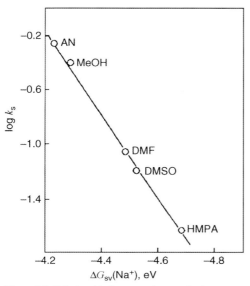

Figure 8.5 Relationship between the standard rate constant (log k_s) for reaction $Na^+ + e^- \rightleftarrows Na(Hg)$ and the solvation energies of Na^+ ion [27a].

(Eq. (5.19)). Although it also occurs in water, it is more marked in aprotic solvents in which small anions are reactive because of their weak solvations.

2. A striking effect of the cation of the supporting electrolyte is shown in Figure 8.6 [30]. Figure 8.6a is for the reductions of alkali metal ions in $HMPA + 0.05 M Et_4NClO_4$: the reduction waves for Cs^+ and Rb^+ are diffusion controlled, but Na^+ and Li^+ are not reduced until the negative end of the potential window. The K^+-wave is lower in height than expected for the diffusion-controlled process and has some kinetic current characteristics.[3] Figure 8.6b indicates the influence of the cation of the supporting electrolyte on the reduction of Na^+. If the cation is Me_4N^+ or Et_4N^+, Na^+ is not reduced at all. If it is Hep_4N^+ or Li^+, on the other hand, Na^+ gives a diffusion-controlled reduction wave. When Bu_4N^+ is the supporting electrolyte cation, Na^+ gives a small reduction current, which behaves as a kinetic current. These effects are closely related to the size of the cations: the Stokes' law radii of the cations in HMPA are in the order $Et_4N^+ < Me_4N^+ \ll Bu_4N^+ \sim Na^+ < Li^+ \sim Hep_4N^+$. If Na^+ and Et_4N^+ are contained in the solution, Et_4N^+ is preferentially attracted to the surface of the DME, which is negatively charged at negative potentials. The Et_4N^+, accumulated on the electrode surface, prevents the Na^+ from approaching the electrode and being reduced. In the solution containing Na^+ and Bu_4N^+, some Na^+ can

---

3) The kinetic current is obtained when an electroactive species is supplied by a chemical reaction occurring near the electrode surface (see Ref. [2] in Chapter 5). With a DME, the kinetic current is proportional to the electrode surface area and, therefore, is independent of the height of mercury head (h). In contrast, the diffusion-controlled limiting current is proportional to $h^{1/2}$.

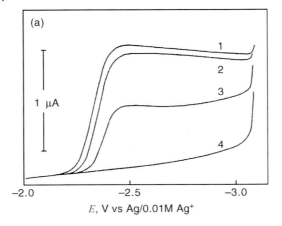

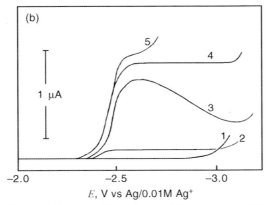

Figure 8.6 Polarographic reductions of alkali metal ions (0.5 mM) in HMPA + 0.05 M Et₄NClO₄ (a) and the influence of the supporting electrolyte on the reduction of 0.5 mM Na⁺ in HMPA (b) [30]. In (a), curve 1 is for Cs⁺, 2 for Rb⁺, 3 for K⁺ and 4 for Na⁺ and Li⁺. In (b), the cation of the supporting electrolyte (0.05 M perchlorate) for curve 1 is Me₄N⁺, Et₄N⁺; 2, Bu₄N⁺; 3, Hex₄N⁺; 4, Hep₄N⁺; and 5, Li⁺.

approach the electrode surface, probably by partial desolvation. Because Hep₄N⁺ and Li⁺ are larger than Na⁺, they do not interfere with the reduction of Na⁺. Interestingly, the well-defined Na⁺ wave, obtained in the HMPA + Hep₄NClO₄ solution, almost disappears if a small amount of Et₄N⁺ (~2 mM) is added. This phenomenon is the so-called electrochemical masking.

Effects of the cation of the supporting electrolyte similar to 2. occur, to a greater or lesser degree, in the reductions of alkali and alkaline earth metal ions in basic aprotic solvents [30a]. In dimethylacetamide (DMA), the reductions of alkaline earth metal ions are electrochemically masked by Et₄N⁺. In DMF and DMSO, the reversibility of the reductions of alkali and alkaline earth metal ions decreases with the decrease in

Table 8.3 Rate constants and transfer coefficients for the polarographic reductions of alkali metal ions in DMF (25 °C) [27b].

Ion	Supporting electrolyte	k_s, cm s^{-1}	α	Ion	Supporting electrolyte	k_s, cm s^{-1}	α
Li$^+$	Et$_4$NClO$_4$	0.00 011	0.52	K$^+$	Et$_4$NClO$_4$	0.65	0.37
	Pr$_4$NClO$_4$	0.00 028	0.77		Pr$_4$NClO$_4$	0.81	0.60
	Bu$_4$NClO$_4$	0.00 047	0.82		Bu$_4$NClO$_4$	1.33	0.62
Na$^+$	Et$_4$NClO$_4$	0.090	0.53	Rb$^+$	Et$_4$NClO$_4$	0.59	0.39
	Pr$_4$NClO$_4$	0.20	0.59		Pr$_4$NClO$_4$	0.82	0.49
	Bu$_4$NClO$_4$	0.32	0.63				

the cationic size of the supporting electrolyte. This effect is apparent from the kinetic data in Table 8.3, which were obtained by Baranski and Fawcett [27b] for the reductions of alkali metal ions in DMF.

Izutsu *et al.* [30] interpreted qualitatively the effect of cationic size of the supporting electrolyte by the double-layer effect,[4] which considered separately the potential at

[4] *Frumkin's double-layer effect:* As shown in Figure 8.7, the electrical double layer in the absence of specific adsorption consists of two parts, i.e. the Helmholtz layer and the diffuse layer (Gouy–Chapman–Stern model). At a negatively charged electrode, the outer Helmholtz plane (OHP) is an imaginary plane passing through the center of the cations of the supporting electrolyte, at their closest approach to the electrode surface. The potential varies linearly in the Helmholtz layer but exponentially in the diffuse layer (curve a). The following table shows, for some solvents, the potentials of zero charge ϕ_s at Hg/0.1 M LiClO$_4$(S) interface and the OHP potential ($\phi_{OHP} - \phi_s$) at the electrode charge density of $-10\,\mu C\,cm^{-2}$ and the electrolyte concentration of 0.1 M [29a].

that the electrode reduction of a univalent cation is accelerated by the double-layer effect. On the other hand, if $z = 0$, $n = 1$, $\alpha \sim 0.5$ and $(\phi_{OHP} - \phi_s) < 0$, $k_{s,app} < k_{s,corr}$ and the electrode reduction of a neutral molecule is decelerated by the double-layer effect. In the study of electrode kinetics, it is usual to get $k_{s,corr}$ by correcting for the double-layer effect (see Table 8.6 for an example).

Correction for the double-layer effect is not always easy. Especially, there are cases when the distance of the nearest approach of the reactant (x_r) should be considered to be different from that of the OHP (x_{OHP}). Then, in the discussion of the double-layer effect, ϕ_{OHP} should be replaced by ϕ_r, the potential at x_r. In the reductions of alkali and alkaline earth metal ions at very negative potentials, the relation of x_r and x_{OHP} is important; if

Solvent S	Water	FA	Ac	AN	DMSO	PC
ϕ_s, V versus Fc$^+$/Fc	—	−0.729	−0.670	−0.622	−0.717	−0.601
$\phi_{OHP} - \phi_s$, V	−0.088	−0.080	−0.121	−0.106	−0.101	−0.093

Usually the electrode reaction is considered to occur when the reactant reaches the OHP; thus, the rate of the electrode reaction is influenced by the value of $(\phi_{OHP} - \phi_s)$. For a reduction Ox^z + e$^- \rightleftarrows$ Red^{z-1}, the experimental standard rate constant, $k_{s,app}$, deviates from the standard rate constant expected for $(\phi_{OHP} - \phi_s) = 0$ (curve b). If the latter rate constant is expressed by $k_{s,corr}$, there is a relation $k_{s,exp} = k_{s,corr} exp[(n\alpha - z)(\phi_{OHP} - \phi_s) \times F/RT]$, where α is the transfer coefficient. If $z = +1$, $n = 1$, $\alpha \sim 0.5$ and $(\phi_{OHP} - \phi_s) < 0$, then $(\alpha - z)(\phi_{OHP} - \phi_s) > 0$ and $k_{s,app} > k_{s,corr}$, showing

$x_r < x_{OHP}$ and x_r is inside the Helmholtz layer, the value of $(\phi_r - \phi_s)$ is very negative and the reduction is much accelerated, compared to the case where $x_r > x_{OHP}$ and x_r is outside the Helmholtz layer.

Though not discussed in this book, the role of nonaqueous solvents in determining the structures and properties of electrical double layer has been the subject of numerous studies since 1920s. For the recent results, see, for example, Trasatti, S. (1987) *Electrochim. Acta,* **32**, 843; Borkowska, Z. (1988) *J. Electroanal. Chem.,* **244**, 1; Bagotskaya, I.A. and Kazarinov, V.E. (1992) *J. Electroanal. Chem.,* **329**, 225.

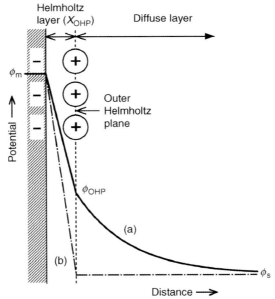

Figure 8.7 Structure of the electrical double layer.

the center of the cation of the supporting electrolyte (the outer Helmholtz plane) and that at the charge center of the reacting metal ions. Fawcett [29b] recently studied the kinetics of the amalgam formation of alkali and alkaline earth metal ions using various solvents and supporting electrolytes. He tried to explain their experimental results by proposing four possible elementary steps (electron transfer, ion transfer, adsorption and metal incorporation) and by considering the double-layer effect for each step. See the original report [29b] for details.

It is impossible in water to electrolytically deposit such active metals as alkali and alkaline earth metals on the electrodes of materials other than mercury. However, it is possible in appropriate aprotic solvents, as discussed in Section 12.7.

In water, the ions of rare earth metals, aluminum and zirconium are hydrolyzed and do not give well-defined polarographic reduction waves. In aprotic solvents, however, the solvolysis reactions do not occur as easily as the hydrolysis reactions in water. Thus, in solvents such as DMSO, DMF and AN, those ions give fairly well-defined polarographic waves. As another example, we cannot measure the reduction of $TiCl_4$ in water but can measure it in AN. This is because $TiCl_4$ is hydrolyzed in water but not solvolyzed in AN [31]. $TiCl_4$ in AN behaves as a Lewis acid and reacts with AN, a Lewis base, to produce a stable neutralization product $TiCl_4(CH_3CN)_2$:

$$TiCl_4 + 2CH_3CN \rightarrow TiCl_4(CH_3CN)_2$$

In 0.1 M Et_4NClO_4, this product gives two waves that correspond to $Ti^{4+} \rightarrow Ti^{3+} \rightarrow Ti^0$. In 0.1 M Et_4NCl, however, it is converted to $TiCl_6^{2-}$ by the reaction $TiCl_4(CH_3CN)_2 + 2Cl^- \rightarrow TiCl_6^{2-} + 2CH_3CN$ and gives three waves that correspond to $Ti^{4+} \rightarrow Ti^{3+} \rightarrow Ti^{2+} \rightarrow Ti^0$. Thus, by a controlled-potential electrolysis in this solution, we can generate Ti^{2+}, which is easily oxidized in air.

8.2.2
Polarography and Voltammetry of Metal Complexes

As described in Section 5.3, DC polarography is useful to study the complex formation of metal ions. If a metal ion M^{n+} reacts with ligand L^b to form a complex $ML_p^{(n+pb)+}$, we can get the value of p and the formation constant K by measuring the half-wave potential for the reduction of the metal ion as a function of the ligand concentration. When a successive complex formation ($M^{n+} \rightleftarrows ML^{(n+b)+} \rightleftarrows ML_2^{(n+2b)+} \rightleftarrows \cdots \rightleftarrows ML_p^{(n+pb)+}$) occurs, we can determine the successive formation constants. In aprotic solvents, the reduction of organic compounds (Q) in the presence of metal ions (M^+) is sometimes used to get the formation constants of ion pairs, $Q^{\bullet-}-M^+$ (see footnote 11). In some cases, the effect of metal ions on the voltammetric oxidation of organic ligands gives information on metal–ligand complexation; thus, the anodic oxidation of cryptands at a rotating gold disk electrode has been used to get the stability of metal cryptates [32].

However, recent interests in polarography and voltammetry of metal complexes in nonaqueous solutions has mainly focused on the so-called organometallic complexes. Many organometallic complexes, which are insoluble in water, are soluble in nonaqueous solvents. In particular, in aprotic solvents, many of the reaction products and intermediates, which are unstable in water and other protic solvents, are stabilized to a considerable extent. Thus, in appropriate aprotic solvents, we can study the elementary process for the reduction and/or oxidation of organometallic complexes. By studying these processes, we can obtain useful information about the redox properties of the complexes, including their electronic properties and the sites of electronic changes. Cyclic voltammetry, which is most useful in such studies, is now popular in the field of coordination chemistry. The following are some examples of the electrode reactions of organometallic complexes.

Organometallic Complexes for Potential Reference Systems
Some redox couples of organometallic complexes are used as potential references. In particular, the ferricinium ion/ferrocene (Fc^+/Fc) and bis(biphenyl)chromium(I)/bis(biphenyl)chromium(0) (BCr^+/BCr) couples have been recommended by the IUPAC as the potential reference in each individual solvent (Section 6.1.3) [15]. Furthermore, these couples are often used as solvent-independent potential reference for comparing the potentials in different solvents [25]. The oxidized and reduced forms of each couple have similar structures and large sizes. Moreover, the positive charge in the oxidized form is surrounded by the bulky ligands. Thus, the potentials of these redox couples are expected to be fairly free of the effects of solvents and reactive impurities. However, these couples do have some problems. One problem is that in aqueous solutions, Fc^+ in water behaves somewhat differently in other solvents [33]; the solubility of $BCr^+BPh_4^-$ is not enough in aqueous solutions, although it increases somewhat at higher temperatures ($>45\,^\circ C$) [26]. The other problem is that the potentials of these couples are influenced to some extent by solvent permittivity; this has been discussed in footnote 9 of Chapter 2. The influence of solvent permittivity can be removed by using the average of the potentials of the

(a)

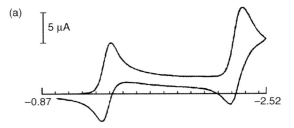

(b)

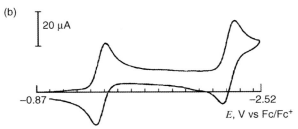

E, V vs Fc/Fc$^+$

Figure 8.8 Cyclic voltammograms of cobalticinium ion in 1 mM $(\eta\text{-}C_5H_5)_2CoPF_6$–0.1 M Bu$_4NBF_4$–AN [34a]. Recorded using a hanging mercury drop electrode (a) and a glassy carbon electrode (b) at 100 mV s^{-1} and at 25 °C.

Cc$^+$/Cc^0 and Cc^0/Cc$^-$ couples, where Cc denotes cobaltocene, because the effects of solvent permittivity are canceled out between the two couples. In practice, the half-wave potentials for Cc$^+$/Cc^0 and Cc^0/Cc$^-$ are measured by cyclic voltammetry (Figure 8.8) and the mean value obtained [34a]. Krishtalik et al. [34b] showed that the Born equation is almost valid for the solvations of Cc$^+$ and Cc$^-$ if $r = 0.37$ nm is used, and the authors proposed to use the formal potential of the Cc$^+$/Cc^0 couple as potential reference, after correcting the effect of solvent permittivity. Except for water, the corrected formal potential of the Cc$^+$/Cc^0 couple is -1.31 V versus Fc$^+$/Fc^0, while the formal potential of the BCr$^+$/BCr^0 couple is -1.13 V versus Fc$^+$/Fc^0 (Table 6.4).

Furthermore, as described in Section 4.1.3, the electrode processes of metallocenes are suitable for studying the role of fast solvation dynamics in the kinetics of electron-transfer reactions.[5] Fawcett et al. [35a–c] determined the standard rate constants (k_s) for the reductions of Cc$^+$ to Co^0 and Cc^0 to Cc^{-1} and obtained near-linear relations between $\log k_s$ and $\log \tau_L$, where τ_L is the longitudinal relaxation time of solvents (see Table 1.3). The result for the reduction of Cc$^+$ to Cc^0 is shown in Figure 8.9, line 1; the slope is somewhat smaller than unity. Fawcett et al. [35d] also studied the oxidation of nickelocene by AC voltammetry at a mercury hemi-spherical ultramicroelectrode (25 μm). They examined the relation between

5) The dynamic solvent effects on the kinetics of electron-transfer processes have been re-viewed in detail in Refs [28a, 37] and concisely in Section 3.6 of Ref. [2].

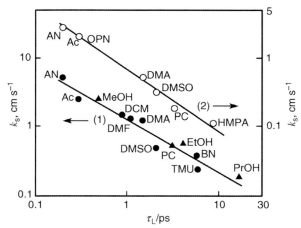

Figure 8.9 Relationship between the standard rate constants and the longitudinal relaxation times of solvents for the reductions of (1) Cc^+ to Cc^0 and (2) benzophenone to its radical anion, both at an Hg electrode, plotted in logarithmic scales using the data in Refs [35a], [69a]. Circles show Debye solvents and triangles non-Debye solvents.

$\{\ln[k_s/(\Delta G_{os}^*)^{1/2}] + \Delta G_{os}^*/RT\}$ and $\ln \tau_L$ (see Eq. (4.11)), by taking the double-layer effect on k_s and the solvent effect on ΔG_{os}^* (outer-shell activation energy) into account and by making correction for them. Here, τ_L varied from 0.2 ps for AN to 2.7 ps for PC. The relation was nearly linear with a unit slope, showing that the process was adiabatic. Bard and coworkers [35e] determined k_s for Fc^+/Fc^0 system in DMSO by adding sucrose to change the medium viscosity (η) and observed rough proportionality between k_s and $(1/\eta)$ as expected from the relation $\eta \propto \tau_L$.[6] It is interesting to compare these results with the results for the reduction of sodium ion to its amalgam, where log k_s was in linear relation with the solvation energies of Na^+ (Figure 8.5).

Organometallic Compounds of Biological Importance

Electrochemical properties of porphyrins and metalloporphyrins in nonaqueous solutions have been studied extensively to elucidate their biological functions. Various books and review articles have appeared on this subject [36]. Metalloporphyrins have three active sites at which electronic changes may occur by electrode reactions. (i) As shown by the cyclic voltammogram in Figure 8.10, a metal-free porphyrin is reduced in two steps, forming its radical anion and dianion, and is also

---

6) The near-linear relations of $\ln k_s$,
$\{\ln[k_s/(\Delta G_{os}^*)^{1/2}] + \Delta G_{os}^*/RT\}$ and $\ln \eta$
against $\ln \tau_L$ have also been confirmed for tris
(1,10-phenanthroline)ruthenium(II) (Ru
$(phen)_3^{2+}$) [35f].

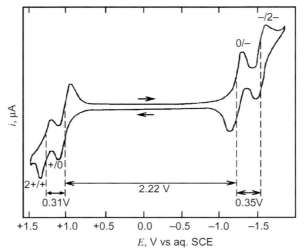

Figure 8.10 Cyclic voltammogram of tetraphenylporphyrin
(TPPH$_2$) at a platinum electrode in a solution of 2.6 mM
TPPH$_2$–0.1 M Bu$_4$NClO$_4$–CH$_2$Cl$_2$ [36a].

oxidized in two steps, forming its radical cation and dication. Metalloporphyrins of
Zn(II) and Mg(II) also give cyclic voltammograms as in Figure 8.10 because Zn(II)
and Mg(II) are electrochemically inactive. Information on the electronic properties of
porphyrins can be obtained from Figure 8.10: for example, the potential difference
between the first reduction step and the first oxidation step is 2.22 V and corresponds
to the energy difference between the lowest unoccupied molecular orbital (LUMO)
and the highest occupied molecular orbital (HOMO) of the porphyrin. (ii) If the
metalloporphyrin contains a metal ion that is electroactive, the electronic change can
occur also at the central metal. (iii) If the two axial ligands, i.e. the fifth and sixth
ligands, are electroactive, electronic change can occur there. For details of electrochemical studies of metalloporphyrins, see Ref. [36].

Stabilization of Low-Valency Complexes
The ligand of a metal complex and the solvent molecule compete with each other, as
Lewis bases, to interact with the central metal ion. At the same time, the metal ion
and the solvent molecule compete with each other, as Lewis acids, to interact with
the ligand. Thus, the behavior of a metal complex is easily influenced by the Lewis
acid–base properties of solvents. In an aprotic solvent, which is of weak acidity, the
interaction of the solvent molecules with the coordinating ligands is very weak and a
low-valency metal complex tends to be stabilized. For example, as described in
Section 4.1.2, tris(bipyridine)iron(II) complex, [Fe(bpy)$_3$]$^{2+}$, in aprotic solvents
gives three reversible waves, each corresponding to one-electron reduction process,
and generates [Fe(bpy)$_3$]$^-$ as the final product (see Eq. (4.8)) [38]. Figure 8.11 shows
the three waves obtained by cyclic voltammetry. However, this does not mean that
all three electrons are accepted by the central Fe(II) to form Fe(−I). From the ESR

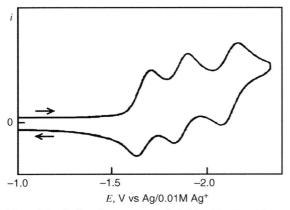

Figure 8.11 Cyclic voltammogram of tris(bipyridine)iron(II) complex at a glassy carbon electrode in 0.05 M Bu_4NClO_4–AN.

signal obtained with the electrogenerated low-valency complexes, it is obvious that the ligand, bpy, is also involved in the reduction processes [38].

Ligand Relaxation by the Supporting Electrolyte

The ligand of a metal complex and the anion of the supporting electrolyte compete with each other, as Lewis bases, to interact with the central metal ion. At the same time, the metal ion and the cation of the supporting electrolyte compete with each other, as Lewis acids, to interact with the ligand. Therefore, the behavior of a metal complex is easily influenced by the supporting electrolyte; for example, tris(acetylacetonato)iron(III), $Fe(acac)_3$, in AN forms $Fe(acac)_3^-$ by one-electron reduction. If the supporting electrolyte is $LiClO_4$, however, Li^+ picks up $acac^-$ from Fe $(acac)_3^-$ [39].

$$Fe(acac)_3^- + Li^+ \rightarrow Fe(acac)_2 + Li(acac)$$

This so-called ligand relaxation process is detected by the positive shift of the polarographic reduction wave of $Fe(acac)_3$. If the cation of the supporting electrolyte is R_4N^+, which is a very weak Lewis acid, this process does not occur.

8.2.3
Polarography and Voltammetry of Anions

The anion that forms either an insoluble mercury salt or a soluble mercury complex gives an anodic mercury dissolution wave in DC polarography. For chloride and hydroxide ions, the reactions of anodic dissolution are given by $2Hg + 2Cl^- \rightleftharpoons Hg_2Cl_2 + 2e^-$ and $Hg + 2OH^- \rightleftharpoons HgO + H_2O + 2e^-$, respectively. In aprotic solvents, these anodic dissolution waves appear at much more negative potentials than in water. It is because, in aprotic solvents, small anions such as Cl^- and OH^- are solvated only weakly and are very reactive, causing extremely small solubility products of Hg_2Cl_2 and HgO. Figure 8.12 shows the potentials at which the anodic

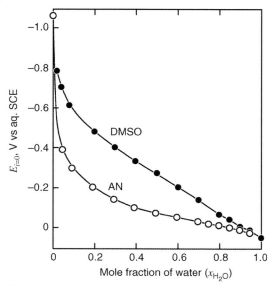

Figure 8.12 The potentials at which the anodic mercury dissolution due to OH^- starts to occur in H_2O–AN and H_2O–DMSO mixtures. Solution: 0.28 mM $Et_4NOH + 0.1$ M Et_4NClO_4 [40].

dissolution wave due to OH^- starts to appear in H_2O–AN and H_2O–DMSO mixed solvents [40]. The large negative shifts in water-poor regions show that OH^- becomes very reactive with the increase in the aprotic nature of the mixed solvents. OH^- in aprotic solvents is so reactive that, interestingly, it can work even as a reducing agent, donating an electron to an electron acceptor by $OH^- \rightarrow OH^{\bullet} + e^-$ [41].

The oxidation processes of various anions, including halide ions, have been studied by voltammetry at solid electrodes [42a, b]. For example, at a platinum electrode, iodide ion (I^-) in various organic solvents is oxidized in two steps, i.e. $3I^- \rightarrow I_3^- + 2e^-$ and $2I_3^- \rightarrow 3I_2 + 2e^-$. From the difference in the standard potentials of the two steps, the dissociation constants of I_3^-, $p\beta = -\log ([I_2][I^-]/[I_3^-])$, have been determined as in Table 8.4. In aqueous solutions, $p\beta$ is ~3 and I_3^- is not stable enough to use this

Table 8.4 Standard potentials of I^-/I_3^- and I_3^-/I_2 couples and dissociation constants of I_3^- [42c].

Solvent	NM	PC	TMS	DMF	AN	DMSO	MeOH
$E^0(I_2/I_3^-)$	+0.36$_5$	+0.39	+0.35	+0.245	+0.16	+0.16	+0.24
$E^0(I_3^-/I^-)$	−0.27$_5$	−0.28	−0.31	−0.39	−0.315	−0.32	−0.13
$p\beta$	7.4	7.6	7.5	7.2	7.3	5.4	4.2

The potentials are V versus Fc/Fc^+. Nürnberg, H.W. (ed.) (1974) *Electroanalytical Chemistry*, John Wiley & Sons, Inc., New York,, p. 386.

method. The electrode oxidation of CN^- has been studied in several solvents using ESR spectroscopy [42c]. From the ultimate formation of relatively stable tricyano-methylenimine radical anion, the following steps have been considered:

$$2CN^- \xrightarrow{-2e^-} 2CN\bullet \longrightarrow (CN)_2 \xrightarrow{CN^-} (NC)_2C{=}N^- \xrightarrow{-e^-, CN^-} (NC)_2C \cdots N - CN^{\bullet -}$$

8.2.4
Electrode Reactions of Dissolved Oxygen, Dissolved Hydrogen, Carbon Dioxide and Solvated Electrons

Dissolved Oxygen
The solubility of oxygen at 1 atm and 25 °C has been determined to be 1.0 mM in water, 2.1 mM in DMSO, 4.8 mM in DMF, 8.1 mM in AN, ~10 mM in hydrocarbons and ~25 mM in fluorocarbons [43a]. In aqueous solutions, the first reduction step of the dissolved oxygen is either two-electron or four-electron process, depending on the catalytic activity of the electrode material and on the solution composition.

$$\text{2-electron process :} \quad O_2 + 2H^+ + 2e^- \rightarrow H_2O_2$$
$$(O_2 + 2e^- \rightleftarrows O_2{}^{2-} \text{ in strongly alkaline solutions})$$
$$\text{4-electron process :} \quad O_2 + 4H^+ + 4e^- \rightarrow 2H_2O$$

In aprotic solvents, however, the first step is one-electron process to produce a superoxide ion $O_2^{\bullet -}$ [44].

$$O_2 + e^- \rightleftarrows O_2^{\bullet -}$$

If a tetraalkylammonium salt is used as supporting electrolyte, this process is either reversible or quasireversible and occurs at around -0.8 V versus aq. SCE in various aprotic solvents and with various electrode materials (Hg, Pt, GC). The formation of $O_2^{\bullet -}$ at room temperature has been confirmed by ESR spectroscopy, directly in pyridine and with the aid of spin-trapping using 5,5-dimethyl-1-pyrroline-N-oxide in DMSO [45]. If a Brønsted acid is added to the solution, the first step is converted to a two-electron process: $O_2^{\bullet -}$ produced in the first step is protonated to form $O_2H^{\bullet}$, which is more reducible than O_2. Thus, $O_2H^{\bullet}$ is further reduced to O_2H^- at the potential of the first step. According to detailed polarographic study in DMSO–H_2O mixtures, about 30% v/v H_2O is needed to convert the polarographic wave from one-electron process to two-electron process [46]. A metal ion M^+ interacts with $O_2^{\bullet -}$ to form an ion pair $M^+{-}O_2^{\bullet -}$ (often insoluble) and shifts the half-wave potential of the first wave in a positive direction [47]. Electrogenerated superoxide $O_2^{\bullet -}$ can act either as a nucleophile or as an electron donor and has been used in organic syntheses [48].

In dipolar aprotic solvents, the second step of the oxygen reduction is $O_2^{\bullet -} \rightarrow O_2{}^{2-}$. However, if H^+ is available from the solvent or from protic impurities, $O_2{}^{2-}$ reacts with H^+ to form O_2H^- or O_2H_2. Sometimes, the O_2H^- or O_2H_2 is further reduced at more negative potentials to give H_2O as the final product. Detailed information concerning the electrochemistry of molecular oxygen can be found in Ref. [43b].

Dissolved Hydrogen

At a platinum electrode that is slightly activated by platinization, the dissolved hydrogen in various solvents is oxidized nearly reversibly by $H_2 \rightleftarrows 2H^+ + 2e^-$. We can determine by cyclic voltammetry the standard potential of this process. If the standard potentials in various solvents are compared using a common potential scale, the Gibbs energies of transfer of H^+ can be obtained [49]. With electrodes other than platinum, it is difficult to observe reversible oxidation of dissolved hydrogen [49b].

Carbon Dioxide

The electrode reduction of CO_2 has considerably been studied in nonaqueous solvents (see review article [50]). Carbon dioxide dissolves more in nonaqueous solvents than in water, the solubilities of CO_2 at 1 atm and 25 °C being 0.25_6 M in AN, 0.18_6 M in DMF, 0.15_9 M in MeOH, 0.14_1 M in PC and 0.13_0 M in DMSO compared to 0.033 M in H_2O [51]. One-electron reduction of CO_2 forms $CO_2^{\bullet-}$ anion radical, but the radical is either adsorbed or not adsorbed at the electrode surface, depending on the electrode materials used. Then, $CO_2^{\bullet-}$ undergoes further inner-sphere processes when it is adsorbed, while $CO_2^{\bullet-}$ undergoes further outer-sphere processes when it is not adsorbed. The details of CO_2 reduction in aqueous and nonaqueous solutions are discussed in Section 12.5, in connection with its future applicability as the carbon resources.

Here, we consider the outer-sphere processes that occur at a mercury (Hg) electrode in DMF–0.1 M Bu_4NClO_4 [52]. The cyclic voltammogram for the reduction of CO_2 is usually irreversible, though somewhat reversible behavior of the $CO_2/CO_2^{\bullet-}$ couple can be observed at a high voltage scan rate (e.g. 4400 V s^{-1}) [52c], and from this, the standard potential of the $CO_2/CO_2^{\bullet-}$ couple has been obtained to be $\sim$ −2.21 V versus aq. SCE or $\sim$−2.6 V versus Fc/Fc$^+$ couple. The number of electrons per molecule for an irreversible wave was close to 1 when $[CO_2] > 5$ mM, in agreement with the formation of oxalate ($2CO_2^{\bullet-} \rightarrow {}^-O_2C–CO_2^-$) as well as the formation of CO and CO_3^{2-} [$CO_2^{\bullet-} + CO_2 \rightarrow {}^\bullet O_2C–CO_2^-$ followed by ${}^\bullet O_2C–CO_2^- + e^- \rightarrow CO + CO_3^{2-}$ (at electrode) or ${}^\bullet O_2C–CO_2^- + CO_2^{\bullet-} \rightarrow CO$ $CO_3^{2-} + CO_2$ (in solution)]. When the DMF solution contained residual water and upon decreasing the CO_2 concentration, the number of electrons per molecule approached 2, indicating that the formation of formate, $HCOO^-$, occurred by $CO_2^{\bullet-} + H_2O \rightarrow HCO_2^{\bullet} + OH^-$ followed by $HCO_2^{\bullet} + e^- \rightarrow HCOO^-$ (at electrode) or $HCO_2^{\bullet} + CO_2^{\bullet-} \rightarrow HCOO^- + CO_2$ (in solution). The product distribution obtained after the preparative-scale electrolysis of $\sim$100 mM CO_2 at an Hg electrode in DMF–0.1 M Bu_4NClO_4 showed a decrease in the percentage of oxalate and an increase in the percentage of CO with the decrease in temperature: i.e. the percentages of oxalate and CO were 80 and 20%, respectively, at 25 °C but 20 and 80%, respectively, at −20 °C. The percentages of oxalate and CO also depended on the CO_2 concentration: at 0 °C, they were 90 and 10%, respectively, at $[CO_2] = 20$ mM but 23 and 77%, respectively, at $[CO_2] = 217$ mM. Interestingly, the electrolysis catalyzed by radical anions of aromatic esters and nitriles exclusively produced oxalate in the DMF medium.

Solvated Electrons

In strongly basic solvents such as HMPA, amines and liquid ammonia, solvated electrons are relatively stable. In these solvents, if the supporting electrolyte is the salt of Li^+ or Na^+, blue solvated electrons, e_{SH}^-, are generated from the surface of the platinum electrode polarized at a very negative potential.

$$e^- (cathode) + SH \rightleftarrows e_{SH}^-$$

This process was shown by voltammetry to be reversible, with the standard potentials equal to -3.44 V versus Ag/0.1 M $AgClO_4$ in HMPA (25 °C) and -2.90 V versus Ag/0.1 M $AgClO_4$ in methylamine (-50 °C) [53a, b]. The stability of solvated electrons in these solvents has been attributed to the capture of electrons into the cavity of the solvent. For the details of electrochemistry of solvated electrons, see Ref. [53c, d].

8.3
Polarography and Voltammetry of Organic Compounds

Polarographic and voltammetric studies of organic compounds are often carried out in nonaqueous solutions for the following reasons: (1) Many organic compounds that are insoluble in water dissolve in appropriate nonaqueous solvents. (2) Polarography and voltammetry in aprotic solvents are useful to study the electronic properties of organic species, including extremely unstable intermediates. This is because, in such solvents, the first steps for the reduction and oxidation of organic compounds are usually one-electron processes that form radical anions and cations, respectively, and the potentials of these processes reflect the LUMO and HOMO of the compounds. Several excellent books on organic electrochemistry are available. Among them, *Organic Electrochemistry* (fourth edition, edited by Lund and Hammerich, 2001) [54] is the most authoritative, covering all aspects of organic electrochemistry. References [55, 56] also cover electrochemistry of organic compounds. The book by Mann and Barnes [57], though somewhat old, deals with electrode reactions of a variety of organic and inorganic species in nonaqueous solutions. Meites and Zuman [58] and Bard [59] have compiled many data of organic electrochemistry. References [4, 5] show how to get electrochemical data and mechanistic information of organic reactions, mainly from cyclic voltammetry. In addition to these books, many useful reviews that discuss topical problems are available (e.g. [60–63]). In this section, some characteristics of the electrode reactions of organic species in nonaqueous solutions are outlined.

8.3.1
Reductions of Organic Compounds

As described in Section 4.1.2, electrophilic organic compounds are reducible at the electrode. Some reducible organic compounds are listed in Table 8.5 with the potentials of the first reduction step in dipolar aprotic solvents. As described in

Table 8.5 Examples of reducible organic compounds and the potentials of their first reduction step in aprotic solvents[a]

Compounds	Solvent	Supporting electrolyte	Potential reference	Potential
Fullerenes				
C_{60}	AN/toluene ($-10\,^{\circ}$C)	Bu_4NPF_6	Fc/Fc^+	-0.98^b
C_{70}	AN/toluene ($-10\,^{\circ}$C)	Bu_4NPF_6	Fc/Fc^+	-0.97^c
Hydrocarbons[d]				
Benzene	DMF (-40 to $-65\,^{\circ}$C)	Me_4NBr	aq. Ag/AgCl	-3.42
Naphthalene	DMF (-40 to $-65\,^{\circ}$C)	Me_4NBr	aq. Ag/AgCl	-2.53
Anthracene	DMF (-40 to $-65\,^{\circ}$C)	Me_4NBr	aq. Ag/AgCl	-2.04
Phenanthrene	DMF (-40 to $-65\,^{\circ}$C)	Me_4NBr	aq. Ag/AgCl	-2.49
Biphenyl	THF	$LiBPh_4$	aq. Ag/AgCl	-2.68
Oxygen-containing compounds				
p-Benzoquinone	DMF	Et_4NClO_4	SCE	-0.54
1,4-Naphthoquinone	DMF	Et_4NClO_4	SCE	-0.60
9,10-Anthraquinone	DMF	Et_4NClO_4	SCE	-0.98
Benzaldehyde (PhCHO)	DMF	Et_4NI	SCE	-1.80
Acetone (MeCOMe)	DMF	Et_4NBr	SCE	-2.84
Acetophenone (PhCOMe)	DMF	Et_4NI	SCE	-1.99
Benzophenone (PhCOPh)	DMF	Et_4NI	SCE	-1.72
Benzyl (PhCOCOPh)	DMSO	Et_4NClO_4	SCE	-1.04

Nitrogen-containing compounds

t-Nitrobutane [(CH$_3$)$_3$NO$_2$]	AN	Bu$_4$NBr	SCE	−1.62
Nitrobenzene (PhNO$_2$)	AN	Pr$_4$NClO$_4$	SCE	−1.15
p-Chloronitrobenzene	AN	Pr$_4$NClO$_4$	SCE	−1.06
p-Bromonitrobenzene	AN	Pr$_4$NClO$_4$	SCE	−1.05
1,2-Dinitrobenzene [C$_6$H$_4$(NO$_2$)$_2$]	AN	Pr$_4$NClO$_4$	SCE	−0.69
Nitrosobenzene (PhNO)	DME	NaNO$_3$	SCE	−0.81
Azobenzene (PhN=NPh)	DMF	Bu$_4$NClO$_4$	SCE	−1.36

Halogenated compounds

Methyl iodide	DMF	Et$_4$NClO$_4$	SCE	−2.10
n-Butyl bromide	DMF	Et$_4$NClO$_4$	SCE	−2.41
n-Butyl iodide	DMF	Et$_4$NClO$_4$	SCE	−2.05
Benzyl bromide	DMF	Et$_4$NClO$_4$	SCE	−1.68
Benzyl chloride	DMF	Et$_4$NClO$_4$	SCE	−1.90
Tetrachloromethane	DMF	Et$_4$NClO$_4$	SCE	−1.13
Chlorobenzene [PhCl]	DMF	Et$_4$NClO$_4$	SCE	−2.7
1,2-Dichlorobenzene (C$_6$H$_4$Cl$_2$)	DMF	Et$_4$NClO$_4$	SCE	−2.5
1,2,3-Trichlorobenzene (C$_6$H$_3$Cl$_3$)	DMF	Et$_4$NClO$_4$	SCE	−2.2
1,2,3,4-Tetrachlorobenzene (C$_6$H$_2$Cl$_4$)	DMF	Et$_4$NClO$_4$	SCE	−1.9
Pentachlorobenzene (C$_6$HCl$_5$)	DMF	Et$_4$NClO$_4$	SCE	−1.6
Hexachlorobenzene (C$_6$Cl$_6$)	DMF	Et$_4$NClO$_4$	SCE	−1.4

[a] From Refs [1, 54, 57–59].
[b] Reduced in six steps (−0.98, −1.37, −1.87, −2.35, −2.85, −3.26 V).
[c] Reduced in six steps (−0.97, −1.34, −1.78, −2.21, −2.70, −3.70 V).
[d] For other aromatic hydrocarbons, see Table 8.8.

Ref. [54], organic compounds undergo various complicated electrode reductions. Here, however, only simple but typical cases are considered; they are reductions by the outer-sphere electron-transfer reactions and the dissociative electron-transfer reactions.

Outer-Sphere Electron-Transfer Reactions

In reductions of this type, the first step is usually one-electron process that produces a radical anion, $Q^{\bullet-}$:

$$Q + e^- \rightleftarrows Q^{\bullet-}$$

$Q^{\bullet-}$ is generally reactive, but its lifetime varies over a wide range, depending on its intrinsic reactivity and on the reactivity of the environment. If $Q^{\bullet-}$ has a delocalized negative charge, it is more or less stable unless the environment is reactive. Here, an environment of low reactivity (low Lewis acidity) is realized in aprotic solvents using tetraalkylammonium salts as supporting electrolyte. If $Q^{\bullet-}$ is stable enough, we can generate it by controlled-potential electrolysis and study its characteristics by measuring its UV/vis absorption spectra or ESR signals (Sections 9.2.1 and 9.2.2). However, study by cyclic voltammetry is by far the most popular; by employing high-voltage scan rates and low-temperature techniques, we can study the behavior of the radical anion even when its lifetime is very short ($10\,\mu s$ or less, Section 8.4). Because Q and $Q^{\bullet-}$ have similar structures and sizes, the first reduction step of organic compounds in aprotic solvents is usually highly reversible. Some data on the standard rate constants (k_s) are provided in Table 8.6. They were obtained by the AC

Table 8.6 Standard rate constants for electrode reductions of organic compounds determined by an AC impedance method.

Compounds	$E^r_{1/2}$, V versus SCE	$k_{s,exp}$, cm s^{-1}	α	$(\phi_{OHP} - \phi_s)$, mV	$k_{s,corr}$ cm s^{-1}	k_{ex}, M^{-1} s^{-1}
Benzonitrile	−2.17	0.61	0.64	−83	4.9	5.5×10^8
Nitrobenzene	−1.05	2.2	0.70	−56	10	3.0×10^7
m-Dinitrobenzene	−0.76	2.7	0.50	−46	6.5	5.2×10^8
p-Dinitrobenzene	−0.55	0.93	0.61	−36	2.2	6.0×10^8
Dibenzofuran	−2.41	2.9	(0.57)	−89	21	1.6×10^9
Dibenzothiophene	−2.37	1.6	(0.57)	−88	12	1.2×10^9
1,4-Naphthoquinone	−0.52	2.1	(0.57)	−35	4.6	4.2×10^8
Anthracene	−1.82	5	0.55	−76	27	1.8×10^9
Perylene	−1.54	5	0.50	−70	20	2.1×10^9
Naphthalene	−2.49	1.0	0.56	−145	23	6.2×10^8
trans-Stilbene	−2.15	1.2	0.58	−139	27	1.0×10^9

$E^r_{1/2}$: reversible half-wave potential, $k_{s,exp}$: experimental standard rate constant, α: transfer coefficient, $(\phi_{OHP} - \phi_s)$: potential at the outer Helmholtz plane, $k_{s,corr}$: standard rate constant after correction for the double-layer effect (see footnote 4), k_{ex}: rate constant for the homogeneous self-exchange electron transfer (see footnote 4 in Chapter 9); obtained with an HMDE in DMF–0.5 M Bu$_4$NClO$_4$ at $22 \pm 2\,°C$, except the last two obtained with a DME in DMF–0.1 M Bu$_4$NI at $30\,°C$.
From Kojima, H. and Bard, A.J. (1975) *J. Am. Chem. Soc.*, **97**, 6317.

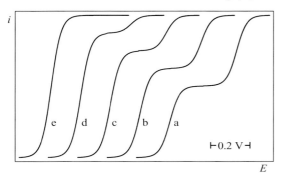

Figure 8.13 Influence of a weak proton donor (e.g. water) on the DC polarogram of the two-step reduction of an organic compound in an aprotic solvent. Curve a: without proton donor; curves b–e: with increasing amounts of proton donor.

impedance method and the experimental results were corrected for the double-layer effect. Some of them approach $\sim$30 cm s^{-1}. Recently, cyclic voltammetry, which uses ultramicroelectrodes and fast scan rates, has often been employed in studying the kinetics of highly reversible electrode reactions, as discussed in Section 8.4.

Many organic compounds in aprotic solvents are reduced in two steps, as shown by curve a in Figure 8.13. The second step is, in principle, the formation of a dianion, Q^{2-}.

$$Q^{\bullet-} + e^- \rightleftarrows Q^{2-}$$

As described in Section 4.1.2, for aromatic hydrocarbons (AH) in the gas phase, the standard potential of the second step is expected to be by 4–5 V more negative than that of the first step. In solutions, however, the experimental standard potential of the second step (E_2^0) is about 0.5–0.8 V more negative than that of the first step (E_1^0) [64]. This difference in the gas phase and in solutions is caused by the fact that the solvation energy of ions depends approximately on the square of the charge number; the solvation energy of the dianion Q^{2-} is four times that of the anion radical $Q^{\bullet-}$, while the solvation energy of the neutral Q^0 is nearly negligible. This phenomenon also occurs for other kinds of compounds. When the solvation of the dianion (Q^{2-}) is extraordinarily strong by such reasons as the decreased size of dianion, the order of E_1^0 and E_2^0 is inverted and E_2^0 becomes more positive than E_1^0. In such a case, the reduction of Q becomes one-step two-electron process and it occurs at a potential equal to $(E_1^0 + E_2^0)/2$. Such a potential inversion occurs, for example, with 3,6-dinitrodurene (3,6-dinitro-1,2,4,5-tetramethyl-benzene) with $(E_1^0 - E_2^0) = -280$ mV; it is considered that for its dianion, the benzene ring is distorted into the boat form [60, 65].

The Q^{2-}-dianions with delocalized charges can remain somewhat stable and give a well-defined reversible CV curve for the second step. However, because Q^{2-} is a much stronger base than $Q^{\bullet-}$, it easily reacts with protons that are originated from the protic impurity such as H_2O or from the solvent itself.

$$Q^{2-} + H^+ \rightarrow QH^-, \quad QH^- + H^+ \rightarrow QH_2$$

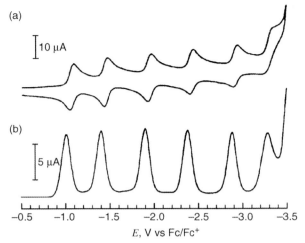

Figure 8.14 Reduction of fullerene, C_{60}, in AN–toluene (1:5) with Bu_4NPF_6 as supporting electrolyte at $-10\,°C$, recorded using (a) cyclic voltammetry at $100\,mV\,s^{-1}$ and (b) differential pulse voltammetry at $25\,mV\,s^{-1}$. $E_{1/2}$: -0.98, -1.37, -1.87, -2.35, -2.85 and $-3.26\,V$ versus Fc/Fc^+ [62].

If this occurs, the second step becomes less reversible or shifts to a more positive potential (see Figure 8.13). Occasionally, QH_2 is further reduced at more negative potentials. Moreover, the second step that appears at negative potentials is easily influenced by the cation of the supporting electrolyte (see footnote 12).

In contrast to the two one-electron steps in aprotic solvents, many organic compounds in water and other protic solvents give one two-electron reduction wave as shown in Figure 8.13, curve e.[7]

$$Q + 2H^+ + 2e^- \rightarrow QH_2$$

The transition from the mechanism in aprotic solvent to that in protic solvent is important and is discussed later.

Some organic compounds undergo multistep one-electron reversible reductions. As shown in Figure 8.14, Buckminsterfullerene (C_{60}) in AN/toluene at $-10\,°C$, for example, gives six one-electron reversible steps in cyclic voltammetry [62, 66]. C_{60} can accept up to six electrons to form diamagnetic C_{60}^{6-} because its LUMO is triply degenerated.[8]

---

7) Some organic compounds (Q) undergo two-step one-electron reduction ($Q \rightarrow Q^{\bullet-} \rightarrow Q^{2-}$) even in aqueous solutions. It occurs when $Q^{\bullet-}$ (semiquinone) is stable and its disproportionation ($2Q^{\bullet-} \rightarrow Q + Q^{2-}$) is difficult (see also footnote 10).

8) All of the six anions are stable on the voltammetric time scale but, if they are generated by controlled-potential electrolysis, only C_{60}^- to C_{60}^{4-} are stable [62a]. Neutral C_{60} is almost insoluble in such solvents as DMF, AN and THF, but its anions dissolve easily. Thus, these anions can be generated from a suspension of C_{60}.

Various factors give influences on the reductions of organic compounds in aprotic solvents. They are discussed below.

1. **Relation between the LUMO and the half-wave potential of the first reduction wave**
   When an organic compound, Q, is reduced, it accepts an electron from the electrode to its lowest unoccupied molecular orbital. Here, the energy of the LUMO of Q corresponds to its electron affinity (EA). If the energies of LUMO (ε_{lu}) for a series of analogous compounds are obtained by the molecular orbital method, there should be a relationship:

$$E_{1/2}(1) = E_{ref}(abs) - \varepsilon_{lu} - \Delta G^{\circ}_{sv}(Q/Q^{\bullet -})$$

where $E_{1/2}(1)$ is the half-wave potential of the first wave, $E_{ref}(abs)$ is the absolute potential of the reference electrode and $\Delta G^{\circ}_{sv}(Q/Q^{\bullet -}) = G^{\circ}_{sv}(Q^{\bullet -}) - G^{\circ}_{sv}(Q)$, i.e. the difference between the solvation energies of Q and $Q^{\bullet -}$.

Two classical results are shown in Figures 8.15 and 8.16. In Figure 8.15, Hoijink obtained near-linear relations between $E_{1/2}(1)$ and ε_{lu} for aromatic hydrocarbons (AHs). Similar near-linear relations have been observed for many organic compounds (quinones, nitorobenzenes, ketones and others). Figure 8.16 shows Peover's results [63]. He measured the spectra of the charge-transfer complexes of hexamethylbenzene (electron donor D) and various aromatic compounds (electron acceptors A) in dichloromethane and obtained the charge-transfer transition energies ($h\nu_{CT}$):

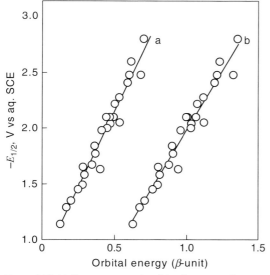

Figure 8.15 Half-wave potentials of the first wave of various conjugated hydrocarbons in 96% dioxane–water and the calculated values of the energy of LUMO obtained by (a) Hückel's and (b) Wheland's approximation. Plotted from the data in Hoijink, G.J. (1955) *Rec. Trav. Chim.*, **74**, 1525.

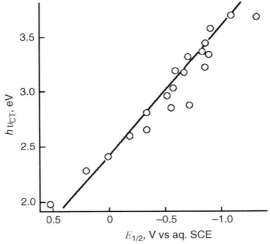

Figure 8.16 The relation between ν_{CT} of the spectra of charge-transfer complexes and the half-wave potentials of electron acceptors [63]. Electron acceptor: derivatives of phthalic anhydride, quinone and nitrobenzene and tetracyanobenzene and tetracyanoethylene.

$$D + A \xrightarrow{h\nu_{CT}} D^+ \cdot A^-$$

$h\nu_{CT}$ is correlated with the ionization potential (IP) of D and the electron affinity of A by

$$h\nu_{CT} = IP - EA + C$$

where C is a constant.

In Figure 8.16, the values of $h\nu_{CT}$ are plotted against the half-wave potentials of the acceptors measured in acetonitrile. A near-linear relation is observed. Peover also obtained, by an electron capture method, the values of EA of aromatic hydrocarbons in the gas phase and confirmed a near-linear relation with the reduction potentials in the solution phase.

It is certain that, in the first reduction step in aprotic solvents, an electron is accepted by the LUMO of the organic compound. However, it was fortunate that this conclusion was deduced from studies that either ignored the influence of solvation energies or used the results in different solvents. Shalev and Evans [67] estimated the values of $\Delta G_{sv}^{\circ}(Q/Q^{\bullet-})$ for 22 substituted nitrobenzenes and 9 quinones from the half-wave potentials measured by cyclic voltammetry. For quinones and some substituted nitrobenzenes, the values of $\Delta G_{sv}^{\circ}(Q/Q^{\bullet-})$ in a given solvent were nearly independent of the EA values. Similar results had been observed for alternant aromatic hydrocarbons (AAH) in AN (Section 8.3.2) [68]. If $\Delta G_{sv}^{\circ}(Q/Q^{\bullet-})$ does not vary with EA, there should be a linear relation of unit slope between $E_{1/2}$ and EA. Shalev and Evans [67], moreover, got a near-linear relation between $\Delta G_{sv}^{\circ}(Q/Q^{\bullet-})$ and EA for some other substituted nitroben-zenes. Here again, the $E_{1/2}$–EA relation should be linear, though the slope deviates from unity.[9]

2. **Effect of dynamic solvent property on the reduction kinetics** As described in Section 4.1.3, the outer-sphere reduction of large organic compounds is suitable to examine the role of dynamic solvent properties in determining the standard rate constant, k_s. Curve 2 in Figure 8.9 is for the reduction of benzophenone in various solvents; a near-linear relation of unit slope is observed between log k_s and log τ_L^{-1}, where τ_L is the longitudinal solvent relaxation time of the subpico- to picosecond region [69a]. For the one-electron reduction of p-dicyanobenzene, the relation between $\{\ln[k_s/(\Delta G_{os}^*)^{1/2}] + \Delta G_{os}^*/RT\}$ and $\ln \tau_L$ (see Eq. (4.11)) has been studied, as in the case of nickelocene in Section 8.2.2, and a roughly linear relation of unit slope has been obtained [35d].

3. **Effect of Lewis acidity of aprotic solvents on half-wave potentials** For an organic compound (Q) in dipolar aprotic solvents, the half-wave potential ($E_{1/2}$) of the first reduction step tends to shift to positive direction with an increase in solvent Lewis acidity (i.e. acceptor number) [69]. It is because, for the redox couple $Q/Q^{\bullet-}$, the reduced form ($Q^{\bullet-}$) is energetically more stabilized than the oxidized form (Q) with increasing solvent acidity. The positive shift in $E_{1/2}$ with solvent acceptor number has been observed with quinones [69c], benzophenone [69a,d] and anthracene [69d]. With fullerene (C_{60}), the positive shift in $E_{1/2}$ with solvent acidity parameter, E_T, has been observed for the reductions of C_{60} to C_{60}^-, C_{60}^- to C_{60}^{2-} and C_{60}^{2-} to C_{60}^{3-} [66c]. However, the positive shift in $E_{1/2}$ is not apparent if the charge in $Q^{\bullet-}$ is highly delocalized, as in the case of perylene and fluoren-9-one [69d].

4. **Effect of Brønsted acids on the reduction mechanisms** Brønsted acids (proton donors, HA) have significant influences on the reduction of organic compounds in aprotic solvents. If a weak Brønsted acid such as water is added step-wise to the electrolytic solution, the height of the first polarographic wave increases at the expense of that of the second wave (Figure 8.13). By the addition of a weak acid, following reactions occur at or near the electrode:

$$Q^{\bullet-} + HA \xrightarrow{k} QH^{\bullet} + A^-; \quad QH^{\bullet} + e^- \longrightarrow QH^-; \quad QH^- + HA \longrightarrow QH_2 + A^-$$

The QH$^{\bullet}$ radical formed by protonation is usually easier to reduce than Q and the first wave approaches a two-electron process. By theoretical analysis of the increase in the first-wave height, we can determine the rate constant (k) for the

---

9) Another interesting result obtained by Shalev and Evans [67] is that the rate constant for the homogeneous self-exchange electron transfer, log k_{ex}, between various Q and $Q^{\bullet-}$ in DMF decreases linearly with the increase in $\Delta G_{sv}^{\circ}(Q/Q^{\bullet-})$. This is because log k_{ex} is mainly governed by the solvent reorganization energy from the configuration around Q to that around $Q^{\bullet-}$, which is expected to increase with the increase in $\Delta G_{sv}^{\circ}(Q/Q^{\bullet-})$. Here, log k_{ex} can be determined from the broadening of the ESR signal of $Q^{\bullet-}$ in the presence of Q (Section 9.2.2).

protonation reaction [70]. However, the regeneration of Q by such reactions as $QH^{\bullet} + Q^{\bullet-} \rightleftarrows QH^- + Q$ and $Q^{\bullet-} + Q^{\bullet-} \rightleftarrows Q^{2-} + Q$ may also contribute to the exaltation of the first wave.

If the Brønsted acid is strong enough, the proton addition to $Q^{\bullet-}$ rapidly occurs and a new wave appears before the first wave because the ratio $[Q^{\bullet-}]/[Q]$ is kept very small. In some cases, QH^+ formed by the reaction of Q and H^+ gives a new wave at much more positive potential than the first wave. Inversely, if the Brønsted acid is very weak, it does not influence the first wave but shifts the second wave to positive direction, because the dianion Q^{2-} is a stronger base than $Q^{\bullet-}$ (Figure 8.13). Possible mechanisms for the reduction of an organic compound in the presence of a Brønsted acid are often represented by the 'square-scheme'. The mechanism actually occurring depends on the strength of the Brønsted acid and the base strength of Q, $Q^{\bullet-}$ and Q^{2-}.

$$
\begin{array}{ccc}
Q & \rightleftarrows\ Q^{\bullet-}\ \rightleftarrows & Q^{2-} \\
\updownarrow & \updownarrow & \updownarrow \\
QH^+ & \rightleftarrows\ QH^{\bullet}\ \rightleftarrows & QH^- \\
\updownarrow & \updownarrow & \updownarrow \\
QH_2^{2+} & \rightleftarrows\ QH_2^{\bullet+}\ \rightleftarrows & QH_2 \quad \text{"Square - Scheme"}
\end{array}
$$

By using Brønsted acids of known pK_a values and by examining their influence on the reduction of Q, we can get some qualitative information about the base strengths of $Q^{\bullet-}$ and Q^{2-}. Recently, some quantitative studies have been carried out on this problem. For example, Niyazymbetov *et al.* [66f] studied the pK_a values of dihydrofullerene, $C_{60}H_2$ and related species by combining cyclic voltammetry, controlled-potential electrolysis and acid–base titrations. They got (or estimated) in DMSO a pK_a of ~8.9 for $C_{60}H^{\bullet}$; pK_{a1} and pK_{a2} of 4.7 and 16 for $C_{60}H_2$; pK_{a1}, pK_{a2} and pK_{a3} of 9, ~9 and ~25 for $C_{60}H_3^{\bullet}$; and pK_{a2}, pK_{a3} and pK_{a4} of 16, ~16 and ~37 for $C_{60}H_4$. To obtain these estimates, they used a relation as in Eq. (8.4):

$$
pK_{a1,C_{60}H^{\bullet}} = pK_{a2,C_{60}H_2} + \frac{E^0_{2,C_{60}} - E^0_{1,C_{60}H^{\bullet}}}{0.059} \tag{8.4}
$$

where $pK_{a1,C_{60}H^{\bullet}}$ and $pK_{a2,C_{60}H_2}$ are for $C_{60}H^{\bullet} \rightleftarrows C_{60}^{\bullet-} + H^+$ and $C_{60}H^- \rightleftarrows C_{60}^{2-} H^+$, respectively, and $E^0_{1,C_{60}H^{\bullet}}$ and $E^0_{2,C_{60}}$ are for $C_{60}H^{\bullet} + e^- \rightleftarrows C_{60}H^-$ and $C_{60}^{\bullet-} + e^- \rightleftarrows C_{60}^{2-}$, respectively. On the other hand, Cliffel and Bard [66g] used voltammetry and near-IR spectrophotometric titrations to obtain $pK_a \sim 3.4$ for $C_{60}H^{\bullet}$ in *o*-dichlorobenzene.

If water is added to the solution of anthracene in DMF, the radical anion of anthracene is protonated as described above and the first wave in DC polarography is gradually converted to a two-electron process as in Figure 8.13. The influence of water on the reduction of anthraquinone is somewhat different from the case of anthracene [70]. By the addition of a large amount of water, the height of the first wave increases, while the second wave decreases in height and shifts to the more positive side; this is similar to the case of anthracene. However, in AC polarography, the two waves for anthraquinone are reversible and their reversibilities do not decrease by the addition of water; this shows that both $Q^{\bullet-}$ and Q^{2-} do not react with the proton from water. Moreover, according to the ESR measurement, $Q^{\bullet-}$ in the bulk of the solution does not disappear by the addition of water. If the exaltation of the first wave is due to the proton addition, $Q^{\bullet-}$ must disappear more rapidly. In order to explain these phenomena, the following disproportionation mechanism was proposed:

$$2Q^{\bullet-} \rightleftarrows Q + Q^{2-}$$

In anhydrous DMF, the equilibrium is almost completely to the left and the first wave is a one-electron process.[10] However, if water is added, the equilibrium slightly shifts to the right, as is apparent from the fact that the second wave approaches the first wave. This is because Q^{2-} interacts strongly with water by hydrogen bonding. As a result, some part of $Q^{\bullet-}$ formed at the first wave returns to Q and is reduced again at the electrode to exalt the first wave.

In some cases, the intramolecular hydrogen bonding stabilizes $Q^{\bullet-}$ and shifts the first wave to positive side. For example, the first wave of *o*-nitrophenol ($E_{1/2} = -0.33$ V) is much more positive than those of *m*- and *p*-nitrophenols ($E_{1/2} = -0.58$ V and -0.62 V) [70]. This is due to the intramolecular hydrogen bonding of the radical anion of *o*-nitrophenol.

5. **Effect of cations** A metal ion (Lewis acid) has an influence on the reductions of organic compounds in aprotic solvents. Namely, the metal ion, M^+, forms ion pairs with $Q^{\bullet-}$ and Q^{2-} and shifts the first and second waves, to a greater or lesser extent, to positive potentials [71]. This influence is pronounced in protophobic aprotic solvents (AN, PC, etc.), in which M^+ is solvated only weakly. The influence increases in the order $K^+ < Na^+ < Li^+$ and it also depends on the Lewis basicity of $Q^{\bullet-}$ and Q^{2-}. The example in Figure 8.17 is for 1,2-naphthoquinone in AN [71b]. Because the radical anion of 1,2-naphthoquinone interacts strongly with metal ions forming a chelate-like structure, the positive shifts of the first and second waves are marked and much larger than the case of 1,4-naphthoquinone.[11]

10) If the first and the second waves are well separated, the difference in their half-wave potentials, $\Delta E_{1/2}$, and the disproportionation constant, $K = [Q][Q^{2-}]/[Q^{\bullet-}]^2$, are related by $\Delta E_{1/2} = -(RT/F)\ln K$.

11) If the ion pair $Q^{\bullet-}-M^+$ is soluble, its formation constant, $K_A = [Q^{\bullet-} - M^+]/([Q^{\bullet-}][M^+])$, is related to the positive shift of the half-wave potential $\Delta E_{1/2}$ by $\Delta E_{1/2} = (RT/F)(\ln K + \ln [M^+])$, where $[M^+]$ is the concentration of free metal ion. The relation is used to determine the K_A value. However, it should be noted that the relation is not applicable if the ion pair is insoluble, as is often the case.

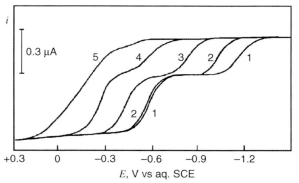

Figure 8.17 Influence of metal ions on the DC polarographic reduction wave of 1,2-naphthoquinone (0.5 mM) in AN–0.05 M Et_4NClO_4 [71b]. Metal ions (5 mM): curve 1, none; 2, K^+; 3, Na^+; 4, Li^+ and 5, Mg^{2+}.

The difference in the size of R_4N^+ ions also has considerable influences:

(i) R_4N^+ ions, which do not form stable ion pairs with $Q^{\bullet-}$, scarcely influence the potential of the first reduction wave. However, they influence the potential of the second wave; small R_4N^+ ions such as Me_4N^+ and Et_4N^+ tend to shift the second wave to positive direction forming ion pairs with Q^{2-}, while larger R_4N^+ ions such as Hex_4N^+ tend to shift the second wave to negative direction by the double-layer effect.[12)]

(ii) The influence of R_4N^+ on the kinetics of the one-electron reduction of organic compounds is worth noting [72]. Petersen and Evans [72a], for example, determined the standard rate constants (k_s) for the reductions of many organic compounds at a mercury electrode in AN, using Et_4NClO_4 and Hep_4NClO_4 as supporting electrolytes. As shown in Figure 8.18, for the reductions at negative potentials, the k_s value was larger with Et_4N^+ than with Hep_4N^+. Generally, the k_s value for the reduction at negative potentials decreases with increasing size of R_4N^+. Interestingly, however, the standard potential and the rate of homogeneous self-exchange electron-transfer reaction are not influenced by R_4N^+. Thus, the influence on log k_s has been attributed to the blocking effect of the compact layer of adsorbed R_4N^+ ions, whose thickness

---

12) The potential of the second wave of organic compounds and dissolved oxygen, $Q^{\bullet-} \rightarrow Q^{2-}$, are often influenced significantly by R_4N^+ of the supporting electrolyte. Small R_4N^+ ions, electrostatically attracted to the negatively charged electrode surface, seem to form ion pairs with Q^{2-} rather than with $Q^{\bullet-}$, causing a positive shift of the second wave. In contrast, with the increase in the size of R_4N^+, the distance of OHP from the electrode increases and makes the potential at the closest approach of $Q^{\bullet-}$ more negative, resulting in the more difficult formation of Q^{2-} and in a negative shift of the second wave. For the double-layer effect that considers the cationic size of the supporting electrolyte, see, for example, Fawcett, R.W. and Lasia, A. (1990) *J. Electroanal. Chem.*, **279**, 243.

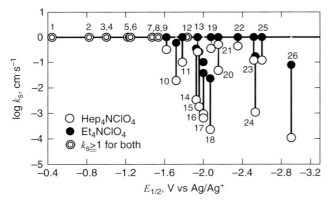

Figure 8.18 Relationship between the half-wave potential and the standard rate constant for the first reduction wave of various organic compounds in AN. The influence of R_4N^+ of the supporting electrolyte [72a]. Compounds: 1, hexafluorobiacetyl; 2, p-benzoquinone; 3, 1,4-naphthoquinone; 5, oxygen; 6, 9,10-anthraquinone; 8, p-nitrotoluene; 9, 4,4'-methoxybenzyl; 10, 2,3-butanedione; 11, nitromesitylene; 13, dicyanobenzene; 14, nitromethane; 15, nitroethane; 16, nitropropane; 18, cyclooctatetraene; 20, 4-cyanopyridine; 21, benzophenone; 22, 4,4'-dimethoxybenzophenone; 25, trans-stilbene; names are abbreviated for 4, 7, 12, 17, 19, 23, 24 and 26. The upper limit of determination of k_s was $1\,cm\,s^{-1}$ (i.e. $\log k_s \sim 0$).

increases with alkyl chain length. The electron transfer occurs by tunneling through the layer.[13] It has been reported that when large R_4N^+ ions such as Bu_4N^+, Hex_4N^+ and Oct_4N^+ are adsorbed onto the electrode surface, the three alkyl chains facing the electrode are bent and the charge centers of R_4N^+ are kept at ~0.37 nm from the surface, irrespective of the size of R [72c].

Dissociative Electron-Transfer Reactions
The reductions of halogenated organic compounds (RX) involve the cleavage of carbon–halogen bonds [73]. Depending on the solvent, supporting electrolyte, electrode material and potential, it is possible to electrogenerate either alkyl radicals ($R^\bullet$) or carbanions (R^-), which then can lead to the formation of dimers (R–R), alkanes (RH) and olefins [R(−H)]:

$$RX + e^- \rightleftarrows RX^{\bullet-}, \quad RX^{\bullet-} \xrightarrow{k} R^\bullet + X^-, \quad R^\bullet + e^- \to R^-$$
$$2R^\bullet \to R - R \,(\text{dimerization}) \quad \text{or} \quad [RH + R(-H)] \,(\text{disproportionation})$$
$$(R^\bullet \text{ or } R^-) + SH \to RH + (S^\bullet \text{ or } S^-) \quad (SH : \text{solvent})$$

13) It was shown in Ref. [72d] that the standard electron transfer rate constants ($\log k_s$) for each nitroalkane, $R'NO_2$, are linearly related to the hydrodynamic diameter of R_4N^+ ($d_{R_4N^+}$). Moreover, it was shown that the $\log k_s$ data for all nitroalkanes are in a single linear relation against ($d_{R_4N^+} + r_{R'}$), where $r_{R'}$ is roughly the radius of R' (Figure 8.19). The distance, ($d_{R_4N^+} + r_{R'}$), may have some relation to the tunneling of electrons from the electrode to the nitro group, although the smallest distance from the electrode to the nitro group is largely independent of the size of R'.

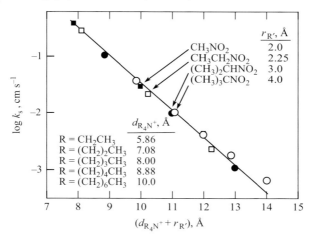

Figure 8.19 Standard electron transfer rate constants for four nitroalkanes, $R'NO_2$, obtained in $AN/0.10 M R_4NClO_4$ at a HMDE [72d]. Data are plotted versus the sum of the hydrodynamic diameter of R_4N^+ ($d_{R_4N^+}$) and the radius of R' ($r_{R'}$).

If a mercury electrode is used, $R^\bullet$ reacts with mercury to form organomercury compounds, whereas R^- tends to form such species as RH and R($-$H).

In aprotic solvents, the radical anion $RX^{\bullet-}$ for aryl halides has been detected as an intermediate. In cyclic voltammetry of aryl halides, though an irreversible two-electron reduction occurs at low scan rate, a reversible one-electron reduction occurs at high scan rate. Thus, it is possible to get the values of the standard potential (E^0) for the $RX/RX^{\bullet-}$ couple and the rate constant (k) for $RX^{\bullet-} \rightarrow R^\bullet$ (therefore, the lifetime of $RX^{\bullet-}$). In Figure 8.20, the relation between E^0 and log k for aryl bromides in DMF is linear with a slope of $\sim$0.5 [61f]. It is apparent that the lifetime of $RX^{\bullet-}$, obtained by $1/k$, increases with the positive shift of E^0. In contrast, the existence of $RX^{\bullet-}$ for alkyl monohalides has never been confirmed. With these compounds, it is difficult to say whether the two processes, i.e. electron transfer and bond cleavage, are step-wise or concerted ($RX + e^- \rightarrow R^\bullet + X^-$). According to Savéant [61f], the smaller the bond dissociation energy, the larger the tendency for the concerted mechanism to prevail over the step-wise mechanism.

Studying the electrochemical reduction of halogenated organic compounds has practical importance, especially for organic syntheses [73]. Moreover, the reductive cleavage of the C$-$X bond can be applied to convert hazardous chlorinated compounds; for example, polychlorinated biphenyl (PCB) is converted to biphenyl by the reduction in DMF at -2.8 V versus SCE [74].

8.3.2
Oxidation of Organic Compounds

As described in Section 4.1.2, nucleophilic organic compounds are oxidized at the electrode. Some oxidizable organic compounds are listed in Table 8.7 with the

Table 8.7 Examples of oxidizable organic compounds and the potentials of their first oxidation step in nonaqueous solvents[a].

Compounds	Solvent	Supporting electrolyte	Potential reference	Potential
Fullerene				
C_{60}	TCE	Bu_4NPF_6	Fc/Fc^+	+1.26
Hydrocarbons				
Benzene	AN	Bu_4NClO_4	SCE	+2.62
Toluene	AN	$NaClO_4$	Ag/Ag^+	+1.98
Naphthalene	AN	$NaClO_4$	Ag/Ag^+	+1.34
Anthracene	AN	$NaClO_4$	Ag/Ag^+	+0.84
9,10-Diphenylanthracene	AN	$NaClO_4$	Ag/Ag^+	+0.86
Biphenyl	AN	$NaClO_4$	Ag/Ag^+	+1.48
Ethylene	AN	Et_4NBF_4	Ag/Ag^+	+2.90
1-Alkenes	AN	Et_4NBF_4	Ag/Ag^+	+2.7–+2.8
2-Alkenes	AN	Et_4NBF_4	Ag/Ag^+	+2.2–+2.3
Cyclohexene	AN	Et_4NBF_4	Ag/Ag^+	+2.05
1,3-Butadiene	AN	Et_4NBF_4	Ag/Ag^+	+2.09
Nitrogen-containing compounds				
Diethylamine (Et_2NH)	AN	$NaClO_4$	Ag/Ag^+	+1.01
Triethylamine (Et_3N)	AN	$NaClO_4$	Ag/Ag^+	+0.66
Aniline ($PhNH_2$)	H_2O	buffer	SCE	+1.04
p-Phenylenediamine ($H_2NC_6H_4NH_2$)	AN	$NaClO_4$	Ag/Ag^+	+0.18
N,N-Dimethylformamide	AN	$NaClO_4$	Ag/Ag^+	+1.21
Pyrrole	AN	$NaClO_4$	Ag/Ag^+	+0.76
Pyridine	AN	$NaClO_4$	SCE	+1.82
Sulfur-containing compounds				
Thiophenol (PhSH)	CH_2Cl_2	CF_3COOH	Ag/Ag^+	+1.65
Dimethyl sulfide (MeSMe)	AN	$NaClO_4$	Ag/Ag^+	+1.41
Methyl phenyl sulfide (MeSPh)	AN	$NaClO_4$	Ag/Ag^+	+1.00
Diphenyl disulfide (PhSSPh)	AN/CH_2Cl_2	$LiClO_4$	Ag/Ag^+	+1.75
Thiophene	AN	$NaClO_4$	SCE	+2.10
Oxygen-containing compounds				
Water (H_2O)	AN	Et_4NClO_4	SCE	+2.8
Methanol (MeOH)	AN	Et_4NClO_4	SCE	+2.50
Phenol (PhOH)	AN	Et_4NClO_4	SCE	+1.55
Anisole ($PhOCH_3$)	AN	Et_4NClO_4	SCE	+1.75
4-Methylphenol ($4\text{-}CH_3PhOH$)	AN	Et_4NClO_4	SCE	+1.35
Benzyl alcohol ($PhCH_2OH$)	AN	Et_4NClO_4	SCE	+2.00

[a]From Refs [1, 54, 57–59].

potentials of the first oxidation step in nonaqueous solvents. By using a solvent of weak basicity and a supporting electrolyte that is difficult to oxidize, we can expand the potential window on the positive side and can measure oxidations of difficult-to-oxidize substances such as benzene. Anodic oxidations of organic compounds usually occur via complicated mechanisms but many of them are of practical importance. Various books and review articles on the electrolytic oxidation of organic

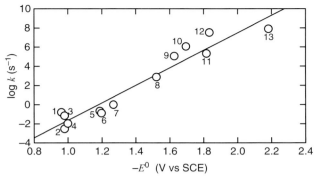

Figure 8.20 Cleavage rate constant k of aryl bromides as a function of their standard potential E^0 in DMF [61g]. 1, 2-isopropyl-4-nitrophenyl; 2, 4-nitrophenyl; 3, 2-methyl-3-nitrophenyl; 4, 2-methyl-4-nitrophenyl; 5, 3-fluorenyl; 6, 1-fluorenyl; 7, 2,6-dimethyl-4-nitrophenyl; 8, 3-benzoylphenyl; 9, 4-benzoylphenyl; 10, 9-anthracenyl; 11, 3-acetylphenyl; 12, 4-acetylphenyl; 13, 1-naphthyl.

compounds are available [54–57, 63]. Here, however, we focus our discussion only on the oxidation of C_{60} and AHs.

A reversible one-electron oxidation CV peak for C_{60} has been obtained at room temperature and at $100 \, \mathrm{mV \, s^{-1}}$ in 1,1,2,2-tetrachloroethane (TCE) containing Bu_4NPF_6 as supporting electrolyte. In other solvents such as AN and DCE, the oxidation was irreversible and contained multielectrons. The HOMO–LUMO energy gap for C_{60} in TCE was 2.32 V [62a].

In solvents of weak basicity (e.g. AN), aromatic hydrocarbons are oxidized, at least in principle, in two steps. In the first step, the compound gives an electron in its highest occupied molecular orbital to the electrode to form a radical cation $(AH^{\bullet +})$,[14] whereas, in the second step, it gives another electron to the electrode to form dication AH^{2+}.

$$AH \rightleftarrows AH^{\bullet +} + e^-$$
$$AH^{\bullet +} \rightleftarrows AH^{2+} + e^-$$

The two-step oxidation really occurs, for example, with 9,10-diphenylanthracene. As shown by the CV curve in Figure 8.21, the two waves are reversible or nearly reversible. The radical cation of 9,10-diphenylanthracene is fairly stable and, as in the case of radical anions, its ESR signals can be measured.

On the contrary, the radical cation of anthracene is unstable. Under normal voltammetric conditions, the radical cation $AH^{\bullet +}$ formed at the potential

14) The potential for the first oxidation step is linearly related to the donor number of the solvents [29a].

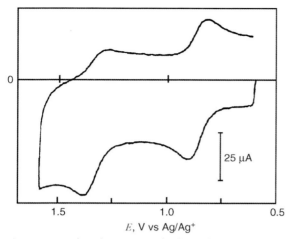

Figure 8.21 Cyclic voltammogram for the oxidation of 9,10-diphenylanthracene (1 mM) at a platinum microelectrode in 0.5 M Bu_4NClO_4–AN. Scan rate: $1000\,V\,s^{-1}$ [73].

of the first oxidation step, undergoes a series of reactions (chemical → electrochemical → chemical → ⋯) to form polymerized species. This occurs because the dimer, trimer, etc. formed from $AH^{\bullet\,+}$, are easier to oxidize than AH. As a result, the first oxidation wave of anthracene is irreversible and its voltammetric peak current corresponds to that of a process of several electrons (Figure 8.22a). However, if fast-scan cyclic voltammetry (FSCV) at an ultramicroelectrode (UME) is used, the effect of the follow-up reactions is removed and a reversible one-electron CV curve can be obtained (Figure 8.22b) [75]. By this method, the half-life of the radical cation of anthracene has been determined to be ∼90 µs.[15] Okazaki et al. [76a] used FSCV of up to $30\,000\,V\,s^{-1}$ at a UME to determine the oxidation and reduction potentials of 9-substituted anthracenes and correlated them with the results of the calculated energy levels of the HOMO and LUMO. Nozaki et al. [76b] used FSCV of up to $300\,000\,V\,s^{-1}$ at a UME to study the kinetics of the dimerization of 9-methoxy-anthracene cation radicals in nitroethane.

The radical cation of 9,10-diphenylanthracene is much more stable than that of anthracene because, with 9,10-diphenylanthracene, the 9- and 10-positions, which are reactive because of the high unpaired electron densities, are masked by phenyl groups and the unpaired electrons are delocalized. The stabilization of the radical

15) Useful methods for detecting short-lived cationic radicals are the fast-scan cyclic voltammetry at a UME (Section 8.4.2) and cyclic voltammetry at low temperatures (Section 8.4.3). It is preferable to prepare electrolytic solutions using solvents of low basicity (AN, DME, CH_2Cl_2, etc.) and to remove water almost completely with the aid of active alumina (Sections 8.1.2 and 10.2) [68].

Table 8.8 Oxidation potentials, reduction potentials, ionization potentials and electron affinities of AAHs[a].

Compound	E_{red}, V	E_{ox}, V	M_{AAH}, V	E_M, V	IP, eV	EA, eV	ΔG°_{sv}, eV
Benzene		2.79,i		(3.10)	9.37		
Naphthalene	−2.50,R	1.84,i	−0.33	(2.17)	8.54	0.074	−2.05
Anthracene	−1.97,R	1.37,r	−0.30	1.67	7.89	0.653	−1.99
Phenanthrene	−2.49,R	1.83,r	−0.33	2.16	8.43	0.273	−1.92
Benzanthracene	−2.02,R	1.44,r	−0.29	1.73	8.04	0.640	−1.97
Triphenylene	−2.48,R	1.88,i	−0.30	(2.17)	8.45	0.251	−1.93
Chrysene	−2.27,R	1.64,r	−0.32	1.95	8.19	0.516	−1.89
Perylene	−1.66,R	1.06,R	−0.30	1.37	7.72	0.956	−2.01
Benzoperylene	−1.91,R	1.35,R	−0.28	1.63	7.85	0.779	−1.91
Pyrene	−2.04,R	1.36,R	−0.34	1.70	7.95	0.664	−1.95
Benzopyrene	−1.84,R	1.16,R	−0.34	1.50	7.75	0.930	−1.91
Mean value			−0.31		(IP + EA) = 8.68		−1.94

[a]From Ref. [68]; the potentials are values in AN and against aq. SCE. $E_M = E_{Ox} - M_{AAH} = M_{AAH} - E_{Red}$. R: reversible; i: irreversible; r: reversible at high potential scan-rate.

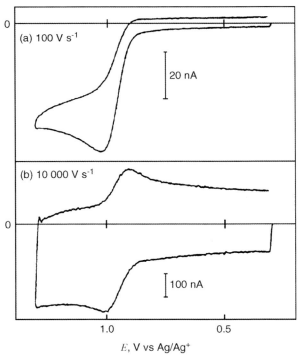

Figure 8.22 Effect of scan rate on the cyclic voltammogram for the oxidation of anthracene (2.4 mM) at a platinum ultramicroelectrode in 0.6 M Et$_4$NClO$_4$–AN. Electrode radius: 5 μm [73].

cation of anthracene also occurs by introducing other substituents such as $-NH_2$ and $-OCH_3$ at positions 9 and 10.

If the solution contains a nucleophile (Nu:$^-$) or a base (B) or consists of a basic solvent, the radical cation (AH$^{\bullet+}$) formed at the first oxidation step reacts as follows [54]:

$$AH^{\bullet+} + Nu:^- \rightarrow H-A^{\bullet} - Nu, \quad AH^{\bullet+} + B \rightarrow A^{\bullet} + BH^+$$

The products of the above two reactions, $H-A^{\bullet}-Nu$ and $A^{\bullet}$, are easier to oxidize than AH and are oxidized to form $H-A^+-Nu$ and A^+, which chemically react with Nu:$^-$ and B to form electrically neutral final products. Thus, in the presence of a nucleophile or a base, aromatic hydrocarbons give a one-step two-electron oxidation wave.[16]

Parker [68] obtained in acetonitrile the oxidation and reduction potentials (E_{ox} and E_{red}) of alternant aromatic hydrocarbons by cyclic voltammetry and examined how those potentials are related with the ionization potential and the electron affinity of the compounds (Table 8.8). As expected, he found linear relations of unit slopes between E_{ox} and IP and between E_{red} and EA. Moreover, he found that E_{ox} and E_{red} of each AAH were symmetrical with respect to a common potential M_{AAH} (-0.31 V versus SCE). The values of ($E_{ox} - M_{AAH}$) and ($E_{red} - M_{AAH}$) are correlated with the values of IP and EA, obtained in the vacuum, by $E_{ox} - M_{AAH} = IP - \phi + \Delta G^{\circ}_{sv+}$ and $E_{red} - M_{AAH} = EA - \phi - \Delta G^{\circ}_{sv-}$, respectively (Figure 8.23). Here, ϕ is the work function of graphite and is equal to 4.34 eV, and ΔG°_{sv+} and ΔG°_{sv-} are the differences in solvation energies for the $0/+1$ and $0/-1$ couples of AAH. Experimentally, ΔG°_{sv+} and ΔG°_{sv-} were almost equal, not depending on the species of AAH, and were equal to -1.94 eV in AN.

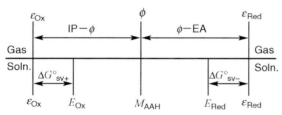

Figure 8.23 Relationship between the oxidation potential (E_{Ox}), the reduction potential (E_{Red}), the ionization potential (IP), the electron affinity (EA) and the solvation energies (ΔG°_{sv+}, ΔG°_{sv-}) for alternant aromatic hydrocarbons [68].

16) The anodic partial fluorination of organic compounds occurs in weak Lewis base solvents (AN, CH_2Cl_2, DME) containing fluoride ions (QH $\xrightarrow{-e^-}$ QH$^{\bullet+}$ $\xrightarrow{-e^- -H^+}{\xrightarrow{F^-}}$ Q-F or QH $\xrightarrow{-2e^- -H^+}$ Q$^+$ $\xrightarrow{F^-}$ Q-F). This occurs because F$^-$ in aprotic solvents is extremely nucleophilic (Table 2.7). In contrast to chemical fluorination that uses hazardous reagents and to anodic fluorination in anhydrous HF, this method can be carried out under mild conditions and without using hazardous reagents. For details, see Ref. [54a], Chapter 25.

There is a relation $\Delta E \equiv E_{red} - E_{ox} = EA - IP - 2\,\Delta G^\circ_{sv}$. Here, we imagine an AAH with $IP = 6.26$ eV and $EA = 2.38$ eV. Then, we get $\Delta E = 0$, and 50% of the AAH dissolved in AN is expected to ionize by disproportionation ($2AAH \rightleftarrows AAH^+ + AAH^-$). This ionization is motivated by the ion solvation energies, in contrast to the fact that in the solid state, electrical conductivity is obtained only with graphite, for which $IP = EA = 4.34$ eV.

8.4
Cyclic Voltammetry for Electrochemical Studies in Nonaqueous Solutions

Cyclic voltammetry is one of the most useful techniques for studying chemistry in nonaqueous solutions. It is especially useful in studying electrode reactions that involve an unstable intermediate or product. By analyzing cyclic voltammograms, we can elucidate the reaction mechanisms and can know the thermodynamic and kinetic properties of the unstable species. Some applications were described in previous sections. Much literature is available on cyclic voltammetry dealing with the theories and practical methods of measurements and data analyses [3–5, 77]. In this section, three useful cyclic voltammetry techniques are outlined.

8.4.1
Digital Simulation in Cyclic Voltammetry

For many inorganic and organic substances, it is rare that the electrode reaction is simply an electron transfer at the electrode surface. In most cases, the electron transfer process is accompanied by preceding and/or following reactions, which are either chemical or electrochemical. For example, for the electrode reduction of substance A, mechanisms described here can be considered.

E mechanism: $A + e^- \rightarrow B$

EE mechanism: $A + e^- \rightarrow B$, $B + e^- \rightarrow C$

EEE mechanism: $A + e^- \rightarrow B$, $B + e^- \rightarrow C$, $C + e^- \rightarrow D$, ...

EC mechanism: $A + e^- \rightarrow B$, $B \rightarrow C$ (or $2B \rightarrow C$, etc.)

ECE mechanism: $A + e^- \rightarrow B$, $B \rightarrow C$ (or $2B \rightarrow C$, etc.), $C + e^- \rightarrow D$

EEdisp mechanism: $A + e^- \rightarrow B$, $B + e^- \rightarrow C$, $2B \rightarrow A + C$

(E = electrochemical reaction, C = chemical reaction and disp = disproportionation)

Moreover, each of the chemical and electrochemical reactions can have different reaction rates and reversibilities. All of them are reflected in cyclic voltammograms [3–5]. If we measure cyclic voltammograms of an electrode reaction, changing parameters such as potential range, voltage scan rate, temperature, electrode material and solution composition and analyze the voltammograms appropriately, we can get various information about the electrode reaction. However, except the cases where the electrode process is very simple, it is not an easy matter to analyze the cyclic voltammograms appropriately.

Digital simulation software, which is now commercially available, is useful in analyzing cyclic voltammograms of complicated electrode reactions [78]. If we assume a possible reaction mechanism and can get simulated CV curves that fit the experimental CV curves, we can confirm the reaction mechanism and obtain thermodynamic and kinetic parameters concerning the electron transfer and chemical processes. By the development of simulation softwares, cyclic voltammetry has become a very powerful technique. On the contrary, without a simulation software, cyclic voltammetry is not as convenient.[17]

8.4.2
Ultramicroelectrodes in Cyclic Voltammetry

The UME described in Section 5.54 displays its ability when used in nonaqueous solutions [79].

1. Because the current that flows at a UME is extremely small (10 pA–100 nA), we can keep the effect of the iR-drop very small and can measure voltammograms even in solutions of high resistance, i.e. in solutions without supporting electrolyte [80] and in solutions of nonpolar solvents such as benzene, toluene and hexane [81]. UMEs have also been used for voltammetry in the gaseous phase [82] and in supercritical fluids [83].

2. Because the current that flows at a UME is small, we can scan the electrode potential at extremely high rates, sometimes reaching 1 000 000 V s^{-1} or even more [84]. By using cyclic voltammetry at such fast scan rates, we can detect intermediates and products with very short lifetimes and study their thermodynamic and kinetic properties [61e, 77c]. We can also determine the rate constants of very rapid electrode reactions, as shown in Table 8.9.

UMEs of ~10 μm in diameter and voltammetric instruments for use with such UMEs are commercially available. Electrodes of smaller dimensions can be prepared

17) *Determination of homogeneous electron transfer rate constants by cyclic voltammetry:* Evans and Gilicinski [72b] used cyclic voltammetry assisted by a simulation method to determine the rate constants for a homogeneous electron-transfer reaction $Ox_1 + Red_2 \underset{k_b}{\overset{k_f}{\rightleftarrows}} Red_1 + Ox_2$.

They measured cyclic voltammograms for the mixtures of Ox_1 and Ox_2. If the above reaction does not occur, the reduction–reoxidation peaks for Ox_1 and Ox_2 should be obtained. However, if the above reaction occurs in the solution near the electrode, the cyclic voltammograms are distorted. Under appropriate conditions, it is possible to determine the homogeneous electron-transfer rate constants, k_f and k_b, by simulating the cyclic voltammograms. They employed this method for $Ox_1 = RNO_2$ (R: = Me, Et, i-Pr and t-Bu) and $Ox_2 =$ terephthalonitrile.

Table 8.9 Examples of the rate constants of electrode reactions and subsequent reactions determined by rapid-scan cyclic voltammetry [61e, 77e].

Electroactive substance	Solvent	Scan rate, V s^{-1}	Electrode, μmφ	k_s, cm s^{-1}[a]
Anthracene	AN	5×10^2–10^4	Au (6.5)	3.5 ± 0.6
		10^2–10^3	Pt (10)	2.6
		2×10^4–2×10^6	Au (5)	3.8 ± 0.5
	DMF	2×10^4–3×10^5	Au (5)	3.3 ± 0.2
Anthraquinone	AN	5×10^2–10^4	Au (6.5)	1.8 ± 0.4
		10^2–10^3	Pt (10)	1.5
Naphthoquinone	AN	5×10^2–10^4	Au (6.5)	0.7 ± 0.1
		5×10^2–10^4	Pt (5)	0.6 ± 0.1
Benzoquinone	AN	5×10^2–10^4	Au (6.5)	0.4 ± 0.1
		5×10^2–10^4	Pt (5)	0.2 ± 0.05
Ferrocene	DMF	5×10^2–10^4	Pt (5)	1 ± 0.4
		5×10^2–10^4	Hg (5.5)	1 ± 0.6
		5×10^2–10^4	Au (5)	3 ± 1
Ru(bpy)$_3$ $(2+/+)$	DMF	5×10^2–10^4	Au (5)	2.5
Ru(bpy)$_3$ $(+/0)$	DMF	5×10^2–10^4	Au (5)	3.5

Reaction	Scan rate, V s^{-1}	Electrode, μmφ	Lifetime, s^b
Dissociation of RX$^{\bullet-}$ anion radical in DMF^c	10^5–4×10^5	Au (5)	10^{-5}–10^{-6}
Dissociation of RX$^{\bullet-}$ anion radical in AN^c	10^3–5×10^5	Au, Pt (5, 10)	10^{-1}–2×10^{-6}
Oxidation of pyrrole in AN (first step of electrolytic polymerization)	10^3–2×10^4	Pt (5, 10)	2×10^{-3}–3×10^{-5}

^{a} Rate constant at the standard potential.
^{b} Reciprocal of the rate constant (k) of RX$^{\bullet-}$ $\rightarrow$ R$^{\bullet}$ + X$^-$.
^{c} RX (haloallene) + e$^-$ $\rightleftharpoons$ RX$^{\bullet-}$ (anion radical), RX$^{\bullet-}$ $\xrightarrow{k}$ R$^{\bullet}$ + X$^-$, R$^{\bullet}$ + SH (solvent) $\rightarrow$ RH + S$^{\bullet}$.

in the laboratory, though it needs considerable skills [85]. In order to use UMEs successfully for high-speed voltammetry in highly resistive solutions, care must be taken concerning the effects of the ohmic drop and the capacitance of the cell system [76b, 85, 86]. Moreover, two types of voltammograms, i.e. curves (a) and (b) in Figure 5.23, should be used appropriately according to the objective of the measurements. For practical experimental techniques, see Refs [76, 78–84]. Some recent topics related to UMEs are described in footnotes.[18),19),20)]

8.4.3
Low-Temperature Electrochemistry and Cyclic Voltammetry [87, 88]

The low-temperature electrochemistry technique is useful in studying electrode reactions involving unstable products or intermediates. Lowering the temperature by 30–40 °C decreases the reaction rate of the unstable species to one-tenth of the original value. It is equivalent to a 10-fold increase in the voltage scan rate. Figure 8.24 shows the effect of temperature on the cyclic voltammogram for the oxidation of

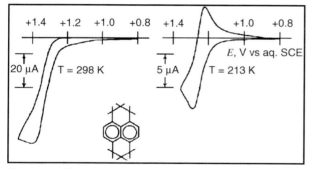

Figure 8.24 Effect of temperature on the cyclic voltammogram for the oxidation of 1,2,3,6,7,8-hexahydropyrene (2.5 mM) at a platinum electrode in 0.1 M Bu_4NClO_4–butyronitrile. Scan rate: 50 mV s^{-1} [87a].

18) In AN without supporting electrolyte, the positive end of the potential window reaches + 6 V versus Ag/Ag^+, and the electrolytic oxidations of the substances with high ionization potentials, including methane, butane, pentane, heptane, rare gases (Ar, Kr, Xe) and oxygen, can be observed [80a]. In SO_2 at −70 °C and without supporting electrolyte, Cs^+, Rb^+, K^+ and Na^+ were oxidized at a platinum UME (10 μm diameter) [80d].

19) The *infinite dilution* half-wave potential of 7,7,8,8-tetracyanoquinodimethane (TCNQ) versus Fc/Fc^+ was obtained by measuring

steady-state UME voltammograms (Figure 5.23b) in AN with 1–70 mM Me_4NPF_6 and by extrapolating the half-wave potential to zero ionic strength [80c].

20) Ultramicroelectrodes of extremely small radius (1–2 nm, called nanodes or nanoelectrodes) [89] are suitable for determining the rate constants of very fast electrode reactions. For example, the standard rate constant, k_s, has been determined to be 220 ± 120 cm s^{-1} for the ferrocenium/ferrocene couple in AN–0.3 M Bu_4NClO_4.

Table 8.10 Solvent-supporting electrolyte couples for low-temperature electrochemistry[a].

Solvent[b]	Fp, °C	Supporting electrolyte	Lowest temperature, °C
Ac	−94.7	0.3 M Et_4NPF_6	−75
AN	−43.8	0.1 M Bu_4NClO_4	−45
BuN	−111.9	0.3 M Bu_4NClO_4	−75
CH_2Cl_2	−94.9	0.5 M Bu_4NClO_4; 0.1M Bu_4NPF_6	−90
DMF	−60.4	0.6 M Bu_4NClO_4; 0.6M Bu_4NPF_6	−60
EtOH	−114.5	0.5 M $LiClO_4$	−103
PrN	−92.8	0.1 M Bu_4NClO_4	−100
THF	−108.4	0.2 M Bu_4NClO_4	−78
BuN/C_2H_5Cl (1 : 1 by volume)		0.2 M Bu_4NClO_4	−185
DMF/toluene (2 : 3 by volume)		0.1 M Et_4NPF_6	−88
$C_2H_5Cl/THF/2$-MeTHF (2 : 0.88 : 0.12 by volume)		0.6 M $LiBF_4$	−173

[a]Prepared from the data in Table 16.2 of Ref. [88].
[b]For abbreviated symbols, see Table 1.1.

1,2,3,6,7,8-hexahydropyrene. At ambient temperatures, it does not give a re-reduction peak. However, at −60 °C, reversible oxidation and re-reduction waves are observed. Low-temperature electrochemistry techniques are often used for studying electrode reactions of organic compounds and metal complexes. Examples of the solvent-supporting electrolyte systems for low-temperature electrochemistry are listed in Table 8.10. Some specific situations occur for voltammetry at low temperatures: increase in solution resistance, decrease in the rate of diffusion of electroactive species, decrease in the Nernstian slope, etc. Appropriate care must be taken for these. The use of a UME is advantageous in overcoming the influence of solution resistance. The reference electrode is usually kept at a constant temperature (e.g. 25 °C), while the temperature of the indicator electrode is varied at much lower temperature ranges. Fortunately, the thermal potential developed between the solution of the reference electrode and that containing the indicator electrode is negligibly small. For the practical aspects of electrochemistry at reduced temperatures, the review by Evans and Lerke [88] is useful.

8.5
Voltammetry of Isolated Nanoparticle Solutions (Nanoelectrochemistry)

Voltammetry in solutions of metal nanoparticles of various sizes is the subject of research and has been reviewed in detail [90]. Here, we consider a typical case of Au nanoparticles stabilized by organothiolate ligand monolayers: this kind of nanoparticle is called *monolayer-protected clusters* (MPCs).

The MPCs are synthesized as follows: Chloroaurate is phase-transferred, using transfer reagent $Oct_4N^+Br^-$, from water to toluene, where organothiol is added and reduces Au^{III} to a colorless Au^I thiolate complex that is not isolated.

$$[\text{Oct}_4\text{N}^+ + \text{AuCl}_4^-]_{\text{water}}$$
$$\rightarrow [\text{Oct}_4\text{N}^+ + \text{AuCl}_4^-]_{\text{toluene}} \xrightarrow{\text{RSH}} [\text{Oct}_4\text{N}^+ + \text{Au}^I\text{SR} + \text{RSSR}]_{\text{toluene}}$$

A reducing agent, BH_4^-, is then added and Au nanoparticles are formed. By selecting the thiol: Au ratio, reaction temperature, etc. we can get MPCs of sizes from 1 to 5 nm, though the raw product is their mixture (polydispersity). The nature of organothiolate dominates the MPC solubilities: those based on alkanethiolate or arylthiolate are soluble in moderately polar to nonpolar media but not in water. The MPC is stable and the raw product can be fractionated into monodisperse samples. Here, partial replacement of thiolate by other functional ligand is also possible.

Three types of voltammetry exist by the size of MPCs: they are *bulk continuum voltammetry, quantized double-layer charging voltammetry* and *voltammetry of molecule-like nanoparticles* [90]. This metal-to-molecule transition of voltammetry is summarized in Figure 8.25.

Bulk continuum voltammetry is for large nanoparticles of >3–4 nm sizes. The relation that distinguishes such nanoparticles from even smaller ones is $\Delta V = ze/C_{\text{CLU}}$, where ΔV is the change in the electrochemical potential of a nanoparticle with a double-layer capacitance C_{CLU}, that is incurred upon transfer of z electrons to/from the nanoparticle. The subscript 'CLU' means an individual metal cluster. If $\Delta V \leq k_B T_{25} (= 25.7\,\text{meV})$, successive electron transfer to/from the nanoparticle will result in a continuum – as opposed to step-wise – change in the nanoparticle's potential. The bulk continuum behavior or voltammetry of smooth current–potential relation is expected for C_{CLU} larger than about 6 aF.

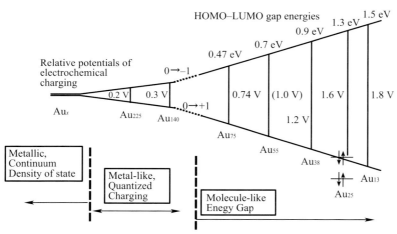

Figure 8.25 Summary of electrochemical behavior by MPC core size, showing estimated energy gaps (HOMO–LUMO gaps) and electrochemical energy gaps (which is the spacing between the first oxidation peak and the first reduction peak for the native nanoparticle) [90].

In quantized double-layer charging voltammetry, the MPCs with smaller sizes (diameter: ~ 2.0 nm for Au_{225} and 1.6 nm for Au_{140}) behave as quantum capacitors. Figure 8.26 (upper) is an example of this type of voltammogram [91]: differential pulse voltammogram (DVP) was measured for hexanethiolate-coated Au_{147} MPCs (177 µM) dissolved in 10 mM bis(triphenylphosphoranylidene) ammonium tetrakis (pentafluorophenyl)borate ($BTPPATPBF_{20}$)–1,2-dichloroethane (DCE) solution. A two-electrode arrangement, a Pt microelectrode as working electrode and a silver wire as both counter and quasireference electrode, was used. Evenly spaced 15 peaks (0.25 V in average space, corresponding to $C_{CLU} = 0.64$ aF) were observed.

By further decrease in nanoparticle size (e.g. ~ 1 nm for Au_{25}), a molecule-like phenomenon appears and an energy gap is observed in the voltammogram. Figure 8.26 (lower) is a voltammogram for 170 µM hexanethiolate-coated Au_{38} in the same solution as in Figure 8.26 (upper). If the charging energy (0.3 V; estimated as the difference between the two oxidation peaks) is corrected for from 1.2 V, which is the gap in peak potentials between the first oxidation (1/0) and the first reduction (0/−1), the HOMO–LUMO gap of 0.9 V is obtained. Though under limited conditions, the monolayer of thiolate ligand of the MPCs works only to stabilize the Au core against aggregation. For the voltammetric behavior at the functionalized MPCs, see Ref. [90].

Nanoelectrochemistry is a branch of electrochemistry that deals with electrochemical phenomena of materials at nanometer size levels. This includes, in addition to voltammetry of solutions of isolated nanoparticle, electrochemistry at nanoscopic electrodes (nanoelectrode), electrochemistry of films of nanoparticles

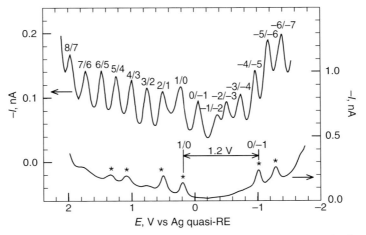

Figure 8.26 Differential pulse voltammetric responses for MPC solutions measured at a Pt microelectrode; as-prepared 177 µM hexanethiolate(C6S)-coated Au_{147} (upper) showing 15 high-resolution quantized double-layer (QDL) peaks and 170 µM C6S-Au_{38} (lower) showing a HOMO–LUMO gap [91]. It can be seen that the as-prepared solution contains a residual fraction of Au_{38} that smears out the charging response at potentials where QDL peaks overlap. The electrode potential was scanned from negative to positive. For the solution and electrodes, see text.

and electrochemistry of single nanopores in membranes [90]. Nanoelectrodes are the electrodes with submicrometer (nanometer) dimensions and useful to investigate the transition from micro-scale to molecular- or atomic-scale of various electrochemical phenomena, including electrodeposition and corrosion. The nanoelectrodes dealt with in footnote 20 of this chapter and in Figure 9.11, which is for single-molecule electrochemistry, are the examples of such electrodes. For the electrochemistry of films of nanoparticles and that of single nanopores in membranes, see Ref. [90].

References

1 Sawyer, D.T., Sobkowiak, A. and Roberts, J.L., Jr (1995) *Electrochemistry for Chemists*, 2nd edn, John Wiley & Sons, Inc., New York.

2 Bard, A.J. and Faulkner, L.R. (2001) *Electrochemical Methods, Fundamentals and Applications*, 2nd edn, John Wiley & Sons, Inc., New York.

3 Bond, A.M. (2002) *Broadening Electrochemical Horizons, Principles and Illustration of Voltammetric and Related Techniques*, Oxford University Press, Oxford.

4 Savéant, J.-M. (2006) *Elements of Molecular and Biomolecular Electrochemistry, An Electrochemical Approach to Electron Transfer Chemistry*, John Wiley & Sons, Inc., NJ.

5 Compton, R.G. and Banks, C.E. (2007) *Understanding Voltammetry*, World Scientific, London.

6 Mann, C.K. (1969) *Electroanalytical Chemistry*, vol. 3 (ed. A.J. Bard), Marcel Dekker, New York, p. 57.

7 Bauer, D. and Breant, M. (1975) *Electroanalytical Chemistry*, vol. 8 (ed. A.J. Bard), Marcel Dekker, New York, p. 281.

8 Badoz-Lambling, J. and Cauquis, G. (1974) *Electroanalytical Chemistry* (ed. H.W. Nürnberg), John Wiley & Sons, Inc., New York, p. 335.

9 (a) Lund, H. (2001) *Organic Electrochemistry*, 4th edn (eds H. Lund and O. Hammerich), Marcel Dekker, New York, Chapter 5; (b) Lund, H. (1991) *Organic Electrochemistry*, 3rd edn (eds H.

Lund and M.M. Baizer) Marcel Dekker, New York, Chapter 6.

10 Kissinger, P.T. and Heinemann, W.R. (eds) (1997) *Laboratory Techniques in Electroanalytical Chemistry* 2nd edn, Marcel Dekker, New York.

11 Aurbach, D. (ed.) (1999) *Non-aqueous Electrochemistry*, Marcel Dekker, New York.

12 Horrocks, B.R. (2003) *Encyclopedia in Electrochemistry: vol. 3, Instrumentation and Electroanalytical Chemistry* (eds A.J. Bard, M. Stratmann and P.R. Unwin), Wiley-VCH Verlag GmbH, Weinheim.

13 (a) Alehasham, S., Chambers, F., Strojek, J.W., Swain, G.M. and Ramesham, R. (1995) *Anal. Chem.*, **67**, 2812; (b) Wu, Z., Yano, T., Tryk, D.A., Hashimoto, K. and Fujishima, A. (1998) *Chem. Lett.*, 503; (c) Fujishima, A., Einaga, Y., Rao, T.N. and Tryk, D.A. (eds) (2005) *Diamond Electrochemistry*, Elsevier, Amsterdam.

14 Butler, J.N. (1970) *Advances in Electrochemistry and Electrochemical Engineering*, vol. 7 (eds P. Delahay and C.W. Tobias), Interscience Publishers, New York, pp. 77–175.

15 Gritzner, G. and Kuta, J. (1984) *Pure Appl. Chem.*, **56**, 461.

16 Ref. [1], pp. 197, 199.

17 Ref. [1], pp. 249–286.

18 Katovic, V., May, M.A. and Keszthelyi, C.P., in Ref. [10], Chapter 18.

19 Frank, S.N. and Park, S.-M., in Ref. [10], Chapter 19.

20 (a) Lines, R., Jensen, B.S. and Parker, V.D. (1978) *Acta Chem. Scand.*, **B32**, 510;

(b) Jensen, B.S. and Parker, V.D. (1975) *J. Am. Chem. Soc.*, **97**, 5211.

21 Galus, Z. (1984) *Pure Appl. Chem.*, **56**, 635.

22 For example, (a) Mann, C.K. and Barnes, K.K. (1970) *Electrochemical Reactions in Non-aqueous Systems*, Marcel Dekker, New York, Chapter 14; (b) Meites, L. and Zuman, P. (eds) (1980–1988) *CRC Handbook Series in Inorganic Electrochemistry*, vols. I–VIII, CRC Press, Boca Raton, FL; (c) Bard, A.J. (ed.) (1973–1980) *Encyclopedia of Electrochemistry of the Elements*, vols. I–XIV (Inorganic Section), Marcel Dekker, New York; (d) Reports from IUPAC Commission on Electroanalytical Chemistry, *Pure Appl. Chem.*, 1977, **49**, 217, 877; 1983, **55**, 1373.

23 Gritzner, G. (1990) *Pure Appl. Chem.*, **62**, 1839.

24 Gutmann, V. (1978) *The Donor–Acceptor Approach to Molecular Interactions*, Plenum Press, New York, p. 121.

25 Gritzner, G. (1986) *J. Phys. Chem.*, **90**, 5478.

26 Gritzner, G. and Hörzenberger, F. (1992) *J. Chem. Soc., Faraday Trans.*, **88**, 3013; 1995, **91**, 3843; 1996, **92**, 1083.

27 (a) Baranski, A.S. and Fawcett, W.R. (1978) *J. Electroanal. Chem.*, **94**, 237; (b) Baranski, A.S. and Fawcett, W.R. (1980) *J. Chem. Soc., Faraday Trans. I*, **76**, 1962.

28 (a) Galus, Z. (1995) Electrochemical reactions in non-aqueous and mixed solvents, in *Advances in Electrochemical Science and Engineering*, vol. 4 (eds H. Gerischer and C.W. Tobias), John Wiley & Sons, Inc., New York, pp. 217–295; (b) Galus, Z. (1991) *Pure Appl. Chem.*, **63**, 1705.

29 (a) Fawcett, W.R. (1989) *Langmuir*, **5**, 661; (b) Fawcett, W.R. (1989) *J. Phys. Chem.*, **93**, 267.

30 (a) Izutsu, K., Sakura, S., Kuroki, K. and Fujinaga, T. (1971) *J. Electroanal. Chem.*, **32**, app.11; (b) Izutsu, K., Sakura, S. and Fujinaga, T. (1972) *Bull. Chem. Soc. Jpn.*, **45**, 445; 1973, **46**, 493, 2148.

31 Kolthoff, I.M. and Thomas, F.G. (1964) *J. Electrochem. Soc.*, **111**, 1065.

32 Ritzler, G., Peter, F. and Gross, M. (1983) *J. Electroanal. Chem.*, **146**, 285.

33 (a) Diggle, J.W. and Parker, A.J. (1973) *Electrochim. Acta*, **18**, 975; (b) Coetzee, J.F. and Istone, W.K. (1980) *Anal. Chem.*, **52**, 53.

34 (a) Stojanovic, R.S. and Bond, A.M. (1993) *Anal. Chem.*, **65**, 56; (b) Krishtalik, L.I., Alpatova, N.M. and Ovsyannikova, E.V. (1991) *Electrochim. Acta*, **36**, 435.

35 (a) Fawcett, W.R. and Foss, C.A. (1991) *J. Electroanal. Chem.*, **306**, 71; (b) Fawcett, W.R. and Foss, C.A. (1991) *Electrochim. Acta*, **36**, 71; (c) Fawcett, W.R. and Opallo, M. (1992) *J. Phys. Chem.*, **96**, 2920; (d) Winkler, K., Baranski, A. and Fawcett, W.R. (1996) *J. Chem. Soc., Faraday Trans.*, **92**, 3899; (e) Zhang, X., Leddy, J. and Bard, A.J. (1985) *J. Am. Chem. Soc.*, **107**, 3719; (f) Winkler, K., McKnight, N. and Fawcett, W.R. (2000) *J. Phys. Chem. B*, **104**, 3575.

36 (a) Kadish, K.M. (1986) *Progress in Inorganic Chemistry*, vol. 34, John Wiley & Sons, Inc., New York, pp. 435–605; (b) Walder, L. (1991) *Organic Electrochemistry* (eds H. Lund and M.M. Baizer), Marcel Dekker, New York, Chapter 21; (c) Ref. [1], Chapter 13; (d) Kadish, K.M. and Caemelbecke, E.V. (2002) *Encyclopedia of Electrochemistry, vol. 9, Bioelectrochemistry* (eds A.J. Bard, M. Stratmann and G.S. Wilson), Wiley-VCH Verlag GmbH, Weinheim, Chapter 6.

37 (a) Weaver, M.J. and McManis, G.E. III (1990) *Acc. Chem. Res.*, **23**, 294; (b) Weaver, M.J. (1992) *Chem. Rev.*, **92**, 463; (c) Fawcett, W.R. and Opallo, M. (1994) *Angew. Chem. Int. Ed.*, **33**, 2131; (d) Miller, C.J. (1995) *Physical Electrochemistry, Principles, Methods, and Applications* (ed. I. Rubinstein), Marcel Dekker, New York, Chapter 2.

38 (a) Tanaka, N. and Sato, Y. (1966) *Inorg. Nucl. Chem. Lett.*, **2**, 359; *Electrochim. Acta*, 1968, **13**, 335; (b) Tanaka, N., Ogata, T. and Niizuma, S. (1973) *Bull. Chem. Soc. Jpn.*, **46**, 3299.

39 Schaap, W.B. (1960) *J. Am. Chem. Soc.*, **82**, 1837.

40 Izutsu, K., Adachi, T. and Fujinaga, T. (1970) *Electrochim. Acta*, **15**, 135.

41 (a) Sawyer, D.T. and Roberts, J.L., Jr (1988) *Acc. Chem. Res.*, **21**, 469; (b) Umemoto, K., Nagase, Y. and Sasaki, Y. (1994) *Bull. Chem. Soc. Jpn.*, **67**, 3245.

42 (a) Popov, A.P. and Geske, D.H. (1958) *J. Am. Chem. Soc.*, **80**, 1340; (b) Benoit, R.L. (1968) *Inorg. Nucl. Chem. Lett.*, **4**, 723; (c) Andreades, S. and Zahnow, E.W. (1969) *J. Am. Chem. Soc.*, **91**, 4181.

43 (a) Barrette, W.C., Jr, Johnson, H.W., Jr and Sawyer, D.T. (1984) *Anal. Chem.*, **56**, 1890; (b) Ref. [1], Chapter 9.

44 Peover, M.E. and White, B.S. (1966) *Electrochim. Acta*, **11**, 1061.

45 Pieta, P., Petr, A., Kutner, W. and Dunsch, L. (2008) *Electrochim. Acta*, **53**, 3412.

46 Fujinaga, T., Izutsu, K. and Adachi, T. (1969) *Bull. Chem. Soc. Jpn.*, **42**, 140.

47 Johnson, L., Pool, K.H. and Hamm, R.E. (1967) *Anal. Chem.*, **39**, 888.

48 Simonet, J. (1991) *Organic Electrochemistry*, 3rd edn (eds H. Lund and M.M. Baizer), Marcel Dekker, New York, p. 1245.

49 (a) Barrette, W.C., Jr and Sawyer, D.T. (1984) *Anal. Chem.*, **56**, 653; (b) Izutsu, K. and Fujimatsu, T., unpublished results.

50 Hori, Y. (2008) *Modern Aspects of Electrochemistry, Number 42* (eds C. Vayenas, R.E. White and M.E. Gamboa-Aldeco), Springer, New York, Chapter 3.

51 Fogg, P.G.T. (ed.) (1992) *Carbon Dioxide in Non-Aqueous Solvents at Pressures Less Than 200 kPa (Solubility Data Series Vol. 50)*, Pergamon, Oxford.

52 (a) Amatore, C. and Savéant, J.M. (1981) *J. Am. Chem. Soc.*, **103**, 5021; (b) Amatore, C., Nadjo, L. and Savéant, J.M. (1984) *Nouv. J. Chim.*, **8**, 565; (c) Gennaro, A., Isse, A.A., Severin, M.-G., Vianello, E., Bhugun, I. and Savéant, J.-M. (1996) *J. Chem. Soc., Faraday Trans.*, **92**, 3963.

53 (a) Kanzaki, Y. and Aoyagui, S. (1972) *J. Electroanal. Chem.*, **36**, 297; 1974, **51**, 19; (b) Harima, Y., Kurihara, H. and Aoyagui, S. (1981) *J. Electroanal. Chem.*, **124**, 103;

(c) Krishtalik, L.I. (1976) *Electrochim. Acta*, **21**, 693; (d) Alpatova, N.M., Krishtalik, L.I. and Pleskov, Y.V. (1987) Electrochemistry of solvated electrons, in *Topics in Current Chemistry*, vol. 138, Springer, Berlin, (review article).

54 (a) Lund, H. and Hammerich, O. (eds) (2001) *Organic Electrochemistry*, 4th edn, Marcel Dekker, New York; (b) Lund, H., and Baizer, M.M.(eds) (1991) *Organic Electrochemistry*, 3rd edn, Marcel Dekker, New York.

55 Sainsbury, M. (2002) *Organic Electrochemistry, Rodd's Chemistry of Carbon Compounds*, 2nd edn, Elsevier Science, Amsterdam.

56 Bard, J.A., Stratmann, M. and Schäfer, H.J. (eds) (2004) *Encyclopedia of Electrochemistry, vol. 8, Organic Electrochemistry*, Wiley-VCH Verlag GmbH, Weinheim.

57 Mann, C.K. and Barnes, K.K. (1970) *Electrochemical Reactions in Nonaqueous Systems*, Marcel Dekker, New York, Chapters 2–12.

58 Meites, L. and Zuman, P. (eds) (1977–1981) *CRC Handbook Series in Organic Electrochemistry*, vols. I–V, CRC Press, Boca Raton, FL.

59 Bard, A.J. (ed.) (1973–1980) *Encyclopedia of Electrochemistry of the Elements*, vols. I–XIV (Organic Section), Marcel Dekker, New York.

60 Evans, D.H. (2008) *Chem. Rev.*, **108**, 2113.

61 For example, (a) Parker, V.D. (1983) *Adv. Phys. Org. Chem.*, **19**, 131; (b) Hammerich, O., Ref. [54a], Chapter 2; (c) Hammerich, O. and Parker, V.D., Ref. [54b], Chapter 3; (d) Wayner, D.D.M. and Parker, V.D. (1993) *Acc. Chem. Res.*, **26**, 287; (e) Andrieux, C.P., Hapiot, P. and Savéant, J.-M. (1990) *Chem. Rev.*, **90**, 723; (f) Savéant, J.-M. (1993) *Acc. Chem. Res.*, **26**, 455; (g) Savéant, J.-M. (1990) *Adv. Phys. Org. Chem.*, **26**, 1; (h) Evans, D.H. (1990) *Chem. Rev.*, **90**, 739.

62 (a) Echegoyen, L. and Echegoyen, L.E. (1998) *Acc. Chem. Res.*, **31**, 593; (b) Ref. [54a], Chapter 7; (c) Reed, C.A. and Bolskar, R.D. (2000) *Chem. Rev.*, **100**, 1075;

(d) Echegoyen, L.E., Herranz, M.Á. and Echegoyen, L. (2006) *Bard-Stratmann Encyclopedia of Electrochemistry, vol. 7a, Inorganic Chemistry* (eds F. Scholz and C.J. Pickett), Wiley-VCH Verlag GmbH, Weinheim, Section 6.1.

63 For example, Peover, M.E. (1967) *Electroanalytical Chemistry*, vol. 2 (ed. A.J. Bard), Marcel Dekker, New York, Chapter 1.

64 (a) Fry, A.J. (2005) *Electrochem. Commun.*, **7**, 602; (b) Fry, A.J. (2006) *Tetrahedron*, **62**, 6558; (c) Ref. [60].

65 Kraiya, C. and Evans, D.H. (2004) *J. Electroanal. Chem.*, **565**, 29.

66 (a) Dubois, D., Kadish, K.M., Flanagan, S., Haufler, R.E., Chibante, L.P.F. and Wilson, L.J. (1991) *J. Am. Chem. Soc.*, **91**, 4364; (b) Dubois, D., Kadish, K.M., Flanagan, S. and Wilson, L.J. (1991) *J. Am. Chem. Soc.*, **91**, 7773; (c) Dubois, D., Moninot, G., Kutner, W., Thomas Jones, M. and Kadish, K.M. (1992) *J. Phys. Chem.*, **96**, 7173; (d) Xie, Q., Perex-Cordero, E. and Echegoyen, L. (1992) *J. Am. Chem. Soc.*, **114**, 3798; (e) Ohsawa, Y. and Saji, T. (1992) *J Chem. Soc. Chem. Commun.*, 781; (f) Niyazymbetov, M.E., Evans, D.H., Lerke, S.A., Cahill, P.A. and Henderson, C.C. (1994) *J. Phys. Chem.*, **98**, 13093; (g) Cliffel, D.E. and Bard, J.A. (1994) *J. Phys. Chem.*, **98**, 8140.

67 Shalev, H. and Evans, D.H. (1989) *J. Am. Chem. Soc.*, **111**, 2667.

68 Parker, V.D. (1974) *J. Am. Chem. Soc.*, **96**, 5656; 1976, **98**, 98.

69 (a) Fawcett, W.R. and Fedurco, M. (1993) *J. Phys. Chem.*, **97**, 7075; (b) Ref. [37c]; (c) Jaworski, J., Lesniewska, E. and Kalinowski, M.K. (1979) *J. Electroanal. Chem.*, **105**, 329; (d) Jaworski, J.S. (1986) *Electrochim. Acta*, **31**, 85; (e) Ref. [29a].

70 (a) Fujinaga, T., Izutsu, K., Umemoto, K., Arai, T. and Takaoka, K. (1968) *Nippon Kagaku Zasshi*, **89**, 105; (b) Umemoto, K. (1967) *Bull. Chem. Soc. Jpn.*, **40**, 1058.

71 (a) Peover, M.E. and Davies, J.D. (1963) *J. Electroanal. Chem.*, **6**, 46; (b) Fujinaga, T., Izutsu, K. and Nomura, T. (1971) *J. Electroanal. Chem.*, **29**, 203.

72 (a) Petersen, R.A. and Evans, D.H. (1987) *J. Electroanal. Chem.*, **222**, 129; (b) Evans, D.H. and Gilicinski, A.G. (1992) *J. Phys. Chem.*, **96**, 2528; (c) Fawcett, W.R., Fedurco, M. and Opallo, M. (1992) *J. Phys. Chem.*, **96**, 9959; (d) Kraiya, C., Singh, P. and Evans, D.H. (2004) *J. Electroanal. Chem.*, **563**, 203.

73 (a) Ref. [54a], Chapter 8; (b) Ref. [61e]; (c) Ref. [61f].

74 Sugimoto, H., Matsumoto, S. and Sawyer, D.T. (1988) *Environ. Sci. Technol.*, **22**, 1182.

75 Howell, J.O. and Wightman, R.M. (1984) *J. Phys. Chem.*, **88**, 3915.

76 (a) Okazaki, S., Oyama, M. and Nomura, S. (1997) *Electroanalysis*, **9**, 1242; (b) Nozaki, K., Oyama, M., Hatano, H. and Okazaki, S. (1989) *J. Electroanal. Chem.*, **270**, 191.

77 For example, (a) Ref. [54a], Chapter 2; (b) Ref. [2], Chapter 12; (c) Brown, E.R. and Sandifer, J.R. (1986) *Physical Methods of Chemistry, vol. 2, Electrochemical Methods*, 2nd edn (eds B.W. Rossiter and J.F. Hamilton), John Wiley & Sons, Inc., New York, Chapter 4; (d) Heinze, J. (1984) *Angew. Chem. Int. Ed. Engl.*, **23**, 831; (e) Wightman, R.M. and Wipf, D.O. (1990) *Acc. Chem. Res.*, **23**, 64.

78 (a) Rudolph, M., Reddy, D.P. and Feldberg, S.W. (1994) *Anal. Chem.*, **66**, 589A; (b) Rudolph, M. (1995) *Physical Electrochemistry* (ed. I. Rubinstein), Marcel Dekker, New York, Chapter 3; (c) Gosser, D.K., Jr (1993) *Cyclic Voltammetry, Simulation and Analysis of Reaction Mechanisms*, Wiley-VCH Verlag GmbH, Weinheim; (d) Maloy, J.T., Ref. [10], Chapter 20; (e) Brits, D. (2005) *Digital Simulation in Electrochemistry, Lecture Notes in Physics*, 3rd edn, Springer, Darmstadt.

79 (a) Wightman, R.M. and Wipf, D.O. (1989) *Electroanalytical Chemistry*, vol. 15 (ed. A.J. Bard), Marcel Dekker, New York, p. 267; (b) Michael, A.C. and Wightman, R.M., Ref. [10], Chapter 12; (c) Amatore, C. (1995) *Physical Electrochemistry* (ed. I. Rubinstein), Marcel Dekker, New York,

Chapter 5; (d) Aoki, K., Morita, M., Horiuchi, T. and Niwa, O. (1998) *Methods of Electrochemical Measurements with Microelectrodes*, IEICE, Tokyo, (in Japanese).

80 (a) Dibble, T., Bandyopadhyay, S., Ghoroghchian, J., Smith, J.J., Sarfarazi, F., Fleischman, M. and Pons, S. (1986) *J. Phys. Chem.*, **90**, 5275; (b) Ciszkowska, M., Stojek, Z. and Osteryoung, J. (1990) *Anal. Chem.*, **62**, 349; (c) Lehman, M.W. and Evans, D.H. (1998) *J. Phys. Chem. B*, **102**, 9928; (d) Jehoulet, C. and Bard, A.J. (1991) *Angew. Chem. Int. Ed. Engl.*, **30**, 836.

81 (a) Bond, A.M. and Mann, T.F. (1987) *Electrochim. Acta*, **32**, 863; (b) Geng, L., Ewing, A.G., Jernigan, J.C. and Murray, R.W. (1986) *Anal. Chem.*, **58**, 852.

82 Brina, R., Pons, S. and Fleischmann, M. (1988) *J. Electroanal. Chem.*, **244**, 81.

83 Niehaus, D., Philips, M., Michael, A.C. and Wightman, R.M. (1989) *J. Phys. Chem.*, **93**, 6232; references cited in Section 13.2.2.

84 Andrieux, C.P., Garreau, D., Hapiot, P. and Savéant, J.M. (1988) *J. Electroanal. Chem.*, **248**, 447.

85 For example, Nomura, S., Nozaki, K. and Okazaki, S. (1991) *Anal. Chem.*, **63**, 2665.

86 Wipf, D.O., Kristensen, E.W., Deakin, M.R. and Wightman, R.M. (1988) *Anal. Chem.*, **60**, 306.

87 (a) Van Duyne, R.P. and Reilley, C.N. (1972) *Anal. Chem.*, **44**, 142, 153, 158; (b) Deming, R.L., Allred, A.L., Dahl, A.R., Herlinger, A.W. and Kestner, M.O. (1976) *J. Am. Chem. Soc.*, **98**, 4132; (c) Nelson, S.F., Clennan, E.L. and Evans, D.H. (1978) *J. Am. Chem. Soc.*, **100**, 4012.

88 Evans, D.H. and Lerke, S.A., Ref. [10], Chapter 16.

89 (a) Penner, R.M., Herben, M.J., Longin, T.L. and Lewis, N.S. (1990) *Science*, **250**, 1118; (b) Heinze, J. (1991) *Angew. Chem. Int. Ed. Engl.*, **30**, 170.

90 Murray, R.W. (2008) *Chem. Res.*, **108**, 2688.

91 Quinn, B.M., Liljeroth, P., Ruiz, V., Laaksonen, T. and Kontturi, K. (2003) *J. Am. Chem. Soc.*, **125**, 6644.

9
Other Electrochemical Techniques in Nonaqueous Solutions

As described in Chapter 8, the current–potential curves in polarography and voltammetry are useful for obtaining mechanistic information on electrode reactions. However, for complicated electrode processes, the information obtained from the current–potential curves is not conclusive enough. To get more conclusive information, it is desirable to confirm the reaction products and/or intermediates by some other techniques. In this chapter, we focus our discussion on such techniques. We deal with electrolytic and coulometric techniques in Section 9.1 and the combinations of electrochemical and nonelectrochemical techniques in Section 9.2.

9.1
Use of Electrolytic and Coulometric Techniques in Nonaqueous Solutions

In the electrogravimetry and coulometry described in Section 5.6, the substance under study is completely electrolyzed in obtaining the analytical information. A complete electrolysis is also carried out in electrolytic syntheses and separations. Electrolytic methods are advantageous in that they need no chemical reagent and in that optimum reaction conditions can easily be obtained by controlling electrode potentials.

The method of complete electrolysis is also important in elucidating the mechanism of an electrode reaction. Usually, the substance under study is completely electrolyzed at a controlled potential, and the products are identified and determined by appropriate methods, such as gas chromatography (GC), high-performance liquid chromatography (HPLC) and capillary electrophoresis. In the GC method, the products are often identified and determined by the standard addition method. If the standard addition method is not applicable, however, other identification/determination techniques such as GC–MS should be used. The HPLC method is convenient when the product is thermally unstable or difficult to vaporize. HPLC instruments equipped with a high-sensitivity UV detector are the most popular, but a more sophisticated system such as LC–MS may also be employed. In some cases, the products are separated from the solvent-supporting electrolyte system

Electrochemistry in Nonaqueous Solutions, Second, Revised and Enlarged Edition. Kosuke Izutsu
Copyright © 2009 WILEY-VCH Verlag GmbH & Co. KGaA, Weinheim
ISBN: 978-3-527-32390-6

by such processes as vaporization, extraction and precipitation. If the products need to be collected separately, a preparative chromatographic method is useful. When the products are reactive with oxygen or water, the above procedures should be carried out under oxygen-free or dry atmospheres. For examples of practical procedures, see Ref. [1].

Methods of identifying/determining the products after complete electrolysis are helpful in arriving at a reliable conclusion, but these methods are applicable only to the products that are stable enough. The primary products at the electrode often undergo slow reactions in solution to final products. Although information about the final products is helpful, it does not necessarily interpret the polarographic or voltammetric results. The information obtained by a rapid electrolysis (see below) or by *in situ* measurements as in Section 9.2 may be more useful in explaining the polarographic or voltammetric results.

The method of complete electrolysis is also used in determining the number of electrons (n) participating in the electrode reaction. Here, the controlled-potential coulometry is generally used, i.e. a known amount of the substance under study is completely electrolyzed at a controlled potential and the quantity of electricity needed is measured with a coulometer. The electrolytic current (faradaic current) decays exponentially with time. However, a small residual current remains even after complete electrolysis, making the end of electrolysis somewhat unclear. Figure 9.1 shows an example of the use of controlled-current coulometry [2]. The substance under study is electrolyzed at a constant current (i_c), and the progress of the electrolysis is surveyed by measuring its cyclic voltammograms (CVs). If the constant current is consumed only for the electrolysis of the substance under study, the peak current linearly decreases with time. Thus, the linear portion is extrapolated to time t, at which the peak current reaches zero. The quantity of electricity ($Q = i_c \times t$) is what is needed to completely electrolyze the substance. In Figure 9.1, about 20 C is necessary to electrolyze 0.1 mmol of the substance, showing that $n = 2$. Here, the

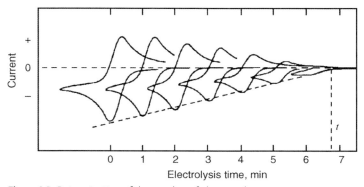

Figure 9.1 Determination of the number of electrons by controlled-current coulometry [2]. The case when 0.1 mmol of 2,3,5,6-tetraphenyl-1,4-dithiin in AN is electrolyzed at constant current (50 mA). The CV curves were measured, from left to right, after 0, 1, 2, 3, 4, 5 and 6 min.

analysis of the CV curves gives $n = 1$. This inconsistency occurs because the product of one-electron reduction in the CV measurement undergoes another one-electron reduction, after a reaction in the solution (ECE mechanism). In reality, this kind of inconsistency often occurs between the result of the fast measurement and that of the slow measurement.

To determine the number of electrons, the flow-coulometric method described in Section 5.6.3 is also useful. The solution of the supporting electrolyte is flowing through the column-type cell for rapid electrolysis (Figure 5.33), and the potential of the carbon fiber working electrode is kept at a value at which the desired electrolysis occurs. If a known amount of the substance under study is injected from the sample injection port, 100% electrolysis occurs in the column and, from the peak-shaped current–time curve, the quantity of electricity and, hence, the number of electrons can be determined. With this cell, the electrolysis is complete within 0.1 s. Moreover, the length of stay in the column can be varied over a wide range, by controlling the flow rate of the supporting electrolyte. Thus, with this cell, the number of electrons in the rapid reaction can be distinguished from that in the slow reaction. The column-type cell for rapid electrolysis is also useful for completely converting a substance in the solution from one oxidation state to another in a short time. This cell is often used in the next section.

9.2
Combination of Electrochemical and Nonelectrochemical Techniques

Recently, it has become popular to study the electrode phenomena by combining electrochemical and nonelectrochemical techniques in various ways. The usefulness of such combined techniques in nonaqueous solutions is shown below with the help of some examples.

9.2.1
Spectroelectrochemistry

Spectroelectrochemistry (SEC) is the field in which electrochemistry is combined with spectroscopy. SEC techniques are useful in studying the electrochemical phenomena that occur both in solutions and at electrode surfaces. Since there are various books and review articles dealing with SEC [3], only a few examples of phenomena in nonaqueous solutions are considered here.

In a typical SEC measurement, an optically transparent electrode (OTE) is used to measure the UV/vis absorption spectrum (or absorbance) of the substance participating in the reaction. Various types of OTE exist, for example (i) a plate (glass, quartz, or plastic) coated either with an optically transparent vapor-deposited metal (Pt or Au) film or with an optically transparent conductive tin oxide film (Figure 5.27) and (ii) a fine micromesh (40–800 wires cm^{-1}) of electrically conductive material (Pt or Au). The electrochemical cell may be either a thin-layer cell with a solution-layer thickness of less than 0.2 mm (Figure 9.2a) or a cell with a solution

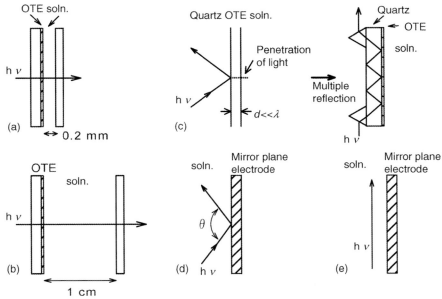

Figure 9.2 Various methods of spectroelectrochemistry.
(a) Transmission method (thin-layer cell), (b) transmission
method (cell of conventional thickness), (c) internal reflectance
method and multiple reflectance method, (d) specular reflectance
method and (e) parallel method.

layer of conventional thickness ($\sim$1 cm, Figure 9.2b). The advantage of the thin-layer
cell is that the electrolysis is complete within a short time ($\sim$30 s). On the other hand,
the cell with conventional solution thickness has the advantage that mass transport in
the solution near the electrode surface can be treated mathematically by the semi-
infinite linear diffusion theory.

SEC with a thin-layer cell is used to determine the formal potential of a redox
system. The absorption curves in Figure 9.3 were obtained with a thin-layer cell
containing a solution of $[Tc^{III}(dmpe)_2Br_2]^+$ (dmpe $= 1,2$-bis(dimethylphosphine)
ethane) in 0.5 M Et_4NClO_4–DMF [4]. On each curve is shown the potential of the OTE
at which the electrolysis was carried out until the redox equilibrium was attained.
$[Tc^{III}(dmpe)_2Br_2]^+$ is not reduced at all at $+0.1$ V versus aqueous Ag/AgCl, but it is
completely reduced to $[Tc^{II}(dmpe)_2Br_2]$ at -0.3 V. Between the two potentials, the
solution after electrolysis contains both $[Tc^{III}(dmpe)_2Br_2]^+$ and $[Tc^{II}(dmpe)_2Br_2]$ in
different ratios. If we express the two species by Ox and Red, respectively, the relation
between C_{Ox} and C_{Red} can be expressed by

$$E_{app} = E^{0\prime} + \frac{RT}{nF} \ln \frac{C_{Ox}}{C_{Red}}$$

where C is the concentration, E_{app} is the potential of electrolysis and $E^{0\prime}$ is the formal
potential of the redox system. If the relation between C_{Ox}/C_{Red} and E_{app} is obtained

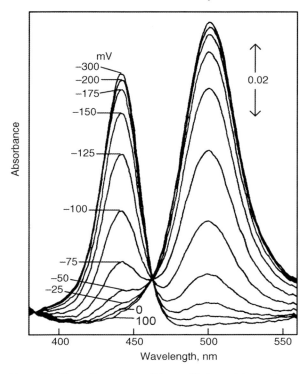

Figure 9.3 Absorption spectra of the equilibrium mixtures of
$[Tc^{III}(dmpe)_2Br_2]^+$ and its reduction product $[Tc^{II}(dmpe)_2Br_2]$,
obtained by the electrolysis of $[Tc^{III}(dmpe)_2Br_2]^+$ in 0.5 M
Et_4NClO_4–DMF in a thin-layer cell at various potentials [4].

from the absorption curves, the values of $E^{0\prime} = -0.108$ V versus aqueous Ag/AgCl
and $n = 0.98$ (practically unity) can be determined. When the formal potential is to
be determined by polarography or voltammetry, the electrode reaction must be
reversible. With this method, however, the reversibility of the reaction is not a
requirement, although Ox and Red must be stable enough. Moreover, the difference
from the potentiometric titration method is that no chemical reagent is needed to
generate Red.

The cell with a conventional thickness (Figure 9.2b) is used to study the reactivity
of an electrode reaction product. For example, substance B, which is generated by
electrode reaction (9.1), may react with substance C existing in the solution by
Eq. (9.2).

$$A \pm ne^- \rightarrow B \quad (+ \text{ for reduction and } - \text{ for oxidation}) \tag{9.1}$$

$$B + C \xrightarrow{k} D \tag{9.2}$$

A potential step is applied to the OTE to switch its potential from the initial value,
at which reaction (9.1) does not occur, to the value of the limiting current region

of reaction (9.1), and then the absorbance $A(t)$ of substance B is measured as a function of time t.[1] If reaction (9.2) does not occur, the $A(t) - t$ relation is expressed by Eq. (9.3).[2]

$$A(t) = 2\varepsilon_B C_A D_A^{1/2} t^{1/2} / \pi^{1/2} \tag{9.3}$$

Here, ε_B is the molar absorption coefficient of substance B, and C_A and D_A are the concentration and the diffusion coefficient, respectively, of substance A. The $A(t) - t$ relation changes when reaction (9.2) occurs. By simulating the expected $A(t) - t$ relation and comparing with the experimental results, the rate constant k can be obtained. This method is useful to study the reaction of radical ions in nonaqueous solutions. For example, the reactions of the cationic radical (DPA$^{\bullet+}$) of 9,10-diphenylanthracene (DPA) with such bases as water and pyridine were studied in Et_4NClO_4–acetonitrile (AN) [5, 6].[3] In this method, the substance to be measured is generated $in\ situ$ by controlled-potential electrolysis and its concentration at every moment is traced. Thus, the reactivity of an unstable substance (lifetime of $\sim$100 μs) can be studied. Various other methods have been developed for studying rapid reactions, e.g. the pulse radiolysis and flash photolysis methods. Compared with such methods, this method has a merit that the reactive substance can be generated at the best condition by the controlled-potential electrolysis.

However, the cell in Figure 9.2b has a disadvantage in that the concentration of the electrogenerated substance decreases with an increase in distance from the OTE surface. Because of this, simulation of the reaction is very difficult, except for the first-order (or pseudo-first-order) reactions. For more complicated reactions, it is desirable that the concentration of the electrogenerated species is kept uniform in the solution. With a thin-layer cell, a solution of uniform concentration can be obtained by complete electrolysis, but it takes $\sim$30 s. Thus, the thin-layer cell is applicable only for slow reactions. For faster reactions, a column-type cell for rapid electrolysis is convenient. Okazaki $et\ al.$ [7] constructed a stopped-flow optical absorption cell using one or two column-type cells (Figure 9.4) and used it to study the dimerization of the radical cations of triphenylamine (TPA$^{\bullet+}$) and the reactions of the radical cation of 9,10-diphenylanthracene (DPA$^{\bullet+}$) with water and alcohols. The solutions of radical cations were prepared by electrolyzing the precursors in the column-type electrolytic cell. Using the stopped-flow cell, the reaction of substance with a half-life of $\sim$1 s could be studied in solutions of uniform concentrations. Recently, Oyama and coworkers [8] developed a method applicable to substances of shorter ($<$30 ms) lives and named it an electron-transfer stopped-flow method. In this method, an unstable cation radical (N$^{\bullet+}$) for spectroscopic detection is generated not by electrolysis but by electron exchange with another long-lived cation radical (M$^{\bullet+}$), whose formal

---

1) An alternative to the potential-step method is to open the electrolytic circuit after keeping the potential for a fixed time at the limiting current region and then to measure the change in absorbance.

2) The current $i(t)$ that flows at time t due to the semi-infinite linear diffusion is expressed by $i(t) = nFC_A D_A^{1/2} / (\pi t)^{1/2}$ for unit area of the smooth OTE. Thus, the quantity of electricity $Q(t)$ that flows by time t is expressed by $Q(t) = 2nFC_A D_A^{1/2} t^{1/2} / \pi^{1/2}$. Compare with Eq. 9.3.

3) Radical ions in solutions are usually beautifully colored.

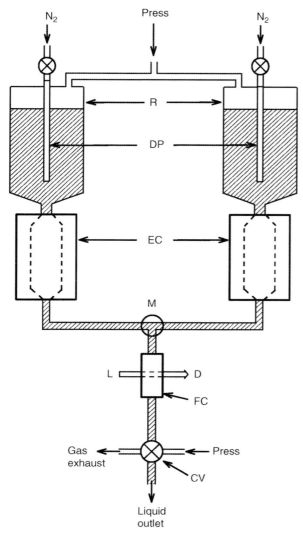

Figure 9.4 A stopped-flow optical absorption cell equipped
with two column-type cells for rapid electrolysis [7]. R, solution
reservoir; DP, N_2-gas bubbling; EC, electrochemical cell for
flow electrolysis; M, mixer; FC, flow-type optical absorption cell;
L, light beam; D, photodetector; CV, control valve.

potential is more positive than that of $N^{\bullet+}$, as shown by $N + M^{\bullet+} \rightarrow N^{\bullet+} + M$, mixing N with $M^{\bullet+}$ using the stopped-flow cell. Contrarily, the solution of $M^{\bullet+}$ is prepared by a batch-wise electrolysis of M in 0.05 M Bu_4NPF_6–AN and diluting it to an appropriate concentration with AN; here, M is either tris(4-bromophenyl)amine (formal potential of $M^{\bullet+}/M$: 0.97 V versus I_3^-, I^-) or tris(2,4-dibromophenyl)amine (formal potential of $M^{\bullet+}/M$: 1.41 V versus I_3^-, I^-)). By this method, the reaction mechanisms of cation radicals of such substances as anthracene, pyrene and their derivatives and aniline, N-methylaniline, N,N-dimethylaniline, diphenylamine and their derivatives have successfully been studied in AN.

In Figure 9.5, curve (a) is the cyclic voltammogram for the oxidation of tri-p-anisylamine (TAA) in 0.1 M Et_4NClO_4–AN, and curves (b) and (c) are the absorbance–potential curve and its derivative curve, respectively. They were obtained using an OTE with a vapor-deposited platinum film [9]. The excellent agreement of curves (a) and (c) shows that the reaction in the CV is purely the oxidation and re-reduction of the TAA. However, the two curves are different in that the peak current for curve (a) is proportional to the square root of the voltage sweep rate, while the peak height of curve (c) is inversely proportional to the square root of the voltage sweep rate.

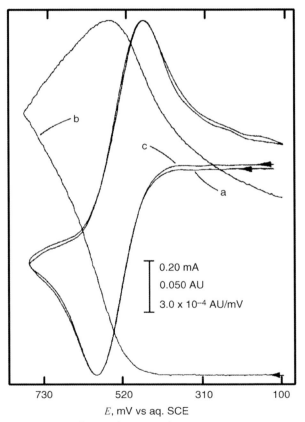

Figure 9.5 Simultaneously measured CV curve (a), absorption–potential curve (b), and its derivative with respect to scan potential (c), for the cyclic linear sweep oxidation of tri-p-anisylamine at a vapor-deposited platinum OTE in 0.1 M Et_4NClO_4–AN [9].

Various other spectroscopic techniques have been used for SEC studies in the solution near the electrode surface. Figure 9.2c is for internal reflectance spectroscopy (IRS). An OTE is on a quartz substrate, and the optical beam approaches the back of the OTE from within the quartz. If the angle of incidence is larger than the critical angle, the beam will be totally internally reflected. However, the light beam slightly penetrates into the solution (typically 50–200 nm). Thus, by using a multiple reflectance method, the absorbance of the electrogenerated substance adjacent to the electrode surface can be measured. Figure 9.2d is for the specular reflectance method, which uses a metal (Au or Pt) electrode. The sensitivity increases when the angle θ approaches 180°. In particular, if an incident laser beam with $\theta = 180°$ is used (Figure 9.2e), it is possible to measure the optical absorption of the solution as a function of the distance from the electrode surface. Furthermore, in addition to the UV/vis absorption methods, infrared absorption and Raman scattering methods, which give useful structural information, are also employed. See books and review articles [3] for the methods of SEC studies.

As described above, SEC methods are useful in studying the reactivity of radical anions and cations in nonaqueous solutions. Related to this, electrochemiluminescence (ECL), which is often caused by the reaction between the radical ions, is interesting [10]. For example, the electrogenerated anionic and cationic radicals of DPA in AN react as follows to give blue light:

$$DPA^{\bullet-} + DPA^{\bullet+} \rightarrow {}^1DPA^* + DPA \quad ({}^1DPA^* \rightarrow DPA + h\nu)$$

$^1DPA^*$ is the exited singlet state of DPA. When the two radical ions are generated at the same electrode, the electrode is alternately kept at the potential to generate the radical anion and at the potential to generate the radical cation. In some cases, the two radical ions are generated at different electrodes and then mixed to cause the ECL. For the ECL by the reaction between $DPA^{\bullet-}$ and the radical cation of N,N,N',N'-tetramethyl-p-phenylenediamine ($TMPD^{\bullet+}$), the process is somewhat more complicated, including triplet–triplet annihilation.

$$DPA^{\bullet-} + TMPD^{\bullet+} \rightarrow {}^3DPA^* + TMPD$$

$$^3DPA^* + {}^3DPA^* \rightarrow {}^1DPA^* + DPA$$

ECL will be discussed in more detail in Section 12.4 because its applications for display devices and for chemical analyses are important.

A method called *photomodulation voltammetry* has been developed for studying the electrochemistry of neutral free radicals [11]. In this method, the light beam from mercury–xenon lamp is modulated in sinusoidal form by using a light chopper and is sent to an electrochemical cell having a microgrid electrode. An organic substance in the solution near the grid-electrode is photodecomposed to form unstable free radicals, the current–potential curves of which are measured by a phase-sensitive voltammetric technique. This method has been used to study the redox properties, including the redox potentials, of very dilute (10^{-7}–10^{-8} M) and short-lived ($\sim$1 ms) free radicals. In a recent study [12], for example, photomodulation voltammetry on benzyl radical ($R^{\bullet}$) was carried out in AN, PC, DMF and DMSO to see the solvent effect. Diphenylacetone (E) was used as a precursor ($E \xrightarrow{h\nu} 2R^{\bullet}$), and, although the radical ($R^{\bullet}$) was rapidly converted to dibenzyl ($2R^{\bullet} \xrightarrow{k} P$), it produced steady voltammetric reduction and oxidation waves by $R^{\bullet} + e^- \rightleftarrows R^-$ and $R^{\bullet} \rightleftarrows R^+ + e^-$, respectively. The results gave the rate constant (k), the formal potentials for the reduction and oxidation processes, and the lifetime and steady concentration of $R^{\bullet}$.

9.2.2
Electrochemical-ESR Method

The ESR spectrum of the electrogenerated substance in nonaqueous solution was first measured in 1958 [13] and the electrochemical-ESR method (EC-ESR) became popular soon after that [14]. The ESR method is very sensitive to substances with unpaired electrons, such as organic and inorganic free radicals, radical anions and cations and paramagnetic metal complexes. In aprotic nonaqueous solvents, such substances are often the products of electrode reactions (Chapter 8), and the ESR method is extremely useful for studying electrode reactions. Moreover, in the ESR study of radical ions, the method of rapid electrolysis (see below) with tetra-alkylammonium salt as supporting electrolyte is best suited for generating radical ions. For example, if a radical anion is prepared by the reduction of its parent compound (Q) with sodium metal, the ESR signal of the radical anion is sometimes distorted by ion pair (Na^+–$Q^{\bullet-}$) formation. Such an ion pair formation does not occur if the electrolytic preparation method is employed using a tetraalkylammonium salt as supporting electrolyte.

The cells for EC-ESR measurements can be classified into two types, i.e. *in situ* and *ex situ* types. Examples of such cells are shown in Figure 9.6. With cell (a), which is the *in situ* type, the working electrode for generating the paramagnetic substance is placed inside the microwave cavity. Thus, even very unstable radical ions can be detected or identified. In some cases, the ESR signal–potential relation can be measured simultaneously with the current–potential curve. However, there are some problems with *in situ* cells; for example, (i) various restrictive conditions (dielectric loss due to solvent, etc.) in the cell design and (ii) the electron exchange between the parent compound (Q) and its radical ion ($Q^{\bullet-}$ or $Q^{\bullet+}$), which tends to increase the line widths of the ESR spectra. With cell (b), which is *ex situ* type, the radical ion is generated outside the cavity and introduced into it. The advantage of this cell is that there is no restriction in the cell design and, if necessary, the parent compound is completely converted to its radical ion before being introduced into the cavity. The problem is the time needed for 100% electrolysis. The cell shown in (b) is equipped with a carbon fiber working electrode for rapid electrolysis, and 100% electrolysis is possible within 0.1 s or less. Recently, EC-ESR cells have been greatly improved [15], and we can select the one best suited for the intended measurement.

The EC-ESR method has been used in various ways to study chemistry in nonaqueous solutions.

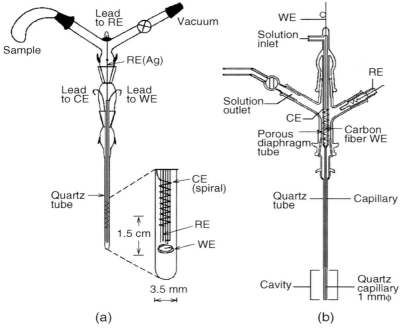

Figure 9.6 Electrochemical-ESR cells. (a) *In situ* generation type (Fernando, K.R. *et al.* (1986) *J. Mag. Reson.*, **68**, 552). (b) *Ex situ* generation type [15].

Detection and Identification of Radicals

The ESR detection and/or identification of the radical ion are very important in getting definite information about electrode processes in nonaqueous solutions. Sometimes, information about the electronic configuration of the radical ion is also obtained. However, in some cases, the radical ion detected by ESR measurement is not the original product from the electroactive species under study. For example, some halogenated nitrobenzenes give the ESR signal of the radical anion of nitrobenzene that is generated by dehalogenation of the original product [16]. The nitro group acts as an intramolecular catalyst in the dehalogenation. Similar dissociative electron transfer ($RX + e^- \rightarrow R^\bullet + X^- \rightarrow$ product) occurs in the electrode reductions of many halogen compounds, and its dynamics is an important subject of research (Section 8.3.1) [17].

Kinetic Studies

If the time course of the concentration of a radical ion can be obtained from the ESR signal, we can use it to obtain kinetic information. For example, Umemoto [18] studied, by means of ESR spectroscopy and polarography, the kinetics of the protonation reactions of the electrogenerated radical anions of anthracene, anthraquinone and benzophenone in water-containing DMF. In the case of anthracene, the two methods gave similar results concerning the rate of the decay of the radical anion ($Q^{\bullet -}$), supporting the following mechanism.

$$Q + e^- \rightleftarrows Q^{\bullet -}$$

$$Q^{\bullet -} + H_2O \rightarrow QH^{\bullet} + OH^- \text{ (rate determining)}$$

$$QH^{\bullet} + e^- \rightarrow QH^- \text{ (at the electrode)}$$

$$QH^{\bullet} + Q^{\bullet -} \rightarrow QH^- + Q \text{ (in the solution)}$$

$$QH^- + H_2O \rightarrow QH_2 + OH^-$$

On the other hand, for anthraquinone and benzophenone, the ESR results showed that the concentration of the radical anion was almost time independent, although, in polarography, the height of the first wave increased, consuming the height of the second wave, as shown in Figure 8.13. This inconsistent result was explained as the establishment of the following disproportionation equilibrium in the solution.

$$2Q^{\bullet -} \rightleftarrows Q + Q^{2-}$$

In polarography, Q formed by the disproportionation is reduced at the electrode and increases the height of the first wave.

Studies of Self-Exchange Electron Transfer in the Solution

If the solution contains both the parent compound (Q) and its radical ion ($Q^{\bullet -}$ or $Q^{\bullet +}$), the self-exchange electron transfer reaction occurs between them ($Q + Q^{\bullet -} \rightleftarrows Q^{\bullet -} + Q$ or $Q + Q^{\bullet +} \rightleftarrows Q^{\bullet +} + Q$), and the line width of the ESR spectrum increases with the increase in the reaction rate. The electron exchange reaction is very fast and the rate constants (k_{ex}) determined by the ESR measurements have values of 10^7–$10^9 \, l \, mol^{-1} \, s^{-1}$ [19].[4]

Studies of Electrode Reactions of Metal Complexes

Metal ions with electron configurations of d^1 to d^9 and some of their complexes produce ESR signals. Metal complexes with ligands of radical anions also give ESR signals. So the EC-ESR method is often used in the study of redox reactions of metal complexes. EC-ESR cells for this purpose are the same as those used for the study of radical ions [21]. Using the ESR method, it is possible to determine the site in the complex at which the electronic change takes place, i.e. whether it is the central metal,

---

4) According to Marcus [20a], the standard rate constant k_s for the electrode reaction $Q + e^- \rightleftarrows Q^{\bullet -}$ and the rate constant, k_{ex}, for the self-exchange electron transfer, $Q + Q^{\bullet -} \rightleftarrows Q^{\bullet -} + Q$, can be correlated by $k_s/A_{el} \cong (k_{ex}/A_{ex})^{1/2}$, where A_{el} and A_{ex} are the pre-exponential factors when k_s and k_{ex} are expressed in the form $k = A\exp(-\Delta G^*/RT)$ (ΔG^* = activation energy). Kojima and Bard [20b] studied this relation using organic compounds as in Table 8.6. The average values of A_{el} and A_{ex}, estimated for the compounds in Table 8.6, were $(5.0 \pm 0.6) \times 10^3 \, cm \, s^{-1}$ and $(2.9 \pm 0.2) \times 10^{11} M^{-1} \, s^{-1}$, respectively. Literature values of k_{ex} determined by ESR method have been listed in Table 8.6. k_s values estimated by using these values and the above relation are between 50 and 425 cm s^{-1} and much larger than $k_{s,corr}$ in Table 8.6. Detailed discussion is given in Ref. [20b].

the ligand or the delocalized molecular orbital (see Section 8.2.2 for an example) [22]. When a metal ion with a rare valence state is formed by an electrode reaction, the ESR method can detect it. For example, it was found by the combined use of cyclic voltammetry, ESR spectroscopy and UV/vis spectroscopy that, when [Hg(cyclam)] $(BF_4)_2$ complex (cyclam $= 1,4,8,11$-tetraazacyclo-decane) was electrolytically oxidized in propionitrile at $-78\,^\circ$C, a complex of Hg(III) ($[Hg(cyclam)]^{3+}$) with a half-life of about 5 s was formed [21]. The formation of the complex of Hg(III) was possible because, in propionitrile and at low temperatures, the oxidation of $[Hg(cyclam)]^{2+}$ occurred more easily than the oxidation of the ligand (cyclam).

9.2.3
Electrochemical Mass Spectroscopy

Mass spectrometry (MS) is based on the generation of gaseous ions from an analyte molecule, the subsequent separation of those ions according to their mass-to-charge (m/z) ratio, and the detection of those ions [23]. The resulting mass spectrum is a plot of the (relative) abundance of the produced ions as a function of the m/z ratio. MS is important in determining the molecular mass, molecular formula and elemental composition and in elucidating the structure of the compound. Moreover, MS is the most sensitive spectrometric method for molecular analysis. Now, online GC–MS and LC–MS are widely used as the most powerful analytical tools.

In conventional electrochemical mass spectrometry (EC-MS), the electrochemical cell is directly coupled to the mass spectrometer and the volatile product or intermediate at the electrode is introduced into the mass spectrometer [24]. The conventional interface between the electrochemical cell and the mass spectrometer is shown in Figure 9.7a. The porous membrane (e.g. Teflon) plays an important role, permeating gases but not liquids through it. Thus, from the mass signal–potential curve measured *in situ* and simultaneously with the current–potential curve,

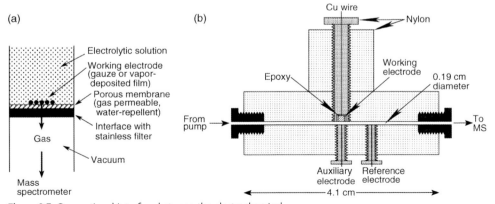

Figure 9.7 Conventional interface between the electrochemical cell and mass spectrometer (a) [24] and the electrochemical cell for the particle beam interface (b) [25].

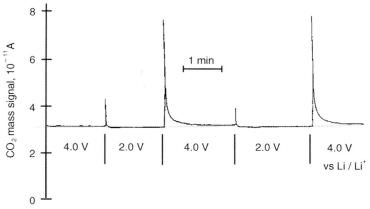

Figure 9.8 Variation with time of the CO_2 ($m/z = 44$) mass signal obtained by switching the potential of porous Pt electrode in 0.1 M $LiClO_4$–PC alternately between 4.0 and 2.0 V versus Li/Li^+ [24a].

it is possible to definitely identify the volatile substance and the potential of its formation. The application to nonaqueous solutions is fairly rare, although there are many applications to aqueous solutions. The result in Figure 9.8 is for the electrolytic oxidation of propylene carbonate (PC) in the electrolyte solution of a lithium battery. It shows the time course of the mass signal for CO_2 ($m/z = 44$), obtained by switching the potential of a porous platinum electrode in 0.1 M $LiClO_4$–PC alternately between 4.0 and 2.0 V (versus Li/Li^+) [24a]. The CO_2 peak was observed after each potential switch, but it decayed quickly because the oxidation product of PC passivates the electrode surface and prevents the generation of CO_2.

Online electrochemical mass spectrometry is also important: the solution in the electrochemical cell is directly sent to the mass spectrometer via interface (electrospray, particle beam, etc.). Thus, the ions formed by the electrochemical reaction can be detected if the ions have a long enough lifetime to enter the ion source of the mass spectrometer and reach the detector, though not all ions that are detected are due to the electrochemical reaction. The use of the method to nonaqueous solutions is not frequent too. As an example, the anodic oxidation of triphenylamine (TPA) in AN–Bu_4NClO_4 has been studied using the particle beam (PB) interface as in Figure 9.7b [25]. The electro-oxidation of TPA generates a $TPA^{\bullet+}$ radical cation ($m/z = 245$), which dimerizes to tetraphenylbenzidine (TPB). TPB is readily oxidized to $TPB^{\bullet+}$ ($m/z = 488$) and TBP^{2+} ($m/z = 244$) at the oxidation potential of TPA. In EC/PB/MS, the direct monitoring of the oxidation of TPB to $TPB^{\bullet+}$ as a function of the electrode potential was achieved via selective ion monitoring of the ion peak at m/z 488. By using the relative intensity ratio of ions at m/z 244 (TPB^{2+}) to 245 ($TPA^{\bullet+}$), the formation of TPB^{2+} as a function of the electrode potential was also monitored. EC/PB/MS showed a maximum rate of formation of $TPB^{\bullet+}$ at $+1.2$ V versus Pd wire, while TPB^{2+} is generated at a maximum rate at $+1.6$ V versus Pd wire.

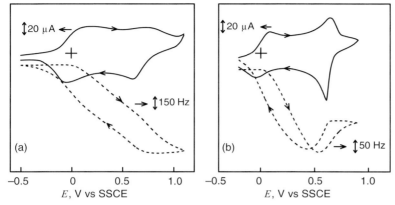

Figure 9.9 CV and frequency–potential curves for the oxidation and re-reduction processes of the electropolymerized polyaniline film [27]. (a) In 0.5 M LiClO$_4$–AN and (b) in aqueous 0.5 M NaClO$_4$ + HClO$_4$ (pH = 1). Voltage sweep rate 5 mV s^{-1}, quantity of film deposition 0.4 C cm^{-2} and SSCE = saturated NaCl calomel electrode.

9.2.4
Use of Electrochemical Quartz Crystal Microbalance

The basic aspects of electrochemical quartz crystal microbalance (EQCM) [26] were outlined in Section 5.5.7. The EQCM is also useful in nonaqueous solutions for studying various phenomena at the electrode surface, including the adsorption and desorption of electrode reaction products and the electrolytic deposition of metals. Figure 9.9 shows the CV curves and the frequency–potential curves for the oxidation and re-reduction of the electropolymerized polyaniline film [27]. Figure 9.9a is in 0.5 M LiClO$_4$–AN and Figure 9.9b is in aqueous 0.5 M NaClO$_4$ + HClO$_4$ (pH = 1). In Figure 9.9a, the comparison of the CV curve and frequency–potential curve shows that the anion (ClO$_4^-$) is doped at the two oxidation steps and the doped anion is released at the two reduction steps (Scheme 9.1). In Figure 9.9b, the mass increases at the first oxidation step, showing the doping of the anion. However, in the second oxidation step, the mass decreases. This has been explained by deprotonation (imine formation) and the accompanying release of the doped anion [27, 28].

9.2.5
Use of Scanning Electrochemical Microscopy

The apparatus and basic principles of scanning electrochemical microscopy (SECM) [29] were described in Section 5.5.7. We consider that the reaction O + e$^-$ R occurs at the tip (see Figure 5.29). If the substrate is conductive, the reduction product, R, at the tip is reoxidized at the substrate and O is regenerated. The

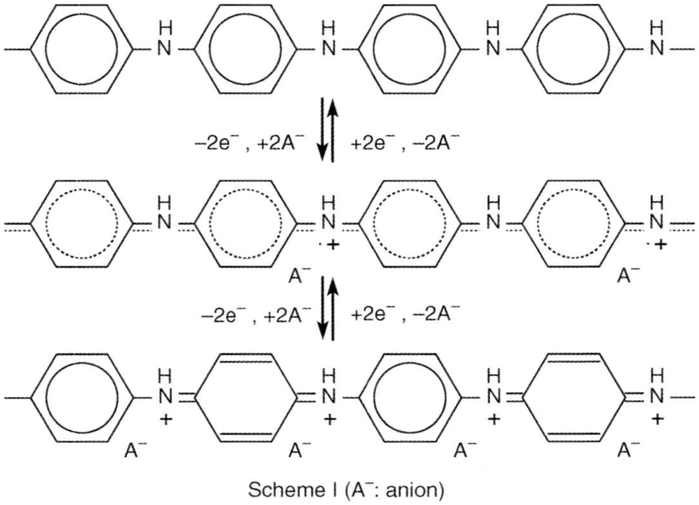

$-2e^-, +2A^- \updownarrow +2e^-, -2A^-$

A^- A^-

$-2e^-, +2A^- \updownarrow +2e^-, -2A^-$

A^- A^- A^- A^-

Scheme I (A⁻: anion)

Scheme 9.1

regenerated O is then reduced again at the tip. The efficiency of this cycle depends on the ratio of (d/a), where a is the radius of the disk-type tip electrode and d is the distance of the tip from the substrate. The tip current i_T is a function of $L \equiv (d/a)$ as approximately described by

$$I_T(L) = i_T/i_{T,\infty} = 0.68 + 0.78377/L + 0.3315 \exp(-1.0672/L)$$

where $i_{T,\infty}$ is the tip current at $(d/a) \to \infty$ and $i_{T,\infty} = 4nFD_OC_Oa$. The tip current at any potential, E, can be expressed by Eq. (9.4) when the tip reaction is reversible and by Eq. (9.5) when the tip reaction is under mixed diffusion-kinetic control.

$$I_T(E, L) = [0.68 + 0.78377/L + 0.3315 \exp(-1.0672/L)]/\theta \tag{9.4}$$

$$I_T(E, L) = [0.68 + 0.78377/L + 0.3315 \exp(-1.0672/L)]/(\theta + 1/\kappa) \tag{9.5}$$

where $\theta = 1 + \exp[nf(E - E^{0\prime})]D_O/D_R$, $\kappa = k_s \exp[-\alpha nf(E - E^{0\prime})]/m_O$, k_s is the standard rate constant, α is the transfer coefficient, $E^{0\prime}$ is the formal potential and $f = F/RT$. m_O is the effective mass transfer coefficient for SECM and

$$m_O = 4D_O[0.68 + 0.78377/L + 0.3315\exp(-1.0672/L)]/(\pi a)$$
$$= i_p/(\pi a^2 nFC_O) \tag{9.6}$$

From Eq. (9.6), $m_O \sim D/d$ if $L \ll 1$. This suggests the usefulness of SECM in the study of rapid heterogeneous electron transfer kinetics, the largest k_s values that can be determined being of the order of D/d. The largest k_s values measurable by voltammetry at UMEs are on the order of D/a. It is easier to get a small L value with SECM than to prepare a UME of very small a value. For example, k_s for the Fc⁺/Fc couple in AN has been determined by SECM to be 3.7 cm s⁻¹ [30].

SECM is also useful for studying homogeneous reactions of the products generated either at the tip or at the substrate. In SECM, almost all of the stable

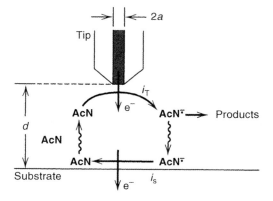

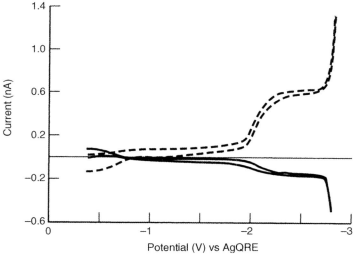

Figure 9.10 SECM study of the dimerization of the acrylonitrile anion radical (AN•⁻) in DMF (see text).

product at the tip is transferred to the substrate if $a/d \leq 2$, meaning that the collection efficiency is practically unity[5] (see Chapter 5, footnote 12 to compare with the collection efficiency at a rotating ring-disk electrode). However, if the product at the tip is lost by a homogeneous reaction, the collection efficiency decreases. The dimerization of acrylonitrile anion radical (shown by AcN•⁻) in DMF is considered in Figure 9.10 [31]. The solution contains acrylonitrile (AcN, 1.5 mM) and Bu_4NPF_6 (0.1 M) in DMF. The substrate is a 60 μm diameter gold electrode held at −1.75 V versus silver quasi-reference electrode and $a = 2.5$ μm and $d = 1.36$ μm.

---

5) The collection efficiency here is obtained as the ratio i_S/i_T, where i_S is the substrate current and i_T is the tip current.

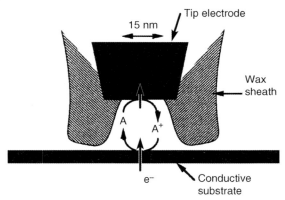

Figure 9.11 Idealized diagram of the trapping of molecule A
that is oxidized at the tip and re-reduced at the conductive
substrate in the system for single-molecule electrochemistry [32].
See the size of the tip electrode.

The potential of the tip was scanned at $100\,\mathrm{mV\,s^{-1}}$, and the voltammogram for the
reduction of AcN (dashed curve) and the substrate current for the reoxidation of
$\mathrm{AcN^{\bullet-}}$ (solid line) were obtained. The mechanism for the homogeneous reaction is

$$2\mathrm{AcN^{\bullet-}} \rightarrow (\mathrm{AcN})_2^{2-}\ \text{(a)} \qquad (\mathrm{AcN})_2^{2-} + 2\mathrm{H^+} \rightarrow (\mathrm{AcNH})_2\ \text{(b)}$$

By studying the dependence of the collection efficiency on d, the rate constant for step
(a) was found to be $6 \times 10^7\,\mathrm{M^{-1}\,s^{-1}}$.

The SECM technique was used for single-molecule electrochemistry [32].
In Figure 9.11, only a small number (1–10) of molecules of electroactive species
A are trapped between the nanometer tip and the conductive substrate. However, the
repeated electron transfers of A produce a current that can be used to detect the
trapped molecules ($\sim$0.6 pA/molecule). Although the system was tested in aqueous
systems, it will also be useful in nonaqueous systems. SECM is also convenient for
studying the kinetics of electron, ion and molecular transfers across the liquid–liquid
interface (ITIES, see Chapter 14) [33]. For example, in the study of electron transfer
kinetics, the tip is immersed in liquid 1 containing $\mathrm{Red_1}$, and liquid 2 containing $\mathrm{Red_2}$
is used as conductive substrate. The oxidation product, $\mathrm{Ox_1}$, at the tip reaches the
liquid–liquid interface and reacts with $\mathrm{Red_2}$ by $\mathrm{Ox_1} + \mathrm{Red_2} \rightarrow \mathrm{Red_1} + \mathrm{Ox_2}$ to
regenerate $\mathrm{Red_1}$ in liquid 1. The information on the electron-transfer kinetics at
the interface can be obtained by measuring the tip current.

References

1 For example, (a) McDonald, R.N., Triebe,
F.M., January, J.R., Borhani, K.J. and
Hawley, M.D. (1980) *J. Am. Chem. Soc.*,
102, 7867; (b) Lawless, J.G., Bartak, D.E.
and Hawley, M.D. (1969) *J. Am. Chem.
Soc.*, **91**, 7121; (c) Moore, W.M.,
Salajegheh, A. and Peters, D.G. (1975)
J. Am. Chem. Soc., **97**, 4954.

2 Hammerich, O. (2001) *Organic Electrochemistry*, 4th edn (eds H. Lund and O. Hammerich), Marcel Dekker, New York, Chapter 2.

3 (a) Kuwana, T. and Winograd, N. (1974) *Electroanalytical Chemistry*, vol. 7 (ed. A.J. Bard), Marcel Dekker, New York, Chapter 1; (b) Heineman, W.R., Hawkridge, F.M. and Blount, H.N. (1984) *Electroanalytical Chemistry*, vol. 13 (ed. J.A. Bard), Marcel Dekker, New York, Chapter 1; (c) McCreery, R.L. (1986) *Physical Methods of Chemistry, vol. 2, Electrochemical Methods*, 2nd edn (eds B.W. Rossiter and J.F. Hamilton), John Wiley & Sons, Inc., New York, Chapter 7; (d) Gale, R.J. (ed.) (1988) *Spectroelectrochemistry, Theory and Practice*, Plenum Press, New York; (e) Plieth, W., Wilson, G.S. and Gutiérrez De La Fe, C. (1998) *Pure Appl. Chem.*, **70**, 1395; (f) Bard, A.J. and Faulkner, L.R. (2001) *Electrochemical Methods, Fundamentals and Applications*, 2nd edn, John Wiley & Sons, Inc., New York, Chapter 17; (g) Crayston, J.A. (2003) *Encyclopedia in Electrochemistry: vol. 3, Instrumentation and Electroanalytical Chemistry*, (eds A.J. Bard, M. Stratmann and P.R. Unwin), Wiley-VCH Verlag GmbH, Weinheim, Section 3.4; (h) Kaim, W. and Klein, A. (eds) (2008) *Spectroelectrochemistry*, RSC, Cambridge.

4 Heineman, W.R. (1983) *J. Chem. Educ.*, **60**, 305.

5 Grant, G. and Kuwana, T. (1970) *J. Electroanal. Chem.*, **24**, 11.

6 Blount, H.N. (1973) *J. Electroanal. Chem.*, **42**, 271.

7 (a) Oyama, M., Nozaki, K., Nagaoka, T. and Okazaki, S. (1990) *Bull. Chem. Soc. Jpn.*, **63**, 33; (b) Oyama, M., Nozaki, K. and Okazaki, S. (1991) *Anal. Chem.*, **63**, 1387.

8 (a) Oyama, M., Higuchi, T. and Okazaki, S. (2000) *Electrochem. Commun.*, **2**, 675; (b) Oyama, M., Matsui, J. and Park, H. (2002) *Chem. Commun.*, 604; (c) Goto, M., Otsuka, K., Chen, X., Tao, Y. and Oyama, M. (2004) *J. Phys. Chem. A*, **108**, 3980; (d) Oyama, M. and Matsui, J. (2004) *Bull.*

Chem. Soc. Jpn., **77**, 953; (e) Oyama, M., Imabayashi, T., Ho, J.-H. and Ho, T.-I. (2006) *Electrochemistry*, **74**, 649 and references therein.

9 Bancroft, E.E., Sidwell, J.S. and Blount, H.N. (1981) *Anal. Chem.*, **53**, 1390.

10 (a) Bard, A.J. (ed.) (2004) *Electrogenerated Chemiluminescence*, Marcel Dekker, New York; (b) Richter, M.M. (2004) *Chem. Rev.*, **104**, 3003; (c) Bocarsly, A.B., Tachikawa, H. and Faulkner, L.R. (1996) *Laboratory Techniques in Electroanalytical Chemistry*, 2nd edn (eds P.T. Kissinger and W.R. Heineman), Marcel Dekker, New York, Chapter 28; (d) Ref. [3f], Chapter 18.

11 (a) Wayner, D.D.M. and Griller, D. (1985) *J. Am. Chem. Soc.*, **107**, 7764; (b) Wayner, D.D.M., McPhee, D.J. and Griller, D. (1988) *J. Am. Chem. Soc.*, **110**, 132; (c) Nagaoka, T., Griller, D. and Wayner, D.D.M. (1991) *J. Phys. Chem.*, **95**, 6264.

12 Grampp, G., Mureşanu, C. and Landgraf, S. (2008) *Electrochim. Acta*, **53**, 3149.

13 Austen, D.E.G., Given, P.H., Ingram, D.J.E. and Peover, M.E. (1958) *Nature*, **182**, 1784.

14 (a) McKinney, T.M. (1977) *Electroanalytical Chemistry*, vol. 10 (ed. A.J. Bard), Marcel Dekker, New York, Chapter 2; (b) Goldberg, I.B. and McKinney, T.M. (1996) *Laboratory Techniques in Electroanalytical Chemistry*, 2nd edn (eds P.T. Kissinger and W.R. Heineman), Marcel Dekker, New York, Chapter 29; (c) Wadhawan, J. and Compton, R.G. (2003) *Encyclopedia of Electrochemistry: vol. 2, Interfacial Kinetics and Mass Transport*, (eds A.J. Bard, M. Stratmann and E.J. Calvo), Wiley-VCH Verlag GmbH, Weinheim, Section 3.2.

15 Okazaki, S. and Nagaoka, T. (1991) *Denki Kagaku*, **59**, 106.

16 (a) Kitagawa, T., Layloff, T.P. and Adams, R.N. (1963) *Anal. Chem.*, **35**, 1086; (b) Fujinaga, T., Deguchi, Y. and Umemoto, K. (1964) *Bull. Chem. Soc. Jpn.*, **37**, 822.

17 (a) Gallardo, I., Guirado, G. and Marquet, J. (2000) *J. Electroanal. Chem.*, **488**, 64 and references therein; (b) Savéant,

J.-M. (1993) *Acc. Chem. Res.*, **26**, 455; (c) Savéant, J.-M. (1994) *J. Phys. Chem.*, **98**, 3716.

18 (a) Fujinaga, T., Izutsu, K., Umemoto, K., Arai, T. and Takaoka, K. (1968) *Nippon Kagaku Zasshi*, **89**, 105; (b) Umemoto, K. (1967) *Bull. Chem. Soc. Jpn.*, **40**, 1058.

19 Layloff, T., Miller, T., Adams, R.N., Fah, H., Horsfield, A. and Procter, W. (1965) *Nature*, **205**, 382.

20 (a) Marcus, R.A. (1963) *J. Phys. Chem.*, **67**, 853; (b) Kojima, H. and Bard, A.J. (1975) *J. Am. Chem. Soc.*, **97**, 6317.

21 Deming, R.L., Allred, A.L., Dahl, A.R., Herlinger, A.W. and Kestner, M.O. (1976) *J. Am. Chem. Soc.*, **98**, 4132.

22 For example, Kadish, K.M., Franzen, M.M., Han, B.C., McAdams, C.A. and Sazou, D. (1991) *J. Am. Chem. Soc.*, **113**, 512. (Factors that determine the site of the nickel porphyrin at which reduction occurs.)

23 Kellner, R., Mermet, J.-M., Otto, M. and Widmer, H.M. (eds) (1998) *Analytical Chemistry*, Wiley-VCH Verlag GmbH, Weinheim, Section 9.4.

24 (a) Bittins-Cattaneo, B., Cattaneo, E., Konigshoven, P. and Vielstich, W. (1991) *Electroanalytical Chemistry*, vol. **17** (ed. A.J. Bard), Marcel Dekker, New York, Chapter 3; (b) Baltruschat, H. (2004) *J. Am. Mass Spectrom.*, **15**, 1693.

25 (a) Regino, M.C.S. and Brajter-Toth, A. (1997) *Anal. Chem.*, **69**, 5067; (b) Zhang, T.

and Brajter-Toth, A. (2000) *Anal. Chem.*, **72**, 2533.

26 For review, (a) Buttry, D.A. (1991) *Electroanalytical Chemistry*, vol. 17 (ed. A.J. Bard), Marcel Dekker, New York, Chapter 1; (b) Ward, M.D. (1995) *Physical Electrochemistry* (ed. I. Rubinstein), Marcel Dekker, New York, Chapter 7.

27 Daifuku, H., Kawagoe, T., Yamamoto, N., Ohsaka, T. and Oyama, N. (1989) *J. Electroanal. Chem.*, **274**, 313.

28 Orata, D. and Buttry, D.A. (1987) *J. Am. Chem. Soc.*, **109**, 3574.

29 (a) Ref. [3f], Chapter 16; (b) Bard, A.J. and Mirkin, M.V. (2001) *Scanning Electrochemical Microscopy*, Marcel Dekker, New York; (c) Bard, A.J., Fan, F.-R. and Mirkin, M. (1995) *Physical Electrochemistry* (ed. I. Rubinstein), Marcel Dekker, New York, Chapter 5; (d) Horrocks, B.R. (2003) *Encyclopedia in Electrochemistry: vol. 3, Instrumentation and Electroanalytical Chemistry*, (eds A.J. Bard, M. Stratmann and P.R. Unwin), Wiley-VCH Verlag GmbH, Weinheim, Section 3.3.

30 Mirkin, M.V., Richards, T.C. and Bard, A.J. (1993) *J. Phys. Chem.*, **97**, 7672.

31 Zhou, F. and Bard, A.J. (1994) *J. Am. Chem. Soc.*, **116**, 393.

32 Fan, F.-R., Kwak, J. and Bard, A.J. (1996) *J. Am. Chem. Soc.*, **118**, 9669.

33 Barker, A.L., Unwin, P.R., Amemiya, S., Zhou, J. and Bard, A.J. (1999) *J. Phys. Chem. B*, **103**, 7260 and references therein.

10
Purification of Solvents and Tests for Impurities

In previous chapters, we dealt with various electrochemical processes in nonaqueous solutions by paying attention to solvent effects on them. Many electrochemical reactions that are not possible in aqueous solutions become possible with the use of suitable nonaqueous or mixed solvents. However, for the solvents to display their advantages, they must be sufficiently pure. Impurities in the solvents often have a negative influence. Usually, commercially available solvents are classified into several grades of purity. Some of the highest grade solvents are pure enough for immediate use, but all others need purification before use. In this chapter, the effects of solvent impurities on electrochemical measurements are briefly reviewed in Section 10.1, popular methods used in solvent purification and tests of impurities are outlined in Sections 10.2 and 10.3, respectively, and finally, practical purification procedures are described for 25 electrochemically important solvents in Section 10.4.

As discussed in Section 1.1.4, some solvents are toxic or hazardous to human health, and the use of such solvents should be avoided. This is especially true for carcinogenic solvents because the formation of malignant tumors has a very long latency period (10 years or more). If the solvent that is most suitable for a given purpose is toxic or hazardous, we have to find an alternative solvent that is harmless. If the use of the toxic or hazardous solvent is unavoidable, a minimum amount should be used very carefully and the waste should be treated safely. The evaporation of volatile organic solvents is known to cause various environmental problems (Section 1.1.4). Nowadays, to protect the environment, it is usual to select environmentally benign solvents. In the field of organic chemistry, hazardous organic solvents are gradually being replaced with alternatives such as supercritical fluids, ionic liquids, immobilized solvents, aqueous systems and solvent-free systems. Supercritical fluids, ionic liquids and immobilized solvents are also promising in electrochemical technologies, as discussed in Chapter 13.

Solvents for electrochemical use have been dealt with in the books of Refs [1–3]. A series of IUPAC reports [4, 5] on the methods of solvent purification and tests for impurities is also useful. The book by Riddick, Bunger and Sakano [6] is the most authoritative, covering the properties of about 500 organic solvents and their purification methods. The book authored by Marcus [7] also contains useful information about solvent properties. For the latest information about the toxicity and

Electrochemistry in Nonaqueous Solutions, Second, Revised and Enlarged Edition. Kosuke Izutsu
Copyright © 2009 WILEY-VCH Verlag GmbH & Co. KGaA, Weinheim
ISBN: 978-3-527-32390-6

hazards of solvents to human health and the environment, the handbooks listed in Ref. [8] are useful, but such information can also be accessed via Internet.

10.1
Effects of Solvent Impurities on Electrochemical Measurements

Small amounts of impurities in solvents usually do not have serious effects on the physical properties of solvents (Section 2.5). However, they often have drastic effects on the chemical properties of solvents, changing the reaction mechanisms or making electrochemical measurements impossible. The extent of the effect of an impurity differs considerably, depending on the properties of the impurity and those of the solvent in which it exists. Impurities that have significant effects on chemical reactions or on electrochemical measurements are called reactive impurities.

Most reactive impurities are acids or bases in a broad sense. Here, an acid is a substance that has proton donor capacity, hydrogen bond donor capacity, electron pair acceptability and electron acceptability. A base is a substance that has proton acceptability, hydrogen bond acceptability, electron pair donor capacity and electron donor capacity. Some reactive impurities have both acidic and basic properties.

The impurities interact with ionic species in the solution and change their reactivities. Basic impurities tend to interact with cations (Lewis acids), while acidic impurities interact with anions (Lewis bases). The impurities and the solvent compete to interact with ions. The interactions between basic impurities and cations are stronger in less basic solvents, while those between acidic impurities and anions are stronger in less acidic solvents. The ions that are small in size but are large in charge number easily interact with impurities, while the ions that are large in size but are small in charge number (NR_4^+, Ph_4As^+, BPh_4^-, etc.) do not easily interact with impurities. Coetzee *et al.* [9] developed a method to characterize small amounts of reactive impurities in solvents (Sections 6.3.5 and 10.3).

Impurities also interact with electrically neutral acidic or basic substances. Basic impurities tend to interact with acidic substances, while acidic impurities interact with basic substances. The former interaction is more extensive in less basic solvents, while the latter interaction is more pronounced in less acidic solvents.

The impurities may interfere with electrochemical measurements in various ways:

1. The impurity gives a signal that disturbs the measuring system. An example is shown in Figure 10.1 [10]. The residual current–potential curves were obtained with a platinum electrode in propylene carbonate (PC) containing various concentrations of water. Because water is amphiprotic, its cathodic reduction and anodic oxidation are easier than those of PC, which is aprotic and protophobic. Thus, the potential window is much narrower in the presence of water than in its absence. Complete removal of water is essential for measuring electrode reactions at very negative or positive potentials.

2. The impurity interacts with a substance participating in the process to be measured. An example is met when we determine the dissociation constant

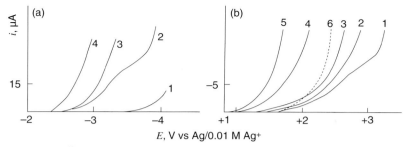

Figure 10.1 Effect of water on the residual current–potential curves at a platinum electrode in PC [10]. (a) Cathodic side in 0.1 M Bu₄NClO₄. Water concentration: curve 1, ~0 M, 2, 0.01 M, 3, 0.03 M, 4, 0.1 M; (b) anodic side in 0.1 M KPF₆. Water concentration: curve 1, 0.01 M, 2, 0.03 M, 3, 0.1 M, 4, 0.5 M, 5, 1 M (curve 6 is in 0.1 M LiClO₄ at a water concentration of 0.001 M).

(pK_a) of a weak acid HA by conductimetry. In the determination, we dissolve HA in the pure solvent and measure the conductivity of the solution containing dilute H^+ and A^- formed by dissociation. If a basic impurity B is contained in the solution, it disturbs the measurement, producing ionic species by the reaction $B + HA \rightarrow BH^+ + A^-$. Because trace amounts of basic impurity are contained even in a purified solvent, conductimetric pK_a determination is practically impossible for HA with $pK_a > 7.5$.

10.2
Procedures for the Purification of Solvents

Commercially available solvents contain impurities from various sources: some come from raw materials, some are produced by manufacturing processes, some are introduced from outside and some are generated by decomposition of the solvents. The most common method of purifying solvents is distillation: solvents of low boiling points are distilled under atmospheric pressure, while those of high boiling point are distilled at reduced pressures under an inert gas such as nitrogen and argon. For the distillation of solvents, the use of high-performance distilling apparatus is highly desirable. It can produce high-purity solvents by one or two distillations and can save solvents and time.

For impurities that are not removed by distillation, some treatment is necessary before distillation. It is usual to make the impurities nonvolatile by treating with such reactive agents as molecular sieves, P_2O_5, $CuSO_4$, active alumina, KOH, CaO, BaO, CaH_2 and metallic sodium. One method is to add a reactive agent to the solvent in a sealed bottle, keep it for some time (several hours to several days) with stirring and then get the supernatant and distill it. Another method is to add a reactive agent into the distilling flask and reflux over it for several hours before distillation. Sometimes the raw solvent is passed through a column packed with a reactive agent before introduction into the distilling flask. These pretreatments are effective to remove water and other impurities that are acidic, basic or reducible substances. However, in some cases, pretreatment causes a decomposition or polymerization of the solvent. For example, if we distill acetonitrile (AN) over a large amount of P_2O_5, a yellowish brown polymer is formed. If we distill PC over CaO, PC is decomposed to give propylene oxide. Activated molecular sieves are versatile and easy to use; because they are moderately active, they can effectively remove both water and basic impurities such as amines, but their tendency to decompose solvents is much less than that in other chemical pretreatments.[1] However, molecular sieves must be kept in contact with the solvent for several days for them to be fully efficient.

Apart from these common pretreatments, special pretreatments are necessary in some cases. For example, a small amount of basic impurity (possibly triethylamine) in PC is not removed even by repeated distillations. However, if we add *p*-toluene-sulfonic acid to neutralize the basic impurity, we can remove it easily by distillation (Section 10.4). As another example, volatile impurities in PC can be removed only by bubbling inert gases (nitrogen or argon) for many hours.

In some cases, fractional freezing is more effective than distillation after chemical pretreatment because thermal decomposition of solvents, caused by distillation, can be avoided. Fractional freezing has been employed in purifying DMSO (mp 18.5 °C), pyridine (mp −41.6 °C) and HMPA (mp 7.2 °C).

The solvent, once purified, may absorb atmospheric water. Moreover, it may decompose to form impurities if exposed to heat, light, oxygen or water. As an extreme case, if HMPA after purification is exposed to air and light for several hours, it generates its peroxide, which produces a large polarographic wave, as in Figure 10.2 [11].

Generally, solvents should be purified just before use, and the purified solvents should be stored in an inert atmosphere in a cool, dark place and should be used as soon as possible.[2] In the preparation of the electrolyte solution and during the measurement, the introduction of moisture from air should be avoided as much as possible, by using a vacuum line or a glove box if necessary.[3] In order to keep the electrolyte solution dry, active molecular sieves can be used. As described in Section 8.1.2, adding active alumina powder directly into the solution in the cell is the most effective way of removing water almost completely, although its unfavorable influence must also be taken into account.

1) The dehydrating abilities of various reactive agents are compared in Burfield, D.R., Lee, K.-H. and Smithers, R.H. (1977) *J. Org. Chem.*, **42**, 142; 1978, **43**, 3966. However, it should be noted that the amounts of reactive agents used were considerably larger than those in conventional use.

2) If the purified solvents are stored in a freezer (−20 to −30 °C), the formation of impurities by solvent decomposition can be suppressed to a considerable extent.

3) For example, 100 ml of air at 25 °C and at 100% humidity contains about 2.5 mg of water.

Therefore, when we handle electrolyte solutions in nonaqueous solvents, we must estimate the amount of water introduced from the air and the extent of its effect on the measurements. The vacuum line techniques and the glove box operations for electrochemical studies in nonaqueous solvents have been dealt with in several books. See, for example, Kissinger, P.T. and Heineman, W.R. (eds) (1996) *Laboratory Techniques in Electroanalytical Chemistry*, 2nd edn, Marcel Dekker, New York, Chapters 18 and 19.

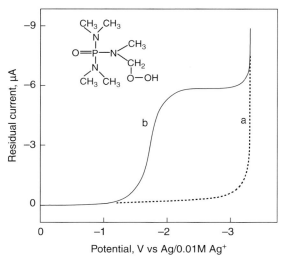

Potential, V vs Ag/0.01M Ag⁺

Figure 10.2 Polarographic residual current–potential curve in HMPA–0.1 M LiClO₄ [11]. (a) In HMPA just after distillation; (b) in HMPA kept in the light for 4 days after distillation. The structural formula shows the peroxide of HMPA.

10.3
Tests for Purity of Solvents

The following methods are often employed for testing the purity of organic solvents:

Karl Fischer (KF) titration This method is used to determine trace amounts of water in organic solvents. The KF reagent is a mixture of iodine, sulfur dioxide and a base in anhydrous methanol. It reacts with water by $I_2 + SO_2 + H_2O \rightarrow 2HI + SO_3$ and $SO_3 + CH_3OH \rightarrow CH_3OSO_3H$, where the base works both as a coordinating solvent and as a proton acceptor. The titration end point is often detected by biamperometric or bipotentiometric methods, and the lower limit of water determined is ~10 ppm. The original KF reagent contained pyridine as the base. Nowadays nonpyridine KF reagents, which use bases other than pyridine (e.g. imidazole), are widely used with automatic volumetric KF titrators, which are commercially available. In the coulometric KF method for which automatic titrators are also available, I_2 is coulometrically generated *in situ* in the titration cell, and water down to 2 ppm can be determined. Hence, the method has an advantage that the amount of the sample solvent needed is small.

Conductimetry This method is useful in detecting and determining ionic impurities. A conductimetric titration, on the other hand, is useful in determining nondissociative acids and bases. For example, if an amine RNH_2 in a solvent is titrated with a solution of weak acid HA, a titration curve as in Figure 10.3 is obtained due to the ionization reaction $RNH_2 + HA \rightarrow RNH_3^+ + A^-$. Thus, the amine can be determined. This method is employed in the purity test of PC (see Section 10.4).

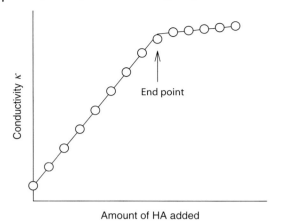

Figure 10.3 Conductimetric titration curve of an amine impurity with a very weak acid (HA).

Potentiometry Conventional nonaqueous pH titrations are useful in detecting and determining acidic and basic impurities. On the other hand, the ion-probe method proposed by Coetzee *et al.* [9] is convenient in characterizing trace amounts of reactive impurities. The principle of the method was described in Section 6.3.5. In Figure 10.4, the method is applied to reactive impurities

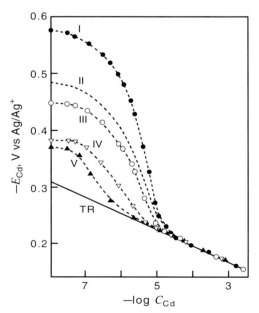

Figure 10.4 Characterization of reactive impurities in commercial acetonitrile by the ion-probe method (use of the Cd^{2+} probe ion and the Cd^{2+}-ISE) [9]. Samples: I, Mallinckrodt AR; II, Aldrich, 99%; III, Burdick and Jackson; IV, Fisher, 99%; V, after purification of I; TR, theoretical response in the absence of impurities.

in commercial AN products. The probe ion and the ISE were Cd^{2+} and a Cd^{2+}-selective electrode (CdS–Ag_2S solid membrane), respectively. The solid line TR is the theoretical relation expected when the ISE responds to Cd^{2+} in the Nernstian way. The total concentrations of impurities, which were reactive with Cd^{2+}, were estimated to be 4×10^{-5}, 8×10^{-6}, 3×10^{-6}, 1×10^{-6} and 3×10^{-7} M for curves I, II, III, IV and V, respectively. It was confirmed by gas chromatography that the main impurities were alkylamines such as ethylamine. By appropriately selecting a probe ion and an ISE, we can characterize various reactive impurities, which are too dilute to be detected by other methods.

Polarography and voltammetry If we measure a residual current–potential curve by adding an appropriate supporting electrolyte to the purified solvent, we can detect and determine the electroactive impurities contained in the solution. In Figure 10.2, the peroxide formed after HMPA purification was detected by polarography. Polarography and voltammetry are also used to determine the applicable potential ranges and how they are influenced by impurities (see Figure 10.1). These methods are the most straightforward for testing solvents to be used in electrochemical measurements.

Gas chromatography (GC) This method is important in detecting and determining organic impurities in solvents. Today, gas chromatography–mass spectrometry (GC–MS) is also used to identify impurities. Water is sometimes determined by GC with a column packed, for example, with Porapak Q. The lower limit of the water determination is $\sim$2 ppm.

Measurement of UV absorption spectra This method is used to detect impurities that show UV absorption. The UV cutoff points for various solvents are listed in Table 10.1. Usually, it is not easy to identify the impurities that cause UV absorption. However, if the impurities are removed by purification, better

Table 10.1 UV cutoff (nm) of high-quality organic solvents.[a]

Solvents	UV cutoff	Solvent	UV cutoff
Acetone (Ac)	330	Ethanol (EtOH)	205
Acetonitrile (AN)	190	Ethylene carbonate (EC)	215
Benzene	280	Methanol (MeOH)	205
Benzonitrile (BN)	300	4-Methyl-2-pentanone (MIBK)	335
1,2-Dichloroethane (DCE)	225	N-Methyl-2-pyrrolidinone (NMP)	260
Dichloromethane	230	Nitromethane (NM)	380
Diethyl carbonate (DEC)	255	Propylene carbonate (PC)	240
1,2-Dimethoxyethane (DME)	220	Pyridine (Py)	305
N,N-Dimethylacetamide (DMA)	270	Sulfolane (TMS)	220
N,N-Dimethylformamide (DMF)	270	Tetrahydrofuran (THF)	220
Dimethyl sulfoxide (DMSO)	265	Water	190
Dioxane	215		

[a]From Refs [3, 6, 7].

results can be obtained in electrochemical measurements. UV/vis spectrometry can also be applied to determine colorless impurities that are acids or bases. For example, colorless amines in the solvent can be determined by adding nitrophenol and measuring the absorbance of the resulting nitrophenolate anion.

There are some impurities that are not detectable by the methods described above but that have undesirable effects on electrochemical measurements. For these impurities, appropriate testing methods must be employed (see Refs [4, 5]).

10.4
Purification Methods for Solvents in Common Use

The procedures described below are for purifying commercial products to a level that is pure enough for ordinary electrochemical measurements. Most of them are based on the reports from the IUPAC Commission on Electroanalytical Chemistry [4, 5].

Acetone (Ac) [5a] Commercial products are usually very pure. The major impurity is water ($\leq 0.5\%$) and minor impurities are methanol, 2-propanol ($\leq 0.05\%$) and aldehyde ($\leq 0.002\%$). An acidic desiccant (e.g. silica gel) and a basic desiccant (e.g. alumina) are not suitable for removing water because they work as a catalyst and cause aldol condensation. Drierite (active calcium sulfate anhydride) has a mild dehydrating ability and is suitable as desiccant. It is added to acetone ($25\,g\,l^{-1}$) and the solvent is shaken for several hours for dehydration. The supernatant is then distilled over Drierite ($10\,g\,l^{-1}$), connection to the outside air being made through a drying tube filled with Drierite. Magnesium perchlorate anhydride should not be used in the drying tube because an explosion may occur if it contacts with acetone vapor. The distillate should be stored in a light-resistant bottle.

Acetonitrile (AN) [12] Commercial products contain such impurities as water, ammonia, acetic acid, propionitrile, acrylonitrile, allyl alcohol, acetone and benzene. To obtain AN for general electrochemical use, AN is passed successively through two columns, one packed with molecular sieves (3A or 4A) and the other with neutral chromatographic alumina. Then the solvent is fractionated over CaH_2 at a high reflux ratio until the boiling temperature stabilizes at $81.6\,°C$ ($760\,mmHg$). The forecut is discarded and the remaining solvent (90%) is collected.

To get AN of higher purity, (i) 100 ml of MeOH is added to 1 l of AN and the mixture is fractionated at a high reflux ratio, until the boiling point rises from 64 to $80\,°C$ and the distillate becomes optically transparent down to 240 nm (removal of benzene), and the distillate is discarded; (ii) 0.3 g of NaH suspension in paraffin oil is added to the remaining solvent, and the solvent is refluxed for 10 min and then rapidly distilled until $\sim$30 ml of residue (yellow or brownish) remains; (iii) the distillate is immediately passed through a column packed with

Table 10.2 Comparison of impurity levels before and after purification of AN [12].

Impurities	Before purification	After purification	Impurities	Before purification	After purification
Water	1.7×10^3 (ppm)	1.0 (ppm)	Allyl alcohol	2.6×10^3 (ppm)	$<1 \times 10^1$ (ppm)
Propionitrile	3.2×10^2	2.5×10^2	Acetone	8.1×10^2	$<3 \times 10^{-2}$
Acrylonitrile	2.4×10^2	$<4 \times 10^{-2}$	Benzene	1.8×10^2	$<1 \times 10^{-2}$

100 g of acidic alumina. The first 50 ml of percolate is discarded and the main fraction is collected. The fraction is suitable for various purposes, but the long-term stability is limited, possibly due to the presence of colloidal alumina. The colloidal alumina is removable by distilling over CaH_2. Table 10.2 shows the effectiveness of the latter procedure. Although propionitrile is not removed, other impurities are removed sufficiently.

Benzene [5b] Commercial products are fairly pure but contain small amounts of sulfur compounds, such as thiophene, and water (<0.05%). To purify a commercial product, 200 ml of it is vigorously shaken with 50 ml of conc. H_2SO_4 for 1 h to remove thiophene. The yellow color of the H_2SO_4 phase shows the presence of thiophene. Benzene, separated from the H_2SO_4 phase, is washed in sequence with water, 0.1 M NaOH and several times with water. It is dried with $CaCl_2$ or Na_2SO_4 and refluxed for 1 h over metallic sodium under nitrogen or argon atmosphere and fractionally distilled under the same atmosphere to collect the 10–80% fraction. *Benzene is toxic and may cause cancer. It should be replaced by a less hazardous solvent such as toluene.*

Benzonitrile (BN) [5c] The purity of commercially available BN is 99% or more, and the major impurities are water, benzoic acid, isonitriles and amines. In getting pure BN, the commercial product is dried with a mild dehydrating agent, e.g. calcium chloride, the supernatant is transferred to a distilling vessel containing P_2O_5 and then it is distilled at a reduced pressure under an inert atmosphere to collect the 10–80% fraction.

1,1- and 1,2-Dichloroethane [5c] The two solvents contain such impurities as water, HCl and chlorinated hydrocarbons. They can be purified by the same method, i.e. they are shaken with conc. H_2SO_4 repeatedly until the H_2SO_4 is not colored. Then, they are washed in sequence with water, saturated aqueous Na_2CO_3 and water. After preliminary drying with calcium chloride, they are distilled over P_2O_5 to collect the 10–80% fraction. These solvents are suspected to be carcinogenic and should be used with care.

Dichloromethane [5c] The main impurities in commercial products are HCl, alcohols, water and decomposition products such as formaldehyde. Reagent grade CH_2Cl_2 is shaken with conc. H_2SO_4 repeatedly until the H_2SO_4 layer is not colored. Then, it is washed in sequence with water, saturated aqueous Na_2CO_3 and water. After preliminary drying with calcium chloride, it is refluxed for 2 h over P_2O_5, and the fraction (10–80%) is collected over new P_2O_5. This CH_2Cl_2 is stored over P_2O_5 and redistilled before use. For the tests of impurities, see Ref. [5e]. This solvent is suspected to be carcinogenic and should be used with care.

N,N-Dimethylformamide (DMF) [13] In the presence of water, DMF is slowly hydrolyzed to HCOOH and $(CH_3)_2NH$. It is also decomposed by light and heat. Thus, the main impurities are water, HCOOH and $(CH_3)_2NH$. Small amounts of HCN, formed by photolysis, and CO, formed by thermal decomposition, are also present. In the purification of a commercial product, it is kept with molecular sieves (4A or 5A) for 1–4 days to dehydrate. Then, it is shaken with BaO for 1–2 days and the supernatant is distilled twice at 20 mmHg under nitrogen. All these operations must be carried out in the absence of light. The distillate should be stored under nitrogen gas and used as soon as possible. DMF has toxic effects, particularly on the liver and kidney, and care should be taken in its handling.

Dimethyl sulfoxide (DMSO) [14] Commercial products contain water and small amounts of dimethyl sulfide and dimethyl sulfone as impurities. In the purification, 5A molecular sieves (activated at 500 °C under argon for 16 h) are added and kept for several days to reduce water to <10 ppm and other low-boiling point impurities to <50 ppm. Then, the solvent is filtered and the filtrate is distilled over CaH_2 at reduced pressure in a nitrogen atmosphere.

Dioxane [5d] Major impurities are water and peroxides. In the purification, the commercial product is passed slowly through a column packed with 3A molecular sieves (20 g l^{-1}, activated at 250 °C for 24 h) to remove water, and then the effluent is passed slowly through a column packed with chromatographic grade basic alumina (20 g l^{-1}, activated at 250 °C for 24 h) to remove peroxides and other impurities. Here, *the absence of peroxides must be confirmed* with titanium tetrachloride or with aqueous potassium iodide–starch reagent. After confirming the absence of peroxides, the effluent is refluxed for several hours over sodium wire (10 g l^{-1}), avoiding atmospheric moisture, adding more sodium from time to time until the surface of the added sodium remains shiny and clean. If a brown crust is formed on the sodium, it shows the presence of excessive amounts of glycol acetal and the solvent should be pretreated with HCl (see Ref. [5f]). Finally the solvent is distilled at a high reflux ratio in a dry argon or nitrogen atmosphere to collect the middle 80%. The distillate is stored in the dark, in filled containers and, when in use, is dispensed by dry argon or nitrogen pressure.

1,2-Ethanediol (ethylene glycol) [5e] Commercial products are pure enough for most purposes. To remove water of ~2000 ppm, the ethanediol is dehydrated with sodium sulfate anhydride and distilled twice at reduced pressure in a dry nitrogen atmosphere to avoid oxidation to aldehyde.

Ethanol (EtOH) [5f] Commercial products are pure enough for most purposes. The impurities are MeOH, PrOH, Ac, water, etc. To reduce ~2000 ppm of water in 'absolute' EtOH, about 5% of hexane or cyclohexane is added to the EtOH and the mixture is fractionated to remove the aqueous azeotrope. By this step, the water content is reduced to <500 ppm. The dehydrated EtOH is then slowly passed through a column of dry molecular sieves (4A) to reduce the water content to ~7 ppm.

Ethylenediamine (en) [15] Among the impurities are water, carbon dioxide, ammonia and polyethylene amines (e.g. diethylenetriamine and triethylenetetramine). In the recommended purification method, commercial product (98%) is shaken for about 12 h with activated molecular sieves (5A, $70 \, g \, l^{-1}$), the supernatant is decanted and shaken for about 12 h with a mixture of CaO ($50 \, g \, l^{-1}$) and KOH ($15 \, g \, l^{-1}$), and then the supernatant is fractionally distilled (reflux ratio 20: 1) in the presence of freshly activated molecular sieves to collect the fraction boiling at 117.2 °C (760 mmHg). The fraction collected is further fractionated from sodium metal. All distillations should be carried out under nitrogen. The purified solvent is stored in a sealed reservoir to prevent contamination of CO_2 and water and dispensed by dry argon or nitrogen pressure. *Care should be taken in handling en because it is toxic.*

Hexamethylphosphoric triamide (HMPA) [16] The major impurities in commercial products are water, dimethylamine and its salt with HCl, the peroxide of HMPA, etc. In purifying commercially available HMPA, it is refluxed over BaO for several hours at ~4 mmHg under N_2 atmosphere and is then distilled (bp 90 °C). The distillate is refluxed over metallic sodium ($\sim 1 \, g \, l^{-1}$) and distilled again at ~4 mmHg under N_2 atmosphere. If the purified HMPA is exposed to air and light, its peroxide is formed rapidly and interferes with electrochemical measurements (see Figure 10.2). It should be stored in a cold dark place in an inert atmosphere or under a vacuum and used within 1–2 days. *Although HMPA has interesting properties, it may cause cancer and heritable genetic damage* [17a]. *It must be handled with extreme care or its use should be avoided.* There are some solvents that share the properties of HMPA but are less harmful (e.g. N,N'-dimethylpropioneurea (DMPU) [17b] and dimethyl-2-imidazolidinone [17c]).

Methanol (MeOH) [5g] Because commercial products are highly pure, they are applicable for most general purposes without purification. To remove water (300–1000 ppm) in commercial MeOH, it is distilled with a good fractionating column under the exclusion of moisture and the water content is reduced to ~100 ppm. Then the distillate is twice distilled over metallic sodium ($1–2 \, g \, l^{-1}$) and the fraction within ±0.01 °C of the boiling point is immediately used. In this case, water content is ~0.5 ppm. *Caution:* Anhydrous magnesium perchlorate is not suitable to dry MeOH, there is a danger of explosion!

N-Methylacetamide (NMA) [18] Major impurities are acetic acid and methylamine. Methods for purifying NMA are of two types: (a) fractional freezing, including zone melting and (b) chemical treatment followed by distillation at reduced pressure. In the zone melting method, an apparatus suitable for substances melting at near-room temperature is used. It consists of 10 Vycor tubes for zone melting (0.7 cm in radius, 78 cm in length and equipped with heating wires and cooling coils), which contain 1 l of NMA and can be treated at one time. By repeating six zone-melting cycles, NMA with a melting point of 30.35–30.55 °C and a relative permittivity of 184.3 (40 °C) can be obtained. For details, see Ref. [18].

N-**Methylpropionamide [19]** The commercial product from Eastman Organic Chemicals is pure enough if distilled once under vacuum. Moreover, this solvent can easily be synthesized from methylamine and propionic acid. Thus, in the recommended method, it is first synthesized and then purified. For details, see Ref. [19].

Nitromethane (NM) [5h] Commercially available NM contains water and ∼0.1 M of nitroethane, 2-nitropropane and propionitrile. Its purification is performed by combining a crystallization method with vacuum distillation. To a 1 : 1 (v/v) mixture of NM and diethyl ether in a Dewar vessel, dry ice powder is added slowly, with stirring, until a temperature below −60 °C is attained. The stirring is continued for another 5 min, and the mixture is transferred to a deep Büchner funnel, with a coarse porosity glass filter, and is cooled with a cooling tube of dry ice temperature. Then, most of the ether is removed by sucking with a tap aspirator for several seconds, the suction is stopped, and the crystal in the funnel is pressed with a glass rod. Dry ice powder is scattered over the crystal and the funnel is sucked for 20 min. Then, the crystal is washed twice with ether cooled to −78 °C. Here, in each wash, the amount of ether used is half the amount of the original mixture, the crystal in the funnel is pressed with the rod, and the dry ice powder is scattered over the crystal. Then, the cooling tube is removed and the crystal in the funnel is liquefied, and the liquid is filtered and transferred to a distilling flask. The liquid is vacuum distilled with a high-performance distilling apparatus: when the distillate of NM begins to come out, the previous fraction is removed, the condenser is dried and then the distillation is continued. The middle 80% fraction is collected, stored in a colored bottle under dry nitrogen and dispensed by nitrogen gas pressure. NM should be distilled with great care because it is explosive.

1- and 2-Propanol (PrOH) [5i] In recent commercially available PrOH, water is the only impurity that affects electrochemical measurements. To remove water, the commercial product is refluxed for 4 h over freshly ignited calcium oxide or for 2 h over magnesium ribbon activated by iodine and fractionally distilled twice with a glass column having high theoretical plates. The middle fraction is collected and stored over CaH_2 under nitrogen. At this stage, the purity determined by GC measurement is 99.94% and the remaining impurity of ∼600 ppm is mainly water. By passing the distillate through a column packed with activated molecular sieves (type A in the K-form), the water content is reduced to ∼20 ppm.

Propylene carbonate (PC) [20] High-quality PC for immediate electrochemical use is now commercially available. The impurities contained in technical grade PC are water, CO_2, propylene oxide, 1,2-propanediol, allyl alcohol and ethylene carbonate (EC). To purify technical grade PC, it is kept with molecular sieves (4A or 5A) for a night and is subjected to two or three fractional distillations at reduced pressure under a nitrogen atmosphere; the distillate collected is transferred to an inert atmosphere glove box. By this procedure, the contents of both water and organic impurities are reduced to <20 ppm. However, commercially available PC often contains a basic impurity (triethylamine) that is difficult to eliminate by distillation. If PC is to be used for studying acid–base equilibria in PC, it is desirable to run the first distillation by adding p-toluenesulfonic acid (twice equivalent of the basic impurity) to make it nonvolatile [20b].[4]

Pyridine (Py) [21] Major impurities are water and amines such as picoline and lutidine. To get pure Py, it is necessary to keep KF reagent grade Py or an equivalent Py over solid KOH ($\sim$20 g kg^{-1}) for 2 weeks, fractionally distill the supernatant liquid over freshly activated molecular sieves 5A and solid KOH (reflux ratio 20: 1) and collect the fraction boiling at 115.3 °C (760 mmHg). To prevent contamination with atmospheric CO_2 and water, the solvent should be stored in a sealed container. *Because Py is toxic, it should be used with adequate ventilation and great care.*

Sulfolane (TMS) [22] Main impurities are water and 3-sulfolene (decomposed to SO_2 and 1,3-dibutadiene at >70 °C), but 2-sulfolene and isopropyl sulfolanyl ether may also be present. For fairly pure commercial products, two vacuum distillations, the second of which is from solid NaOH, can give a distillate applicable for most general purposes. The water level is <5 mM. The distillate should be stored in the dark under a nitrogen atmosphere. For a method of purifying fairly contaminated products, see Ref. [22].

Tetrahydrofuran (THF) [5d] The main impurities are water and peroxides. The same purification procedure as that for dioxane can be applied.

Toluene [5b] Commercial toluene contains water and methylthiophene as major impurities and can be purified in the same way as benzene.

2,2,2-Trifluoroethanol [5e] This solvent is strongly acidic and the major impurities are water and trifluoroacetic acid. In its purification, it is dried over potassium carbonate anhydride for a night and then subjected to fractional distillation (reflux ratio 10: 1) under atmospheric pressure, the middle fraction being collected.

---

4) Coetzee *et al.* (*Electroanalysis*, 1993, **5**, 765) characterized reactive impurities in γ-BL and PC by ion-probe method. They detected in PC triethylamine and chloride ion, both above 10^{-4}M.

References

1 Mann, C.K. (1969) *Electroanalytical Chemistry*, vol. 3 (ed. A.J. Bard), Marcel Dekker, New York, p. 57.

2 (a) Lund, H. (2001) *Organic Electrochemistry*, 4th edn (eds H. Lund and O. Hammerich), Marcel Dekker, New York, Chapter 5; (b) Lund, H. (1991) *Organic Electrochemistry*, 3rd edn (eds H. Lund and M.M. Baizer), Marcel Dekker, New York, Chapter 6.

3 Sawyer, D.T., Sobkowiak, A. and Roberts, J.L., Jr (1995) *Electrochemistry for Chemists*, 2nd edn, John Wiley & Sons, Inc., New York, Chapter 7.

4 Coetzee, J.F. (ed.) (1982) *Recommended Methods for Purification of Solvents and Tests for Impurities*, Pergamon Press, Oxford.

5 (a) Coetzee, J.F. and Chang, T.-H. (1986) *Pure Appl. Chem.*, **58**, 1535 (acetone); (b) Kadish, K.M., Mu, X. and Anderson, J.E. (1989) *Pure Appl. Chem.*, **61**, 1823 (benzene and toluene); (c) Kadish, K.M. and Anderson, J.E. (1987) *Pure Appl. Chem.*, **59**, 703 (benzonitrile,1,1-and 1,2-dichloroethane, and dichloromethane); (d) Coetzee, J.F. and Chang, T.-H. (1985) *Pure Appl. Chem.*, **57**, 633 (dioxane and THF); (e) Marcus, Y. (1990) *Pure Appl. Chem.*, **62**, 139 (1,2-ethanediol and 2,2,2-trifluorothanal); (f) Marcus, Y. (1985) *Pure Appl. Chem.*, **57**, 860 (ethanol); (g) Marcus, Y. and Glikberg, S. (1985) *Pure Appl. Chem.*, **57**, 855 (methanol); (h) Coetzee, J.F. and Chang, T.-H. (1986) *Pure Appl. Chem.*, **58**, 1541 (nitromethane); (i) Marcus, Y. (1986) *Pure Appl. Chem.*, **58**, 1411 (1- and 2-propanol).

6 Riddick, J.A., Bunger, W.B. and Sakano, T.K. (1986) *Organic Solvents, Physical Properties and Methods of Purification*, 4th edn, John Wiley & Sons, Inc., New York.

7 (a) Marcus, Y. (1998) *The Properties of Solvents*, John Wiley & Sons, Inc., New York; (b) Marcus, Y. (2002) *Solvent Mixtures, Properties and Selective Solvation*, Marcel Dekker, New York.

8 (a) Wypych, G. (ed.) (2000) *Handbook of Solvents*, ChemTec Publishing, Toronto; (b) Ash, M. and Ash, I. (2003) *Handbook of Solvents*, 2nd edn, Synapse Information Resources, New York; (c) Henning, H. (ed.) (1993) *Solvent Safety Sheets: A Compendium for the Working Chemist*, Royal Society of Chemistry, Cambridge; (d) Refs [6] and [7].

9 Coetzee, J.F., Deshmukh, B.K. and Liao, C.-C. (1990) *Chem. Rev.*, **90**, 827.

10 Courtot-Coupez, J. and L'Her, M. (1970) *Bull. Soc. Chim. Fr.*, 1631.

11 Gal, J.Y. (1972) Thesis, University of Limoges.

12 Coetzee, J.F. and Martin, M.W., Ref. [4], p. 10.

13 Juillard, J.,Ref. [4], p. 32.

14 Karakatsanis, C.G. and Reddy, T.B., Ref. [4], p. 25.

15 Asthana, M. and Mukherjee, L.M., Ref. [4], p. 47.

16 Fujinaga, T. and Izutsu, K. and Sakura, S., Ref. [4], p. 38.

17 (a) Sleere, N.V. (1976) *J. Chem. Educ.*, **53**, A12; (b) Seebach, D. (1985) *Chem. Britain*, **21**, 632; (c) Chipperfield, J.R. (1999) *Non-aqueous Solvents*, Oxford University Press, Oxford, p. 60.

18 Knecht, L.A., Ref. [4], p. 50.

19 Hoover, T.B., Ref. [4], p. 55.

20 (a) Fujinaga, T. and Izutsu, K., Ref. [4], p. 19; (b) Izutsu, K., Kolthoff, I.M., Fujinaga, T., Hattori, M. and Chantooni, M.K., Jr (1977) *Anal. Chem.*, **49**, 503.

21 Asthana, M. and Mukherjee, L.M.,Ref. [4], p. 44.

22 Coetzee, J.F.,Ref. [4], p. 16.

11
Selection and Preparation of Supporting Electrolytes

In polarography and voltammetry, supporting electrolytes play various roles, i.e. to make the electrolytic solution conductive, to eliminate migration current that may flow in its absence, and to control reaction conditions by varying acid–base property and complexing ability of the solution and by changing the double-layer structure at the electrode. The indifferent electrolyte in potentiometry is also important: it adjusts ionic strength of the solution and gives appropriate reaction conditions to the solution. The selection of supporting electrolytes and indifferent electrolytes must be carefully carried out, just as the selection of solvents. In this chapter, some problems connected to the selection and preparation of supporting electrolytes are discussed. Some of the discussion here will also apply to ionic liquids (IL), which are dealt with in Sections 13.3 and 14.2.

11.1
Selection of Supporting Electrolytes for Electrochemical Measurements

A supporting electrolyte for use in polarography and voltammetry should fulfill the following conditions: (i) it should be soluble in the solvent under study and should dissociate into ions to give enough conductivity to the solution, (ii) it should be resistant to oxidation and reduction and should give a wide potential window, (iii) it does not have an unfavorable effect on the electrode reaction to be measured, (iv) pure product is commercially available at a reasonable cost and (v) it should neither be toxic nor dangerous to handle. Although these conditions are examined one by one in this section, the conclusion is that the supporting electrolytes for organic solvents are fewer in number than those for aqueous solutions. Supporting electrolytes often used in nonaqueous solutions are listed in Table 11.1. More information is available in Ref. [1].

11.1.1
Solubility and Conductivity of Supporting Electrolytes

As described in Section 2.1, the solubility of a crystalline electrolyte is determined by the difference between the lattice Gibbs energy and the solvation energy of

Electrochemistry in Nonaqueous Solutions, Second, Revised and Enlarged Edition. Kosuke Izutsu
Copyright © 2009 WILEY-VCH Verlag GmbH & Co. KGaA, Weinheim
ISBN: 978-3-527-32390-6

Table 11.1 Examples of supporting electrolytes for use in organic solvents.

Solvents	Examples of supporting electrolytes[a]
HOAc	NaOAc, NH$_4$OAc, LiCl, HCl, H$_2$SO$_4$, HClO$_4$, NaClO$_4$, Bu$_4$NClO$_4$, Bu$_4$NBF$_4$
MeOH	NH$_4$Cl, LiCl, HCl, KOH, KOMe, NaClO$_4$, R$_4$N$^+$ salts
en	LiCl, NaNO$_3$, R$_4$N$^+$ salts
DMF	R$_4$N$^+$, Li$^+$ salts of ClO$_4^-$, BF$_4^-$, PF$_6^-$; LiCl, NaClO$_4$
NMP	R$_4$N$^+$, Li$^+$ salts of ClO$_4^-$, BF$_4^-$, PF$_6^-$; LiCl, NaClO$_4$
HMPA	R$_4$N$^+$, Li$^+$ salts of ClO$_4^-$, BF$_4^-$, PF$_6^-$; LiCl, NaClO$_4$
Py	R$_4$N$^+$ salts of ClO$_4^-$, BF$_4^-$; LiCl, LiClO$_4$, LiNO$_3$, NaI
DMSO	R$_4$N$^+$, Li$^+$, Na$^+$, NH$_4^+$ salts of ClO$_4^-$, BF$_4^-$, PF$_6^-$, Cl$^-$, NO$_3^-$
AN	R$_4$N$^+$, Li$^+$ salts of ClO$_4^-$, BF$_4^-$, PF$_6^-$
PC	R$_4$N$^+$, Li$^+$ salts of ClO$_4^-$, BF$_4^-$, PF$_6^-$; NaClO$_4$
TMS	R$_4$N$^+$, Li$^+$ salts of ClO$_4^-$, BF$_4^-$, PF$_6^-$; NaClO$_4$, NH$_4$PF$_6$
NM	Bu$_4$N$^+$, Li$^+$ salts of ClO$_4^-$, BF$_4^-$, PF$_6^-$
NB	Bu$_4$NClO$_4$
THF, DME	Bu$_4$NClO$_4$, NaClO$_4$, LiClO$_4$
CH$_2$Cl$_2$	Bu$_4$N$^+$, Hex$_4$N$^+$ salts of ClO$_4^-$, BF$_4^-$, PF$_6^-$
Benzene	Bu$_4$NBF$_4$, Hex$_4$NClO$_4$

[a]When a mercury electrode is used, alkali metal salts are not appropriate because the potential window is narrow on the negative side.

the electrolyte. For a given electrolyte, the solubility increases with the increase in the solvation energy of the electrolyte, which, in principle, equals the sum of the cationic and anionic solvation energies.

The solubilities of alkali metal halides in various solvents are shown in Table 11.2 [2]. In the table, water can dissolve all of the listed halides. Because water has a high permittivity and moderate acidic and basic properties, the hydration energies of the halides are large enough. Polar protic solvents such as MeOH, HCOOH, FA and NMF can also dissolve many of the halides to a considerable extent. However, in polar protophobic aprotic solvents such as AN and Ac, halides other than NaI are virtually insoluble. It is because both the cations and the anions

Table 11.2 Solubilities of alkali metal halides in various solvents (g/100 g solvent, 25 °C).

Salts	H$_2$O	MeOH	HCOOH	AN	Ac	FA	NMF	DMF
LiCl	55	41.0	27.5	0.14	0.83	28.2	23.0	11(28)
NaCl	36	1.4	5.2	0.0003	0.000042	9.4	3.2	0.04
KCl	36	0.53	19.2	0.0024	0.000091	6.2	2.1	0.017–0.05
RbCl	94	1.34	56.9	0.0036	0.00022			
CsCl	192	3.01	130.5	0.0084	0.00041			0.0052
NaF	4	0.03		0.003	0.0000025			0.0002
NaCl	36	1.4	5.2	0.0003	0.000042	9.4	3.2	0.04
NaBr	46	17.4	19.4	0.040	0.011	35.6	28	3.2(10.3)
NaI	184	83.0	61.8	24.9	28.0	57(85)	61	3.7(6.4)

Table 11.3 Solubilities of tetraalkylammonium salts and conductivities of their solutions (25 °C).

Salts	Solubilities, mol l^{-1}				Conductivities, mS cm^{-1} (concentration, mol l^{-1})			
	AN	DMF	DME	THF	AN	DMF	DME	THF
Et$_4$NClO$_4$	1.13	1.00	<0.01	<0.01	38.5(0.6)	19.2(0.6)	—	—
Et$_4$NBF$_4$	1.69	1.24	<0.01	<0.01	55.5(1.0)	26.3(1.0)	—	—
Et$_4$NCF$_3$SO$_3$	3.10	2.58	—	0.08	41.7(1.0)	20.8(1.0)	—	—
Bu$_4$NClO$_4$	2.05	2.29	1.10	1.48	27.0(0.6)	13.0(0.6)	3.2(1.0)	2.7(1.0)
Bu$_4$NBF$_4$	2.21	2.34	1.70	2.02	32.3(1.0)	14.5(1.0)	4.4(1.0)	2.7(1.0)
Bu$_4$NCF$_3$SO$_3$	2.50	2.25	—	2.35	23.3(1.0)	10.9(1.0)	—	3.3(1.0)
Bu$_4$NBr	1.99	1.57	<0.1	0.14	20.8(0.6)	9.4(0.6)	—	—

are solvated only weakly in such solvents. In DMF, which is a polar protophilic aprotic solvent, the solubilities are often intermediate as expected. Because the supporting electrolytes are usually used in the concentration range of 0.05–1 M, the use of alkali metal halides is rather limited in polar aprotic solvents. When alkali metal salts are to be used as supporting electrolyte in aprotic solvents, lithium or sodium perchlorate, tetrafluoroborate, hexafluorophosphate and trifluoromethane-sulfonate are the candidates because they are fairly soluble.

The electrolytes containing large hydrophobic ions, such as R_4N^+ (R = Bu or larger alkyl), Ph_4As^+ and BPh_4^-, are often insoluble in water because those ions are energetically unstable in water that forms three-dimensional networks (Section 2.2.1). In nonaqueous solvents that do not form such networks, large hydrophobic ions are energetically more stable than in water (Table 2.4) and their salts are considerably soluble. In aprotic solvents, tetraalkylammonium perchlorates (R_4NClO_4), especially Et_4NClO_4 and Bu_4NClO_4, were used for many years as the most popular supporting electrolytes.[1] Et_4NClO_4 and Bu_4NClO_4 are commercially available in high purity, have wide potential windows and rarely unfavorably influence electrode reactions (but see Section 11.1.3). Bu_4NClO_4, which is soluble only slightly in water, is highly soluble in many aprotic solvents, including low-permittivity solvents such as DME ($\varepsilon_r = 7.2$) and THF ($\varepsilon_r = 7.6$) (Table 11.3). However, if the solvent permittivity is even lower as in benzene ($\varepsilon_r = 2.3$) and CH$_2$Cl$_2$ ($\varepsilon_r = 1.6$), Hex$_4$NClO$_4$ and Hep$_4$NClO$_4$ are convenient as supporting electrolytes. The disadvantage of tetraalkylammonium perchlorates is that they are explosive and may explode if they are heated or shocked. They must be handled with enough care and, if possible, their use should be avoided.

Recently, tetraalkylammonium tetrafluoroborates (R_4NBF_4), hexafluoropho-sphates (R_4NPF_6) and trifluoromethanesulfonates ($R_4NCF_3SO_3$) have been widely

---

1) Me$_4$NClO$_4$ and Et$_4$NClO$_4$ are not soluble enough in MeOH and EtOH. On the other hand, R$_4$NClO$_4$ salts with R higher than Pr are difficult to dissolve in water.

used as substitutes for tetraalkylammonium perchlorates (R_4NClO_4). Table 11.3 also shows the solubilities of R_4NClO_4, R_4NBF_4 and $R_4NCF_3SO_3$ in four solvents and the electrical resistivities of their solutions [3]. The solubilities and conductivities for R_4NBF_4 and $R_4NCF_3SO_3$ are near to or larger than those for R_4NClO_4. Especially, Bu_4NBF_4 and $Bu_4NCF_3SO_3$ are highly soluble and give considerable conductivities even in DME and THF. Although not included in the table, R_4NPF_6 are also highly soluble and give solutions of high conductivity. Moreover, all these electrolytes have wide potential windows (see below). They are commercially available and, if necessary, they are easily prepared or purified (Section 11.2). In some cases, novel types of lithium salts ($LiN(SO_2CF_3)_2$, $LiC(SO_2CF_3)_3$, etc.), developed for lithium batteries may also be of use (see Section 12.1). Contrarily, the use of tetraalkylammonium halides is rather limited because they often narrow the potential windows (see below) and unfavorably influence electrochemical reactions.

11.1.2
Potential Windows and Supporting Electrolytes

The potential window of an electrolytic solution is determined by the reduction and oxidation of the supporting electrolyte as well as by those of the solvent. Because all these potentials depend on the indicator electrode used, the following discussion on the relation between the potential window and the supporting electrolyte is dealt with for each of the mercury, platinum and carbon electrodes.

(i) **Mercury electrodes** In an aprotic solvent containing a tetraalkylammonium salt as supporting electrolyte, the reduction of R_4N^+ to R_4N-amalgam determines the negative limit of the potential window.[2]

$$R_4N^+ + e^- + nHg \rightleftharpoons R_4N(Hg)_n$$

When $R_4N^+ \geq Bu_4N^+$, the potential limit is usually between -3.0 and -3.3 V versus Fc/Fc^+ in many aprotic solvents.[3] When $R_4N^+ = Et_4N^+$ or Me_4N^+; however, it shifts to positive direction by 0.1–0.2 V.

In an aprotic solvent containing an alkali metal salt as supporting electrolyte, the reduction of the alkali metal ion to its amalgam determines the negative end of the potential window.

$$M^+ + e^- + Hg \rightleftharpoons M(Hg)$$

---

2) In aprotic solvents, the amalgam of R_4N is somewhat stable at temperatures well below 0 °C, forming a film on the surface of mercury electrode, and its reoxidation reaction can be observed by cyclic voltammetry. However, at ambient temperatures or in the presence of water and other protic solvents, the amalgam soon decomposes.

3) The negative potential limit at a mercury electrode in 0.1 M Bu_4NClO_4 (25 °C, V versus Fc/Fc^+) is -2.55 in MeOH, -3.14 in THF, -3.25 in γ-BL, -3.32 in PC, -1.6 in FA, -3.12 in NMF, -3.35 in DMF, -3.3 in DMA, -3.26 in NMP, -3.2 in TMU, -3.26 in AN, -1.41 in NM, -2.36 in CH_2Cl_2, -2.51 in 1,2-DCE, -3.35 in DMSO and -3.0 in HMPA. From Gritzner, G. (1990) *Pure Appl. Chem.*, **62**, 1839.

Because alkali metals are stabilized by amalgam formation, the reductions of alkali metal ions occur at 0.5–1.0 V more positive potential than those of R_4N^+ ions. Therefore, for polarographic measurements at negative potentials, alkali metal salts are not suitable. Moreover, alkali metal ions often have unfavorable influences on electrode reactions (Section 11.1.3).

When the anion of the supporting electrolyte is ClO_4^-, BF_4^-, PF_6^- or $CF_3SO_3^-$, the positive limit of the potential window is determined by the anodic dissolution of mercury ($Hg \rightarrow Hg^{2+}$ or Hg_2^{2+}). However, when it is a halide ion, the potential of the anodic mercury dissolution shifts to negative direction because the halide ion forms a precipitate with Hg_2^{2+}. The negative shift is more marked in aprotic solvents than in water (Section 8.2.3). Therefore, halides are not suitable as supporting electrolyte in aprotic solvents.

(ii) **Platinum electrodes** In protic solvents, the reduction of the solvent to form molecular hydrogen ($SH + e^- \rightarrow 1/2H_2 + S^-$) determines the negative potential limit.

In aprotic solvents, however, the potential on the negative side is often limited by the reduction of the cation of the supporting electrolyte. When the cation is R_4N^+, it is reduced to its radical ($R_4N^\bullet$), which immediately decomposes. When the cation is an alkali metal ion, it is usually reduced to deposit as its metal on the electrode. Lithium and sodium ions in basic solvents are strongly solvated, and their reductions occur at more negative potentials than those of R_4N^+ ions. As an extreme, in the solution of lithium or sodium ion in HMPA (strongly basic), deep-blue solvated electrons are generated from the platinum electrode into the solution at the negative end of the potential window. In some cases, the negative region of the potential window is wider at a platinum electrode than at a mercury electrode (see footnote 3 and Table 8.1). However, at a platinum electrode, the negative potential region is narrowed significantly by the presence of small amounts of protic impurities including water. The influence is especially marked in protophobic aprotic solvents such as AN and PC (Figure 10.1).

The potential limits on the positive side are also shown in Table 8.1. Halide ions (X^-) are not suitable for measurements at positive potentials because they are easily oxidized to X_2 or X_3^- at a platinum electrode (Section 8.2.3). On the contrary, perchlorate ions are not easily oxidized. Thus, when a perchlorate is used as supporting electrolyte, the positive potential limit is determined by the oxidation of the solvent if it is a strong electron donor (DMF, DMSO or HMPA), but by the oxidation of ClO_4^- if the solvent is a weak electron donor (AN, PC, NM or TMS). Here, however, it has been shown by ESR measurement that the oxidation product of ClO_4^- in AN gradually reacts with AN ($ClO_4^- \rightarrow ClO_4^\bullet + e^-$, $ClO_4^\bullet + CH_3CN$ $HClO_4 + {}^\bullet CH_2CN$, $2{}^\bullet CH_2CxN \rightarrow NCCH_2-CH_2CN$) [4]. For the measurements at positive potentials, the salts of BF_4^- and PF_6^- are preferable to the salts of ClO_4^- and $CF_3SO_3^-$. For instance, the positive potential limit at a platinum electrode in AN is $+2.48$ V (versus $Ag/0.1M\ Ag^+$ and at $10\ mA\ cm^{-2}$) for ClO_4^-, $+2.91$ V for BF_4^- and $+3.02$ V for PF_6^- [5]. As shown in Table 8.1, the positive potential limit in NM is $+2.3$ V (versus Fc/Fc^+) for ClO_4^- and $+2.9$ V for BF_4^-. Here, if the solution of BF_4^- in NM is pre-electrolyzed at a positive potential, the potential limit shifts to $+3.8$ V. This large positive shift has been attributed to the elimination of trace residual water by pre-electrolysis, indicating the serious influence of water on potential windows.

(iii) **Carbon electrodes** In protic solvents, the negative potential region at a glassy carbon (GC) electrode is a little wider than that at a platinum electrode because the hydrogen overpotential is larger at a GC electrode. In aprotic solvents, however, the potential window at a GC electrode is not much different from that at a platinum electrode. The potential window at a GC electrode in PC (1 mA cm^{-2}) is $+2.6$ to -3.2 V (versus Fc/Fc^+) for 0.1 M $LiClO_4$, $+3.6$ to -3.2V for 0.1 M KPF_6, $+0.4$ to -3.0 V for 0.1 M Et_4NCl and $+3.6$ to -3.0 V for 0.1 M Bu_4NPF_6 [6]. Ue *et al.* [7] studied the potential window at a GC electrode using various supporting electrolytes in PC and using 0.65 M Et_4NBF_4 in various solvents. In 0.65 M $Et_4NBF_4 + \gamma$-BL, the width of the potential window reached 8.2 V. Recently, a boron-doped diamond thin-film electrode has been attracting attention for its wide potential windows and low residual currents (see footnote 8 in Chapter 5), although its application in nonaqueous solvents is still scarce [8, 9]. Its potential window in 0.1 M Bu_4NPF_6–AN/toluene mixture is said to be $\sim$0.5 V wider than that at a platinum electrode.

If the supporting electrolyte and the electrode material are chosen appropriately, the potential window in such protophobic aprotic solvents as AN, NM, PC and TMS easily exceeds 6 V (Table 8.1, see also footnote 18 in Chapter 8). In aqueous solutions, the potential window never exceeds 4.5 V, even when a mercury electrode is used on the negative side and a diamond electrode on the positive side. This difference is important not only for electrochemical measurements but also for electrochemical technologies of, for example, rechargeable batteries and supercapacitors. For more information on the potential windows in nonaqueous solutions, see Ref. [10].

11.1.3
Influences of Supporting Electrolytes on Electrode Reactions in Nonaqueous Solutions

The supporting electrolytes influence polarographic and voltammetric measurements in the manner as described below:

(a) The cation or anion of the supporting electrolyte reacts at the electrode: at a mercury electrode, the negative side of the potential window is narrowed by an alkali metal ion that is reduced to its amalgam, while the positive side by a halide ion that facilitates the anodic dissolution of mercury. At a platinum or carbon electrode, a halide ion is anodically oxidized and narrows the positive side of the potential window. These influences of halide ions are also observed in water but more markedly in aprotic solvents. With tetraalkylammonium salts of ClO_4^-, BF_4^-, PF_6^- and $CF_3SO_3^-$, these influences can be avoided.

(b) The cation or anion of the supporting electrolyte interacts with the electroactive substances: a typical example is when a halide ion forms a complex with an electroactive metal ion and shifts the reduction potential of the metal ion to negative direction. Sometimes a halide ion forms a precipitate with an electroactive metal ion. These influences also occur in water but are more pronounced in aprotic solvents. With tetraalkylammonium salts of ClO_4^-, BF_4^-, PF_6^- and $CF_3SO_3^-$, these influences rarely occur.

(c) The cation or anion of the supporting electrolyte interacts with the product or intermediate of the electrode reaction: (i) In Section 8.2.2, a lithium ion shifted the reduction wave of tris(acetylacetonato)iron(III), $[Fe(acac)_3]$, in AN to positive direction, by liberating $acac^-$ from the reduction product, $Fe(acac)_3^-$. (ii) In Section 8.3.1, lithium and sodium ions shifted the two reduction waves of 1,2-naphthoquinone (Q) in AN to positive direction, by forming ion pairs with $Q^{\bullet-}$ and Q^{2-}. These influences are remarkable with small cations (strong Lewis acids) such as Li^+ and in protophobic aprotic solvents such as AN. On the contrary, they are not significant with large cations (weak Lewis acids) such as R_4N^+ and in protophilic aprotic solvents such as DMF and DMSO.

(d) The cation of the supporting electrolyte influences the electrical double layer: A typical example is the influence of the cation of the supporting electrolyte on the polarographic reduction of Na^+ in HMPA [11]. As shown in Figure 8.6b, Na^+ in HMPA gives a diffusion-controlled wave when the cation of the supporting electrolyte is Li^+ or Hep_4N^+ and a small (kinetic-controlled) reduction wave when the cation is Bu_4N^+, but not reduced at all when the cation is Et_4N^+ or Me_4N^+. This influence is the so-called electrochemical masking: Et_4N^+ and Me_4N^+, with small solvated radii, are preferentially attracted electrostatically onto the negatively charged electrode surface. The resulting double-layer structure works unfavorably for the reduction of Na^+, which is strongly solvated and much larger in size than Et_4N^+ and Me_4N^+. These interfering effects of small R_4N^+ ions on the reduction of metal ions in strongly basic solvents were discussed in Section 8.2.1 [11, 12].[4] On the contrary, the double-layer structure obtained with large R_4N^+ ions tends to work unfavorably on the reduction of the radical anion of an organic compound ($Q^{\bullet-} \rightarrow Q^{2-}$), as described in Section 8.3.1. By the same reason, the second reduction wave of dissolved oxygen ($O_2^{\bullet-} \rightarrow O_2^{2-}$) in HMPA was at -1.9 V (versus Ag/0.01 M Ag^+) with Et_4NClO_4 (0.05 M) and -2.3_5 V with Bu_4NClO_4, but it did not appear until the negative end of the potential window (~ -3.0 V) with Hex_4NClO_4 [13]. In addition to these,

---

4) The influence of this type is specific to aprotic solvents. In aqueous solutions and at very negative potentials, water molecules as electron acceptor are strongly attracted to the negatively charged electrode surface. A tetraalkylammonium ion (R_4N^+), which is specifically adsorbed near the potential of zero charge, tends to be desorbed at very negative potentials, being replaced by water molecules. In aprotic solvents, however, solvent molecules are not strongly attracted onto the negatively charged electrode surface because they are weak as electron acceptor. Thus, R_4N^+ tends to be *directly* attracted to the electrode and the tendency is more pronounced with a smaller R_4N^+.

Petersen and Evans [14], Evans and Gilicinski [15] and Fawcett *et al.* [16] found that the standard rate constants (k_s) in AN and PC for the reduction of organic compounds at negative potentials decreased with the increase in the size of R_4N^+ (Section 8.3.1). The decrease in k_s with the increase in the size of R_4N^+ was interpreted by the decrease in electron tunneling rate with the increasing thickness of the compact layer of adsorbed R_4N^+ ions.

Although tetraalkylammonium salts are most frequently used as supporting electrolyte in aprotic solvents, it should be noted that even tetraalkylammonium ions give significant influences on electrode reactions. An appropriate R_4N^+ should be selected for each measurement.

The above is the general guideline for the selection of the supporting electrolytes in nonaqueous solutions. The supporting electrolytes that are relatively versatile are as follows (R = Et or Bu):

For electrode reductions $NaClO_4$ (for Pt electrode), $LiClO_4$ (for Pt electrode), R_4NX, R_4NClO_4, R_4NBF_4, R_4NPF_6, $R_4N(CF_3SO_3)$.

For electrode oxidations $LiClO_4$, $LiBF_4$, $LiPF_6$, R_4NClO_4, R_4NBF_4, R_4NPF_6, $R_4N(CF_3SO_3)$.

All these electrolytes are neutral in Brønsted acid–base properties. Although rather exceptional, an acid, a base or a pH buffer may be added to the supporting electrolyte of neutral salts. The acid–base system to be selected depends on the purpose of the measurement. We often use trifluoromethanesulfonic acid (CF_3SO_3H) as a strong acid; acetic acid, benzoic acid or phenol as a weak acid; an amine or pyridine as a weak base and tetraalkylammonium hydroxide (R_4NOH) as a strong base. Examples of buffer systems are the mixtures of picric acid and its R_4N salt and amines and their $HClO_4$ salts. Here, we should note that the acid–base reactions in aprotic solvents considerably differ from those in water, as discussed in Chapter 3.

11.2
Methods for Preparing and Purifying Supporting Electrolytes

Tetraalkylammonium salts are frequently used as supporting electrolyte in non-aqueous solutions, although alkali metal salts such as $NaClO_4$, $LiClO_4$, $LiBF_4$ and $LiPF_6$ are also used in some cases. Nowadays, various tetraalkylammonium salts for use as supporting electrolyte are commercially available as pure products. We only need to dry them before use. Here, however, the methods to synthesize and purify some tetraalkylammonium salts are described. To synthesize and purify those electrolytes is not difficult, and the readers can try themselves if necessary, although they must be careful in handling substances that are explosive, corrosive or harmful to humans and to the environment.

1. **Tetraalkylammonium bromides** Et_4NBr: To the mixture of 1 mol of triethylamine (Et_3N) and 1 mol of ethyl bromide (EtBr) in a flask, add AN of the same volume as the mixture and reflux it for several hours. Collect the crystal of Et_4NBr formed,

recrystallize it several times from EtOH and then dry it under vacuum at 100 °C.

Bu$_4$NBr: To the mixture of 1 mol of tributylamine (Bu$_3$N) and 1 mol of butyl bromide (BuBr) in a flask, add AN of the same volume as the mixture and reflux it for 36 h. Obtain the Bu$_4$NBr crystal by removing AN with the aid of a rotating evaporator, recrystallize it several times from ethyl acetate and then dry it under vacuum at 80 °C.

2. **Tetraalkylammonium perchlorates** Et$_4$NClO$_4$: Dissolve 5.3 g of Et$_4$NBr (25 mmol) in ~8 ml of water, add 2.1 ml of 70% HClO$_4$ (~26 mmol) to precipitate Et$_4$NClO$_4$ and filter after cooling to collect the precipitate. Recrystallize the precipitate three times with water and dry it at 70 °C under vacuum. *Because Et$_4$NClO$_4$ is explosive, treat a little amount of it at a time and do not keep it at a temperature above 70°C.*

Bu$_4$NClO$_4$: Dissolve 8.4 g of Bu$_4$NBr (25 mmol) in 18 ml of water, add 2.1 ml of 70% HClO$_4$ (~26 mmol) to it to precipitate Bu$_4$NClO$_4$, collect the precipitate by filtration and wash it with cold water. Recrystallize it three times from an ethyl acetate–pentane mixture or from EtOH and dry it at 70 °C under vacuum. *Because Bu$_4$NClO$_4$ is an explosive, treat a little amount of it at a time and do not keep it above 70°C.*

Hex$_4$NClO$_4$: Dissolve commercially available Hex$_4$NI or Hex$_4$NBr in water and add dilute aqueous HClO$_4$ to it to precipitate Hex$_4$NClO$_4$. Collect the precipitate by filtration and wash it with cold water. Then, recrystallize it three times from ethyl acetate and dry it at 60 °C under vacuum. *Because Hex$_4$NClO$_4$ is explosive, treat a little amount of it at a time and do not keep it above 60°C.*

3. **Tetraalkylammonium tetrafluoroborates** [3] Et$_4$NBF$_4$: Dissolve 5.3 g of Et$_4$NBr (25 mmol) in ~8 ml of water, add 3.6 ml of 48–50% HBF$_4$ (~26 mmol) to it and concentrate by heating. Then, dilute it with ethyl ether and filter it to get crude precipitate of Et$_4$NBF$_4$. Recrystallize the precipitate twice from a MeOH–petroleum ether mixture to get pure crystal of Et$_4$NBF$_4$. Before use, powder the crystal and dry at 80–100 °C for several days under vacuum, mp 377–378 °C.

Bu$_4$NBF$_4$: Dissolve 8.4 g of Bu$_4$NBr (25 mmol) in ~18 ml of water, add 3.6 ml of 48–50% HBF$_4$ (~26 mmol) to it and stir the resulting mixture at 25 °C for 1 min. Collect the crystalline salt on a filter, wash it with water until the washings are neutral and then dry it. Recrystallize the crude Bu$_4$NBF$_4$ three times from ethyl acetate–pentane to get pure crystal of Bu$_4$NBF$_4$. Before use, dry the crystal by the procedure described above, mp 162–162.5 °C.

4. **Tetrabutylammonium hexafluorophosphate** [17] Bu$_4$NPF$_6$: Dissolve 100 g of Bu$_4$NI (0.27 mol) in a minimum amount of acetone, and add, with stirring, an acetone solution that contains 50 g (0.31 mol) of NH$_4$PF$_6$. It is important to maintain at least a 5% excess of PF$_6^-$. Remove the precipitate of NH$_4$I by filtration and slowly add water to the filtrate to precipitate Bu$_4$NPF$_6$. Collect the precipitate on a filter and wash it several times with water (removal of ammonium salts). Precipitate the crude product again from a 5% solution of NH$_4$PF$_6$ to remove iodide ion completely. Recrystallize the precipitate from EtOH–water and dry it for 10 h under vacuum at 100 °C. The yield is ~80%.

5. **Tetraalkylammonium trifluoromethanesulfonates** [3] $Et_4N(CF_3SO_3)$: Add 5 g of CF_3SO_3H (33 mmol) to 50 g of 10% Et_4NOH (34 mmol), stir it for 30 min and dry it using a rotating evaporator to get crude $Et_4N(CF_3SO_3)$. Recrystallize it three times from THF and dry at vacuum. An alternative procedure is to recrystallize the crude product from acetone–ethyl ether (1:9) and to dry it for 24 h at 60 °C at vacuum and in the presence of P_2O_5, mp 160–161 °C.

   $Bu_4N(CF_3SO_3)$: Dissolve 9.7 g of Bu_4NBr (30 mmol) in 30 ml of water and, with vigorous stirring, slowly add to it 4.5 g of CF_3SO_3H (30 mmol). After cooling to room temperature, filter the mixture and wash the precipitate with water. Recover the product remaining in the filtrate by extracting with dichloromethane and add it to the above precipitate. Recrystallize the precipitate twice from dichloromethane–ether and dry at vacuum, mp 111–112.5 °C. As an alternative, the method to neutralize Bu_4NOH with CF_3SO_3H can be employed [18].

6. **Other tetraalkylammonium salts** Other tetraalkylammonium salts are also used in electrochemical measurements; they are tetraalkylammonium nitrates, picrates, carboxylates, sulfonates, etc. They can be prepared in the laboratory by neutralizing the corresponding acid in water with R_4NOH just to the equivalence point, removing water and then drying. If necessary, the products are recrystallized. Some tetraalkylammonium salts form hydrates and are difficult to completely dehydrate. For practical information, see, for example, Ref. [19].

7. **Air- and water-stable ionic liquids** Some of the air- and water-stable ionic liquid may be used as supporting electrolytes in nonaqueous solvents. Many of the ILs are commercially available, but they can be synthesized by the general route as follows [20]: The first step is the alkylation of the organic base using haloalkane, $B + R-X \rightarrow [B-R]^+X^-$. Here, B = N-methylimidazole, N-methylpyrrolidine, pyridinium, etc., R = methyl, ethyl, propyl, etc. and X = chloride, bromide, iodide, trifluoromethanesulfonate, etc. The second step is the anion exchange in the water/organic solvent system, $[B-R]^+X^- + M^+A^- \rightarrow [B-R]^+A^-$ (organic phase) $+ M^+X^-$ (aqueous phase). Here, M = H or alkali metal (Li, Na, etc.) and A = $[BF_4]$, $[PF_6]$, etc. The ionic liquid is extracted into the organic phase, leaving acid or metal salt in aqueous phase. The ionic liquid is washed with water repeatedly to remove the halide impurities. The more hydrophilic the ionic liquid, the more difficult it is to purify because extraction of halide with water is accompanied by a considerable loss of ionic liquid to aqueous phase.

References

1 For example, (a) Mann, C.K. (1969) *Electroanalytical Chemistry*, vol. 3 (ed. A.J. Bard), Marcel Dekker, New York, p. 57; (b) Lund, H. (1991) *Organic Electrochemistry*, 3rd edn (eds H. Lund and N.M. Baizer), Marcel Dekker, New York, p. 304; (c) for new types of electrolytes, Barthel, J., Gores, H.-J., Neueder, R. and Schmid, A. (1999) *Pure Appl. Chem.*, **71**, 1705 and references therein.

2 Burgess, J. (1978) *Metal Ions in Solution*, Ellis Horwood, Chichester, p. 220.

3 (a) House, H.O., Feng, E. and Peet, N.P. (1971) *J. Org. Chem.*, **36**, 2371; (b) Rousseau, K., Ferrington, G.C. and Dolphin, D. (1972) *J. Org. Chem.*, **37**, 3968.

4 Maki, A.H. and Geske, D.H. (1959) *J. Chem. Phys.*, **30**, 1356.

5 Fleischmann, M. and Pletcher, D. (1968) *Tetrahedron Lett.*, 6255.

6 Gross, M. and Jordan, J. (1984) *Pure Appl. Chem.*, **56**, 1095.

7 Ue, M., Ida, K. and Mori, S. (1994) *J. Electrochem. Soc.*, **141**, 2989.

8 For review, Xu, J., Granger, M.C., Chen, Q., Strojek, J.W., Lister, T.E. and Swain, G.M. (1997) *Anal. Chem. News Features*, **69**, 591A.

9 For the application in nonaqueous solvents (a) Alehasham, S., Chambers, F., Strojek, J.W., Swain, G.M. and Ramesham, R. (1995) *Anal. Chem.*, **67**, 2812; (b) Wu, Z., Yano, T., Tryk, D.A., Hashimoto, K. and Fujishima, A. (1998) *Chem. Lett.*, 503.

10 Aurbach, D. and Gofer, Y. (1999) *Nonaqueous Electrochemistry* (ed. D. Aurbach), Marcel Dekker, New York, Chapter 4.

11 (a) Izutsu, K., Sakura, S., Kuroki, K. and Fujinaga, T. (1971) *J. Ekectroanal. Chem.*, **32**, app. 11; (b) Izutsu, K., Sakura, S. and Fujinaga, T. (1972) *Bull. Chem. Soc. Jpn.*, **45**, 445; 1973, **46**, 493, 2148.

12 (a) Baranski, A.S. and Fawcett, W.R. (1978) *J. Electroanal. Chem.*, **94**, 237; (b) Baranski, A.S. and Fawcett, W.R. (1980) *J. Chem. Soc., Faraday Trans. I*, **76**, 1962.

13 Fujinaga, T. and Sakura, S. (1974) *Bull. Chem. Soc. Jpn.*, **47**, 2781.

14 Petersen, R.A. and Evans, D.H. (1987) *J. Electroanal. Chem.*, **222**, 129.

15 Evans, D.H. and Gilicinski, A.G. (1992) *J. Phys. Chem.*, **96**, 2528.

16 Fawcett, W.R., Fedurco, M. and Opallo, M. (1992) *J. Phys. Chem.*, **96**, 9959.

17 Fry, A.J. (1996) *Laboratory Techniques in Electroanalytical Chemistry*, 2nd edn (eds P.T. Kissinger and W.R. Heinemann), Marcel Dekker, New York, Chapter 14.

18 (a) Fujinaga, T. and Sakamoto, I. (1976) *J. Electroanal. Chem.*, **67**, 201, 235; 1977, **85**, 185; (b) Fujinaga, T. and Sakamoto, I. (1980) *Pure Appl. Chem.*, **52**, 1387.

19 Kolthoff, I.M. and Chantooni, M.K., Jr (1966) *J. Phys. Chem.*, **70**, 856.

20 (a) Beyersdorff, T., Schubert, T.J.S., Welz-Biermann, U., Pitner, W., Abbott, A.P., McKenzie, K.J. and Ryder, K.S. (2008) *Electrodeposition from Ionic Liquids* (eds F. Endres, A.P. Abbott and D.R. MacFarlane), Wiley-VCH Verlag GmbH, Weinheim, Chapter 2; (b) Gordon, C.M. and Muldoon, M.J. (2008) *Ionic Liquids in Synthesis*, 2nd edn, vol. 1 (eds P. Wasserscheid and T. Welton), Wiley-VCH Verlag GmbH, Weinheim, Section 2.1; (c) Wagner, M. and Hilgers, C. (2008) *Ionic Liquids in Synthesis*, 2nd edn, vol. 1 (eds P. Wasserscheid and T. Welton), Wiley-VCH Verlag GmbH, Weinheim, Section 2.2.

12
Use of Nonaqueous Solutions in Modern Electrochemical Technologies

Recently, applications of nonaqueous solutions are increasing in the field of modern electrochemical technologies. In this chapter, examples of such applications of nonaqueous solutions are outlined. The technological aspects of nonaqueous electrochemistry have been dealt with in books [1] and review articles [2].

12.1
Batteries Using Nonaqueous Solutions – Lithium Batteries

(a) **Primary lithium batteries** [1] Primary lithium batteries, which are commercially available since 1970s, are still used as power sources for cameras, watches, calculators and various other portable electronic devices. As a typical example, the schematic diagram of the coin-type Li/MnO_2 battery is shown in Figure 12.1. It consists of an anode (negative electrode), a cathode (positive electrode), an electrolyte solution and a porous separator. The anode is Li metal, the cathode is MnO_2 and the electrolyte solution is ~ 1 M $LiClO_4$ or $Li(SO_3CF_3)$ in a nearly 1: 1 mixture of propylene carbonate (PC) and 1,2-dimethoxyethane (DME). The porous separator is an unwoven cloth of polyolefins (e.g. polypropylene). The voltage of this cell is about 3.0 V and the reactions during the discharge are as follows:

At the anode	$Li \rightarrow Li^+ + e^-$
At the cathode	$Li^+ + Mn(IV)O_2 + e^- \rightarrow Mn(III)O_2(Li^+)$
Total reaction	$Li + Mn(IV)O_2 \rightarrow Mn(III)O_2(Li^+)$

At the anode, metallic lithium dissolves as lithium ion (Li^+) and at the cathode, Li^+ diffuses into the crystal lattice of manganese dioxide.

Electrolyte solutions of various aprotic organic solvents are used in primary lithium batteries. Among the organic solvents are alkyl carbonates (PC ($\varepsilon_r = 64.4$), ethylene carbonate (EC, $89.6_{40°C}$), dimethyl carbonate (DMC, 3.1), diethyl carbonate (DEC, 2.8)), ethers (DME (7.2), tetrahydrofuran (THF, 7.4), 2-Me-THF (6.2), 1,3-dioxolane (DIOX, 7.1)) and esters (methyl formate (MF, $8.5_{20°C}$), γ-BL

Electrochemistry in Nonaqueous Solutions, Second, Revised and Enlarged Edition. Kosuke Izutsu
Copyright © 2009 WILEY-VCH Verlag GmbH & Co. KGaA, Weinheim
ISBN: 978-3-527-32390-6

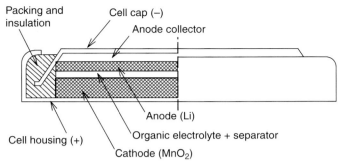

Packing and insulation

Cell cap (−)

Anode collector

Cell housing (+)

Anode (Li)

Organic electrolyte + separator

Cathode (MnO₂)

Figure 12.1 A coin-type Li/MnO$_2$ primary battery.

(39.1)), where the values in parentheses show the solvent permittivity at 25 °C, unless otherwise stated. The solvents of high permittivity and high viscosity are often mixed with the solvents of low permittivity and low viscosity to get highly conductive electrolyte solutions. Among the lithium electrolytes, on the other side, are LiClO$_4$, LiBF$_4$, LiPF$_6$, LiAsF$_6$, LiOSO$_2$CF$_3$, LiN(SO$_2$CF$_3$)$_2$ and LiC (SO$_2$CF$_3$)$_3$ (see Ref. [2a] for other new types of lithium electrolytes).[1] They are stable, dissolve and dissociate easily and give high conductivities. As cathode active materials, polycarbon fluoride (CF)$_n$ and manganese dioxide are typical.

Lithium is the lightest metal, and the Li$^+$/Li electrode has a very negative potential. Thus, the primary lithium batteries have high emfs ($\sim$3.5 V) and working voltages ($\sim$3.0 V), high energy densities, long lives ($\sim$10 years) and wide working temperature ranges (-40 to $+70$ °C). For detailed performance data, see Ref. [1b].

(b) **Secondary lithium batteries** [1, 3] The reaction of the Li$^+$/Li electrode in the primary lithium battery can proceed inversely. Thus, the lithium battery is rechargeable, if a reversible positive electrode is employed.[2] However, the Li$^+$/Li electrode has a problem as the lithium metal loses its smooth surface by repeated charging/discharging cycles, forming dendritic deposits, which may cause an internal electrical short circuit between the anode and the cathode. This difficulty can be overcome by replacing the liquid electrolyte (solvent + salt) with an ion-conducting gel-polymer electrolytes (GPEs) (polymer + solvent + salt) [4].

1) Conductivities ($\times 10^{-3}$ S cm^{-1}) for 1 M solutions in 1:1 PC–DME at 25 °C: LiBF$_4$ 9.46, LiCF$_3$SO$_3$ 6.12, Li(CF$_3$SO$_2$)$_2$N 12.6, LiClO$_4$ 13.5, LiAsF$_6$ 14.8, LiPF$_6$ 15.3 (Webber, A. (1991) *J. Electrochem. Soc.*, **138**, 2586). LiF is insoluble in DME but, in DME containing 1 M tris(pentafluorophenyl)borane (B(C$_6$F$_5$)$_3$, an anion receptor), LiF is soluble up to 1 M and gives a conductivity of 6.8×10^{-3} S cm^{-1} at 25 °C. It is because F$^-$ is combined with B (C$_6$F$_5$)$_3$ to form a big anion with a delocalized charge. (Sun, X., Lee, H.S., Lee, S., Yang, X.Q., Mcbreen, J. (1998) *Electrochem. Solid State Lett.*, **1**, 239).

2) For reversible positive electrodes, channel- or layer-structured materials such as MnO$_2$, MoO$_3$, V$_2$O$_3$, MoS$_2$ and TiS$_2$ and conducting polymers are used. An example of the battery with a positive electrode of conducting polymer is Li/LiClO$_4$–PC/polypyrrole. In the charging process, the oxidation of the polymer and the insertion of ClO$_4^-$ from the electrolyte solution occur concurrently. In the discharging process, the doped ClO$_4^-$ is released from the polymer. The performance of the Li/polymer battery, however, is limited by such factors as slow charging/discharging rate, self-discharge and low energy content.

In the lithium-ion secondary battery, which was put on the market in 1990, the difficulty of the Li^+/Li electrode was avoided by the use of a graphite negative electrode (anode) (C_y), which works as a host for Li^+ ions by intercalation. The active materials for the positive electrode (cathode) are $LiCoO_2$, $LiNiO_2$, $LiMn_2O_4$ and $LiFePO_4$, which also work as a host for Li^+ ions. Electrolyte solutions used in the primary lithium batteries can be used, in principle, also in lithium-ion secondary batteries. But the majority of manufacturers use electrolyte solutions composed of EC, $LiPF_6$ and one or more linear carbonates from DME, DEC and ethylmethyl carbonate (EMC), although one manufacturer uses (γ-BL + $LiBF_4$) as an electrolyte solution. The linear carbonate(s) is added to increase the fluidity and reduce the melting point of the electrolyte solution. A latest comprehensive review is available for nonaqueous liquid electrolytes for lithium-based rechargeable batteries [5].

A schematic diagram of a lithium-ion battery is shown in Figure 12.2. The cell reaction for positive electrode of $LiCoO_2$ is as follows:

$$LiCoO_2 + C_y \xrightarrow{\text{first charging}} Li_{1-x}CoO_2 + Li_xC_y \underset{\text{charging}}{\overset{\text{discharging}}{\rightleftharpoons}} Li_{1-x+\Delta x}CoO_2 + Li_{x-\Delta x}C_y$$

In the charging process, Co(III) in $LiCoO_2$ is oxidized to Co(IV). This battery has a working voltage of ~3.7 V and a cycle number of >1000. Lithium-ion batteries of

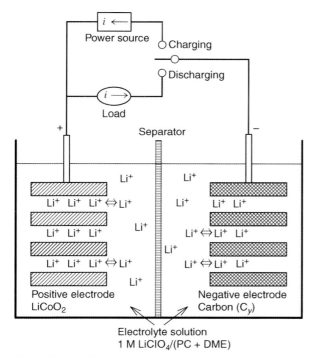

Figure 12.2 A lithium-ion secondary battery.

small to moderate sizes are widely used as power sources for computers, telephones, cameras, camcorders, etc. Moreover, the applications of large-size lithium-ion batteries for electric vehicle (EV) or hybrid electric vehicle (HEV) and for electricity storage are now really beginning.

Because of the importance of high-performance secondary batteries, various efforts have been made to improve the technologies of the lithium-ion batteries. The current state-of-the-art electrolyte systems (EC, $LiPF_6$ and linear carbonate(s) from DME, DEC and EMC) are not perfect and need improvement, at least, in the following four points: (1) The linear carbonates are highly flammable with flash points usually below 30 °C. The flammable solvents may cause a fire, and the seriousness of the hazard is proportional to the size of the battery. Flame-retardant or nonflammable lithium-ion electrolytes are needed especially for vehicle traction batteries. (2) The usable temperature range of the above electrolyte systems is limited between −20 and 50 °C. EC (mp: 36.4 °C) is responsible for the lower temperature limit and $LiPF_6$ for the higher temperature limit. This temperature range should be widened, especially for batteries of space and vehicle traction uses. (3) Formation of insoluble films (Li_2O, lithium alkyl carbonate) on the surface of the anode (and cathode also) more or less causes the loss of lithium ions. During the charging/discharging process, the insoluble film formed in EC tends to protect the surface from further formation of the film, although the insoluble film formed in PC tends to cause, due to the PC cointercalation, the disintegration of a graphene structure by an exfoliation process. EC is superior to PC in this respect, but a more improved film formation is desirable. (4) Better ionic transport is desirable. Although the impedances at the anode/electrolyte and cathode/electrolyte interfaces are more important than the bulk ion conductivity, the impedances at the two interfaces tend to decrease with the increase in the bulk ion conductivity.

In the investigations for improving the state-of-the-art electrolyte solutions, two methods have been applied: one is to add additives to the electrolyte solutions and the other is to use new solvent/electrolyte systems. To get low flammability by additives, gas-evolving ingredients (e.g. pyrocarbonates that release CO_2) or phosphorous-based organic molecules (e.g. phosphate esters) are used. An example is to add a substituted cyclic phosphorimide, hexamethoxycyclophosphazene, in a level ≤10% to $EC/DMC/LiPF_6$. On the other hand, to get low flammability by new solvents, fluorinated esters and fluorinated carbonates may be used, alone or as a cosolvent. Using fluorinated solvents improves, in addition to flammability, the low-temperature performances and the stabilities of insoluble films on both electrodes. Details are not described here because this topic has thoroughly been reviewed in Ref. [5].

There are two types of lithium secondary batteries using polymers, one is based on solid polymer electrolytes (SPEs) and the other is based on GPE. SPEs, consisting of pure polymers (e.g. poly(ethylene oxide), PEO) that dissolve lithium salts, have low conductivities and their applications are limited only to batteries of low current densities. In GPEs, conventional electrolyte solutions may be used, but their spilling from batteries can be avoided by gel formation with polymers (PEO or poly(vinylidene fluoride-co-hexafluoropropylene) (PVdF–HFP); see Ref. [4] for details).

Plastic sheet-type lithium-ion batteries have been developed for EV and HEV [6]; for example, the cathode and anode active materials were $LiCoO_2$ and synthetic graphite material, respectively. These materials were fabricated into thin films with Super P carbon black, a binder of PVdF–HFP, and a plasticizer of dibutylphthalate (DBP). The separator was made of PVdF–HFP copolymer, silicate powder and DBP that was later removed by extraction. The electrolyte was a $LiPF_6$ solution in EC–DMC–EMC. The cathodes ($23.6\,cm \times 13.0\,cm$) were fabricated by laminating two cathode films containing 65–73% active material by weight, with an aluminum current collector between the films. The anodes ($24.0\,cm \times 13.3\,cm$) were also prepared by laminating two films containing 65% (or 73%) active material, with an expanded metal copper current collector between the films. Unit bi-cell laminates having a cathode/separator/anode/separator/cathode layer structure were fabricated by heat-laminating the component sheets together. The bi-cells were prepared by extracting DBP using either ether or methanol followed by drying, adding the electrolyte, and packaging.

Recently, anode materials other than carbon have been developed for lithium-ion batteries. Promising examples among them are silicon nanowires synthesized on the stainless steel substrates [7a] and nanostructured Ni_3Sn_4 intermetallic compound electrodeposited on the Cu nanorods current collector [7b].[3] In the latter case, the first activation step, $Ni_3Sn_4 + 17.6Li^+ + 17.6e^- \rightarrow 4Li_{4.4}Sn + 3Ni$, is followed by the processes $Li_{4.4}Sn \rightarrow Sn + 4.4Li^+ + 4.4e^-$ (charge) and $Sn + 4.4Li^+ + 4.4e^- \rightarrow Li_{4.4}Sn$ (discharge).

On the use of ionic liquids to rechargeable lithium batteries, see Chapter 13.

12.2
Capacitors Using Nonaqueous Solutions

12.2.1
Electrochemical Double-Layer Capacitors and Pseudocapacitors [8]

Electrochemical (or electric) double-layer capacitors (EDLCs) are, alone or coupled with pseudocapacitors, called supercapacitors or ultracapacitors. They are now

3) Nanosized (1–100 nm) particles, rods, wires, fibers, tubes and films are used in nanotechnology. At these nanosized substances, many processes occur more favorably than at the normal-sized substances. In addition to the high-specific surface area, various other factors can be considered for this. The applications of nanotechnologies are widespread, but the cases that are electrochemical and in nonaqueous media are not so many. The use of carbon nanotubes as the anode of lithium-ion batteries and as the electrodes of non-aqueous EDLCs is a promising example of such cases. The use for EDLC can produce high power density of $\sim 30\,kW\,kg^{-1}$, compared to $4\,kW\,kg^{-1}$ of the most advanced EDLC currently available commercially. The application of nanosized particles of $LiCoO_2$, $LiMn_2O_4$, etc. as cathode materials of lithium-ion batteries is also interesting. Another example is the use of conductive polymer nanotubes for electrochromic devices and for supercapacitors, as described in footnote 9. As media for electrodepositions or syntheses of nanosized metals, alloys, semiconductors and conducting polymers, ionic-liquids and liquid–liquid interfaces are sometimes used conveniently (see Chapters 13 and 14).

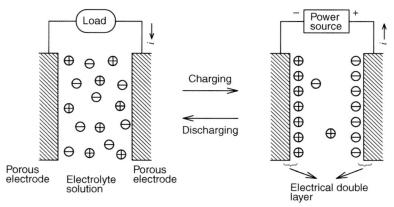

Figure 12.3 An electrochemical double-layer capacitor.

attracting much attention as new power sources complementary to rechargeable batteries. The EDLCs are based on the double-layer capacitance at carbon electrodes of high specific areas, while the pseudocapacitors are based on the pseudocapacitance of the films of redox oxides (RuO_2, IrO_2, etc.) or redox polymers (polypyrrole, polythiophene, etc.).

The principle of EDLC is shown in Figure 12.3. The capacitor, consisting of two carbon electrodes and an electrolyte solution between them, is charged/discharged with a voltage in the double-layer region, i.e. well below the width of the potential window. Because the activated carbon electrodes have large capacitance (usually $100–150\ F g^{-1}$), the electricity stored in the charging process is considerable and the capacitor in the discharging process can work as a power source. The energy density (*e.d.*) of the EDLC (or capacitors in general) is expressed by $e.d. = {}^1/_2 C (\Delta V)^2$, where C is the capacitance and ΔV the charging voltage. For nonaqueous electrolyte solutions with potential windows larger than 5 V (typically 0.5–1.0 M Et_4NBF_4 in PC), the value of ΔV reaches 3.5–4.0 V.[4] It is a big advantage over aqueous electrolyte solutions (e.g. H_2SO_4), for which ΔV is 1.0–1.5 V. The EDLCs with nonaqueous electrolyte dominate the market, though the EDLCs with aqueous solutions have merit over them as they can obtain a higher current. According to Ue [9], PC is the best solvent for EDLC if the electrolytic conductivity, electrochemical window, liquid range and resistance toward hydrolysis are considered. On the other hand, Et_4NBF_4 is the best electrolyte if electrolytic conductivity, electrochemical window and double-layer capacitance are considered. The merits and demerits of EDLCs and secondary batteries have been compared in detail [10]. The merits of the EDLCs are the long cycle lives (10^6 or more) and the possibility of rapid charging/discharging. They are because the processes at the EDLCs are mainly capacitative, and the surfaces of

4) The EDLCs, which use all-solid-state ion-conducting polymer (e.g. poly(ethyene oxide)/ $LiClO_4$) or polymer gel electrolyte, have also been developed [4].

the carbon electrodes are not influenced by the repeated charging/discharging processes. Recently, carbon materials with much higher surface area have been reported: carbon nanotubes are a typical example [11, 12] (see footnote 3).

Pseudocapacitors use metal oxides or redox polymers as electrodes. In pseudo-capacitors of metal oxides (e.g. RuO_2), aqueous electrolyte solutions such as H_2SO_4 are usually used [13]. Contrarily, in pseudocapacitors of redox polymers, nonaqueous electrolyte solutions are often employed. Typically, Et_4NBF_4–PC is used for capacitors with poly(thiophene), poly(3-methylthiophene) and poly[3-(4-fluorophenyl)thio-phene] electrodes [14]. The charging/discharging processes of such capacitors are mainly faradaic in nature (Section 12.3) [4, 15].

Recently, highly effective energy storage was achieved with cells called asymmetric hybrid capacitors: a combination of an EDLC electrode and a battery-like electrode, which works as a pseudocapacitative, forming a capacitor [10, 16]. For example, with a combination of an activated carbon electrode (anode) and a graphitic carbon electrode (cathode) in 1 M Et_4NBF_4/PC, energy density of over 20 Wh kg^{-1} could be obtained, in contrast to less than 4 Wh kg^{-1} for conventional EDLC [16b]. Here, the intercalation of anions makes the graphitic electrode pseudocapacitative.

12.2.2
Aluminum Electrolytic Capacitors

As shown in Figure 12.4, aluminum electrolytic capacitors usually consist of an aluminum foil with a thin film (dielectric) of anodically formed aluminum oxide, an aluminum foil, an electrolyte solution and a separator. The whole 'sandwich' is compactly rolled and packed in a container. The electrolytic capacitors are in wide use because of their small sizes, high capacitances and low prices. However, the characteristics of electrolytic capacitors deteriorate with time. Recently, owing to

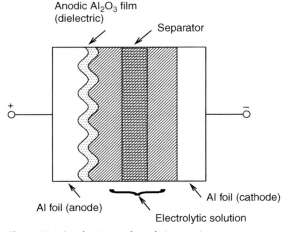

Figure 12.4 An aluminum electrolytic capacitor.

the rapid developments of high-performance electronic instruments, the needs have increased for high-quality electrolytic capacitors, having long lives at elevated temperatures and low impedances at high frequencies. These requirements are met by the use of nonaqueous electrolyte solutions such as 25 wt% triethylmethylammonium hydrogen maleate in γ-BL and 25 wt% tetramethylammonium hydrogen phthalate in γ-BL [9, 17]. The capacitors filled with these organic electrolyte solutions can keep their high performances for a long time.[5] However, the quaternary ammonium systems have disadvantage of a lower sustaining voltage, which limits the nominal voltage up to 50 V. Recently, the automotive industry is demanding high-performance aluminum electrolytic capacitors usable up to 100 V. This demand has been satisfied by adding colloidal silica into the above-mentioned quaternary ammonium electrolyte solutions [18].

12.3
Conducting Polymers and Electrochemistry in Nonaqueous Solutions [19]

Conducting polymers, such as polyacetylenes, polyaniline, polypyrrole and polythiophene (Figure 12.5), are promising new materials. They have a high degree of π-orbital conjugation and are capable of being electrochemically oxidized or reduced by the withdrawal or injection of electrons. When the polymer is in the oxidized state, anions are doped to the polymer to keep electroneutrality (p-doping). Inversely, when

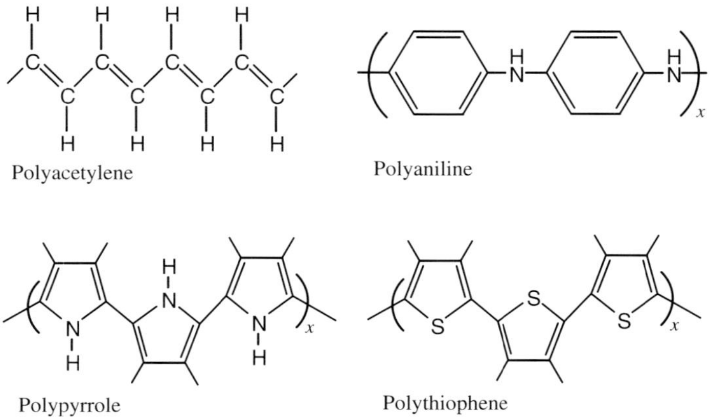

Polyacetylene   Polyaniline

Polypyrrole   Polythiophene

Figure 12.5 Structures of conducting polymers.

5) The problem with the tetraalkylammonium hydrogen phthalate or maleate solution in γ-BL is that the solution has no pH buffer action (Section 6.2.2). In rare occasions, the solution near the cathode of the capacitor locally becomes very high in pH by the cathodic reaction, damages the container and leaks out from it. Recently, the tetraalkylammonium salt has been replaced with the alkylimidazolium salt to avoid such a rise in pH.

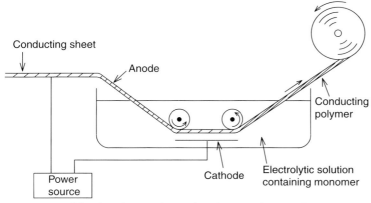

Figure 12.6 Cell for electrolytic synthesis of conducting polymer (roll-type).

the polymer is in the reduced state, cations are doped (n-doping).[6] The doped and undoped polymers have completely different properties. The most important is that the doped polymer has a high conductivity, which is comparable to that of metal, while the undoped polymer is practically an insulator. Conducting polymers are closely related to the electrochemistry in nonaqueous solutions, from their syntheses to their applications, as outlined below.

1. **Syntheses of conducting polymers** Conducting polymers are often synthesized by electrolytic polymerization [19, 20]. The electrolytic solution, consisting of the monomer, the supporting electrolyte and the solvent, is taken into the cell, and the electrolysis is carried out by applying a voltage between the working electrode (Pt, Au, C, etc.) and the counter electrode. Usually, a film of the conducting polymer is formed during anodic polarization. An example of the cell for synthesizing polymer sheet continuously is shown in Figure 12.6. The sheet of anode material is rolled after being covered with the film of the deposited conducting polymer.

   In order to get high-quality polymer films, it is essential to select best conditions of the electrolytic solution, electrode potential, temperature, etc. For synthesizing the films of polypyrrole, polythiophene, polyazulene and their derivatives, the aprotic solvents with weak nucleophilicity (basicity) are suitable and acetonitrile (AN) is used most often.[7] In strongly nucleophilic solvents such as pyridine, the polymer films are not formed. In the moderately nucleophilic solvents such as

---

6) Although conjugated polymers can be either n-doped or p-doped – and thus, be capable of behaving either as negative or positive electrodes – the majority of applications are confined to the p-doped positive side.

7) Recently, the mechanisms of pyrrole electropolymerization have been reviewed in Ref. [20b]. By the anodic reaction, an electron is withdrawn from the pyrrole monomers and cationic radicals are formed. The cationic radicals undergo a series of

chemical–electrochemical–chemical reactions and as a result, the polymerization proceeds. If the cationic radicals of the monomers or oligomers react with a nucleophilic solvent or solute, the chain reaction is terminated and the polymerization process is stopped. Thus, strongly nucleophilic (basic) solvents and strongly nucleophilic anions $(F^-, Cl^-, OH^-, RO^-, CH_3COO^-,$ etc.) should be avoided in the polymerization processes.

Table 12.1 Applicability of conducting polymers.

(a) Conductor; (b) (variable) resistor; (c) electromagnetic shielding; (d) heat absorber; (e) electronic devices (diode, transistor, FET, etc.); (f) pseudocapacitor; (g) solar cell; (h) thermoelectromotive element; (i) electrochromic display; (j) thermochromism, solvatochromism; (k) photomemory; (l) polymer battery; (m) fuel cell; (n) sensor; (o) electrode; (p) electric contact; (q) heating element; (r) electric field relaxation; (s) conducting polymer gel; (t) catalyst, photocatalyst; (u) filter, separation membrane, adsorber; (v) actuator; (w) neuroelement

DMF and DMSO, the films are formed only when the solution is buffered by adding protonic acid. To get high-quality polypyrrole films in aqueous solutions, the electrolytic solution must be highly concentrated. The most suitable medium differs from one polymer to another.[8] For example, for polyaniline, the best electrolytic solution is aqueous sulfuric acid. The conducting polymer films formed by electrolytic polymerization are usually in ∼30% oxidized state (by the withdrawal of electrons) and are doped with anions (p-doping). Because the anions give significant influences on the conductivity of the polymer, the selection of the anion of the electrolytic solution is also very important (see footnote 7). Moreover, by endowing the dopant anion with an active site having an oxidizing, reducing or complexing ability, it is possible to prepare functional polymer films.

2. **Characterization of conducting polymers** Electrochemical methods play an important role in characterizing the functions of conducting polymers [15, 20b, 21]. Cyclic voltammetry (CV) is the most useful; by the measurements of cyclic voltammograms, we can know the potentials, the currents and the quantities of electricity for the oxidation–reduction processes of the conducting polymers. By combining the CV technique with the *in situ* measurements of optical absorptions and mass changes, we can get more knowledge about the doping and undoping processes. An example of the use of an electrochemical quartz crystal microbalance (EQCM) was shown in Figure 9.9. By using the EQCM, the mass changes during the oxidation–reduction processes can be quantitatively measured as frequency–potential curves, simultaneously with the current–potential curves. The cyclic voltammogram in Figure 9.9a shows that the oxidation of polyaniline in 0.5 M $LiClO_4$–AN proceeds in two steps. The frequency– potential curve, on the other hand, shows that, in both steps, the doping of $ClO_4{}^-$ occurs.

3. **Application of conducting polymers** Doped conducting polymers behave like metals while undoped conducting polymers behave like insulators or semiconductors. The reversible conversion between the metallic and the insulating (or semiconducting) states can be realized simply by switching the potential of the conducting polymer. Conducting polymers have a wide applicability, as listed in Table 12.1, including applications as insulating (or semiconducting) material,

8) Ionic liquids are very suitable in synthesizing conducting polymers (see Pringle, J.M., Forsyth, M. and MacFarlane, D.R. (2008) *Electrodeposition from Ionic Liquids* (eds F. Endres, A.P. Abbott and D.-R. Macfarlane), Wiley-VCH Verlag GmbH, Weinheim, Chapter 7). Details are described in Chapter 13

applications as metallic material, applications of reversible conversions between insulating–metallic states and applications of the specific functions of the dopant counter ions.[9]

The use of conducting polymers as electrodes in lithium secondary batteries and supercapacitors (pseudocapacitors) was discussed above. From the standpoint of electrochemistry in nonaqueous solutions, the use as electrochromic materials is also interesting. Conducting polymers usually have different colors between the doped and undoped states. The color change of the transmitted light by the undoped ↔ doped conversion is red ↔ blue for the polythiophene film, yellow ↔ blue for the polypyrrole film and light yellow ↔ green for polyaniline film. In principle, the color switching is possible by the use of a cell composed of an optically transparent indium-tin oxide (ITO) glass electrode coated with a conducting polymer film,[10] a bare ITO glass electrode and an electrolyte solution, as shown in Figure 12.7, and by applying a pulse voltage between the two electrodes to switch the conducting polymer between doped and undoped states. For the recent status of the applications of conductive polymers to electrochromic displays, mirrors and windows, see Ref. [23].

12.4
Electrochemiluminescence (ECL)

Electrogenerated chemiluminescence or electrochemiluminescence (ECL) is a means of converting electrical energy into radiative energy. Reactive intermediates are formed from stable precursor(s) at the surface of electrode(s), and these intermediates then react to form excited states that emit light. For details of ECL, a book [24] and a review article [25] are available.

Two types of ECL exist, annihilation ECL and coreactant ECL. In annihilation ECL, light can be emitted from substance R by the following mechanisms: $R - e^- \rightarrow R^{\bullet+}$

---

9) Usually, it takes $1-2$ s for conducting polymers to switch between doped and undoped states, owing to the slow diffusion of counterions into the polymers. However, faster switching is possible if conducting polymers of long, thin-walled nanotubes are used because the time necessary for diffusion is shortened. For example, at the nanotubes of poly(3,4-ethylenedioxythiophene) (PEDOT), which were electrochemically synthesized using a porous alumina template at the monomer concentration of 20 mM in 0.1 M LiClO$_4$–acetonitrile and at 1.5 V versus Ag/AgCl, the switching rate was <10 ms [22a]. This rate was fast enough for color-switching electrochromic devices to play movies at 24 frames/s (necessary switching rate ≤40 ms) and the coloration was strong. Moreover, supercapacitors based on the fast responsive PEDOT nanotube can provide high power density without a significant loss of its energy density. See [22b] for details.

10) Examples of electrochromic materials other than conducting polymers: WO$_3$ (transparent ↔ blue), IrO$_2$ (transparent ↔ dark gray), viologen (colorless ↔ violet and green) and anthraquinone (colorless ↔ red).

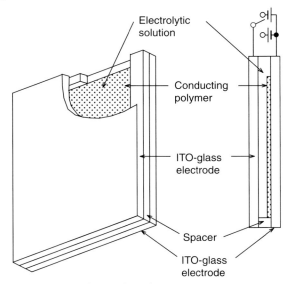

Figure 12.7 A color-switching device.

(oxidation at the electrode), $R + e^- \rightarrow R^{\bullet-}$ (reduction at the electrode), $R^{\bullet-} + R^{\bullet+} \rightarrow R^{\bullet} + R$ (excited-state formation), $R^{\bullet} \rightarrow R + hv$ (light emission). Here, R is such substance as metal chelate (e.g. $[Ru(bpy)_3]^{2+}$) or organic compound (e.g. rubrene (5,6,11,12-tetraphenyltetracene)). For $[Ru(bpy)_3]^{2+}$ in acetonitrile–Bu_4NBF_4, $E^0(R^{\bullet+}/R) = +1.2$ V versus SCE, $E^0(R/R^{\bullet-}) = -1.4$ V versus SCE, and light at 610 nm is emitted. On the other hand, for rubrene in benzonitrile–Bu_4NClO_4, $E^0(R^{\bullet+}/R) = +0.95$ V versus SCE, $E^0(R/R^{\bullet-}) = -1.37$ V versus SCE, and light at ~540 nm is emitted. Here, CV is a useful method for determining the potentials at which $R^{\bullet+}$ and $R^{\bullet-}$ are generated and for knowing their stability and reactivity. In emitting light, one electrode or two electrodes situated close to each other may be used as the working electrode(s). When one electrode is used, it is held for some time (say from microseconds to seconds) at the potential at which $R^{\bullet+}$ is produced, the potential is then switched to a value at which $R^{\bullet-}$ is generated and then returned back to the initial value. This sequence is repeated as many times as desired. During the sequence, both $R^{\bullet+}$ and $R^{\bullet-}$ exist in the diffusion layer near the electrode and can react to emit light. Instead of using R, it is possible to use two different precursors (A and D) to generate ECL: $A + e^- \rightarrow A^{\bullet-}$, $D - e^- \rightarrow D^{\bullet+}$, $A^{\bullet-} + D^{\bullet+} \rightarrow A^{\bullet} + D$ (or $A + D^{\bullet}$), $A^{\bullet}$ (or $D^{\bullet}$) $\rightarrow A$ (or D) $+ hv$. For example, 9,10-diphenylanthracene (DPA) and N,N,N',N'-tetramethyl-p-phenylenediamine (Wurster's Blue (TMPD)) in DMF serve as A and D, respectively. Here, $E^0(A/A^{\bullet-}) = -1.89$ V versus SCE, $E^0(D^{\bullet+}/D) = +0.24$ V versus SCE and $A^{\bullet}$

is an excited single-state ^{1}DPA$^\bullet$, which is generated by A$^{\bullet-}$ + D$^{\bullet+}$ → ^{3}A$^\bullet$ + D and ^{3}A$^\bullet$ + ^{3}A$^\bullet$ → ^{1}A$^\bullet$ + A (^{3}A$^\bullet$: triplet state). This triplet route seems to occur in many ECL reactions with different precursors.

When two working electrodes situated close to each other are used, the potential of one electrode is adjusted to generate R$^{\bullet+}$ (or D$^{\bullet+}$) and that of the other to generate R$^{\bullet-}$ (or A$^{\bullet-}$). At a rotating ring-disk electrode, R$^{\bullet-}$ or A$^{\bullet-}$ is generated at the disk while R$^{\bullet+}$ or D$^{\bullet+}$ is generated at the ring. A ring of light appears on the inner edge of the ring electrode.

Annihilation ECL can be observed with a conventional apparatus, although cells, electrodes and experimental procedures must be modified to electrogenerate two intermediates and to detect light emission. For organic systems such as polyaromatic hydrocarbons, water and oxygen are harmful because they can quench ECL. Thus, solvent/supporting electrolyte systems must be treated with care. Many of the annihilation systems require pure nonaqueous solvents because the potential window in water is too narrow to generate two intermediates (R$^{\bullet+}$ and R$^{\bullet-}$ or D$^{\bullet+}$ and A$^{\bullet-}$).

Contrarily, coreactant ECL needs potential step or sweep in only one direction and allows the use of aqueous solutions. A coreactant is a species that, upon oxidation or reduction, produces an intermediate that can react with an ECL luminophore to produce excited states. Usually, this occurs upon bond cleavage of the coreactant to form strong reductants or oxidants. For example, oxalate ion ($C_2O_4^{2-}$) produces strong reductant $CO_2^{\bullet-}$ upon oxidation in aqueous solution: $C_2O_4^{2-}$ − e$^-$ → $(C_2O_4^{\bullet-})$ → $CO_2^{\bullet-}$ + CO_2. At the same potential, luminophore D (e.g. $[Ru(bpy)_3]^{2+}$) can be oxidized by D − e$^-$ → D$^{\bullet+}$. D$^{\bullet+}$ and $CO_2^{\bullet-}$ then react to form D$^\bullet$ that emits light (D$^{\bullet+}$ + $CO_2^{\bullet-}$ → D$^\bullet$ + CO_2 and D$^\bullet$ → D + $h\nu$). Oxalate is often referred to as 'oxidative–reductive' coreactant because of its ability to form a strong reducing agent upon electrochemical oxidation [25]. Another example of an oxidative–reductive system is the commercially important $[Ru(bpy)_3]^{2+}$/TPrA system (TPrA = tri-*n*-propylamine). Two reactions, $[Ru(bpy)_3]^{2+}$ − e$^-$ → $[Ru(bpy)_3]^{3+}$ and TPrA − e$^-$ → (TPrA$^{\bullet+}$) → TPrA$^\bullet$ + H$^+$ occur at the same potential. And light is emitted by $[Ru(bpy)_3]^{3+}$ + TPrA$^\bullet$ (or TPrA$^{\bullet+}$) → $[Ru(bpy)_3]^{\bullet2+}$ + products, $[Ru(bpy)_3]^{\bullet2+}$ → $[Ru(bpy)_3]^{2+}$ + $h\nu$. 'Reductive–oxidative' coreactants are also used. For example, in the case of peroxydisulfate ($S_2O_8^{2-}$), reduction produces a strong oxidant, $SO_4^{\bullet-}$. Two reactions, $[Ru(bpy)_3]^{2+}$ + e$^-$ → $[Ru(bpy)_3]^+$ and $S_2O_8^{2-}$ + e$^-$ → $SO_4^{\bullet-}$ + SO_4^{2-}, occur at the same potential and $[Ru(bpy)_3]^+$ and $SO_4^{\bullet-}$ react to emit light. Electrochemical apparatus for coreactant ECL may be similar to those used in annihilation ECL, but there is no need to work with nonaqueous systems.

Many annihilation and coreactant ECL systems now exist and these have been reviewed in detail [24, 25].

ECL is applied mainly in two ways, for the analytical purposes and for the display devices. In the applications for analytical purposes, coreactant ECL is mostly used and aqueous solutions may be employed. Because ECL emission-intensity is usually proportional to the concentrations of the emitter and the coreactant, ECL can be used in the analyses of various species.[11]

In the applications as display devices, ECL is emitted either by the emitters in solution or by the emitters accumulated at the electrode surface. Transparent electrodes, usually ITO, are used as the working electrode(s). The examples of emitters accumulated at the electrode surface [27] are tris(4-vinyl-4'-methyl-2,2'-bipyridyl)ruthenium(II) ($[Ru(vbpy)_3]^{2+}$), poly(vinyl-9,10-diphenylanthracene) (PVDPA), etc. $[Ru(vbpy)_3]^{2+}$ gives orange color (lifetime $\sim$20 min) when the potential is pulsed between $+1.5$ and -1.5 V (versus SSCE) in AN while PVDPA gives blue color (lifetime 5–10 min) when the potential is pulsed between $+1.6$ and -2.0 V versus SCE in PC, AN and THF. Many other examples exist and highly efficient ECL devices reaching 1000–2000 cd m^{-2} have been reported. The examples of emitters for ECL in solutions are $[Ru(bpy)_3]^{2+}$ and rubrene, and they are often used without supporting electrolyte. Various efforts have been made to enhance the ECL from the emitters in solution [28].

12.5
Electrochemical Reduction of CO$_2$ in Nonaqueous Solutions [29–31]

In both aqueous and nonaqueous solutions, CO_2 first undergoes a one-electron reduction to $CO_2^{\bullet-}$. The $CO_2^{\bullet-}$ radical anion, however, is very reactive and reacts in various ways depending on the electrode material, reaction medium, temperature, CO_2 concentration, etc. The complicated nature of the reduction of CO_2 has been studied in detail because electrochemical reduction of CO_2 can pave the way for fixing CO_2 as organic materials, solving simultaneously the problems of energy and environment. The reduction of CO_2 occurs at negative potentials; acidic aqueous solutions are not appropriate for the reduction because it competes with the generation of hydrogen gas. Moreover, CO_2 is converted to CO_3^{2-} in basic aqueous

---

11) For example, ECL from $[Ru(bpy)_3]^{2+}$ has been used to analyze the concentrations of coreactants such as oxalate and peroxydisulfate to the concentration levels as low as 10^{-13} M. In these examples, ECL is measured in the presence of high, predetermined concentration of ECL emitter ($[Ru(bpy)_3]^{2+}$). These types of experiments can be used to assay compounds that act as coreactants including a variety of amines. The ECL assays of amines are useful because amine groups are prevalent in many biologically and pharmacologically important compounds including alkylamines, antibiotics, antihistamines, opiates, nicotinamide and the reduced form of NADH. Because these compounds contain no chromophore, they cannot undergo luminescence unless an ECL-active compound is present. ECL is often used in clinical diagnostic assays. Here, ECL emitters are used as labels in affinity-binding assays that attach the ECL emitter to the analyte of interest. The label is physically linked to one of the binding partners in the assay and provides the means for detecting the coupling of the binding partner to the analyte [25]. Commercial instruments are available for ECL assays of antibodies, antigens and DNA. Assays that have been developed for these systems include alpha-fetoprotein, digoxin, thyrotopin, protein, steroidal hormones, cytokines, various antibodies and others. The use of flow injection, liquid chromatography and capillary electrophoresis in these applications has been reviewed in [26].

solutions. Thus, in water, the reduction of CO_2 is observable only in neutral solutions. Nonaqueous aprotic media have wide potential regions on the negative side and are suitable for the observation of the CO_2 reduction. CO_2 dissolves more conveniently in nonaqueous solvents than in water (Section 8.2.4). Therefore, a considerable number of studies have been carried out in nonaqueous solutions. Many review articles have been published concerning electrochemical reduction of CO_2 in nonaqueous and aqueous solutions [29, 30], of which Ref. [29] is the most comprehensive and updated. The reduction of CO_2 in nonaqueous solvents (especially in DMSO and at a gold electrode) has also been studied from the standpoint of developing electrochemical sensors for CO_2 [31].

The $CO_2^{\bullet-}$ radical anion formed by $1e^-$ reduction is either adsorbed or not adsorbed on metal electrodes (Table 12.2) [29, 30]. Metal electrodes at which $CO_2^{\bullet-}$ is not adsorbed are Pb, Tl and Hg electrodes in aprotic nonaqueous media and Cd, Sn, In, Pb, Tl and Hg electrodes in aqueous media, while those at which $CO_2^{\bullet-}$ is adsorbed are Au, Ag, Cu, Zn, Cd, Sn and In electrodes in aprotic nonaqueous media and Au, Ag, Cu and Zn electrodes in aqueous media. Savéant et al. [30a] studied the reduction of CO_2 at a mercury electrode ($CO_2^{\bullet-}$ not adsorbed) in 0.1 M Bu_4NClO_4–DMF. The cyclic voltammogram was usually irreversible, somewhat reversible behavior of the $CO_2/CO_2^{\bullet-}$ couple was observed only at a high scan rate (e.g. $4400\,V\,s^{-1}$). But, from the reversible behavior, the standard potential of the $CO_2/CO_2^{\bullet-}$ couple was estimated to be about -2.21 V versus aqueous SCE. The number of electrons per molecule for the irreversible reduction was close to 1 when $[CO_2] > 5$ mM, in agreement with the formation of oxalate ($2CO_2^{\bullet-} \rightarrow {}^-O_2C-CO_2^-$) as well as the formation of CO and CO_3^{2-} [$CO_2^{\bullet-} + CO_2 \rightarrow {}^{\bullet}O_2C-CO_2^-$ followed by ${}^{\bullet}O_2C-CO_2^- + e^- \rightarrow CO + CO_3^{2-}$ (at electrode) or ${}^{\bullet}O_2C-CO_2^- + CO_2^{\bullet-} \rightarrow CO + CO_3^{2-} + CO_2$ (in solution)]. When the DMF solution contained residual water and the CO_2 concentration decreased, the number of electrons per molecule approached 2, indicating that the formation of formate, $HCOO^-$, occurred by $CO_2^{\bullet-} + H_2O \rightarrow HCO_2^{\bullet} + OH^-$ followed by $HCO_2^{\bullet} + e^- \rightarrow HCO_2^-$ or $HCO_2^{\bullet} + CO_2^{\bullet-} \rightarrow HCO_2^- + CO_2$. In general, in aprotic nonaqueous media and at the electrodes at which $CO_2^{\bullet-}$ is not adsorbed, one path is the formation of oxalate by the reaction of two $CO_2^{\bullet-}$ radical anions, though, in the presence of H^+ ion, some part of the oxalate is further reduced to glyoxylic acid or glycolic acid. In the other path, $CO_2^{\bullet-}$ reacts with CO_2 to form ${}^{\bullet}O_2C - CO_2^-$, but the C–C bond is broken by accepting an electron or by reacting with another $CO_2^{\bullet-}$ to form CO and CO_3^{2-}.[12] In aprotic nonaqueous media and at the electrodes at which $CO_2^{\bullet-}$ is adsorbed, $CO_2^{\bullet-}$(ads) reacts with CO_2 to form ${}^{\bullet}O_2C-CO_2^-$(ads), which is further reduced to CO and CO_3^{2-}. In aqueous media and at the electrodes at which $CO_2^{\bullet-}$ is not adsorbed, $CO_2^{\bullet-}$ and H_2O react to form $HCO_2^{\bullet}$ radical, but the radical is reduced to $HCOO^-$

---

12) The reduction products of CO_2 in PC–Et_4NClO_4 can be classified into the following three groups by the catalytic activity of the electrode materials: $(COOH)_2$ is selectively generated at Hg, Tl, and Pb, CO is selectively generated at Au, Ag, Cu, Zn, In, Sn, Pt, Pd and Ni and $(COOH)_2$ and CO are simultaneously generated at Cr, Mo, Ti and Fe [30b].

Table 12.2 Influence of electrode materials and solvents on the reduction path of CO_2.

(I) $CO_2^{\cdot-}$ not adsorbed on metal electrodes:

Nonaqueous (aprotic) media (Pb, Tl, Hg electrodes)

$$CO_2 \xrightarrow{+e^-} CO_2^{\cdot-} \xrightarrow{+CO_2^{\cdot-}} \boxed{(COO^-)_2} \left(\xrightarrow{+H^+,+e^-} \boxed{HCO-COO^-} + OH^- \atop \xrightarrow{+H^+,+e^-} \boxed{H_2COH-COO^-} \right)$$

$$CO_2 \xrightarrow{+e^-} CO_2^{\cdot-} \xrightarrow{+CO_2} \cdot O_2C-CO_2^- \xrightarrow{+e^- \text{ (at electrode)}} \boxed{CO} + CO_3^{2-}$$
$$\xrightarrow{+CO_2^{\cdot-} \text{ (in solution)}} \boxed{CO} + CO_3^{2-} + CO_2$$

Aqueous media (Cd, Sn, In, Pb, Tl, Hg electrodes)
(or nonaqueous media containing water)

$$CO_2 \xrightarrow{+e^-} CO_2^{\cdot-} \underset{}{\overset{+H_2O}{\rightleftarrows}} HCO_2^{\cdot} + OH^-$$
$$\xrightarrow{+e^- \text{ (at electrode)}} \boxed{HCOO^-}$$
$$\xrightarrow{+CO_2^{\cdot-} \text{ (in solution)}} \boxed{HCOO^-}$$

(II) $CO_2^{\cdot-}$ adsorbed on metal electrodes:

Nonaqueous (aprotic) media (Au, Ag, Cu, Zn, Cd, Sn, In electrodes)

$$CO_2 \xrightarrow{+e^-} CO_2^{\cdot-}(ads) \xrightarrow{+CO_2} \cdot O_2C-CO_2^-(ads) \xrightarrow{+e^-} CO(ads) + CO_3^{2-}$$
$$\xrightarrow{} \boxed{CO}$$

Aqueous media (Au, Ag, Cu, Zn electrodes)

$$CO_2 \xrightarrow{+e^-} CO_2^{\cdot-}(ads) \xrightarrow{+H_2O} CO(OH)(ads) (+OH^-) \xrightarrow{+e^-} CO(ads) (+OH^-)$$
$$\xrightarrow{} \boxed{CO}$$

(only at Cu electrode)

$$\xrightarrow{+H^+,+e^-} CH_2^{\cdot}(ads)(+H_2O) \xrightarrow{+H^+,+e^-} \boxed{CH_4}$$
$$\xrightarrow{+CH_2^{\cdot}(ads)} \boxed{C_2H_4}$$
$$\xrightarrow{+CH_2^{\cdot}(ads),+H_2O} \boxed{C_2H_5OH}$$

by the reduction either at the electrode or in solution (by $CO_2^{\cdot-}$). In the aqueous media and at the electrode at which $CO_2^{\cdot-}$ is adsorbed, $CO_2^{\cdot-}$(ads) reacts with H_2O to form $CO(OH)$(ads), which is further reduced to CO. Here, the phenomena at the Cu electrode are somewhat different from those at other electrodes: CO is adsorbed with a moderate strength and is further reduced to $CH_2^{\cdot}$(ads), from which methane (CH_4), ethylene (C_2H_4) and ethanol (C_2H_5OH) are formed. Methanol is a nonaqueous solvent but protic and behaves like water. In MeOH and at a Cu electrode, hydrocarbon such as CH_4 and C_2H_4 are formed in addition to CO and $HCOCH_3$

(formed by esterification of HCOOH with CH_3OH). The carbon atom and the hydrogen atom of CH_4 and C_2H_4 have been proven to come from the CO_2 molecule and CH_3OH, respectively.

The distribution of products obtained also depends on temperature. For example, oxalate and CO are formed at an Hg electrode in DMF–0.1 M Bu_4NClO_4. The preparative-scale electrolysis at $\sim$100 mM CO_2 showed a decrease in the percentage of oxalate and an increase in that of CO with the decrease in temperature: i.e. the percentages of oxalate and CO were 80 and 20%, respectively, at 25 °C but 20 and 80%, respectively, at −20 °C. In the same electrode and solution, the percentages of oxalate and CO also depended on the CO_2 concentration: at 0 °C, they were 90 and 10%, respectively, at $[CO_2] = 20$ mM, but 23 and 77%, respectively, at $[CO_2] = 217$ mM [30a].

Related to the fixation of CO_2, 'electrochemical carbon', which is prepared by converting carbon halides (e.g. polytetrafluoroethylene) to carbon, is attracting attention for its technological applicability. Nonaqueous electrolyte solutions are often used in the electrochemical carbonization processes [32]. Although not dealt with in this book, the use of nonaqueous electrolyte solutions is really very popular in the field of electrochemical organic syntheses [33].

12.6
Use of Acetonitrile in Electrowinning and Electrorefining of Copper [34]

The solvation of Cu^{2+} in AN is relatively weak as expected from the weak basicity of AN. The solvation of Cu^+ in AN, on the other hand, is extremely strong because of the back-donation of the d^{10}-electron of Cu^+ to the CN-group. Thus, if Cu^{2+} in AN comes in contact with metallic copper (Cu^0), the reaction (comproportionation) $Cu^{2+} + Cu^0 \rightleftarrows 2Cu^+$ occurs to form Cu^+ (Chapters 2 and 4). The preferential solvation of AN to Cu^+ also occurs in (H_2O + AN) mixtures, and the equilibrium $Cu^{2+} + Cu^0 \rightleftarrows 2Cu^+$ lies very far to the right. The equilibrium constant is 10^{11} M in the 1: 1 (molar ratio) H_2O–AN mixture and 10^{20} M in anhydrous AN, in contrast to 10^{-6} M in pure water. As a result, the potentials of Cu^{2+}/Cu^+ and Cu^+/Cu^0 couples vary as shown in Figure 12.8 [35]. Parker [34] used this phenomenon in the electrowinning and electrorefining of copper.

The process for electrowinning of copper is schematically shown in Figure 12.9. If copper(I) sulfate in AN–H_2O–H_2SO_4 solution is electrolyzed using a platinum electrode as anode and a copper electrode as cathode, one-electron processes occur at the two electrodes ($Cu^+ \rightarrow Cu^{2+}$ at the anode and $Cu^+ \rightarrow Cu^0$ at the cathode). Compared to the conventional electrowinning from the aqueous acidic solution of copper(II) sulfate (water oxidation at the anode and $Cu^{2+} \rightarrow Cu^0$ at the cathode), the electric power consumed is only about 10%, and high-quality copper can be obtained. It is of course necessary to return Cu^{2+}, generated at the anode, to Cu^+. But various methods are applicable to it, e.g. through contact with coarse copper.

In the electrorefining of copper, copper(I) sulfate in the AN–H_2O–H_2SO_4 solution is electrolyzed using a coarse copper electrode as anode and a pure copper electrode

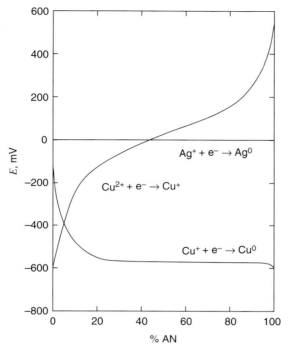

Figure 12.8 The potentials of Cu^{2+}/Cu^+ and Cu^+/Cu^0 couples in AN–H$_2$O mixtures containing 0.01 M H$_2$SO$_4$ (values versus Ag$^+$/Ag couple) [35].

as cathode. The reaction at the anode is $Cu^0 \rightarrow Cu^+$ while that at the cathode is $Cu^+ \rightarrow Cu^0$. Compared to the conventional process that uses aqueous acidic solutions of copper(II) sulfate, this method is advantageous in that the quantity of electricity is one half and the electric power is also small. Moreover, the low-quality

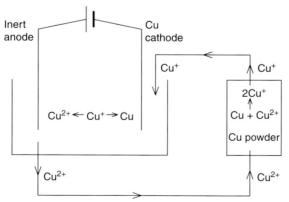

Figure 12.9 Electrowinning of copper from AN–H$_2$O–H$_2$SO$_4$ solution (electrolysis at 0.6 V and ~0.01 A cm^{-2}) [34].

AN used in this method is available at low price and in large quantity as a by-product of chemical industries.

12.7
Electrodeposition of Metals and Semiconductors from Nonaqueous Solutions

Some metals, which are difficult to deposit electrolytically from aqueous solutions, can be deposited from appropriate nonaqueous solutions. They are, for example, alkali metals, magnesium and aluminum. They are usually deposited at more negative potentials than the reduction of water. The electrodeposition of lithium metal from electrolyte solutions in aprotic solvents[13] is an important process in the lithium secondary batteries and the mechanistic study has been carried out extensively (Section 12.1). The electrodepositions of magnesium and calcium metals have also been studied for developing secondary batteries. So far, however, the two metals cannot be deposited from common polar aprotic solvents containing normal salts (perchlorates, trifluoromethanesulfonates, etc.). The only solution for the deposition of magnesium is the ether (THF) solution of Grignard reagent [RMgCl (R = Me, Et, Bu), EtMgBr]. In the solution, magnesium can be deposited smoothly and uniformly at high plating efficiency. For calcium, the electrodeposition of this type is impossible because Grignard-type reagent cannot be obtained with calcium. The aluminum electrodeposition (mainly electroplating) has a great commercial significance. By the use of appropriate electrolytic-bath compositions, highly pure aluminum can be deposited efficiently. Aluminum can be electroplated on a variety of metal surfaces including steel, nonactive metals and active metals (e.g. Mg, Al). The electroplated aluminum can further be anodized to obtain hard, corrosion-resistant, electrically insulating surfaces for various applications. Among the useful bath compositions are $AlCl_3 + LiAlH_4$ in ethers (diethyl ether, THF), $AlBr_3 + KBr$ in aromatic hydrocarbons (toluene, xylene) and $AlCl_3 + LiCl$ in dimethylsulfone $[(CH_3)_2SO_2]$. For more details on electrodepositions of metals from nonaqueous solutions, see Ref. [36].

The formation of amorphous silicon films by electrodeposition from nonaqueous solutions has also been studied [37, 38]. For example, a flat homogeneous silicon film of about 0.25 μm thick can be deposited from 0.2 M $SiHCl_3$–0.03 M Bu_4NBr–THF bath on the cathode of Pt, Au, Cu, GC, ITO, etc. though small amounts of impurities (O, C, Cl) are contained. Their use in photovoltaic or photoelectrochemical solar cells is promising, though there are still many problems to be solved.

Electroplating from nonaqueous solutions is useful when the surface of a substrate, which easily reacts with water, should be protected with a thin metal film.

---

13) For the aprotic solvents used in depositing active metals, a 'moderate' basicity is required. If the basicity is too weak, the metal salts are not soluble enough. On the other hand, if the basicity is too strong, the strongly solvated metal ions cannot be reduced. Especially in HMPA, which is strongly basic, Li^+ and Na^+ are not electrodeposited, and solvated electrons are generated from the electrode surface instead.

For instance, a superconductor, $Ba_2YCu_3O_7$, has a strong oxidizing ability and its surface easily reacts with air. In order to protect the surface of the $Ba_2YCu_3O_7$ with a metal film, a method to electroplate such metals as Ag, Cu, Pb and Sn from acetonitrile has been studied [39]. In aqueous solutions, the reduction of Cu^{3+} in $Ba_2YCu_3O_7$ narrows the applicable potential range and only Ag can be deposited.

Room-temperature ionic liquids are also useful for electrodeposition of active metals and semiconductors, as described in Section 13.3.3. For more details, see Ref. [40].

Although not electrode processes, solvents such as DMF and formamide can act as reductants for silver and gold salts.[14] Recently the reduction of Ag^+ in DMF ($HCONMe_2 + 2Ag^+ + H_2O \rightarrow 2Ag^0 + Me_2NCOOH + 2H^+$) has been used to form thin films of silver nanoparticles on glass surfaces or stable dispersions of silver nanoparticles in solutions [41].

References

1 For example, (a) Aurbach, D. (ed.) (1999) *Nonaqueous Electrochemistry*, Marcel Dekker, New York; (b) Barthel, J. and Gores, H.-J. (1994) *Chemistry of Nonaqueous Solutions: Current Progress* (eds. G. Mamantov and A.I. Popov), Wiley-VCH Verlag GmbH, Weinheim, Chapter 1.

2 For example, (a) Barthel, J., Gores, H.-J., Neueder, R. and Schmid, A. (1999) *Pure Appl. Chem.*, **71**, 1705; (b) Gores, H.-J. and Barthel, J.M.G. (1995) *Pure Appl. Chem.*, **67**, 919.

3 (a) Yamamoto, O. and Wakihara, M. (eds) (1998) *Lithium Ion Batteries: Fundamentals and Performance*, Wiley-VCH Verlag GmbH, Weinheim; (b) Van Schalkwijk, W.A. and Scrosati, B. (eds) (2002) *Advances in Lithium-Ion Batteries*, Kluwer Academic/Plenum, New York; (c) Balbuena, P.B. and Wang, Y. (eds) (2004) *Lithium-Ion Batteries: Solid-Electrolyte Interphase*, Imperial College Press, London; (d) Nazri, G. and Pistoia, G. (eds) (2004) *Lithium Batteries: Science*

and Technology, Kluwer Academic, Norwell.

4 Osaka, T., Komaba, S. and Liu, X., Ref. [1a], Chapter 7.

5 Xu, K. (2004) *Chem. Rev.*, **104**, 4303.

6 (a) Han, K.N., Seo, H.M., Kim, J.K., Kim, Y.S., Shin, D.Y., Jung, B.H., Lim, H.S., Eom, S.W. and Moon, S.I. (2001) *J. Power Sources*, **101**, 196; (b) Lackner, A.M., Sherman, E., Braatz, P.O. and Margerum, J.D. (2002) *J. Power Sources*, **104**, 1.

7 (a) Chan, C.K., Peng, H., Liu, G., Mcilwrath, K., Zhang, X.F., Huggins, R.A. and Cui, Y. (2008) *Nature Nanotech.*, **3**, 31; (b) Hassoun, J., Panero, S., Simon, P., Taberna, P.L. and Scrosati, B. (2007) *Adv. Mater.*, **19**, 1632.

8 For example, Conway, B.E. (1999) *Electrochemical Superconductors: Scientific Fundamentals and Technological Applications*, Kluwer Academic/Plenum Publishers, New York.

9 Ue, M. (2007) *Electrochemistry*, **75**, 565.

10 Ref. [8], Chapter 2.

14) This property of DMF makes the potential of the Ag^+/Ag reference electrode in DMF unstable (Section 6.1.2).

11 Naoi, K. and Simon, P. (2008) *Electrochem. Soc. Interface*, Spring, 34.

12 (a) Futaba, D.N., Hata, K., Yamada, T., Hiraoka, T., Hayamizu, Y., Kakudate, Y., Tanaike, O., Hatori, H., Yumura, M. and Iijima, S. (2006) *Nature Materials*, **5**, 987; (b) Zhou, H., Xhu, S., Hibino, M. and Honma, I. (2003) *J. Power Sources,* **122**, 219.

13 For example, (a) Ref. [8], Chapter 11; (b) Sugimoto, W., Iwata, H., Yasunaga, Y., Murakami, Y. and Takasu, Y. (2003) *Angew. Chem. Int. Ed.*, **42**, 4092.

14 For example, Mastragostino, M., Paraventi, R. and Zanelli, A. (2000) *J. Electrochem. Soc.*, **147**, 3167.

15 Ref. [8], Chapter 12.

16 For example, (a) Amatucci, G.G., Badway, F., Pasquier, A.D. and Zhang, T. (2001) *J. Electrochem. Soc.*, **148**, A930; (b) Yoshio, M., Nakamura, H. and Wang, H. (2006) *Electrochem. Solid-State Lett.*, **9**, A561.

17 Ue, M., Asahina, H. and Mori, S. (1995) *J. Electrochem. Soc.*, **142**, 2266 and references therein.

18 Ue, M., Tamamitsu, K., Tsuji, T., Sato, T. and Takeda, M. (2005) *J. Surf. Fin. Soc. Jpn.*, **56**, 286.

19 For example, Skotheim, T.A. and Reynolds, J. (eds) (2007) *Handbook of Conducting Polymers*, 3rd edn, vols. 1, 2, CRC Press, New York.

20 For example, (a) Heinze, J. (2001) *Organic Electrochemistry*, 4th edn (eds H. Lund and O. Hammerich), Marcel Dekker, New York, Chapter 32; Diaz, A.F. (1991) *Organic Electrochemistry*, 3rd edn (eds H. Lund and M.M. Baizer), Marcel Dekker, New York, Chapter 33; (b) Sadki, S., Schottland, P., Brodie, N. and Sabouraud, G. (2000) *Chem. Soc. Rev.*, **29**, 283.

21 (a) Scrosati, B. (1995) *Solid State Eletrochemistry* (ed. P.G. Bruce), Cambridge University Press, Cambridge, Chapter 9; (b) *Electrochim. Acta*, 1999, **44**, 1845–2163 (A special issue for electroactive polymer films).

22 (a) Cho, S.I., Kwon, W.J., Choi, S.-J., Kim, P., Park, S.-A., Kim, J., Son, S.J.,

Xiao, R., Kim, S.-H. and Lee, S.B. (2005) *Adv. Mater.*, **17**, 171; (b) Cho, S.I. and Lee, S.B. (2008) *Acc. Chem. Res.*, **41**, 699 and references therein.

23 *Electrochim. Acta*, 1999, **44**, 2969–3272 (A special issue for electrochromic materials and devices).

24 Bard, A.J. (ed.) (2004) *Electrogenerated Chemiluminescence*, Marcel Dekker, New York.

25 Richter, M.M. (2004) *Chem. Rev.*, **104**, 3003.

26 Danielson, N.D., Ref. [24], Chapter 9.

27 Buda, M., Ref. [24], Chapter 10.

28 (a) Kado, T., Takenouchi, M., Okamoto, S., Takashima, W., Kaneto, K. and Hayase, S. (2005) *Jpn. J. Appl. Phys.*, **44**, 8161; (b) Nishimura, K., Hamada, Y., Shibata, K. and Fu Yuki, T. (2002) *IEICE Trans. Electron. (Jpn. edn)*, **J85-C**, 1108.

29 Hori, Y. (2008) *Modern Aspects of Electrochemistry*, Number 42 (eds C. Vayenas, R.E. White and M.E. Gamboa-Aldeco), Springer, New York, Chapter 3.

30 (a) Gennaro, A., Isse, A.A., Severin, M.-G., Vianello, E., Bhugun, I. and Savéant, J.-M. (1996) *J. Chem. Soc., Faraday Trans.*, **92**, 3963; (b) Ikeda, S., Takagi, T. and Ito, K. (1987) *Bull. Chem. Soc. Jpn.*, **60**, 2517; (c) Ito, K. (1990) *Electrochemistry*, **58**, 984; (d) Hoshi, N., Murakami, T., Tomita, Y. and Hori, Y. (1999) *Electrochemistry*, **67**, 1144.

31 (a) Welford, P.J., Brookes, B.A., Wadhawan, J.D., McPeak, H.B., Hahn, C.E.W. and Compton, R.G. (2001) *J. Phys. Chem. B*, **105**, 5253; (b) Hahn, C.E.W., McPeak, H. and Bond, A.M. (1995) *J. Electroanal. Chem.*, **393**, 69.

32 Kavan, L. (1997) *Chem. Rev.*, **97**, 3061.

33 *Electrochim. Acta*, 1997, **42**, 1931–2270 (A special issue for electrochemistry in organic synthesis).

34 Parker, A.J. (1976) *Electrochim. Acta*, **21**, 671 and references therein.

35 Parker, A.J. (1972) *Proc. Roy. Aust. Chem. Inst.*, **39**, 163.

36 Aurbach, D., Ref. [1a], Chapter 6; Ref. [1b].

37 Brenner, A. (1967) *Advances in Electrochemistry and Electrochemical*

Engineering, vol. V (ed. C.W. Tobias),
John-Wiley & Sons, Inc., New York.

38 (a) Takeda, Y., Kanno, R., Yamamoto, O.,
Rama Mohan, T.R., Chia-Hao, L. and
Kröger, F.A. (1981) *J. Electrochem. Soc.*, **128**,
1221; (b) Gobet, J. and Tannenberger, H.
(1988) *J. Electrochem. Soc.*, **135**, 109; Ref. [1b].

39 Rosamilia, J.M. and Miller, B. (1989)
J. Electrochem. Soc., **136**, 1053.

40 Endres, F., Abbott, A.P. and Macfarlane,
D.R. (eds) (2008) *Electrodeposition from
Ionic Liquids*, Wiley-VCH Verlag GmbH,
Weinheim. (In this book, the
electrodepositions of metals, alloys,
semiconductors, conducting polymers,
etc. are reviewed in detail.)

41 Pastoriza-Santos, I. and Liz-Marzán, L.M.
(2000) *Pure Appl. Chem.*, **72**, 83.

Part Three
Electrochemistry in New Solvent Systems

Electrochemistry in Nonaqueous Solutions, Second, Revised and Enlarged Edition. Kosuke Izutsu
Copyright © 2009 WILEY-VCH Verlag GmbH & Co. KGaA, Weinheim
ISBN: 978-3-527-32390-6

13
Electrochemistry in Clean Solvents

13.1
Introduction

Following the principles of green chemistry, many researchers in academic and industrial fields are searching for new solvents or solvent systems that can reduce or eliminate the intrinsic hazards associated with traditional solvents [1]. In some cases new substances are being designed and developed for use as solvents, while in some other cases, the best known substances are finding new applications as solvents. Of course, using no solvent at all is the ultimate solution for eliminating hazards associated with solvents.

The following are the strategies for the alternative green solvents [1a]:

1. **Replacing organic solvents by water or aqueous solutions** Among the reactions or processes for which organic solvents seemed necessary, some have been found possible, under appropriate conditions, to use water or aqueous solutions as solvent. This is quite desirable because water is the most popular and safe solvent.

2. **Solvent-free systems** The complete elimination of solvents is the ultimate goal in solving the hazardousness of the solvent usage. This needs that the reagent for a particular process can also serve as solvent. See Ref. [2] for many organic syntheses carried out under solvent-free conditions.

3. **Reduced hazard organic solvents** Next-generation organic solvents with a reduced degree of hazard and toxicity are now actively being pursued.

4. **Supercritical and subcritical fluids** Supercritical carbon dioxide, water, etc. are used as solvents with reduced hazard in a wide range of applications, from reaction and separation medium to cleaning solvent. Although somewhat different from the supercritical fluids (SCFs), the liquids near the critical points (subcritical fluids) can also be used for similar purposes.

5. **Ionic liquids** Most ionic liquids (ILs) or room-temperature molten salts are nonvolatile, nonflammable and almost harmless to human health and the environment. They dissolve many organic and inorganic substances and are good solvents. Their electrochemical applications are widespread.

Electrochemistry in Nonaqueous Solutions, Second, Revised and Enlarged Edition. Kosuke Izutsu
Copyright © 2009 WILEY-VCH Verlag GmbH & Co. KGaA, Weinheim
ISBN: 978-3-527-32390-6

6. **Immobilized solvents** Solvent molecules, which are volatile and cause hazardous influences, can become harmless if they are properly bound to a polymeric backbone but still maintaining the solvent property.

7. **Fluorous solvents** Perfluorinated or very highly fluorinated solvents are called fluorous solvents. They are typically nonpolar and immiscible with organic solvents and water; thus, three phases (fluorous, organic and aqueous) may be separated at room temperature though fluorous and organic phases may be converted to one homogeneous phase when heated. Their boiling points are near to those of parent hydrocarbons, but they are nonflammable and can easily be recovered and reused. They are widely used as solvents for extractions as well as for reactions.

Among the alternative green solvents described above, supercritical fluids and ionic liquids are electrochemically most interesting, and they are dealt with in Sections 13.1 and 13.2, respectively. The electrochemistry in fluorous solvents is just at its beginning [3]. For example, to carry out voltammetric measurements in perfluoromethylcyclohexane, a fluorous solvent of $\varepsilon_r \approx 1$, two strategies have been used: one is to add another fluorous solvent, benzotrifluoride of $\varepsilon_r = 9.2$, in 1: 1 ratio in order to get a mixture with a modest ε_r value and the other is to use, as a fluorophilic supporting electrolyte, a salt consisting of an imidazolium cation with a fluorous ponytail and a lipophilic anion, tetrakis(3,5-trifluoromethyl)phenyl borate (Scheme 13.1). With the solvent and electrolyte, voltammetry was possible at moderate scan rate with electrodes of millimeter size and at faster scan rate with ultramicroelectrodes.

As described in Chapter 12, some reduced hazard organic solvents and immobilized solvents are used electrochemically in association with the lithium-ion batteries.

13.2
Supercritical Fluids

13.2.1
General Aspects of Supercritical Fluids

A substance is supercritical when its temperature and pressure are above the values of the critical point (Figure 13.1). Examples of critical temperature (T_c, °C) and pressure (P_c, bar) are shown in Table 13.1. See Ref. [4] for general affairs of SFCs.

The properties of SFCs are, as shown in Table 13.2, somewhere in between those of a liquid and a gas. The solubilities of solutes in SCFs are considerable because they have relatively large densities. For supercritical water (scW), the dielectric constant is

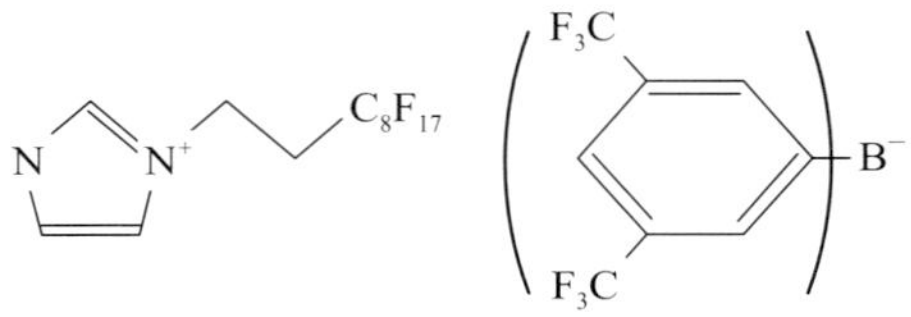

Scheme 13.1 Supporting electrolyte used in (perfluoromethylcyclohexane + benzotrifluoride).

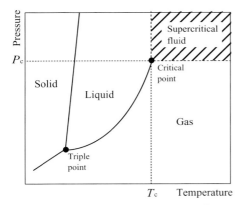

Figure 13.1 Temperature–pressure relation for pure substances.

low (ε_r: 5 or less, Figure 13.2[1]) and even nonpolar organic solute, which is insoluble in ordinary liquid water, can dissolve. On the other hand, the viscosities are near to the gas state, and the rates of mass transfer are much higher than in liquids. Moreover, in SCFs, the solubilities, mass transfer rates and reaction rates of solutes can be controlled by adjusting the applied pressure. The polarity of SCFs ranges from nonpolar region (e.g. supercritical CO_2 (scCO$_2$)) to polar region (e.g. scW)[2]; the polarity of each SCF can be modified by adding small amounts of cosolvent.

Table 13.1 Examples of critical temperatures and pressures.

Solvent	T_c, K (T_c, °C)	P_c, bar
Carbon dioxide	304 (31)	7.4
Water	647 (374)	221
Ammonia	405 (132)	114
Sulfur dioxide	431 (158)	77.8
Methanol	513 (240)	81.0
Ethanol	514 (241)	61.5
Acetonitrile	546 (273)	48.3
Difluoromethane (HFC32)	351 (78)	5.8
1,1,1,2-Tetrafluoroethane (HFC134a)	374 (101)	4.1

1) In Figure 13.2, the value of autoprotolysis constant of H_2O (pK_W) at 25 MPa is also shown as the function of temperature. In high-temperature water (200–300 °C), pK_W is 12 or less. However, the pK_W suddenly increases at temperatures over critical point.

2) The Kamlet–Taft parameters (α for hydrogen bond donor (HBD) acidity, β for hydrogen bond acceptor (HBA) basicity and π^* for dipolarity/polarizability) or related parameters have been determined in some supercritical fluids. For scCO$_2$ [5], α value at 318 °C varied with density, from 0 at above ~0.4 g cm^{-3} to 0.19 at 0.25 g cm^{-3}. The value

0.19 is a little less than that of acetonitrile (0.23). The β value was at most of the order of hexane. The π^* value varied from negative value at lower densities to zero at higher densities, showing that the dipolarity/polarizability is also very weak. For scHFC 32 and 134a [6], the α, β and π^* values show that they have moderate HBD acidity, HBA basicity and dipolarity/polarizability. For scW, only π^* value has been reported. Though the water near critical point (360 °C) had π^* value near 0.5, that for scW was near to zero or minus, showing the low polarity of scW.

Table 13.2 Comparison of properties of supercritical fluids and those of gases and liquids.

State	Condition	Density, g cm^{-3}	Viscosity, g cm^{-1} s^{-1}	Diffusion coefficient, cm^2 s^{-1}
Gas	1 atm, 25 °C	$0.6-2 \times 10^{-3}$	$1-3 \times 10^{-4}$	$1-4 \times 10^{-1}$
Liquid	1 atm, 25 °C	0.6–2	$0.2-3 \times 10^{-2}$	$0.2-2 \times 10^{-5}$
Supercritical	T_c, P_c	0.2–0.5	$1-3 \times 10^{-4}$	$0.5-4 \times 10^{-3}$
fluid	T_c, $4P_c$	0.4–0.9	$3-9 \times 10^{-4}$	$0.5-4 \times 10^{-3}$

SCFs have been used for many years in a variety of ways. Some SCFs serve as 'green solvents' to replace organic solvents that are hazardous to human health and the environment. Especially, scW and scCO$_2$ are important as such SCFs.

In the case of scW, as described above, nonpolar organic substances can dissolve in high concentrations owing to the low dielectric constant. This fact and the high critical temperature ($T_c = 374.2$ °C) make possible the supercritical water oxidation that decomposes the toxic organic substances and hazardous wastes and makes them innocuous. If an oxidizing agent (O$_2$ or H$_2$O$_2$) coexists, considerable parts of PCB and dioxins are decomposed to H$_2$O, CO$_2$ and HCl (Figure 13.3). ScW is also useful for chemical recycling of various waste plastics; the plastics are decomposed to raw materials and fuels. In these processes, subcritical water also works more or less in the same way. Supercritical and subcritical waters are useful to decompose biomass wastes and get resources from them: the hydrolysis in subcritical water gives useful substances that are contained in the biomasses, the hydrothermal reactions in subcritical and supercritical waters give fuel gas (hydrogen and methane), and the supercritical water oxidation gives CO$_2$, H$_2$O and N$_2$ and heat that can be used as energy sources.

The scCO$_2$ is widely applied because it is supercritical near room temperatures, does not flame, does not decompose substances and yet dissolves much organic substances. Examples of the applications of scCO$_2$ are as follows:

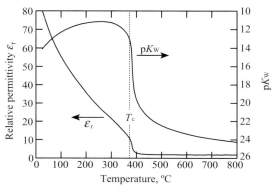

Figure 13.2 Temperature dependence of relative permittivity (ε_r) and autoprotolysis constant (pK$_w$) of water at a pressure of 25 MPa (Tester, J.W. *et al.* (1993) *ACS Symp. Ser.*, **518**, 35).

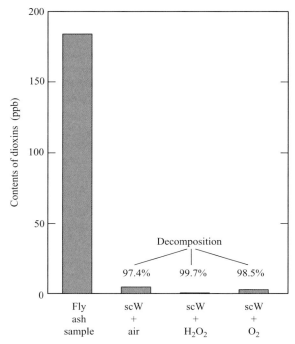

Figure 13.3 Decomposition of dioxins by scW plus oxidizing agents at 400 °C, 30 MPa and reaction time of 30 min [4d].

1. **For separation or extraction** $scCO_2$ is used for the decaffeination of green coffee beans, the extraction of hops for beer production and the production of essential oils and pharmaceutical products from plants. It is also used to remove dioxins from soils and to separate and recover uranium and plutonium from nuclear fuels. $ScCO_2$ is also an extractant of organic substances from ionic liquids [8].

2. **For chemical reactions and syntheses** $scCO_2$ is used to produce various chemicals and polymers with high yield and high-energy efficiency and under benign conditions. Examples are the synthesis of polyacrylic acid in $scCO_2$ and the synthesis of carbonate using $scCO_2$ as raw materials.

3. **For manufacturing and processing of micro- and nanomaterials** $scCO_2$ is suitable for preparing micro- and nanoparticles and fine-porosity materials (e.g. silicaaerogel).

4. **For cleaning** $scCO_2$ is used for dry cleaning, replacing perchloroethylene and other undesirable solvents. It is also used to clean semiconductors and important parts of precision tools.

5. **For the supercritical fluid chromatography (SFC)**

The critical points of alcohols are between CO_2 and water and, thus, supercritical alcohols can decompose substances by the extent in between $scCO_2$ and scW.

For example, PET bottles are decomposed by supercritical methanol to dimethyl terephthalate and ethylene glycol; by using ethylene glycol and terephthalic acid, obtained by further hydrolysis of dimethyl terephthalate, a resin is reproduced to get new PET bottles.

13.2.2
Electrochemical Aspects of Supercritical Fluids

In 1981, Silvestri *et al.* [9] used supercritical HCl and NH_3 to study the anodic dissolution of Fe and Ag. Since then, electrochemical studies in SCFs have been carried out to a considerable extent.

Bard and coworkers [10] carried out voltammetric works in supercritical and near-critical water. High temperature water is very corrosive, attacks most metals and dissolves quartz and Pyrex glass, leading to equipment failure and solution contamination. So they got an improved electrochemical cell as follows [10d]. They replaced the alumina tube reactor by an oxidized titanium tube to produce a more rugged reactor. All metal fittings were replaced by titanium ones to decrease corrosion problems. The working electrodes were Pt disk UMEs (radius: 14 μm) fabricated by sealing Pt–Ir wires in a high PbO-containing glass. This glass had the advantage over borosilicate glass in showing lower conductivity at temperatures above 230 °C, better corrosion resistance and a coefficient of thermal expansion well matched to Pt. They used this improved electrochemical cell to get well-defined steady-state cyclic voltammograms for the oxidation of iodide ion and hydroquinone in 0.2 M $NaHSO_4$ solutions over a temperature range 23–385 °C. For both processes, diffusion coefficients, determined from the limiting currents, agreed well with those calculated from the Stokes–Einstein equation and viscosity values for water at the relevant temperatures and pressures (Figure 13.4). The facts that no special change was observed before and after the critical point show that no large change occurred in the size or shape of these species and that the changes in water for diffusion can simply be treated in terms of bulk viscosity changes. Electrochemical studies are useful to understand various phenomena (e.g. supercritical water oxidation) occurring in supercritical and near-critical water. Moreover, because near-critical water and scW can dissolve many nonpolar organic compounds, electrochemical syntheses in high-temperature water may become a promising procedure. Bard and coworkers also studied electrochemistry in near-critical and supercritical ammonia [11], sulfur dioxide [12] and acetonitrile [13].

$scCO_2$ is important in that it is nontoxic, nonflammable, environmentally benign, commercially available in high quality and has easily attainable critical parameters. However, it is practically nonpolar ($\varepsilon_r \sim 1.6$), and polar solutes and electrolytes are difficult to dissolve in it. Abbott and Harper [14] tried voltammetric measurement of bis(tetradodecylammonium) nickel maleonitrile by dissolving a hydrophobic electrolyte, $Decyl_4NBF_4$, in $scCO_2$, but they did not get a well-defined wave. Voltammetric measurements in $scCO_2$ are usually carried out by adding some polar modifier (e.g. water, acetonitrile) to increase the polarity. Wightman and coworkers [15] studied, using UMEs, the voltammetric oxidation of ferrocene in $scCO_2$ containing H_2O as a modifier and Hex_4NPF_6 as a supporting electrolyte. When $[H_2O] \sim 55$ mM,

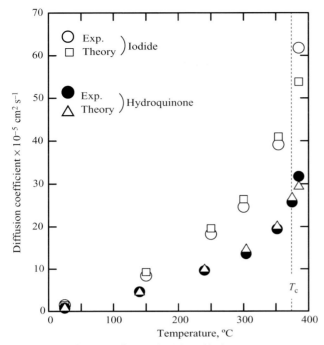

Figure 13.4 Diffusion coefficient of iodide and hydroquinone as a function of temperature in water [10d].

ferrocene gave a well-defined oxidation voltammogram, though its oxidation was impossible if $[H_2O] < 28$ mM. When the concentration of Hex_4NPF_6 was increased, a phase separation occurred between the $scCO_2$ and Hex_4NPF_6, which complicated the electrode process. Li and Prentice [16] used a high-pressure, near-critical CO_2 + H_2O + EtOH solution, with LiCl as electrolyte and a copper cathode, to synthesize methanol by direct electrochemical reduction of CO_2.

$$CO_2 + 6H^+ + 6e^- \rightarrow CH_3OH + H_2O$$

Methanol was produced at 80 °C with a CO_2 pressure of 68 bar. The best current efficiency of 40% is too low to be commercially feasible, but this synthesis uses an environment friendly dense-phase CO_2 for the solvent system, and the CO_2 is also the raw material of synthesis. Conductive polymers have been synthesized in $scCO_2$ by several workers [17]. The nanostructured nickel with particle sizes of sub-100 nm was electroplated in emulsion of $scCO_2$ and electrolyte solution by Yoshida *et al.* [18]. Platinum/ruthenium nanoparticles were decorated on carbon nanotubes in $scCO_2$, and the PtRu/carbon nanotube nanocomposite was shown to have high activity for the electrooxidation of MeOH in 1 M H_2SO_4 + 2 M MeOH [19]. The results proved the feasibility of processing bimetallic catalysts in $scCO_2$ for fuel cell applications.

Supercritical difluoromethane (HFC32), trifluoromethane (HFC23) and 1,1,1,2-tetrafluoroethane (HFC134a) are somewhat more polar than scCO$_2$ and have readily attainable critical parameters. They are used as a voltammetric solvent alone or by mixing with CO$_2$. Olsen and Tallman [20] measured in supercritical chlorodifluoromethane and HFC23 the oxidation wave of ferrocene (Fc^0 → Fc$^+$) and the reduction wave of cobaltocenium ion (Cc$^+$ → Cc^0). The difference between their half-wave potentials was 1.28 V, in good agreement with 1.31 V obtained in various nonaqueous solutions (Section 8.2.2). Abbott and coworkers carried out a series of studies [21–25]: (1) By using HFC134a and HFC32 in liquid and supercritical states, oxidation waves of Cs$^+$ and Xe were obtained [21]. The potential window had an unprecedented width of 9.4 V in HFC134a/0.1 M Bu$_4$NClO$_4$, which was liquid at 30 °C and 10 bar. (2) The effect of electrolyte concentration on the viscosity of HFC32 was measured by using a modified quartz resonator [22]. Ion solvation caused a significant structuring of the solvent and an appreciable increase in solution viscosity. Voltammetry was used to measure the changes of viscosity. (3) The dissociation constants of ferrocene carboxylic acid and its Bu$_4$N salt were studied in scHFC32, and the changes in the redox potentials of ferrecene derivatives were related to the dissociation constants of the electroactive species [23]. (4) The capacitance of a Bu$_4$NBF$_4$/scHFC32 electrolyte at a Pt electrode was shown to depend strongly on fluid density [24]. The capacitance–potential plots at high pressures exhibited a double-layer response similar to that in the liquid state, but, when the pressure was lowered toward the critical value, the diffuse layer collapsed and a pure Helmholtz response was observed. (5) The reduction of CO$_2$ in a mixed SCF (CO$_2$/HFC134a, $x_{\text{HFC134a}} = 0.3$) was studied using Pt and Pb electrodes [25]. At 60 °C and 260 bar, the faradaic efficiencies (%) of (COOH)$_2$, CO and HCOOH formations were 41.6, 14.6 and 0, respectively, at Pt and 17.5, 42.0 and 0, respectively, at Pb. The highly efficient (COOH)$_2$ formation at Pt, compared to <5% in aprotic solutions, was interpreted in terms of the adsorption of CO$_2^{\bullet-}$ onto the Pt electrode (Section 12.5).

The use of SCFs in electrochemical technologies is still scarce but will become more popular, considering their wide applications in many fields. Electrochemical *in situ* monitoring of solutes in SCFs will be another type of applications.

13.3
Ionic Liquids

Ionic liquid or room-temperature ionic liquids (RTILs) are the salts that are liquids near room temperatures, although often defined as the salts melting below 100 °C. Many of them are nonvolatile, nonflammable, less toxic, chemically and thermally stable and good solvents for both organic and inorganic materials. From electrochemical point of view, some of them have good electrical conductivity and wide potential windows. Hydrophobic ILs form interfaces with water (W) and are applicable to chemical separations, extractions and electrochemistry at the IL/W interfaces, just like the cases between two immiscible electrolyte solutions (ITIES).

The electrochemistry at the IL/W interfaces will be discussed in the next chapter. Some books [26] and review articles [27] are available on the electrochemical use of ILs.

ILs can be divided into two groups: chloroaluminate (or more generally haloaluminate) ILs and nonchloroaluminate ILs. Chloroaluminate ILs are the mixtures of $AlCl_3$ and organic chloride (R^+Cl^-). R^+ include 1,3-dimethylimidazolium $([MMIm]^+)$, 1-ethyl-3-methylimidazolium $([EMIm]^+)$, 1-propyl-3-methylimidazolium $([PMIm]^+)$, 1-butyl-3-methylimidazolium $([BMIm]^+)$, 1-propyl-2,3-dimethylimidazolium $([PMMIm]^+)$, N-butylpyridinium $([BP]^+)$, etc. (see Scheme 13.2 for the structure of skeletons). The reaction $2AlCl_3 + Cl^- \rightarrow AlCl_4^- + AlCl_3 \rightarrow Al_2Cl_7^-$ occurs between $AlCl_3$ and Cl^-. If we express the mole fraction of $AlCl_3$ by N $(= [AlCl_3]/([AlCl_3] + [RCl]))$, the ILs of $N = 0.5$ are neutral with an anion $AlCl_4^-$, the ILs of $N > 0.5$ are acidic with anions $Al_2Cl_7^-$ and $AlCl_4^-$ and the ILs of $N < 0.5$ are basic with anions $AlCl_4^-$ and Cl^-. Chloroaluminate ILs are usually hygroscopic.

Nonchloroaluminate ILs are the salt between a cation R^+ and an anion A^-, where the examples of R^+ for aprotic ILs are $[MMIm]^+$, $[EMIm]^+$, $[PMIm]^+$, $[BMIm]^+$, 1-hexyl-3-methylimidazolium $([HMIm]^+)$, $[BP]^+$, 1-butyl-3-methylpyridinium $([BMP]^+)$, 1,1-dimethylpyrrolidinium $([MMPyr]^+)$, trimethylethylammonium $([Me_3EtN]^+)$, trimethylpropylammonium $([Me_3PrN]^+)$, tetraethylammonium $([Et_4N]^+)$, tetrabutylammonium $([Bu_4N]^+)$, tetrabutylphosphonium $([Bu_4P]^+)$ and trimethylsulfonium $([Me_3S]^+)$, and those of A^- are $[BF_4]^-$, $[PF_6]^-$, $[CF_3CO_2]^-$, $[CF_3SO_3]^-$ (TfO^-), $[C_2F_5SO_3]^-$, $[(CF_3SO_2)_2N]^-$ (NTf_2^-) and $[F(HF)_{2.3}]^-$. Some are very hydrophobic and immiscible with water as discussed in the next chapter. There are protic ILs too, the cations R^+ of which being primary, secondary and tertiary ammonium cations, 1-alkylimidazolium cations, 1-alkyl-2-alkylimidazolium cations, etc. [28].

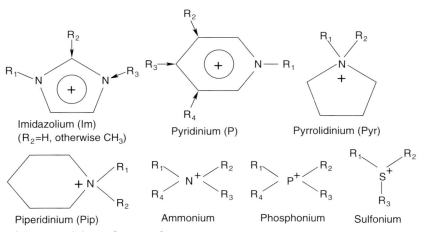

Scheme 13.2 Skeleton of cations of ILs.

13.3.1
Physical, Chemical and Electrochemical Properties of ILs

Dielectric constants [29] Static dielectric constants have been determined for some ILs. They are 15.2 for [EMIm][CF$_3$SO$_3$], 12.8 for [EMIm][BF$_4$], 11.7 for [BMIm][BF$_4$], 11.4 for [BMIm][PF$_6$] and 8.9 for [HMIm][PF$_6$] compared to 20.56 for acetone, 12.9 for pyridine, 10.37 for 1,2-dichloroethane and 8.93 for dichloromethane.

Viscosity [30] Viscosity is a measure of resistance of fluid to flow. Viscosities of ILs (η) at room temperature are from about 10 cP to over 500 cP. These values are much higher than 0.34 cP of AN, 0.89 cP of W and 2.53 cP of PC, and are comparable to 16.1 cP of ethylene glycol and 934 cP of glycerol. Some of the data are shown in Table 13.3. The viscosities are much influenced by temperature. However, the large dispersions of some data in Table 13.3 mainly result from impurities; especially, Cl$^-$-ion and water that are easily contained in ILs much influence the viscosities. Chloride concentration between 1.5 and 6 wt.% raised the viscosity between 30 and 600%. Two wt% of water reduced the viscosity of [BMIm][BF$_4$] by more than 50%. For nonchloroaluminate ILs of the same cation, the viscosity increases with the anion in the order NTf$_2$$^-$ < BF$_4$$^-$ < CF$_3$CO$_2$$^-$ < CF$_3$SO$_3$$^-$ < C$_2$F$_5$SO$_3$$^-$ < C$_2$F$_5$CO$_2$$^-$ < CH$_3$CO$_2$$^-$ < CH$_3$SO$_3$$^-$ < C$_4$F$_9$SO$_3$$^-$. For ILs between substituted imidazolium cations and NTf$_2$$^-$, the viscosity increases in the order [EMIm]$^+$ < [EEIm]$^+$ < [BEIm]$^+$ < [BMIm]$^+$ < [PMMIm]$^+$. For chloroaluminate [EMIm]Cl-AlCl$_3$ at 303 K, the viscosity is nearly constant (14–18) for [EMIm]Cl below 50 mol%, but it increases for [EMIm]Cl above 50 mol% and becomes over 190 cP at 67 mol%. This increase is explained to be due to the hydrogen bonding between Cl$^-$ and the imidazolium ring. For chloroaluminate ILs with the same anion, the viscosity increases with the size of the cation but tends to decrease if the cation is highly asymmetrically substituted. In general, the addition of small amount of cosolvent (e.g. acetonitrile) decreases the viscosity of ILs significantly.

Conductivity [31] The conductivity depends on the number of available charge carriers and their mobility. Very high conductivities are expected for ILs because they are composed entirely of ions. But the actual conductivities of common ILs are comparable to the highest nonaqueous solvent/electrolyte systems ($\sim$10 mS cm^{-1}), and they are much smaller than those of concentrated aqueous electrolyte solutions. The smaller than expected conductivities of ILs can be attributed to ion pairing or ion aggregations, in addition to the large sizes of constituent ions and the high viscosities of ILs. Some of the conductivity (κ) and molar conductivity (Λ) data are also shown in Table 13.3. Here, the conductivity of ILs with F (HF)$_{2.3}$$^-$-anion is exceptionally high and is $\sim$100 mS cm^{-1}. The conductivity of ILs is influenced significantly by the impurities, especially by Cl$^-$-ions and water. Small amount of Cl$^-$-ions tends to decrease the conductivity while small amount of water increases it. The relation between the molar conductivity and the

Table 13.3 Viscosities, conductivities and molar conductivities of ILs at 298 or 293 K (with*).

ILs	η, cP	κ^a	Λ^b
Chloroaluminate ILs			
[MMIm]Cl-AlCl₃ (N = 0.66)	17	15.0	4.26
[EMIm]Cl-AlCl₃ (N = 0.66)	14	15.0	4.46
[EMIm]Cl-AlCl₃ (N = 0.50)	18	23.0	4.98
[EMIm]Cl-AlCl₃ (N = 0.40)	47	6.5	1.22
[BMIm]Cl-AlCl₃ (N = 0.66)	19	9.2	3.04
[BMIm]Cl-AlCl₃ (N = 0.50)	27	10.0	2.49
[MP]Cl-AlCl₃ (N = 0.67)	21	8.1	2.23
[EP]Cl-AlCl₃ (N = 0.67)	18	10.0	2.91
[BP]Cl-AlCl₃ (N = 0.67)	21	6.7	2.18
PhMe₃NCl-AlCl₃ (N = 0.67)	33		
BenPr₃NCl-AlCl₃ (N = 0.67)	226		
Nonchloroaluminate ILs			
[MIm][NTf₂]	81	7.2	
[EIm][PF₆]	550		
[EIm][OTf]	58		
[EIm][NTf₂]	54		
[EMIm][BF₄]	32–43	14	2.24
[EMIm][CH₃SO₃]	160	2.7	0.45
[EMIm][TfO]	43	9.2	1.73
[EMIm][NTf₂]	28–34	9.2	2.38
[EMIm][F(HF)₂.₃]	6	100	15.5
[BMIm][BF₄]	92–219	3.5	0.65
[BMIm][PF₆]	173–450	1.5	
[BMIm][CH₃CO₂]	440		
[BMIm][CF₃CO₂]	70	3.2	0.67
[BMIm][TfO]	90	3.7*	0.83*
[BMIm][NTf₂]	47–69	3.9*	1.14*
[HMIm][PF₆]	585		
[HMIm][NTf₂]	68	2.2	
PrMe₃N[NTf₂]	72	3.3	0.88
HeEt₃N[NTf₂]	167	0.67	0.25
[BP][BF₄]	103	1.9	0.35
[BP][NTf₂]	56.8	2.2	0.63
[PMPyr][NTf₂]	63	1.4	0.39
[BMPyr][NTf₂]	85	0.07	0.02

Skeleton for cations: Im, imidazolium; P, pyridinium; Pyr, pyrrolidinium. Substituents: M and Me, methyl; E and Et, ethyl; B, butyl; Ben, benzyl; He, hexyl; Ph, phenyl; Pr, propyl. Anions: NTf₂, N(CF₃SO₂)₂; TfO, CF₃SO₃.
[a] mS cm⁻¹.
[b] mS cm⁻¹ mol⁻¹. From Ref. [25b].

viscosity nearly follows the Walden's rule, i.e. $\Lambda\eta \approx$ constant. Thus, for ILs of high viscosity, the molar conductivity is small. With ILs of the same anions, the value of $\Lambda\eta$ tends to decrease with the increase of cationic size while with ILs of the same cation, no clear relations can be found between the anionic size.

Acid–base properties and polarity Acceptor numbers (*AN*s) have been determined for some aprotic ILs, by using a linear relation between *AN*s of various molecular solvents and Raman spectrum of diphenylcyclopropenone [32]. *AN*s for ILs are 27.7 for [BMIm][PF$_6$], 26.9 for [BMIm][BF$_4$] and [BMIm]Cl, 25.2 for [BMIm][NTf$_2$], 27.5 for [BMMIm][BF$_4$], 22.2 for [PrMe$_3$N][NTf$_2$], 23.9 for [Et$_2$Me (MeOEt)N][BF$_4$] and 20.3 for [Et$_2$Me (MeOEt)N][NTf$_2$]. These values compare with 27.1 for *t*-BuOH, 22.2 for *t*-PeOH and 20.5 for nitromethane, and these are somewhat larger than those for conventional aprotic solvents (*AN* ∼20 or smaller). The Kamlet–Taft parameters (α, β and π*) for nonchloroaluminate aprotic ILs have also been determined, as shown in Table 13.4 [33]. Here, α is a measure of hydrogen bond acidity, β a measure of hydrogen bond basicity and π* a measure of dipolarity/polarizability. The values in Table 13.4 were obtained from the UV/vis spectra of a single set of dyes (Reichardt's dye, *N*,*N*-diethyl-4-nitroaniline and 4-nitroaniline)[3] though they are usually obtained by averaging the values for several sets of dyes. For aprotic ILs, the α values are largely determined by the nature of the cation though there is also a smaller anion effect. The α values for [BMIm] and [HMIm] ILs are close to the value of *t*-BuOH and a little higher than those for ILs of other cations. The [BMMIm] IL has lowest α values, reflecting the loss of the proton on the 2-position of the ring. Usually, the hydrogen bond acidity of ILs ranges from aprotic to a little higher than aprotic. The β values of ILs are sensitive to anions, but generally the values are smaller than the value of acetone and close to or even smaller than that of acetonitrile, which is a very weak base. All of the π* values for ILs are higher than those for nonaqueous molecular solvents but lower than that for water, showing that the polarities of ILs are moderately high. About the Kamlet–Taft parameters of ILs, the temperature effects on them [33e] and the values in IL/organic solvent mixtures [33d] have been studied. Kamlet–Taft parameters have also been determined for the protic ILs [28] and some of them are shown in Table 13.4. Except for [H*MIm][TfO] (H*: hydrogen), α values for protic ILs are higher than those for aprotic ILs. The autoprotolysis constant (*K*$_{SH}$) of [EtH$_3$N][NO$_3$], a protic IL, has been obtained using a rapidly responding pH-ISFET to be $10^{-10.0}$ mol^2 dm^{-6}, indicating that the neutral [EtH$_3$N][NO$_3$] involves HNO$_3$ and EtNH$_2$ molecules as much as 1.0×10^{-5} M or pH is 5.0 at 298 °C [34a]. A method utilizing the rapidly responding pH-ISFET has been developed for the characterization of ILs and patented [34b]. The rapidity of pH-ISFET response in molecular solvents has been dealt with in Chapter 6.

---

3) $E_T(30)$ (in kcal mol^{-1}) is obtained by 25891/ λ_{max} (nm), where λ_{max} is the wavelength of the maximum of the charge transfer absorption band of the Reichardt's dye. π* is obtained by measuring the wavelength of maximum absorbance, ν_{max} in kK (kilokeyser, 10^{-3} cm^{-1}), of the dye *N*,*N*-diethyl-4-nitroaniline from the relation π$^* = (\nu_{max} - \nu_0)/s$, where ν_0 (27.52 kK) is the regression value for a reference solvent system and *s* (−3.182) is the susceptibility of intensity of spectral absorption due to changing solvent dipolarity/polarizability. α is then obtained by α = {$E_T(30)$ − 14.6(π* − 0.23) − 30.31}/16.5. β is obtained by measuring the relative difference of solvatochromism between 4-nitroaniline (1) and *N*,*N*-diethyl-4-nitroaniline (2) by β = {1.035ν (2)$_{max}$ − ν(1)$_{max}$ + 2.64}/2.8, where ν(2)$_{max}$ and ν(1)$_{max}$ are the wavelengths of maximum absorbance.

Table 13.4 Kamlet–Taft parameters in ILs and in some molecular solvents.[a]

ILs	α	β	π*
Aprotic ILs			
[4MBP][NTf₂]	0.51	0.29	0.98
[OP][NTf₂]	0.51	0.28	0.99
[2MOP][NTf₂]	0.48	0.35	0.95
[3MOP][NTf₂]	0.50	0.33	0.97
[4MOP][NTf₂]	0.50	0.33	0.97
[3MBP][TfO]	0.40	0.46	1.02
[3MBP][NTf₂]	0.43, 0.54	0.25, 0.28	0.95, 0.97
[3MHP][NTf₂]	0.54	0.31	0.98
[BMIm][BF₄]	0.63	0.38	1.05
[BMIm][SbF₆]	0.64	0.15	1.04
[BMIm][PF₆]	0.63	0.21	1.03
[BMIm][TfO]	0.63	0.46	1.01
[BMIm][NTf₂]	0.62	0.24	0.98
[HMIm][TfO]	0.67	0.52	0.98
[HMIm][NTf₂]	0.65	0.25	0.98
[OMIm][NTf₂]	0.60, 0.46	0.28	0.97
[BMMIm][BF₄]	0.40	0.36	1.08
[BMMIm][NTf₂]	0.38	0.24	1.01
[HMMIm][NTf₂]	0.45	0.26	0.99
Protic ILs			
[EtNH⁺₃][NO₃]	1.10	0.46	1.12
[PrNH⁺₃]formate	0.87	0.75	0.73
[PrNH⁺₃][NO₃]	0.88	0.52	1.17
[Pr₂NH⁺₂][SCN]	0.97	0.39	1.16
[Bu₃NH⁺][NO₃]	0.84	—	0.97
[H⁺MIm][TfO]	0.259	0.650	0.971
Molecular Solvents			
Water	1.12	0.14	1.33
Methanol	1.05	0.61	0.73
2,2,2-Trifluoroethanol	1.04, 1.51	0.01, 0.00	1.20, 0.73
DMSO	0.00	0.76	1.00
AN	0.35	0.37	0.80
Ac	0.20	0.54	0.70
2-Butanone	0.06	0.48	0.67
Dichloromethane	0.04	−0.01	0.79
Toluene	−0.21	0.08	0.53
Hexane	0.07	0.04	−0.12
Cyclohexane	0[b]	0[b]	0[b]

Substituents: O, octyl; H⁺, hydrogen, for others, see Table 13.3.
[a]Ref. [32] for aprotic ILs and Ref. [27] for protic ILs.
[b]By definition.

Solvation of ions Recent study by Raman spectroscopy in [BMIm][NTf₂] [35] showed that the numbers of NTf_2^- bound to Li^+, Na^+, K^+ and Cs^+ are 1.95, 2.88, 3.2 and 3.9, respectively. Because NTf_2^- is a bidentate ligand, this means that the coordination numbers of metal ions are 4, 6, 6 and 8, respectively, for

Li^+, Na^+, K^+ and Cs^+. It was also shown that the ion–ion interaction between the alkali metal ions and NTf_2^- is mainly electrostatic and that *cis*-NTf_2^- is favored in the first solvation sphere of the small Li^+ ion, although this tendency diminishes with an increase in the radii of metal ions. Review articles are available on structural studies of ILs by EXAFS [36a] and of alkylimidazolium-based ILs with fluoroanions [36b].

Electrochemical potential windows [37] Potential windows have been measured using GC, Pt, tungsten (W), etc. as working electrodes. The reference electrodes used are $Al_2Cl_7^-$, $AlCl_4^-|Al$(wire), I_3^-, $I^-|Pt$(wire) and $AgBF_4$(satd) or $AgTaF_6$(satd)$|Ag$(wire), the reactions of which are $4Al_2Cl_7^- + 3e^- \rightleftarrows Al + 7AlCl_4^-$, $I_3^- + 2e^- \rightleftarrows 3I^-$ and $Ag^+ + e^- \rightleftarrows Ag$, respectively. A quasireference electrode (Pt or Ag wire) and $Li^+|Li$ reference electrode are also frequently used. The potentials of some reference electrodes against the Fc^+/Fc system (solvent-independent reference) are shown in Table 13.5.

Some of the potential windows for chloroaluminate ILs are shown in Figure 13.5 using the potential of Fc^+/Fc system as reference. In the acidic chloroaluminate ILs ($N > 0.5$), the cathodic limit is the reduction of $Al_2Cl_7^-$ and the anodic limit is the oxidation of $AlCl_4^-$. In the basic chloroaluminate ILs ($N < 0.5$), the cathodic limit is the reduction of the cation R^+ and the anodic limit is the oxidation of Cl^-. In neutral chloroaluminate ILs ($N = 0.5$), the cathodic limit is the reduction of R^+ and the anodic limit is the oxidation of $AlCl_4^-$, and the width is about 4.4 V for [EMIm]Cl-$AlCl_3$ and 3.6 V for [BP]Cl-$AlCl_3$.

The potential windows for nonchloroaluminate ILs are shown in Figure 13.6 against Fc^+/Fc reference system. Generally, the cathodic limit is the reduction of the cation while the anodic limit is the oxidation of the anion. However, when the anion is easily reduced, the cathodic limit may be the reduction of the anion. Contrarily, when the cation is easily oxidized, the anodic limit may be the oxidation of the cation. The former case occurs with [Et$_4$N][TSAC] (TSAC = 2,2,2-trifluoro-*N*-(trifluoromethylsulfonyl)acetamide) and the latter case with EMIm and *p*-dimethylamino-*N*-butylpyridinium (DMABP) systems. The widest potential windows are obtained with ILs of aliphatic quaternary ammonium ions and fluoroanions: the widths of nearly 6.0 V are comparable

Table 13.5 Potentials of some reference electrodes against Fc^+/Fc system.[a]

	Ag(I)/Ag	Al(III)/Al	I_3^-/I^-	EMIm decomp.
[EMIm][AlCl$_3$]	+0.44	−0.40	—	−2.5$_0$
[EMIm][BF$_4$]	+0.51	−0.49	—	−2.5$_9$
[EMIm][NTf$_2$]	+0.51	—	−0.21	−2.5$_0$
[Me$_3$HexN][NTf$_2$]	+0.49	—	−0.07	—

[a]From Ref. [37].

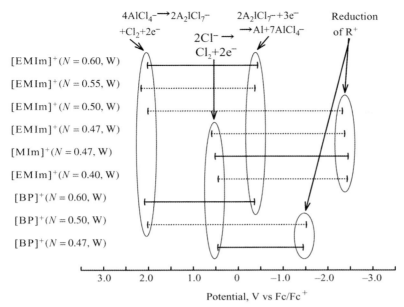

Figure 13.5 Potential windows for chloroaluminate ILs [37].

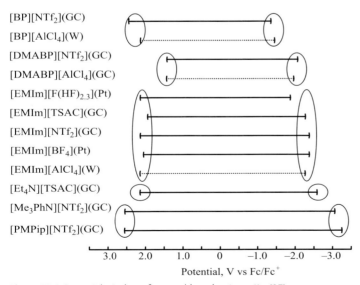

Figure 13.6 Potential windows for nonchloroaluminate ILs [37].
DMABP = p-dimethylamino-N-butylpyridinium and
TSAC = $CF_3SO_2-N-COCF_3$.

with those in aprotic solvents. Water impurity reduces the width of potential window very much.[4]

13.3.2
Voltammetry in ILs

Voltammetric measurements in room-temperature ionic liquids can be performed in similar ways as those in nonaqueous solvents, though reference electrodes should be selected appropriately and, in the case of chloroaluminate ILs, the effect of moisture should be avoided.

1. **Reduction of dissolved oxygen** Dissolved oxygen in ILs is reduced to superoxide ($O_2 + e^- \rightarrow O_2^{\bullet-}$), just as in aprotic solvents: $O_2^{\bullet-}$ generated in [Im]Cl-AlCl$_3$ (GC electrode) [38a] is not stable but that in [BMIm][PF$_6$] (GC electrode) [38b] is stable enough to give a cyclic voltammogram for reverse reaction ($O_2^{\bullet-} \rightarrow O_2 + e^-$). Buzzeo *et al.* [38c] used a microdisk gold electrode of about 10 μm in diameter and studied the reduction of O_2 in [EMIm][NTf$_2$] and [HeMe$_3$N][NTf$_2$] by cyclic voltammetry. For [EMIm][NTf$_2$] and [HeMe$_3$N][NTf$_2$] with the viscosities of 32.1 and 220 cP, respectively, the solubilities of O_2 were 3.9 and 11.6 mM and the diffusion coefficients of O_2 were 8.3×10^{-6} and 1.5×10^{-6} cm^2 s^{-1}. The diffusion coefficients of $O_2^{\bullet-}$, determined by simulation using these values, were 2.7×10^{-6} cm^2 s^{-1} in [EMIm][NTf$_2$] and 4.7×10^{-8} cm^2 s^{-1} in [HeMe$_3$N][NTf$_2$]. The fact that the diffusion coefficients for both O_2 and $O_2^{\bullet-}$ are much different between the two ILs can be attributed to the big difference in viscosity. Superoxide ion, $O_2^{\bullet-}$, generated from IL, e.g. [BMIm][PF$_6$], is a safe and mild oxidizing agent and has several potential applications in green chemistry [40]. For example, it can destroy hexachlorobenzene (HCB), which have a serious environmental problem.

2. **Voltammetry of metallocene derivatives** In nonaqueous molecular solvents, ferrocenium ion/ferrocene (Fc$^+$/Fc) system is used as a reference for electrode potentials because its potential is nearly solvent independent. Such a use is made also in ILs. The oxidation of ferrocene has been studied voltammetrically by various investigators [41]. In [BMIm][PF$_6$] at 22 °C, ferrocene was found to be poorly soluble and need solid-state voltammetric experiment, involving mechan-

---

4) The structure at the IL-electrified metal interface has been studied for imidazolium-based ILs by sum frequency generation (SFG), electrochemical impedance spectroscopy (EIS) and vibrational Stark shift [39]. The vibrational Stark shift and EIS results showed that ions organize in a Helmholtz-like layer at the interface, where the potential drop occurs over the range of 3–5 Å from the metal surface into the liquid. Further, the SFG results im- plied that the 'double-layer' structure is potential dependent; at potentials positive of the potential of zero charge (PZC), anions are adsorbed to the surface and the imidazolium ring is repelled to orient more along the surface normal, compared to the potentials negative of the PZC, at which the cation is oriented more parallel to the surface plane and the anions are repelled from the surface.

ical attachment of finely grinded solid to the electrode surface [41a]. The reaction can be expressed as follows:

$$[Fe(Cp)_2]_{(solid)} + [BMIm][PF_6]_{(solvent)} \rightleftharpoons Fe(Cp)_2][PF_6]_{(solid)}$$
$$+ e^- + [BMIm]^+_{(solvent)}$$

From the solid-state voltammogram, the formal potential for Fc at the Au electrode was determined to be $+1.332$ V versus $[Co(Cp)_2]^{+/0}$, the value that compares with $+1.335$ V versus $[Co(Cp)_2]^{+/0}$ in AN. In this IL, voltammetric studies have also been carried out about ferrocene derivatives and cobaltocenium ion ($[Co(Cp)_2]^+$); $[Co(Cp)_2]^+$ is reduced at GC and Au electrodes in two reversible steps, $[Co(Cp)_2]^+ + e^- \rightleftharpoons [Co(Cp)_2]^0$ and $[Co(Cp)_2]^0 + e^- \rightleftharpoons [Co(Cp)_2]^-$. The potential difference between the two steps was ~ 800 mV. Recently, the behaviors of Fc and Cc$^+$ were studied by double potential step chronoamperometry in several nonchloroaluminate ILs, and the diffusion coefficients of Fc, Fc$^+$, Cc and Cc$^+$ were obtained [41b]. Some of the data for Fc and Fc$^+$ are shown in Table 13.6. The solubility of Fc in [BMIm][PF$_6$] was reported to be poor in Ref. [41a], but the solubility in this study (38.2 mM) was found to be not poor but moderate. Care was taken concerning the water content of ILs in [41a], but the product as received was used in [41b]. The discrepancy of the solubility may have come from the care taken of the IL purity. The rate constants for the oxidation of a series of ferrocene derivatives in [EMIm][NTf$_2$] have been determined using the high-speed channel electrode method at a Pt electrode and at 23 °C [41c]. For ferrocene, the diffusion coefficient is $D = 3.35 \times 10^{-7}$ cm^2 s^{-1} and the rate constant is $k_s = 0.21 \pm 0.06$ cm s^{-1}. In AN, $k_s = 0.99 \pm 0.22$ cm s^{-1} and D was about two orders of magnitude larger. However, a paper reporting the concentration dependence of the diffusion coefficient of ferrocene in [BMIm][BF$_4$] has been published [40d]: D_{Fc} ($\times 10^{-7}$ cm^2 s^{-1}) determined by cyclic voltammetry is 1.3 at 4.3 mM, 1.8 at 8.6 mM, 2.1 at 12 mM, 3.3 at 17 mM, 4.5 at 22 mM, 5.7 at 26 mM and 6.8 at 32 mM though the exact reason for this is not clarified yet.

Table 13.6 Diffusion coefficients and solubilities of ferrocene in various ILs.[a]

ILs	Viscosity[b], cP	$D_{Fc}, \times 10^{-7}$ cm^2 s^{-1}	$D_{Fc^+}, \times 10^{-7}$ cm^2 s^{-1}	D_{Fc}/D_{Fc^+}	Solubility, mM
[EMIm][NTf$_2$]	34	5.34	3.90	1.37	62.3
[BMIm][NTf$_2$]	52	3.77	2.65	1.43	77.4
[BMPry][NTf$_2$]	89	2.31	1.55	1.50	91.2
[BMIm][OTf]	90	2.36	1.80	1.31	58.3
[BMIm][BF$_4$]	112	1.83	1.30	1.85	27.5
[BMIm][NO$_3$]	266	0.83	0.63	1.32	51.4
[BMIm][PF$_6$]	371	0.59	0.31	1.91	38.2

[a]From Ref. [41b], at 26 ± 1 °C.
[b]Viscosity values at 20 °C.

3. **Reduction of metal ions** The reductions of alkali metal ions have been studied at mercury electrodes [42]. Boxall and Osteryoung [42a] determined formal potentials and diffusion coefficients of lithium, sodium and potassium ions in [BMIm][PF$_6$], using a hanging mercury drop electrode. The diffusion coefficients measured by chronoamperometry were 1.9×10^{-8}, 1.4×10^{-8} and 4.5×10^{-8} cm^2 s^{-1} for Li$^+$, Na$^+$ and K$^+$, respectively. The formal potentials (versus Fc/Fc$^+$) of -2.96 V for Na$^+$/Na and -3.35 V for K$^+$/K were obtained from the data by normal pulse voltammetry (NPV), but that of -2.45 V for Li$^+$/Li from the chronopotentiometric data because current maximum was observed in NPV. Since [BMIm][PF$_6$] is reduced at about -2.5 V versus Fc/Fc$^+$, only Li$^+$ could be reduced before IL reduction at nonmercury electrodes. The electrochemistry of cesium was investigated in [Bu$_3$MeN][NTf$_2$] by using cyclic staircase voltammetry, rotating disk voltammetry and chronoamperometry. The reduction of Cs$^+$ at mercury was quasireversible with $k_s = 9.8 \times 10^{-5}$ cm s^{-1}, $\alpha = 0.36$ and the diffusion coefficient $D = 1.04 \times 10^{-8}$ cm^2 s^{-1} at 30 °C. Bulk deposition/stripping experiments conducted at a rotating mercury film electrode gave an average recovery of 97% of the electrodeposited Cs.

The voltammetry and kinetics of the Ag$^+$/Ag system (commonly used as a reference electrode in both molecular solvents and ILs) were studied in [BMPyr][NTf$_2$], using a 10 μm diameter Pt electrode [43]. For AgOTf, AgNTf$_2$ and AgNO$_3$, voltammogram characteristics for 'deposition/stripping' process were obtained. The diffusion coefficients of AgOTf, AgNTf$_2$ and AgNO$_3$, obtained by potential step chronoamperometry, were 1.05, 1.17 and 5.00×10^{-7} cm^2 s^{-1}, respectively. The standard electrochemical rate constants (k_s) for the three were 2.0, 1.5 and 0.19×10^{-4} cm s^{-1}, respectively, which are too small for ideal reference electrode, although these Ag$^+$/Ag electrodes can serve as a reference electrode if the areas of the Ag electrodes are large enough, using a potentiostat through which only small current flows.

Thorium compound, [Th(NTf$_2$)(HNTf$_2$)]·2H$_2$O, where both NTf$_2^-$ and HNTf$_2$ are coordinated to the metal center, has been studied voltammetrically in [BuMe$_3$N][NTf$_2$], and the result showed that Th(IV) is reduced to Th(0) in a single step [44]. Cyclic voltammograms showed that an insoluble product is formed at the electrode surface, which was attributed to the formation of ThO$_2$ by reaction with water. The formal potential of the reduction of Th(IV) to Th(0) was -2.20 V versus Fc$^+$/Fc, which showed that the Th(IV) reduction in [BuMe$_3$N][NTf$_2$] is easier than that in LiCl–KCl eutectic (400 °C), water and nonaqueous solvents.

4. **Reduction of organic substances** Many organic substances in ILs are reduced to anion radicals by 1e$^-$ processes. A typical example is the reduction of benzaldehyde in [BMPyr][NTf$_2$] [45]: at a 10 μmϕ Pt microdisk electrode and at 1000 V s^{-1}, two reversible CV waves are obtained, the first of which at -1.73 V versus Fc$^+$/Fc is 1e$^-$ reduction to anion radical and the second at -2.33 V versus Fc$^+$/Fc is another 1e$^-$ reduction to dianion. Two waves are also observed in [BMIm][NTf$_2$]; however, they are quasireversible and the first wave shifts to -0.75 V versus

Fc$^+$/Fc and the second to -1.85 V versus Fc$^+$/Fc. This positive shift in potential is because of the effect of imidazolium cation: π–π interactions or hydrogen bonds of imidazolium cation to the anion radical and dianion stabilize them.

In the reduction of nitrobenzene in [BMMIm][NTf$_2$], a 1e$^-$-reversible wave appears at -1.2 V (versus Ag quasireference electrode) and a 2e$^-$-irreversible wave at -1.85 V (versus Ag) [46]. The diffusion coefficient is 1.95×10^{-7} cm^2 s^{-1}. A series of aromatic nitro compounds (4-nitrotoluene, 2-nitro-*m*-xylene, 2-nitro-mesitylene, nitropentamethylbenzene, 2,4,6-tritertbutylnitrobenzene) also undergo 1e$^-$ reduction at GC and Au electrodes in [BMIm][NTf$_2$] and [BuEt$_3$N][NTf$_2$], although the rates of electron transfers are slower than that in AN [47]. It is noted that the formal potentials of these compounds in [BMIm][NTf$_2$], measured versus Fc$^+$/Fc reference, are usually more positive by 0.15–0.2 V than those in [BuEt$_3$N][NTf$_2$]. This is also the effect of the acidity of the imidazolium cation. 1,2- and 1,4-Dinitrobenzene exhibit two 1e$^-$-waves in AN but a single 2e$^-$-wave in [BMIm][BF$_4$]. This phenomenon was originally interpreted to be caused by a strong ion pairing between the imidazolium cation and the dinitrobenzene dianion [48], but now the stabilization of the dianion is considered to be due to the π–π interaction or hydrogen bonding of imidazolium cation to the dianion.

5. **Oxidation of organic substances** In ILs, just as in molecular solvents of weak acidity (e.g. AN), typical oxidation of organic compounds proceeds in two 1e$^-$-steps, the first being the formation of cation radical and the second the formation of dication. This occurs, for example, with N,N,N',N'-tetraalkyl-*para*-phenylenediamines (TAPD; alkyl(A) = butyl(B) or methyl(M)) in [C$_n$MIm][NTf$_2$] ($n = 2, 4, 8$ and 10) and [HexMe$_3$N][NTf$_2$] [49].

$$\text{TAPD} \rightarrow \text{TAPD}^{\bullet+} + e^-, \qquad \text{TAPD}^{\bullet+} \rightarrow \text{TAPD}^{2+} + e^-$$

Voltammograms, obtained with an Au microdisk electrode (12.5 μm in diameter) and at 20 °C gave the first wave at around $+0.1$ V and the second at around $+0.7$ V (versus Ag wire). The diffusion coefficients, $10^7 D_{\text{TMPD}}$ and $10^7 D_{\text{TBPD}}$ (cm^2 s^{-1}), were 2.62 and 1.23 in [EMIm][NTf$_2$] ($\eta = 32.1$ cP), 1.87 and 0.80 in [BMIm][NTf$_2$] ($\eta = 57.6$ cP), 1.12 and 0.40 in [C$_8$MIM][NTf$_2$] ($\eta = 119.3$ cP), 1.13 and 0.52 in [C$_{10}$MIm][NTf$_2$] ($\eta = 152.8$ cP) and 0.70 and 0.24 in [HexMe$_3$N][NTf$_2$] ($\eta = 220.0$ cP), compared to 218 and 127 in AN-0.1M Bu$_4$NClO$_4$.

The radical cation of anthracene in [EMIm][NTf$_2$], [BMIm][NTf$_2$] and [BMIm][PF$_6$] was unstable and an irreversible oxidation wave was observed at a scan rate of 0.1–100 V s^{-1}; because the water content in the ILs was of the order of 1500 ppm, a reaction of water with the radical cation occurred and gave a 2e$^-$-oxidation wave. A reversible cyclic voltammogram was obtained at a scan rate above 3000–5000 V s^{-1}. From the reversible behavior at E^0 of about 0.9 V versus Fc$^+$/Fc, the values of $10^7 D$ (cm^2 s^{-1}) and k_s (cm s^{-1}) were obtained to be 2.8 and 0.1–0.2 in [BMIm][NTf$_2$], 5.3 and 0.6–0.7 in [EMIm][NTf$_2$] and 5.6 and 0.2–0.3 in [BMIm][PF$_6$], compared to 250 and 3–5 in (AN/Et$_4$NPF$_6$) [50].

13.3.3
Applications of ILs in Electrochemical Technologies

1. **Electrodeposition of metals and semiconductors** Ref. [26c] has the title 'Electrodeposition from Ionic Liquids' and Chapters 4–6 deal with electrodeposition of metals, alloys and semiconductors, respectively [51]. Some other books [52] and review articles [53] also deal with this topic.

   The electrodepositions of alkali and alkaline earth metals have been studied in neutral chloroaluminate ILs, to use the ILs in rechargeable batteries. However, the reductions of these metals occurred at more negative potential than the decomposition of imidazolium cations. The depositions of Na and Li metals and Na–Li alloys (dendrite free) were possible by using chloroaluminate ILs of quaternary ammonium cations in the presence of such additives as HCl and $SOCl_2$ because these extended the cathodic limit and produced passivation films on the deposited alkali metals [54]. Electrodepositions of such metals as Ag, Al, Bi, Co, Cr, Cu, Fe, Ga, Hg, Ni, Pb, Pd, Pt, Sb, Sn and Zn and such alloys as Al–Ag, Al–Co, Al–Cr, Al–Cu, Al–Fe, Al–Nb, Al–Ni, Al–Pt and Al–Ti have been carried out from acidic chloroaluminate ILs, while electrodepositions of such metals as Au, Cd, Hg, In, Pd, Sb, Sn, Te, Tl and Zn have been carried out from basic or neutral chloroaluminate ILs. However, the chloroaluminate ILs are very reactive with moistures, and this lessens the practical importance of the electrodeposition from these ILs.

   The electrodepositions from nonchloroaluminate ILs can be divided in two groups, one from chlorozincate and the other from the usual nonchloroaluminate ILs. Chlorozincate IL is a mixture of [EMIm]Cl and $ZnCl_2$ [53b]. It is basic if the concentration of $ZnCl_2$ is less than 33.3 mol%, while it is acidic if the concentration of $ZnCl_2$ is more than 33.3 mol%. The predominant ionic species in basic ILs are $[EMIm]^+$, $ZnCl_4^{2-}$ and Cl^-, but those in acidic ILs are $[EMIm]^+$, $ZnCl_4^{2-}$, $ZnCl_3^-$, $Zn_2Cl_5^-$ and $Zn_3Cl_7^-$. Metallic Zn is deposited only from acidic ILs. The metals and alloys deposited from the chlorozincate ILs are Cd, Co, Cu, Fe, Te; Zn–Cd, Zn–Co, Zn–Co–Dy, Zn–Cu, Zn–Fe, Zn–Ni and Zn–Te(semiconductor). As the examples of depositions from the usual nonchloroaluminate ILs are Ge(semiconductor) from $[BMIm][PF_6]$, Cd, Cu, Sb and In–Sb from [EMIm]Cl-$[EMIm][BF_4]$ and Ag, Co, Cu, Li, Mg, Ni, Pd [55], Si(semiconductor) and Zn from NTf_2 salts. Lithium can be electrodeposited from NTf_2 salts of aliphatic and alicyclic ammonium cations. Magnesium can virtually be electrodeposited from 1:2 mixture of Grignard reagent (2 M phenyl-magnesium chloride in THF) and 1-butyl-1-methylpyrrolidium NTf_2 ($[BMPyr][NTf_2]$) IL at $100\,°C$ [56].

   Here, it is interesting to note that nanomaterials of particle sizes or film thicknesses of 1–100 nm can easily be electrodeposited for various metals, alloys and semiconductors from chloroaluminate and nonchloroaluminate ILs though the electrodeposition of such materials is difficult from aqueous solutions. For example, nanocrystalline Cu and Al can be electrodeposited from green [BMPyr][TfO] and $[BMPyr][NTf_2]$, respectively, on conventional solid electrodes [57]. Generally, the obtained Cu and Al deposits are shiny, dense and adherent with very fine crystallites of average sizes of 30–40 nm. The [BMPyr]$^+$ cation might act as a grain refiner, leading to nanosized deposits. For the reviews of electrodeposition of nanoscale materials from ILs, see Ref. [58].

2. **Electrochemical organic synthesis**
    (a) **Electrochemical polymerization [59]** Electrochemical synthesis of electrically conductive polymers such as polyarene, polypyrrole, polythiophene and polyaniline has been performed in chloroaluminate ILs. However, the polymer films are decomposed rapidly by the corrosive products, e.g. HCl, generated by the hydrolysis of these moisture-sensitive ILs. Electropolymerization has also been carried out in nonchloroaluminate ILs. For example, Fuchigami *et al.* employed [EMIm][OTf] ($\eta = 42.7$ cP) for electropolymerization of pyrrole. The polymerization proceeded much faster in this IL than in the conventional media such as aqueous and acetonitrile solutions. In the viscous IL, since the diffusion is slow, radical–radical coupling, further oxidation of oligomers and polymer deposition favorably occur because the reaction products are accumulated in the vicinity of electrode surface and the polymerization rate is increased. Electropolymerization of thiophene is also accelerated in this IL, although that of aniline is decelerated. The polypyrrole film formed in this IL was very smooth and no grains were observed. Moreover, the polypyrrole and polythiophene films prepared in this IL showed much higher electrochemical capacities and electroconductivities. This can be attributed to the extremely high concentration of dopant anions, which results in a much higher doping level. The polymers thus prepared are suitable for high-performance electrochemical capacitors and other electrochemical devices.

    (b) **Other examples of electrochemical organic synthesis**
        (i) Electrosynthesis of organic carbonates from CO_2 and alcohols has been established in CO_2-saturated [BMIm][BF$_4$], followed by the addition of an alkylating agent [60].

$$ROH + CO_2 \xrightarrow{+e^-, + R'I} ROCO_2R'$$

The synthesis was carried out under mild (1 atm CO_2 and 55 °C) and safe conditions, the use of volatile and toxic solvents and catalysts being avoided. The IL used for the reaction was recyclable. The primary and secondary alcohols were converted in good yields, whereas tertiary alcohol and phenol were not reactive.

        (ii) The preparative electrocatalytic homocoupling of PhBr and PhCH$_2$Br, yielding in biphenyl and 1,2-diphenylethane, respectively, was performed using NiCl$_2$(bipy) complex in reusable media, [BMIm][NTf$_2$] [61].

$$2Ph(CH_2)_xBr \xrightarrow{+2e^-, \quad cat.NiCl_2(bpy)} Ph(CH_2CH_2)_xPh$$

        (iii) Partial fluorination has been carried out conveniently in ILs [62]. Momota used liquid fluoride salts, R$_4$NF·xHF (R = Me, Et and *n*-Pr, $x > 3.5$), as

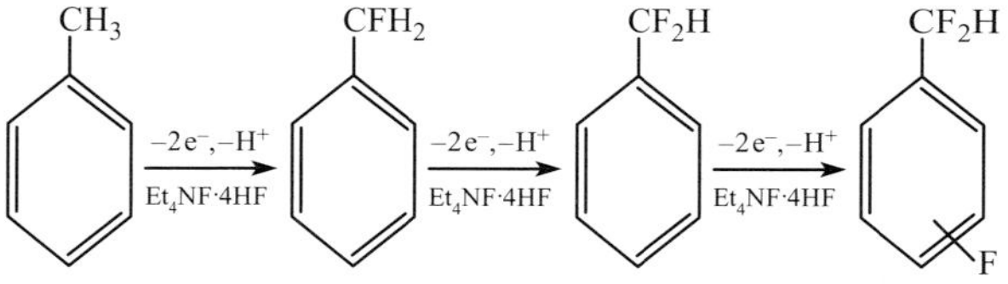

Scheme 13.3

nonviscous, highly conductive and anodically stable ILs and used for anodic partial fluorination of arenes such as benzene, mono-, di- and trifluorobenzene, chlorobenzene, bromobenzene, toluene and quinolines (see Scheme 13.3). For various examples of partial fluorination, see Fujigami's review [63].

3. **Use in electrochemical devices** ILs have been used to various electrochemical devices, e.g. double-layer capacitors, rechargeable lithium batteries and electrochromic windows, mainly because of their safe nonvolatile and nonflammable properties.

   (a) **Double-layer capacitors** Desirable properties for the electrolytes of double-layer capacitors are high double-layer capacitance, electrical conductivity and decomposition voltage, wide operational temperature range and good safety. For commercial double-layer capacitors using activated carbon electrodes, nonaqueous media such as 0.5–1 M Et_4NBF_4–PC or aqueous solutions such as 3.7–4.5 M (30–35 wt%) H_2SO_4 have been used. Nonaqueous solvents have advantages of high decomposition voltage, wide operational temperature range and noncorrosive property but suffer from disadvantages of low electric conductivity, air-tight packaging to avoid moisture, high environmental impact and high cost. Ue [64] used such ILs as [EMIm][BF_4], [EMIm][TaF_6], [EMIm][TfO], [EMIm][NTf_2], [EMIm][$(C_2F_5SO_2)_2$N] and [EMImM][F(HF)$_{2.3}$] and compared their performances with those of nonaqueous 1 M Et_3MeNBF_4/PC and aqueous 4.5 M H_2SO_4/H_2O. Some ILs could perform comparably to Et_3-MeNBF$_4$/PC system but only at higher temperatures; the performance at temperatures below $-10\,°C$ was much inferior. Ue [64] used new IL, [EMIm][$BF_3C_2F_5$] (mp $= -1\,°C$), and found that the electric conductivity was comparable to 1 M Et_3MeNBF_4/PC down to $-35\,°C$, and the capacitor with this IL worked well at low temperatures. Searching for new ILs that show excellent low-temperature characteristics will be the best way to succeed in using ILs in the double-layer capacitors. There are two other ways to get excellent low-temperature characteristics. One is to use polymer ILs, which immobilizes ILs by polymers to overcome the disadvantage of liquid electrolytes. The other is, as often carried out, to mix organic solvents and ILs.

   (b) **Rechargeable lithium batteries** [65] ILs for rechargeable lithium batteries must be liquid in a wide temperature range, wide in potential window, high in conductivity, thermally stable, nonvolatile and nonflammable and chemically stable. To prepare Li battery with chloroaluminate ILs, $SOCl_2$ should be added to make Li metal deposition possible. An example is Li/[EMIm]Cl-AlCl$_3$-LiCl-SOCl$_2$/graphite, but this battery does not work well. For Li batteries with nonchloroaluminate ILs of alkylimidazolium cation, Li electrode should be replaced by, for example, $Li_4Ti_5O_{12}$ electrode that has more positive potential because imidazolium cation is decomposed at more positive potentials than Li^+/Li electrode. With the battery, Li$_4$Ti$_5$O$_{12}$/[EMIm][BF$_4$]-Li[BF$_4$]/LiCoO$_2$, however, cell voltage was reduced to about 2 V, which was too low for Li battery. If IL of nonimidazolium cation, such as 1-ethyl-2-methylpyrazolium ([EMPy]$^+$) and N-methyl-N-propylpiperidinium ([PMPip]$^+$), are used, the Li electrode can be used. The batteries with these cations, Li/[EMPy][BF$_6$]-Li[BF$_4$]/LiMn$_2$O$_4$ and Li/[PMPip][NTf$_2$]-Li[NTf$_2$]/LiCoO$_2$, can give emfs of nearly 3.9 V and fairly good charge–discharge characteristics. However, the Li batteries using ILs are still inferior in performance to those using nonaqueous solvents.

(c) **Other electrochemical devices** Lu *et al.* [66] used [BMIm][BF$_4$] and [BMIm][PF$_6$] to construct electrochemical mechanical actuators, electrochromic windows and numeric displays made from three types of π-conjugated polymers: polyaniline, polypyrrole and polythiophene. The π-conjugated polymers in ILs were electrochemically cycled with enhanced lifetimes (up to 1 million cycles) without failure and with fast cycle switching speeds (100 ms). Brazier *et al.* [67] studied electrochromic devices by using [BMPyr][NTf$_2$] to make two Li-ion conducting elements and by interposing them between transparent film electrodes such as WO$_3$ and Li-charged V$_2$O$_5$. The Li-ion conducting elements were [BMPyr][NTf$_2$] - Li[NTf$_2$] (liquid) and [BMPyr][NTf$_2$]-P(EO)-Li[NTf$_2$] (solid). During WO$_3$ coloration, these showed higher optical contrast than a conventional nonaqueous electrolyte such as PC-Li[NTf$_2$].

13.3.4
Risks of ILs and Their Decomposition

ILs are considered to be more green substitutes of conventional volatile organic solvents. This is mainly because ILs are believed to be nontoxic and nonflammable due to their nonvolatile properties. Certainly, air pollution is reduced by using ILs, but recently, the risks of ILs to the aquatic systems and soils have been pointed out. Many reports are now being published every year on this subject, and two reviews have appeared in 2007 [68].

To test the effect of ILs in aquatic ecosystems, *Daphnia magna* are used because they are the most popular live food for aquarium fish. The acute and chronic toxicity of imidazolium cation-based ILs was first examined by Bernot and coworkers [69]: the median lethal concentration (LC50) was used as an indicator for the acute toxicity. As shown in Table 13.7, the toxicities of imidazolium ILs (anions: Cl$^-$, Br$^-$, PF$_6^-$ and BF$_4^-$) are analogous to ammonia and phenol, which are fairly toxic. These ILs, if leak out, may be more damaging to aquatic ecosystems than conventional volatile organic solvents. *Daphnia magna* were also used to know the effects of ILs based on

Table 13.7 IL ecotoxicity to *Daphni.*[a]

Cation type	R$_1$	R$_2$	Anion	a: LC50/b: 48hEC50 (mg l^{-1})
R$_1$R$_2$Im$^+$	C$_4$H$_9$	CH$_3$	[PF$_6$]$^-$	19.9(a), 24(b)
	C$_4$H$_9$	CH$_3$	Cl$^-$	14.8(a), 6.5(b)
	C$_4$H$_9$	CH$_3$	Br$^-$	8.03(a)
	C$_4$H$_9$	CH$_3$	[BF$_4$]$^-$	10.68(a)
	C$_{12}$H$_{25}$	CH$_3$	Cl$^-$	0.0043(b)
	C$_{16}$H$_{33}$	CH$_3$	Cl$^-$	0.0034(b)
	C$_{18}$H$_{37}$	CH$_3$	Cl$^-$	0.0017(b)
R$_1$P$^+$	C$_4$H$_9$		Cl$^-$	20(b)
(R$_1$)$_3$R$_2$P$^+$	C$_4$H$_9$	C$_2$H$_5$	(EtO)$_2$PO$_2$$^-$	11(b)
	C$_6$H$_{13}$	C$_{14}$H$_{29}$	Cl$^-$	0.072(b)
(R$_1$)$_3$R$_2$N$^+$	C$_8$H$_{17}$	CH$_3$	[NTf$_2$]$^-$	0.2(b)
	{C$_2$H$_4$O(C$_2$H$_4$O)$_4$Me}$_2$C$_{14}$H$_{29}$	CH$_3$	[MeSO$_3$]$^-$	1(b)
Na$^+$			[PF$_6$]$^-$	9344(a), >100(b)
			[MeSO$_3$]$^-$	>100(b)
			[BF$_4$]$^-$	4765

[a]From Ref. [69].

imidazolium, pyridinium, phosphonium and ammonium cations, and the results have been shown using EC50 (medium effective concentration) values, as also summarized in Table 13.7 [70]. All tested ILs had EC50 values below 100 mg l^{-1}, imidazolium cations being most toxic and, irrespective of the cationic type, the toxicities of (C$_8$–C$_{18}$)-substituted salts are much severer than those of (C$_1$–C$_4$)-substituted salts. There are also studies on microorganism toxicology, cytotoxicology and animal tests. In all these cases, more or less toxic effects are observed; the increase in lipophilicity of ILs by elongation of alkyl side chains increases the toxicities [68b].

ILs, more or less, can be distilled and decomposed at 300 °C under reduced pressure [71]. ILs can easily be disposed of by incineration at temperatures above 600 °C, at which even the toughest IL will give up [72]. But, because of the high stability of ILs, their disposal or leakage may cause their accumulation in the environment. Although IL cations are generally not subjected to abiotic hydrolysis, IL anions may undergo hydrolysis. AlCl$_4$$^-$ of chloroaluminate ILs are hydrolyzed and give HCl that is corrosive. Of nonchloroaluminate ILs, PF$_6$$^-$ ion is hydrolyzed in the presence of acid and gives toxic HF [73]. Other fluorinated anions may, more or less, be hydrolyzed; even the most stable NTf$_2$$^-$ has some tendency to hydrolyze, as detected by pH change [74].

If ILs are accumulated in the environment, it is desirable to biologically degrade them. The biodegradation of 1-alkyl-3-methylimidazolium, *N*-alkylpyridinium and *N*-alkyl-4-(dimethylamino)-pyridinium halides (mainly Cl$^-$ salts) has recently been studied [75]. It was found that significant primary biodegradation occurred for (eco)toxicologically unfavorable compounds carrying long alkyl side chain (C6 and

C8). In contrast, for (eco)toxicologically more recommendable imidazolium ILs with short alkyl (<C6) and short functionalized side chain, no biological degradation was found. The introduction of different functional groups into the side chain moiety did not improve biological degradation. After an incubation period of 24 days for the 1-methyl-3-octylimidazolium cation, different biological transformation products carrying hydroxyl, carbonyl and carboxyl groups were identified. Furthermore, shortened side chain moieties were identified indicating the degradation of the octyl side chain via β-oxidation. In the report [75], an electrochemical wastewater treatment was proposed for nonbiodegradable ILs, as part of an alternative disposal strategy: 1-butyl-3-methylimidazolium cation was completely destroyed within 4 h using an electrolysis double cell equipped with electrodes made of iridium oxide (anode), stainless steel (cathode) and a boron-doped diamond-coated bipolar electrode. The products formed electrochemically were easily accessible to biological degradation. The electrochemical decomposition of ILs has also been reported in [76]: when large voltages were applied, [BMPyr][NTf$_2$] formed decomposition products such as methylpyrrolidine, octanes, octenes, 2-butanol, dibutylmethylamine and butylpyrrolidine. On the other hand, the electrochemical breakdown of [BMIm][BF$_4$] formed 1-butyl-3-methylimidazolium radicals that react with each other mainly to disproportionate.

References

1 (a) Anastas, P.T. (2002) *Clean Solvents: Alternative Media for Chemical Reactions and Processing* (eds M.A. Abraham and L. Moens), Oxford University Press, Oxford, Chapter 1; (b) Nelson, W.M. (2003) *Green Solvents for Chemistry: Perspective and Practice*, Oxford University Press, Oxford; (c) Kenton, F.M. (2009) *Alternative Solvents for Green Chemistry (RSC Green Chemistry Series)*, Royal Society of Chemistry, Cambridge.

2 (a) Tanaka, K. (2003) *Solvent-free Organic Synthesis*, Wiley-VCH Verlag GmbH, Weinheim, (b) Tanaka, K. and Toda, F. (2000) *Chem. Rev.*, **100**, 1025.

3 LeSuer, R.J. and Geiger, W.E. (2006) *J. Electroanal. Chem.*, **594**, 20.

4 For example (a) Kiran, E. and Debenedetti, P.G. (eds) (2000) *Supercritical Fluids: Fundamentals and Applications*, Kluwer Academic Publishers, New York, (b) Sunol, A.K. and Sunol, S.G. (2001) *Handbook of Solvents* (ed. G. Wypych), ChemTec Publishing, Toronto, Section 21.1;

(c) Rooney, D.W. and Seddon, K.R. (2001) *Handbook of Solvents* (ed. G. Wypych), ChemTec Publishing, Toronto, Section 21.2; (d) Arai, Y., Sako, T. and Takebayashi, Y. (2002) *Supercritical Fluid: Molecular Interactions, Physical Properties, and New Applications*, Springer, Darmstadt (e) Weingärtner, H. and Frank, E.U. (2005) *Angew. Chem. Int. Ed.*, **44**, 2672.

5 Ikushima, Y., Saito, N., Arai, M. and Arai, K. (1991) *Bull. Chem. Soc. Jpn.*, **64**, 2224.

6 Abbott, A.P., Corr, S., Durling, N.E. and Hope, E.G. (2003) *J. Phys. Chem. B*, **107**, 10628.

7 Minami, K., Mizuta, M., Suzuki, M., Aizawa, T. and Arai, K. (2006) *Phys. Chem. Chem. Phys.*, **8**, 2257.

8 Brennecke, J.F., Blanchard, L.A., Anthony, J.L., Gu, Z., Zarraga, I. and Leighton, D.T. (2002) *Clean Solvents, ACS Symposium Series 819* (eds M. Abraham and L. Moens), American Chemical Society, Washington, DC, Chapter 7.

9 Silvestri, G., Gambino, S., Filardo, G., Cuccia, C. and Guarino, E. (1981) *Angew. Chem. Int. Ed. Engl.*, **20**, 101.

10 (a) McDonald, A.C., Fan, F.-R.F. and Bard, A.J. (1986) *J. Phys. Chem.*, **90**, 196; (b) Flarsheim, W.M., Tsou, Y.-M., Trachtenberg, I., Johnston, K.P. and Bard, A.J. (1986) *J. Phys. Chem.*, **90**, 3857; (c) Flarsheim, W.M., Bard, A.J. and Johnston, K.P. (1989) *J. Phys. Chem.*, **93**, 4234; (d) Liu, C.-Y., Snyder, S.R. and Bard, A.J. (1997) *J. Phys. Chem. B*, **101**, 1180.

11 (a) Crooks, R.M., Fan, F.-R.F. and Bard, A.J. (1984) *J. Am. Chem. Soc.*, **106**, 6851; (b) Crooks, R.M. and Bard, A.J. (1987) *J. Phys. Chem.*, **91**, 1274; (c) Crooks, R.M. and Bard, A.J. (1988) *J. Electroanal. Chem.*, **240**, 253.

12 Cabrera, C.R., Garcia, E. and Bard, A.J. (1989) *J. Electroanal. Chem.*, **260**, 457.

13 Crooks, R.M. and Bard, A.J. (1988) *J. Electroanal. Chem.*, **243**, 117.

14 Abbott, A.P. and Harper, J.C. (1996) *J. Chem. Soc., Faraday Trans.*, **92**, 3895.

15 Niehaus, D., Philips, M., Michael, A. and Wightman, R.M. (1989) *J. Phys. Chem.*, **93**, 6232.

16 Li, J. and Prentice, G. (1997) *J. Electrochem. Soc.*, **144**, 4284.

17 (a) Anderson, P.E., Badiani, R.N., Mayer, J. and Mabrouk, P.A. (2002) *J. Am. Chem. Soc.*, **124**, 10284; (b) Yan, H., Sato, T., Komago, D., Yamaguchi, A., Oyaizu, K., Yuasa, M. and Otake, K. (2005) *Langmuir*, **21**, 12303; (c) Jikei, M., Saitoh, S., Yasuda, H., Itoh, H., Sone, M., Kakimoto, M. and Yoshida, H. (2006) *Polymer*, **47**, 1547.

18 Yoshida, H., Sone, M., Mizushima, A., Abe, K., Tao, X.T., Ichihara, S. and Miyata, S. (2002) *Chem. Lett.*, **31**, 1086.

19 Lin, Y., Cui, X., Yen, C.H. and Wai, C.M. (2005) *Langmuir*, **21**, 11474.

20 Olsen, S.A. and Tallman, D.E. (1994) *Anal. Chem.*, **66**, 503; 1996, 68, 2054.

21 Abbott, A.P., Eardley, C.A., Harper, J.C. and Hope, E.G. (1998) *J. Electroanal. Chem.*, **457**, 1.

22 Abbott, A.P., Hope, E.G. and Palmer, D.J. (2005) *Anal. Chem.*, **77**, 6702.

23 Abbott, A.P. and Durling, N.E. (2001) *Phys. Chem. Chem. Phys.*, **3**, 579.

24 Abbott, A.P. and Eardley, C.A. (1999) *J. Phys. Chem. B*, **103**, 6157.

25 Abbott, A.P. and Eardley, C.A. (2000) *J. Phys. Chem. B*, **104**, 775.

26 Especially useful are (a) Ohno, H. (ed.) (2005) *Electrochemical Aspects of Ionic Liquids*, John-Wiley & Sons, Inc., New York; (b) Wasserscheid, P. and Welton, T. (eds) (2008) *Ionic Liquids in Synthesis*, 2nd edn, vols. 1–2, Wiley-VCH Verlag GmbH, Weinheim; (c) Endres, F., MacFarlane, D. and Abbott, A. (eds) (2008) *Electrodeposition from Ionic Liquids*, Wiley-VCH Verlag GmbH, Weinheim.

27 For example, (a) Buzzeo, M.C., Evans, R.G. and Compton, R.C. (2004) *ChemPhysChem*, **5**, 1106; (b) Galiński, M., Lewandowski, A. and Stępniak, I. (2006) *Electrochim. Acta*, **51**, 5567; (c) Hapiot, P. and Lagrost, C. (2008) *Chem. Rev.*, **108**, 2238 and other review articles cited therein.

28 Greaves, T.L. and Drummond, C.J. (2008) *Chem. Rev.*, **108**, 206.

29 Welton, Y., Ref. [26b], Section 3.5.

30 Mantz, R.A. and Trulove, P.C.,Ref. [26b], Section 3.2.

31 Trulove, P.C. and Mantz, R.A., Ref. [26b], Section 3.6.

32 Fujisawa, T., Fukuda, M., Terazima, M. and Kimura, Y. (2006) *J. Phys. Chem. A*, **110**, 6164.

33 (a) Crowhurst, L., Mawdsley, P.R., Perez-Arlandis, J.M., Stater, P.A. and Welton, T. (2003) *Phys. Chem. Chem. Phys.*, **5**, 2790; (b) Crowhurst, L., Falcone, R., Lancaster, N.L., Llopis-Mestre, V. and Welton, T. (2006) *J. Org. Chem.*, **71**, 8847; (c) Mellein, B.R., Aki, S.N.V.K., Ladewski, R.L. and Brennecke, J.F. (2007) *J. Phys. Chem. A*, **111**, 131; (d) Mellein, B.R. and Brennecke, J.F. (2007) *J. Phys. Chem. B*, **111**, 4837; (e) Lee, J.-M., Ruckes, S. and Prausnitz, J.M. (2008) *J. Phys. Chem. B*, **112**, 1473; (f) Ref. [3].

34 (a) Kanzaki, R., Uchida, K., Hara, S., Umebayashi, Y., Ishiguro, S. and Nomura,

S. (2007) *Chem. Lett.*, **36**, 684; (b) Nomura, S., Iwamoto, K., Shibata, M., Umebayashi, Y., Ishiguro, S. and Kanzaki, R. (2008) Japanese patent P2008–64578A.

35 Umebayashi, Y., Yamaguchi, T., Fukuda, S., Mitsugi, T., Takeuchi, M., Fujii, K. and Ishiguro, S. (2008) *Anal. Sci.*, **24**, 1297.

36 (a) Hardacre, C. (2005) *Annu. Rev. Mater. Res.*, **35**, 29; (b) Matsumoto, K. and Hagiwara, R. (2007) *J. Fluorine Chem.*, **128**, 317.

37 Matsumoto, H., Ref. [26a], Chapter 4.

38 (a) Carter, M.T., Hussey, C.L., Strubinger, S.K.D. and Osteryoung, R.A. (1991) *Inorg. Chem.*, **30**, 1149; (b) AlNashef, I.M., Leonard, M.L., Kittle, M.C., Matthews, M.A. and Weidner, J.W. (2001) *Electrochem. Solid-State Lett.*, **4**, D16; (c) Buzzeo, M.C., Klymenko, O.V., Wadhawan, J.D., Hardacre, C., Seddon, K.R. and Compton, R.G. (2003) *J. Phys. Chem. A*, **107**, 8872.

39 Baldelli, S. (2008) *Acc. Chem. Res.*, **41**, 421.

40 AlNashef, I.M., Matthews, M.A. and Weidner, J.W. (2003) *Ionic Liquids as Green Solvents, Progress and Prospects (ACS Symposium Series 856)* (eds R.D. Roger and K.R. Seddon), American Chemical Society, Washington, DC, p. 509.

41 (a) Hultgren, V.M., Mariotti, A.W.A., Bond, A.M. and Wedd, A.G. (2002) *Anal. Chem.*, **74**, 3151; (b) Rogers, E.I., Silvester, D.S., Poole, D.L., Aldous, L., Hardacre, C. and Compton, R.G. (2008) *J. Phys. Chem. C*, **112**, 2729; (c) Fietkau, N., Clegg, A.D., Evans, R.G., Villagrán, C., Hardacre, C. and Compton, R.G. (2006) *Chem Phys Chem*, **7**, 1041; (d) Eisele, S., Schwarz, M., Speiser, B. and Tittel, C. (2006) *Electrochim. Acta*, **51**, 5304.

42 (a) Boxall, D.L. and Osteryoung, R.A. (2002) *J. Electrochem. Soc.*, **149**, E185; (b) Chen, P.-Y. and Hussey, C.L. (2004) *Electrochim. Acta*, **49**, 5125.

43 Roger, E.I., Silvester, D.S., Ward-Jones, S.E., Aldous, L., Hardacre, C., Russell, A.J., Davies, S.G. and Compton, R.G. (2007) *J. Phys. Chem. C*, **111**, 13957.

44 Bhatt, A.I., Duffy, N.W., Collison, D., May, I. and Lewin, R.G. (2006) *Inorg. Chem.*, **45**, 1677.

45 Brooks, C.A. and Doherty, A.P. (2005) *J. Phys. Chem. B*, **109**, 6276.

46 Silvester, D.S., Wain, A.J., Aldous, L., Hardacre, C. and Compton, R.G. (2006) *J. Electroanal. Chem.*, **596**, 131.

47 Lagrost, C., Preda, L., Volanschi, E. and Hapiot, P. (2005) *J. Electroanal. Chem.*, **585**, 1.

48 Fry, A.J. (2003) *J. Electroanal. Chem.*, **546**, 35.

49 Evans, R.G., Klymenko, O.V., Hardacre, C.H., Seddon, K.R. and Compton, R.G. (2003) *J. Electroanal. Chem.*, **556**, 179.

50 Lagrost, C., Carrié, D., Vaultier, M. and Hapiot, P. (2003) *J. Phys. Chem. A*, **107**, 745.

51 (a) Schubert, T., Zein El Abedin, S., Abbott, A.P., McKenzie, K.J., Ryder, K.S. and Endres, F., Ref. [26c], Chapter 4 (deposition of metals); (b) Sun, I.-W. and Chen, P.-Y., Ref. [26c], Chapter 5 (deposition of alloys); (c) Borisenko, N., Zein El Abedin, S. and Endres, F., Ref. [26c], Chapter 6 (deposition of semiconductors).

52 (a) Katayama, Y., Ref. [26a], Chapter 9; (b) Endres, F. and Zein El Abedin, S., Ref. [26b], Section 6.2.

53 (a) Endres, F. (2002) *ChemPhysChem*, **3**, 144; (b) Zein El Abedin, S. (2006) *ChemPhysChem*, **7**, 58.

54 Doyle, K.P., Lang, C.M., Kim, K. and Kohl, P.A. (2006) *J. Electrochem. Soc.*, **153**, A1353.

55 Bando, Y., Katayama, Y. and Miura, T. (2007) *Electrochim. Acta*, **53**, 87.

56 Cheek, G.T., O'Grady, W.E., Zein El Abedin, S., Moustafa, E.M. and Endres, F. (2008) *J. Electrochem. Soc.*, **155**, D91.

57 Zein El Abedin, S., Pölleth, M., Meiss, S.A., Janek, J. and Endres, F. (2007) *Green Chem.*, **9**, 549.

58 (a) Hempelmann, R., Natter, H., Ref. [26c], Chapter 8; Endres, F., Zein El Abedin, S., Ref. [26c], Chapter 9; (b) Freyland, W., Zell, C.A., Zein El Abedin, S. and Endres, F. (2003) *Electrochim. Acta*, **48**, 3053.

59 (a) Pringle, J.M., Forsyth, M. and MacFarlane, D.R., Ref. [26c], Chapter 7; (b) Fuchigami, T., Ref. [26a], p. 101.

60 Zhang, L., Niu, D., Zhang, K., Zhang, G., Luo, Y. and Lu, J. (2008) *Green Chem.*, **10**, 202.

61 Mellah, M., Gmouh, S., Vaultier, M. and Jouikov, V. (2003) *Electrochem. Commun.*, **5**, 591.

62 (a) Momota, K., Mukai, K., Kato, K. and Morita, M. (1998) *Electrochim. Acta*, **43**, 2503; (b) Momota, K., Mukai, K., Kato, K. and Morita, M. (1998) *J. Fluorine Chem.*, **87**, 173, and references therein.

63 Fuchigami, T., Ref. [26a], p. 93.

64 Ue, M., Ref. [26a], p. 205.

65 Sakaeba, H. and Matsumoto, H., Ref. [26a], p. 173.

66 Lu, W., Fadeev, A.G., Qi, B., Smela, E., Mattes, B.R., Ding, J., Spinks, G.M., Mazurkiewicz, J., Zhou, D., Wallace, G.G., MacFarlane, D.R., Forsyth, S.A. and Forsyth, M. (2002) *Science*, **297**, 983.

67 Brazier, A., Appetecchi, G.B., Passerini, S., Surca Vuk, A., Orel, B., Donsanti, F. and Decker, F. (2007) *Electrochim. Acta*, **52**, 4792.

68 (a) Ranke, J., Stolte, S., Störmann, R., Arning, J. and Jastorff, B. (2007) *Chem. Rev.*, **107**, 2183; (b) Zhao, D., Liao, Y. and Zhang, Z. (2007) *Clean*, **35**, 42.

69 Bernot, R.J., Brueske, M.A., Evans-White, M.A. and Lambert, G.A. (2005) *Environ. Toxicol. Chem.*, **24**, 87.

70 Wells, A.S. and Coombe, V.T. (2006) *Org. Process Res. Dev.*, **10**, 794.

71 Earle, M.J., Esperança, J.M.S.S., Gilea, M.A., Canongia Lopes, J.N., Rebelo, L.P.N., Magee, J.W., Seddon, K.R. and Widegren, J.A. (2006) *Nature*, **439**, 831.

72 Maase, M., Ref. [26b], p. 685.

73 Swatloski, R.P., Holbrey, J.D. and Rogers, R.D. (2003) *Green Chem.*, **5**, 361.

74 Baker, G.A. and Baker, S.N. (2005) *Aust. J. Chem.*, **58**, 174.

75 Stolte, S., Abdulkarim, S., Arning, J., Blomeyer-Nienstedt, A.-K., Bottin-Weber, U., Matzke, M., Ranke, J., Jastorff, B. and Thöming, J. (2008) *Green Chem.*, **10**, 214.

76 Kroon, M.C., Buijs, W., Peters, C.J. and Witkamp, G.-J. (2006) *Green Chem.*, **8**, 241.

14
Electrochemistry at the Liquid–Liquid Interfaces

Electrochemistry at the liquid–liquid interfaces is considered by dividing them into two parts: interfaces between two immiscible electrolyte solutions (Section 14.1) and interfaces between ionic liquids (IL) and water (W) (Section 14.2).

14.1
Interfaces Between Two Immiscible Electrolyte Solutions (ITIES)

In conventional polarography and voltammetry, the indicator electrodes used are electronic conductors such as metals, carbons or semiconductors. In an electrode reaction, an electron transfer (ET) occurs at the electrode/solution interface. Recently, however, it became possible to measure an ion transfer (IT), an electron transfer or a couple of ion–electron transfers at the interface between two immiscible electrolyte solutions (ITIES) by means of voltammetry or polarography. Many books [1] and review articles [2] have been published on this subject. Here, only an outline is provided.

14.1.1
Fundamentals of Electrochemistry at the ITIES

Typical examples of the immiscible liquid–liquid interfaces are water (W)/nitrobenzene (NB), W/1,2-dichloroethane (DCE) and W/o-nitrophenoloctylether (o-NPOE). Here, the solubilities (in mol dm^{-3}, 298 K) of W in NB, DCE and o-NPOE are 0.2, 0.11 and 4.61×10^{-2}, respectively, while those of NB, DCE and o-NPOE in W are 1.5×10^{-2}, 8.5×10^{-2} and 2.01×10^{-6}, respectively [2d]. Thus, in reality, W/o-NPOE is almost immiscible but W/NB and W/DCE are slightly miscible. Here, W is saturated with organic solvent (O) and O with W, but in this section, they are simply shown by W and O, respectively. Figure 14.1 shows the density profile that was obtained by molecular dynamics of water/DCE interface [3]. The interface is very sharp with a thickness of less than 10 Å, though it loses its sharpness to a considerable extent when an external electric field is applied to the interface.

Electrochemistry in Nonaqueous Solutions, Second, Revised and Enlarged Edition. Kosuke Izutsu
Copyright © 2009 WILEY-VCH Verlag GmbH & Co. KGaA, Weinheim
ISBN: 978-3-527-32390-6

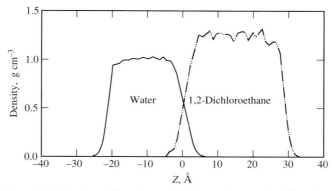

Figure 14.1 Density profile at the water/DCE interface calculated by molecular dynamics [3].

Two examples of the cell used in measurements are shown in Figure 14.2. The indicator electrode is a flat electrode for (a), while it is a dropping electrolyte electrode for (b). In both cases, the cells are usually equipped with four other electrodes, two of which are counter electrodes for supporting the current flow, while the other two are reference electrodes for measuring the potential of the indicator electrode. In practice, various other types of indicator electrodes are used: among them are a gelled electrolyte electrode, a polymer membrane electrode, a micro liquid–liquid interface electrode and an array of these electrodes.

Curve 1 in Figure 14.3 is the current–potential relations at a dropping electrolyte electrode with a cell construction, $Ag|AgCl|SCl(W)|SY(O)|(LiCl + MgSO_4)(W)|AgCl|Ag$, where O is NB and SY is tetrabutylammonium tetraphenylborate (Bu_4NPh_4B). The interface $SCl(W)|SY(O)$ is depolarized because S^+ is contained in both solvents.

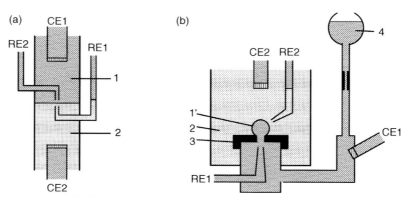

Figure 14.2 Cells for ITIES measurements: (a) cell with a stationary interface for voltammetry and (b) cell with a dropping electrolyte electrode for polarography. (1) aqueous electrolyte phase; (1′) aqueous electrolyte drop; (2) organic phase; (3) Teflon capillary; (4) reservoir of aqueous electrolyte. CE and RE show the counter and reference electrodes.

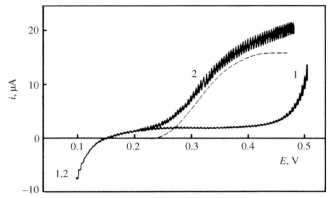

Figure 14.3 ITIES polarogram obtained with a dropping electrolyte electrode. Curve 1: in 0.05 M LiCl + 1 M MgSO₄ (water) and 0.05 M Bu₄NBPh₄ (NB); curve 2: in 1 + 0.5 mM Me₄NCl (water). Dashed curve: curve 2 − curve 1. The potential E is referred to the Galvani potential difference between 0.05 M Bu₄NCl (water) and 0.05 M Bu₄NBPh₄ (NB).

However, the interface SY(O)|(LiCl + MgSO₄)(W) is polarized: LiCl (0.05 M) and MgSO₄ (1 M) are almost insoluble in NB, while Bu₄NPh₄B (0.05 M) is almost insoluble in W. Thus, LiCl, MgSO₄ and Bu₄NPh₄B work as supporting electrolytes and only a small current (*residual current*) flows within the potential windows.

The transfer of ion i^z occurs near the standard ion transfer potential (standard Galvani potential difference), $\Delta_O^W \phi_t^0(i)$, which is related to the standard Gibbs energy of transfer, $\Delta G_t^\circ(i, O \rightarrow W)$, and the standard chemical potentials of i^z in W and O, $\mu_{i,W}^0$ and $\mu_{i,O}^0$, by

$$\Delta_O^W \phi_t^0(i) = -\Delta G_t^\circ(i, O \rightarrow W)/zF = (\mu_{i,O}^0 - \mu_{i,W}^0)/zF$$

The individual values of these quantities are not accessible thermodynamically. But if an extrathermodynamic assumption is introduced, these quantities can be obtained by the measurements of solubility, partition, potentiometry or voltammetry (see Chapter 2). A popular extrathermodynamic assumption is $\Delta G_t^\circ(Ph_4As^+, O \rightarrow W) = \Delta G_t^\circ(Ph_4B^-, O \rightarrow W)$ or $\Delta_O^W \phi_t^0(Ph_4As^+) = -\Delta_O^W \phi_t^0(Ph_4B^-)$. Voltammetrically, this corresponds to using Ph₄AsPh₄B as the supporting electrolyte in O and taking the center of the current–potential curves for the transfers of Ph₄As⁺ and Ph₄B⁻ as potential zero [4]. The $\Delta_O^W \phi_t^0(i)$ values from NB, DCE and o-NPOE to W are shown in Figure 14.4 [5]. The hydrophobicity of the cation increases by going to lower part of the figure, while that of the anion increases by going to its upper part. From this, it is apparent that, for curve 1 in Figure 14.3, the left limit corresponds to the transfer of Bu₄N⁺ from NB to W and the right limit to the transfer of Ph₄B⁻ from NB to W.

The width of the potential window depends considerably on solvents and the electrolytes. Figure 14.5a shows cyclic voltammograms at the 0.1 M tetraoctylam-

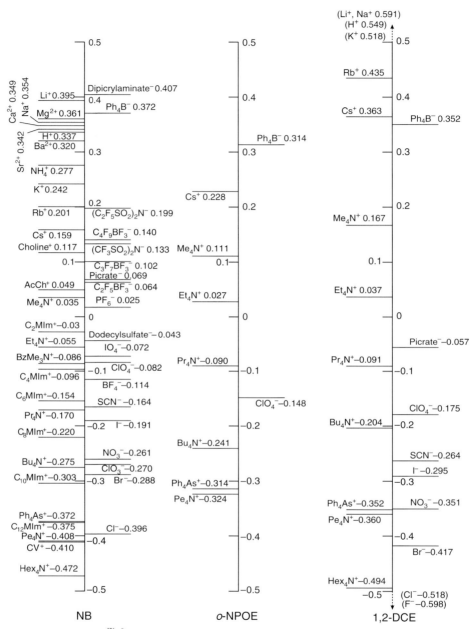

Figure 14.4 $\Delta_O^W \phi_t^0(i)$ values from NB, DCE and *o*-NPOE to W at 25 °C. M and Me, methyl; C_2 and Et, ethyl; Pr, propyl; C_4 and Bu, butyl; Pe, pentyl; C_6 and Hex, hexyl; C_8, octyl; C_{10}, decyl; C_{12}, dodecyl; Ph, phenyl; Bz, benzyl; Im, imidazolium; AcCh, acetylcholine; CV, crystal violet.

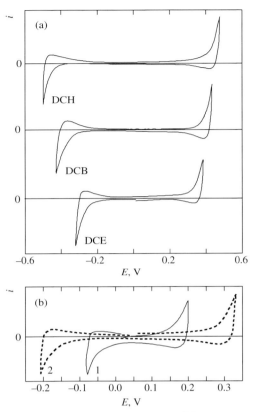

Figure 14.5 (a) Cyclic voltammograms of the 0.1 M tetraoctylammonium tetrakis(4-chlorophenyl)borate (DCH, DCB, or DCE)|0.05 M Li$_2$SO$_4$(W) interfaces [6a] and (b) those of 0.1 M LiCl(W)|0.01 M Bu$_4$NPh$_4$B (curve 1) or tetrapentylammonium tetrakis[3,5-bis (trifluoromethyl)pentyl]borate (curve 2) (o-NPOE) interfaces [6b]. The potentials are referred to the Ph$_4$AsPh$_4$B assumption.

monium tetrakis(4-chlorophenyl)borate(DCH, DCB or DCE)|0.05 M Li$_2$SO$_4$(W) interfaces, where DCH is 1,6-dichlorohexane, DCB is 1,4-dichlorobutane and DCE is 1,2-dichloroethane [6a]. The width of the potential window increases in the order DCE < DCB < DCH. Figure 14.5b shows cyclic voltammograms for 0.1 M LiCl in W and 0.01 M Bu$_4$NPh$_4$B (curve 1) or tetrapentylammonium tetrakis[3,5-bis(trifluoromethyl)pentyl]borate (curve 2) in o-NPOE [6b]. The potential window of curve 2 is much wider than that of curve 1.

Ion Transfer at the W/O Interface

When the two solutions contain ion i^z of activities $a_{i,O}$ and $a_{i,W}$ (concentrations $C_{i,O}$ and $C_{i,W}$), the ion transfer potential $\Delta_O^W \phi_t(i)$ is expressed by

$$\Delta_O^W \phi_t(i) = \Delta_O^W \phi_t^0(i) + \frac{RT}{zF} \ln \frac{a_{i,O}}{a_{i,W}} = \Delta_O^W \phi_t^{0'}(i) + \frac{RT}{zF} \ln \frac{C_{i,O}}{C_{i,W}} \qquad (14.1)$$

Here $\Delta_O^W \phi_t^{0'}(i)$ is the formal ion transfer potential. If the value of $\Delta_O^W \phi_t^{0'}(i)$ is within the potential window, a current–potential curve for ion transfer is obtained. If the IT process is diffusion controlled and the diffusion occurs through the diffusion layers of $\delta_{i,O}$ and $\delta_{i,W}$ thickness, the current–potential curve due to the transfer of i^z can be expressed as follows:

$$E = E_{1/2} + \frac{RT}{zF} \ln \frac{i_{l,p} - i}{i - i_{l,n}} \qquad E_{1/2} = \Delta_O^W \phi_t^{0'}(i) + \frac{RT}{zF} \ln \frac{D_{i,W}\delta_{i,O}}{D_{i,O}\delta_{i,W}} + \text{const} \qquad (14.2)$$

where E is the potential of W-side against O-side, i_{lp} and i_{ln} are the limiting currents at the positive and negative sides of the wave, $D_{i,W}$ and $D_{i,O}$ the diffusion coefficients of i^z in W and O, and 'const' depends on the potential of the reference electrode, Ag|AgCl|Cl$^-$(W), and the liquid junction potential at SCl(W)|SY(O). i_{lp} and i_{ln} can be expressed by the following:

$$i_{i,p} = zFD_{i,W}C_{i,W}/\delta_{i,w} \qquad i_{i,n} = zFD_{i,O}C_{i,O}/\delta_{i,O} \qquad (14.3)$$

The shapes of the current–potential curves are similar to those in Figure 5.6. Such curves can nearly be obtained with a dropping electrolyte electrode and with an ultramicroelectrode. In Figure 14.3, curve 2 was obtained with a dropping electrolyte electrode for the IT of 0.5 mM Me$_4$N$^+$ in W to NB. The limiting current is expressed by the Ilkovič equation and is proportional to the concentration. With a flat electrode, a single-sweep voltammetry and a cyclic voltammetry give current–potential curves as shown in Figures 5.19 and 5.21, respectively.

Here, if ion i^z in W forms at the interface a complex with neutral ligand L in O or an ion pair with ion $j^{z'}$ in O, then the IT process will be facilitated. In case if the formation constants are appropriate, the IT process that originally occurs outside the potential window will occur within the potential window. For example, Figure 14.6 shows cyclic voltammograms of base electrolytes (1, 1') and 1 mM 2-phenylethylammonium ion (2, 2') in the absence (a) and presence (b) of 0.01 M dibenzo-18-crown-6 in NB [7]. Base electrolytes are 0.01 M HCl in W and 0.01 M tetraphenylarsonium 3,3'-*como*-bis (undecahydro-1,2-dicarba-3-cobalta-*closo*-dodecaborate) in NB. These processes are not only useful to get the information about the complexation or ion pair formation but also conveniently applicable to increase the target ions for chemical analyses.

Electron Transfer at the W/O Interface

We consider an electron transfer reaction between a hydrophilic redox couple (Ox$_1$/Red$_1$) in W and a hydrophobic redox couple (Ox$_2$/Red$_2$) in O.

$$\text{Red}_2(O) + \text{Ox}_1(W) \rightleftarrows \text{Red}_1(W) + \text{Ox}_2(O) \qquad (14.4)$$

Here, the number of electrons is equal to n in both cases, namely, Ox$_1$(W) + $ne^- \rightleftarrows$ Red$_1$(W) and Ox$_2$(O) + $ne^- \rightleftarrows$ Red$_2$(O). If this reaction system is in equilibrium, the Galvani potential difference ($\Delta_O^W \phi_{ET}$) at the O/W interface is given by the following equation:

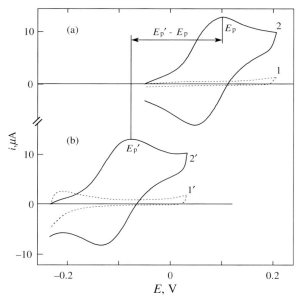

Figure 14.6 Cyclic voltammograms of base electrolyte (1, 1') and 1 mM 2-phenylethylammonium ion (2, 2') in W in the absence (a) and presence (b) of 10 mM dibenzo-18-crown-6 in NB. Sweep rate 0.02 V s^{-1}.

$$\Delta_O^W \phi_{ET} = \Delta_O^W \phi_{ET}^0 + \frac{RT}{nF} \ln \frac{a_{Red_1,W} a_{Ox_2,O}}{a_{Red_2,O} a_{Ox_1,W}} = \Delta_O^W \phi_{ET}^{0'} + \frac{RT}{nF} \ln \frac{C_{Red_1,W} C_{Ox_2,O}}{C_{Red_2,O} C_{Ox_1,W}}$$

(14.5)

where $\Delta_O^W \phi_{ET}^0$ and $\Delta_O^W \phi_{ET}^{0'}$ are the standard and formal Galvani potential differences, respectively, and $\Delta_O^W \phi_{ET}^{0'} = E_{Ox_2/Red_2}^{0'} - E_{Ox_1/Red_1}^{0'}$ ($E_{Ox/Red}^{0'}$: formal redox potential). When $\Delta_O^W \phi_{ET}^{0'}$ is within the potential window, a wave due to the ET reaction of Eq. (14.4) can be observed. The general equation for the wave can be expressed as

$$E = A + \frac{RT}{nF} \ln \frac{(i_{l,Ox_2} - i)(i_{l,Red_1} - i)}{(i_{l,Ox_1} - i)(i_{l,Red_2} - i)}$$

(14.6)

$$A = E^{0'} + \frac{RT}{nF} \ln \frac{(D_{Ox_1,W}/\delta_{Ox_1,W})(D_{Red_2,O}/\delta_{Red_2,O})}{(D_{Ox_2,O}/\delta_{Ox_2,O})(D_{Red_1,W}/\delta_{Red_1,W})}$$

$E^{0'}$ is the formal potential of the reaction (14.4) versus the reference electrode in W-side and $i_{l,i}$ is the limiting current proportional to the concentration of species i. When Red$_2$ in O and Ox$_1$/Red$_1$ couple in W are present and the concentrations of Ox$_1$/Red$_1$ are much larger than that of Red$_2$, the above equation reduces to

$$E = A + \frac{RT}{nF} \ln \frac{i_{l,Red_1}}{i_{l,Ox_1}} + \frac{RT}{nF} \ln \frac{i}{i_{l,Red_2} - i}$$

(14.7)

This means that the same wave form is obtained as that of the IT process.

The truly heterogeneous ET process can be obtained when a sufficiently hydrophobic redox species is used in the O-phase. Lutetium biphthalocyanine (LuPc$_2$) is considered to be such a redox species.

Couples of Ion/Electron Transfers at the W/O Interphasial Region

The system of Fc in NB and Fe(CN)$_6$$^{3-}$ in W was first considered to be an example of the electron transfer at the W/O interface. But the mechanism of the system has been shown to be as follows [8]:

$$Fc(NB) \rightleftarrows Fc(W) \tag{14.8}$$

$$Fc(W) + Fe(CN)_6^{3-}(W) \rightleftarrows Fc^+(W) + Fe(CN)_6^{4-}(W) \tag{14.9}$$

$$Fc^+(W) \rightleftarrows Fc^+(NB) \tag{14.10}$$

Here, the ET reaction (14.9) occurs homogeneously in the W-side and the electric current flows by the IT reaction (14.10). The partition of Fc into W, Eq. (14.8), is not favored, but the depth of the reaction field for the homogeneous ET reaction in W (a few hundred micrometers) is much larger than that of the heterogeneous ET reaction at the W/O interface (several hundred picometers). Thus, the IT mechanism, Eq. (14.10), is the major process for the current flow. Various phenomena, which were first considered to be the ET processes, are now recognized to be the couples of ion/electron transfers.

14.1.2
Practical Applications of Electrochemistry at the ITIES

Voltammetry including single-sweep, cyclic and differential pulse methods, polarography at a dropping electrolyte electrode and amperometry for the IT processes are applicable to the analyses of various ionic species; especially, the applications to redox-inactive biological and pharmaceutical species are promising. They are also useful to get fundamental data of the IT processes, which can be used in the separation chemistry, in the elucidation of biological membrane processes and in the fabrication of ion-selective electrodes. Extraction of ions from aqueous phase 1 (W1) to aqueous phase 2 (W2) by the application of potential difference across the W1/LM/W2 (LM: liquid membrane) interfaces is also useful [9]. The facilitation of the IT processes by complexations and ion pair formations increases the ionic species that can be analyzed or extracted.

As an analytical application, an electrochemical flow cell has been constructed for rapid and coulometric IT at the W/O interface (Figure 14.7) [10]. The cell was composed of a porous Teflon tube (1 mm in inner diameter), an Ag/AgCl wire (0.8 mm in diameter) inserted into the tube, a Pt wire placed outside the tube, O-phase in which the tube was immersed and a tetraphenylborate ISE (connected to an Ag/AgCl reference electrode) in O-phase. The W-phase containing a species of interest was flowed through the narrow gap between the tube and the Ag/AgCl wire. The potential was applied by using the Ag/AgCl wire and the ISE, and the current due

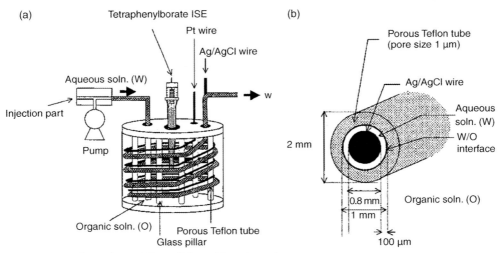

Figure 14.7 (a) Electrochemical flow cell for rapid, coulometric IT at the W/O interface [10]. (b) A porous Teflon tube (1 mm in inner diameter) and an Ag/AgCl wire (0.8 mm in diameter) inserted into the tube.

to the IT was measured by using the Ag/AgCl and Pt wire electrodes. The IT process was complete within the time of flow of the sample through the cell (about 40 s). The cell was used for flow coulometric analyses (p. 151) of K^+, Mg^{2+} and Ca^{2+} with excellent results.

Charge transfer processes that occur, including an intermediate or a product adsorbed at the interfacial region, are of practical interest:

1. Formation of a silicate membrane by the sol–gel process at the liquid–liquid interface is possible by this process [11]. The W-phase contains a precursor ($Si_4O_8(OH)_4^{4-}$) and the O-phase the template ion (e.g. trimethyloctadecylammonium ion, $[TODA]^+$). The template ion, being transferred to the interface and adsorbed there, reacts with the precursor to form a solid.

$$4[TODA]^+(O) + Si_4O_8(OH)_4^{4-}(W) \rightleftharpoons [TODA]_4Si_4O_8(OH)_4(solid)$$

(14.11)

2. Electropolymerization at the W/O interface has been used to prepare the polymer films of polypyrrole, polythiophene, etc. [12]. The monomer, which is in DCE with Ph_4AsBPh_4, is adsorbed at the interface and reacts with an oxidizing agent (Ce^{4+}, Fe^{3+}, etc.) in W to form a polymer film.

3. Electrodepositions of noble metal particles at the interface are also useful. Since Au particles were deposited by the ET reaction between $AuCl_4^-$ in DCE and Fe$(CN)_6^{4-}$ in W [13], electrodeposited particles of various noble metals such as Pd, Pt, Pd/Pt and Ag have been prepared on macro-, micro- or nanoscopic ITIES and membrane-supported ITIES [14]. As a reducing agent, $Fe(CN)_6^{4-}$ in W is used when noble metal complex is in DCE and a ferrocene derivative (dimethylferro-

cene, butylferrocene, etc.) in DCE is used when noble metal complex is in W. The W/O interface is smooth and free of defects that act as a preferential nucleation sites; thus, nanoscale metal particles are usually formed at the interface.

Phase transfer catalysis has been used in organic synthesis to perform reactions in O when some of the reactants are present in W: an example is the substitution reaction involving alkylchloride (RCl), $RCl(O) + NaCN(W) = RCN(O) + NaCl$ (W) [15]. This reaction proceeds only after the addition of tetraalkylammonium chloride, R_4NCl, as a catalyst. It is considered that R_4NCl is partitioned between the two phases and a distribution potential is set up that drives CN^- or electron to the target substance. Polarization studies at the ITIES are useful for the systematic search of suitable catalysts.

14.1.3
Three-Phase Electrochemistry

Three-phase electrochemistry involves three phases in contact: one is the working (indicator) electrode (graphite, Pt or Au) and the other two are liquids that are immiscible (usually O and W). During the electrode process, ions move across the boundary between O and W. Two types of cells exist (Figure 14.8): one uses a flat working electrode and the other a cylindrical working electrode. In the first, a semidroplet of O is attached to the flat graphite disk working electrode surface, which, in turn, is immersed into the W-phase (Figure 14.8a) [16]. In the second, a cylindrical microelectrode (Pt or Au) is immersed vertically across the O/W interface (Figure 14.8b) [17a] or inserted horizontally into a droplet of O-phase in the W-phase

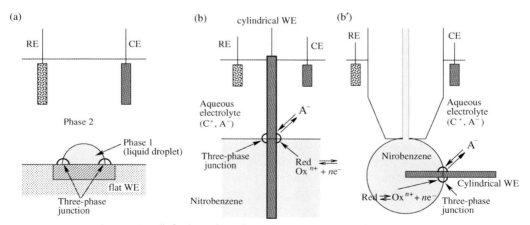

Figure 14.8 Cells for three-phase electrochemistry: (a) a drop of O attached to the flat graphite disk electrode surface is immersed into the W-phase, (b) a cylindrical microelectrode (Pt or Au) is immersed vertically across the O/W interface and (b′) into a drop of O-phase in the W-phase, a cylindrical Pt microelectrode is inserted.

(Figure 14.8b') [17b]. In all cases, two electrodes (reference and auxiliary) are used besides the working electrode, and usually an electroactive (redox) substance is added to O while a supporting electrolyte is added to W. This is in contrast to the conventional electrochemistry at the ITIES, in which four electrodes (two references and two auxiliaries (counters)) are used besides the electrode of W/O interface and the supporting electrolytes are added to both W and O. The case that employs a semi-droplet O-phase has been reviewed in detail [16]. In the case that employs the cylindrical microelectrode, the ET process occurs in the O-phase and simultaneously the IT process occurs from W-phase to O-phase.

$$\text{Red}(O) + nA^-(W) \rightleftarrows Ox^{n+}(O) + nA^-(O) + ne^- \tag{14.12}$$

The formal potential of the process can be expressed by

$$E_f = E^{0'}_{Ox/Red(O)} + \Delta^O_W \phi^0_t(A^-) - \frac{RT}{F}\ln a_{A^-(W)} + \frac{RT}{F}\ln\left(\frac{a_{Red(O)}}{2}\right) \tag{14.13}$$

where $E^{0'}_{Ox/Red(O)}$ is the formal potential of $\text{Red}(O)/Ox^{n+}(O)$ couple, $\Delta^O_W \phi^0_t(A^-)$ is the standard Galvani potential difference for transfer of A^- from W to O and a is the activity. Thus, from the measurements of the formal potential, the values of $\Delta^O_W \phi^0_t(A^-)$ can be obtained. The ET process occurs in the O-side, but the width of the reaction layer from the interface depends on the $\Delta^O_W \phi^0_t(A^-)$ values. With anion A^- of high hydrophlicity, the reaction layer is very thin.

14.2
Interfaces Between Ionic Liquids and Water

The electrochemistry at the interface between hydrophobic ionic liquid and water is a new research field and various phenomena, which are similar to or different from the electrochemistry at the ITIES, can be observed. A useful review article is available on this topic [18].

14.2.1
Fundamentals

The mutual solubility of IL and W plays an important role in determining the electrochemical behavior at the IL/W interface. But, as shown in Table 14.1, it varies considerably depending on the ILs. If the ions of the IL dissolved in W do not form an ion pair, the solubility of the IL in W is determined by the solubility product K^W_{sp} of the IL. Here, if the IL is composed of cation C^+ and anion A^-, it is expressed by

$$\ln K^W_{sp} = -\frac{\Delta G_t(C^+, IL \to W) + \Delta G_t(A^-, IL \to W)}{RT} \tag{14.14}$$

where $\Delta G_t(C^+, IL \to W)$ and $\Delta G_t(A^-, IL \to W)$ are the Gibbs energies of transfer of C^+ and A^- and are the measures of the hydrophobicity of ions. But they cannot have a universal significance because each of them somewhat depends on the counter ion

Table 14.1 Mutual solubilities of water with hydrophobic ionic liquids (IL) and organic solvents (O).[a]

IL or O	Solubility of W in IL or O		Solubility of IL or O in W		$\Delta_{IL}^{W}\phi'$, V
	wt%	mol kg^{-1}	wt%	mol kg^{-1}	
Moderately hydrophobic ILs					
[EMIm][(CF$_3$SO$_2$)$_2$N]	1.94	1.08	1.81	4.60×10^{-2}	0.09
[BMIm][(CF$_3$SO$_2$)$_2$N]	1.48	8.20×10^{-1}	0.716	1.72×10^{-2}	0.02
[HMIm][(CF$_3$SO$_2$)$_2$N]	1.05	5.83×10^{-1}	0.237	5.30×10^{-3}	
[OMIm][(CF$_3$SO$_2$)$_2$N]	0.87	4.83×10^{-1}	8.87×10^{-2}	1.87×10^{-3}	−0.04
[OMIm][(C$_2$F$_5$SO$_2$)$_2$N]	0.6	3×10^{-1}		1.1×10^{-3}	−0.01
Highly hydrophobic ILs					
[Hex$_4$N][(CF$_3$SO$_2$)$_2$N]	0.3	1.6×10^{-1}		7×10^{-6}	−0.17
[C$_{18}$Iq][TFPB]^b	0.31	1.7×10^{-1}		4×10^{-5}	
[TOMA][TFPB]^b	0.20	1.1×10^{-1}		1×10^{-5}	
Organic solvents					
DCE^c	0.15	8.4×10^{-2}	0.81	8.3×10^{-2}	
NB	0.26	1.5×10^{-1}	0.21	1.6×10^{-2}	
NPEO		4.6×10^{-2}		2.0×10^{-6}	

^{a}From [17b]. Temperature: 25 °C. [C$_{18}$Iq] = *N*-octadecylisoquinolinium, [TOMA] = trioctylmethylammonium and [TFPB] = tetrakis[3,5-bis(trifluoromethyl)phenyl]borate. For $\Delta_{IL}^{W}\phi'$, see text.
^{b}Temperature 56 °C.
^{c}Temperature 20 °C.

composing the IL. However, the polarities of ILs of different types are very similar to each other and comparable to those of lower aliphatic alcohols. Thus, it seems reasonable to consider that the hydrophobicities of C$^+$ and A$^-$ of the IL resemble the hydrophobicities of C$^+$ and A$^-$ in organic solvent of similar polarities. Here, we select NB as such organic solvent (O): the standard ion transfer potential, $\Delta_{NB}^{W}\phi_t^0(i)$, which is equal to $-\Delta G_t^{\circ}(i, NB \rightarrow W)/(zF)$ (z is the charge of i^z), is considered instead of the standard ion transfer potential $\Delta_{IL}^{W}\phi_t^0(i)$. The values of $\Delta_{NB}^{W}\phi_t^0(i)$ shown in Figure 14.4 are referred to the Ph$_4$AsPh$_4$B assumption, i.e. Ph$_4$AsPh$_4$B is used as the supporting electrolyte and the center of symmetry of the current–potential curve was considered $\Delta_{NB}^{W}\phi = 0$.

When the absolute values of $\Delta_{IL}^{W}\phi_t^0(i)$ for C$^+$ and A$^-$ are different, the difference in the solubilities of C$^+$ and A$^-$ causes the distribution potential difference across the IL/W interface, $\Delta_{IL}^{W}\phi$, to compensate for the difference in solubilities. Here, $\Delta_{IL}^{W}\phi = (\Delta_{IL}^{W}\phi_t^0(C^+) + \Delta_{IL}^{W}\phi_t^0(A^-))/2$. Some $\Delta_{IL}^{W}\phi'$ values, estimated using $\Delta_{NB}^{W}\phi_t^0(i)$ instead of $\Delta_{IL}^{W}\phi_t^0(i)$, are also shown in Table 14.1; the positive value shows that W-side is positive with respect to the IL-phase and vice versa. For example, $\Delta_{IL}^{W}\phi'$ for [Hex$_4$N][NTf$_2$] is very negative (−0.17 V) due to the high hydrophobicity of Hex$_4$N$^+$; it suggests that the extraction of cations from W to IL is suppressed and that the IL surface carries considerable negative charges.

When W contains electrolytes, the partitioning of ions from W to IL affects $\Delta_{IL}^{W}\phi$. But this occurs only when relatively hydrophobic ions are contained in W. When W contains only hydrophilic ions, their dissolution in IL is negligible, and $\Delta_{IL}^{W}\phi$ is still determined by the partitioning of the ions constituting the IL: $\Delta_{IL}^{W}\phi$ stays constant irrespective of the type and concentration of the hydrophilic electrolyte in W.

14.2.2
Salt Bridges and Reference Electrodes

As an application of the constancy of $\Delta_{IL}^{W}\phi$, Kakiuchi et al. proposed an IL salt bridge [19]. The $\Delta_{IL}^{W}\phi$ values for a $[C_8MIm][NTf_2]$ salt bridge, gelled with poly (vinylidene fluoride-co-hexafluoropropylene), were constant when immersed in aqueous solutions containing from 2×10^{-5} to 2 M KCl, were stabilized within 1 min and were stable within ± 1 mV for at least 3 months. The constant $\Delta_{IL}^{W}\phi$ values were also obtained for HCl, LiCl and NaCl solutions of concentrations <0.2 M. If this salt bridge is inserted between two aqueous solutions, such as $W_1|IL|W_2$, the $\Delta_{IL}^{W}\phi$ at $W_1|IL$ and $IL|W_2$ cancel out and the total $\Delta_{IL}^{W}\phi$ value between W_1 and W_2 is negligibly small, even when the diffusion potential at $W_1|W_2$ is considerably large. Thus, this salt bridge can be used for measurements of purely aqueous systems.

This salt bridge is conceptually different from a concentrated aqueous KCl salt bridge, in which the working principle is the diffusion potential, i.e. a nonequilibrium property. In the case of $[C_8MIm][NTf_2]$ salt bridge, $\Delta_{IL}^{W}\phi$ is defined by the distribution potential that is an equilibrium property. The drawback of this salt bridge is that the $\Delta_{IL}^{W}\phi$ value is influenced if the aqueous solution contains a hydrophobic ion.

Kakiuchi et al. [20] also proposed an Ag/AgCl reference electrode based on $[C_8MIm][NTf_2]$ saturated with AgCl. Both gelled and nongelled $[C_8MIm][NTf_2]$ gave Ag/AgCl electrodes with stable potentials against aqueous KCl solutions between 0.05 mM and 2 M, making possible the miniaturized and solid-state reference electrodes. However, $[C_8MIm][NTf_2]$ dissolves five times more AgCl than in W and, moreover, it dissolves in W up to 1.8×10^{-3} M and W dissolves in it up to 0.9 wt%. Thus, it is desirable to separate the electrode compartment and the salt bridge, such as $Ag|AgCl|AgCl\text{-satd.}[C_8MIm][NTf_2]\vdots[C_8MIm][NTf_2]\vdots$ aqueous solution.

14.2.3
Potential Windows and Voltammograms

Voltammetry that measures ion transfer at a polarized IL/W interface began very recently [21].

The wider the potential window of the IL/W interface the larger the hydrophobicity of the ions constituting the IL, i.e. the smaller the solubility of the IL into W, and the larger the hydrophilicity of the supporting electrolyte in W [18]. When the hydrophilicity of the supporting electrolyte in W is large enough, the potential window of IL, C^+A^-, is determined by the standard ion transfer potential, $\Delta_O^{W}\phi_{C^+}^0$ and $\Delta_O^{W}\phi_{A^-}^0$. For example, the difference between $\Delta_{NB}^{W}\phi_{Hex_4N^+}^0$ and $\Delta_{NB}^{W}\phi_{NTf_2^-}^0$ is 0.605 V and the actual potential window is about 0.3 V. Of course, the location of the potential window is of concern, in addition to its width. For ILs between N-octadecylisoquinolinium

($C_{18}Iq^+$) or trioctylmethylammonium ($TOMA^+$) and tetrakis[3,5-bis(trifluoro-methyl)phenyl]borate ($[TFPB^-]$), the potential windows are about 0.85 V at 56 °C, if the supporting electrolyte in W is 0.1 M LiCl. In this case, the potential limits on both sides are due to the transfers of Li^+ and Cl^- from W to IL. The widest potential window hitherto obtained is 1.1 V at 60 °C in IL of tetraheptylammonium (Hep_4N^+) and $[TFPB^-]$ and W containing $MgSO_4$ [22]. This is wider than the potential window at the conventional ITIES and enables electrochemical studies of various charge transfer processes at the IL/W interface.

Figure 14.9 shows the voltammograms for the facilitated transfer of alkali metal ions in W with dibenzo-18-crown-6 (DB18C6) in $[C_{18}Iq][TFPB]$ [23]. The cell

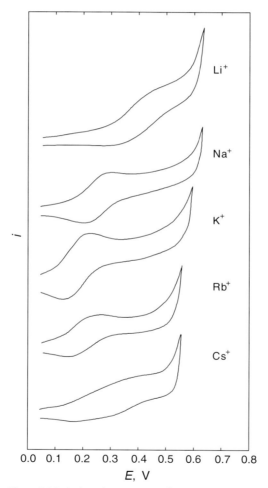

Figure 14.9 Cyclic voltammograms for the transfer of various ions, facilitated by dibenzo-18-crown-6 (DB18C6). IL = [C₁₈Iq][TFPB] (see Table 14.1). The concentration of the ions in W is 100 mM and of DB18C6 in IL is 20 mM [23].

construction is Ag|AgCl|100 mM Li$_2$SO$_4$, 1 mM C$_{18}$IqCl(W)|y mM DB18C6, ([C$_{18}$Iq][TFPB])|x mM MCl(W)|AgCl|Ag, where M$^+$ is either Li$^+$, Na$^+$, K$^+$, Rb$^+$ or Cs$^+$, and the typical values of x and y are 100 and 10, respectively. The stoichiometry of the complexes in [C$_{18}$Iq][TFPB] is 1 : 1 for Li$^+$, Na$^+$, K$^+$ and Rb$^+$, while for the Cs$^+$ transfer both 1 : 1 and 1 : 2 complexes are likely to be formed. The logarithms for the formation constants log K_f are 5.0, 7.0, 8.2 and 7.3 for Li$^+$, Na$^+$, K$^+$ and Rb$^+$, respectively. Because [C$_{18}$Iq][TFPB] has high viscosity (55.7 cP at 56 °C), cyclic voltammograms must be measured at the interface formed at the tip of a micropipette with a radius of about 20 μm.

References

1 For example, (a) Volkov, A.G. and Deamer, D.W. (eds) (1996) *Liquid–Liquid Interfaces, Theory and Methods*, CRC Press, New York;(b) Volkov, A.G., Deamer, D.W., Tanelian, D.L. and Markin, V.S. (1998) *Liquid Interfaces in Chemistry and Biology*, John Wiley & Sons, Inc., New York; (c) Watarai, H., Teramae, N. and Sawada, T. (eds) (2005) *Interfacial Nanochemistry: Molecular Science and Engineering at Liquid–Liquid Interfaces*, Kluwer Academic/Plenum Publishers, New York.

2 For example, (a) Girault, H.H.J. and Schiffrin, D.J. (1989) *Electroanalytical Chemistry*, vol. 15 (ed. A.J. Bard), Marcel Dekker, New York, Chapter 1; (b) Samec, Z. and Kakiuchi, T. (1995) *Advances in Electrochemical Science and Engineering*, vol. 4 (eds H. Gerischer and C.W. Tobias), Wiley-VCH Verlag GmbH, Weinheim, Chapter 5; (c) Kihara, S. and Maeda, K. (1994) *Prog. Sur. Sci.*, **47**, 1; (d) Samec, Z. (2004) *Pure Appl. Chem.*, **76**, 2147.

3 (a) Benjamin, I.,Ref. [1a], Chapter 9; (b) Benjamin, I. (1996) *Chem. Rev.*, **96**, 1449; (c) For the W/NB interface, Michael, D. and Benjamin, I. (1998) *J. Electroanal. Chem.*, **450**, 335.

4 Valent, O., Koryta, J. and Panoch, M. (1987) *J. Electroanal. Chem.*, **226**, 21.

5 Kakiuchi, T. and Nishi, N. (2006) *Electrochemistry*, **74**, 942; Ref. [2d].

6 (a) Katano, H., Tatsumi, H. and Senda, M. (2004) *Talanta*, **63**, 185; (b) Samec, Z.,

Trojánek, A. and Langmaier, J. (1998) *J. Electroanal. Chem.*, **444**, 1.

7 Dvořák, O., Mareček, V. and Samec, Z. (1991) *J. Electroanal. Chem.*, **300**, 407.

8 Osakai, T. and Hotta, H., Ref. [1c], Chapter 8.

9 Shirai, O., Kihara, S., Yoshida, Y. and Matsui, M. (1995) *J. Electroanal. Chem.*, **389**, 61.

10 Yoshizumi, A., Uehara, A., Kasuno, M., Kitatsuji, Y., Yoshida, Z. and Kihara, S. (2005) *J. Electroanal. Chem.*, **581**, 275.

11 (a) Mareček, V. and Jänchenová, H. (2003) *J. Electroanal. Chem.*, **558**, 119; (b) Jänchenová, H., Štulík, K. and Mareček, V. (2006) *J. Electroanal. Chem.*, **591**, 41.

12 (a) Mareček, V., Jänchenová, H., Stibor, I. and Budka, J. (2005) *J. Electroanal. Chem.*, **575**, 293; (b) Vignali, M., Edwards, R.A.H., Serantoni, M. and Cunnane, V.J. (2006) *J. Electroanal. Chem.*, **591**, 59.

13 Cheng, Y. and Schiffrin, D.J. (1996) *J. Chem. Soc., Faraday Trans.*, **92**, 3865.

14 For example, (a) Trojánek, A., Langmaier, J. and Samec, Z. (2007) *J. Electroanal. Chem.*, **599**, 160; (b) Santos, H.A., García-Morales, V., Murtomäki, L., Manzanares, J.A. and Kontturi, K. (2007) *J. Electroanal. Chem.*, **599**, 194; (c) Platt, M. and Dryfe, R.A.W. (2007) *J. Electroanal. Chem.*, **599**, 323.

15 Koryta, J. and Vanýsek, P. (1981) *Advances in Electrochemistry and Electrochemical Engineering*, vol. 12 (eds H. Gerisher and C.W. Tobias), Wiley-Interscience, New York, p. 113.

16 For example, Banks, C.E., Davies, T.J., Evans, R.G., Hignett, G., Wain, A.J.,

Lawrence, N.S., Wadhawan, J.D., Marken, F. and Compton, R.G. (2003) *Phys. Chem. Chem. Phys.*, **5**, 4053.

17 (a) Bak, E., Donten, M. and Stojek, Z. (2007) *J. Electroanal. Chem.*, **600**, 45; (b) Donten, M., Bak, E., Gniadek, M., Stojek, Z. and Scholz, F. (2008) *Electrochim. Acta*, **53**, 5608.

18 (a) Kakiuchi, T. (2007) *Anal. Chem.*, **79**, 6442; (b) Kakiuchi, T. (2008) *Anal. Sci.*, **24**, 1221.

19 Kakiuchi, T. and Yoshimatsu, T. (2006) *Bull. Chem. Soc. Jpn.*, **79**, 1017.

20 Kakiuchi, T., Yoshimatsu, T. and Nishi, N. (2007) *Anal. Chem.*, **79**, 7187.

21 (a) Quinn, B.M., Ding, Z., Moulton, R. and Bard, A.J. (2002) *Langmuir*, **18**, 1734; (b) Kakiuchi, T. and Tsujioka, N. (2003) *Electrochem. Commun.*, **5**, 253; (c) Katano, H. and Tatsumi, H. (2003) *Anal. Sci.*, **19**, 651.

22 Ishimatsu, R., Nishi, N. and Kakiuchi, T. (2007) *Chem. Lett.*, **36**, 1166.

23 Nishi, N., Murakami, H., Imakura, S. and Kakiuchi, T. (2006) *Anal. Chem.*, **78**, 5805.

Index

Electrochemistry in Nonaqueous Solutions, Second, Revised and Enlarged Edition. Kosuke Izutsu
Copyright © 2009 WILEY-VCH Verlag GmbH & Co. KGaA, Weinheim
ISBN: 978-3-527-32390-6